LOGIC FOR AN OVERCAST TUESDAY

LOGIC FOR AN OVERCAST TUESDAY

ROBERT J. RAFALKO

California State University, Bakersfield

Wadsworth Publishing Company

Belmont, California

A Division of Wadsworth, Inc.

Philosophy Editor Ken King
Editorial Assistant Michelle L. Palacio
Production Editor Donna Linden
Managing Designer Donna Davis
Print Buyer Barbara Britton
Designer Wendy Calmenson
Copy Editor Lura Harrison
Photo Researcher Lindsay Kefauver
Technical Illustrator Interactive Composition Corporation, Pleasant Hill, California
Compositor Weimer Typesetting, Indianapolis, Indiana
Cover Designer Donna Davis
Cover Painting Klee, Paul. *Fire at Evening.* 1929. Oil on cardboard, 13⅜ x 13¼″. Collection, The Museum of Modern Art, New York. Mr. and Mrs. Joachim Jean Aberbach Fund. Photograph © 1990 The Museum of Modern Art, New York.
Signing Representative Charlie Delmar
Acknowledgments: p. 532, *The Thinker* by Auguste Rodin, Musée Rodin, Paris (Marburg / Art Resource, NY © 1989 ARS, New York / SPADEM); p. 594, *Guernica* by Pablo Picasso, The Prado Museuem, Madrid (Giraudon / Art Resource, NY © 1989 ARS, New York / SPADEM).

Printed in the United States of America 19

1 2 3 4 5 6 7 8 9 10—94 93 92 91 90

Library of Congress Cataloging-in-Publication Data

Rafalko, Robert J.
Logic for an overcast Tuesday / Robert J. Rafalko.
p. cm.
ISBN 0-534-12552-2
1. Logic. 2. Critical thinking. I. Title.
BC71.R34 1990 89-48118
160—dc20 CIP

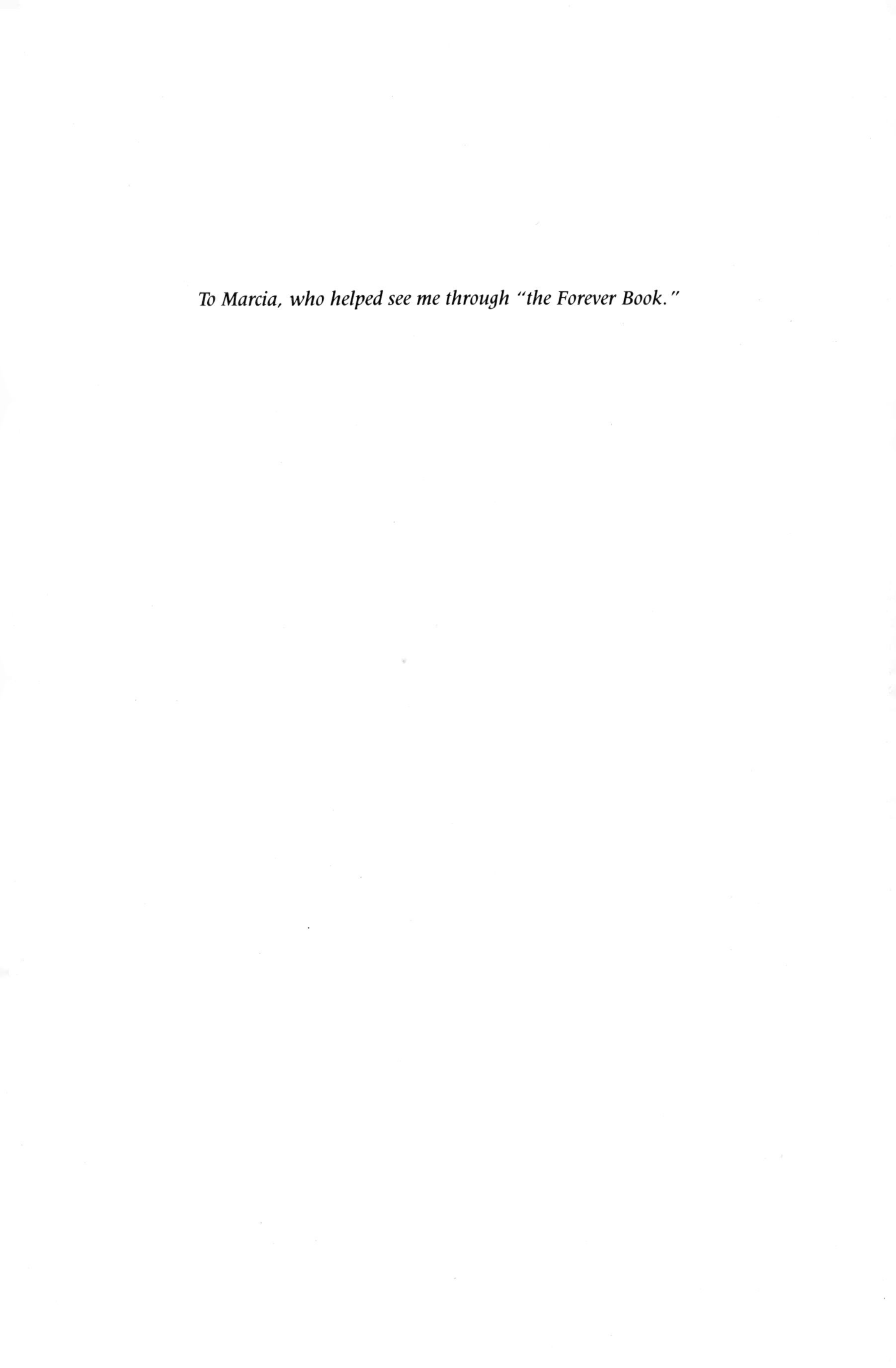

To Marcia, who helped see me through "the Forever Book."

CONTENTS

Module 21 Tips for Testing Venn Diagrams 204

Module 22 Appendix to Unit Two: A Complete Guide to Venn Diagrams 224

UNIT THREE

SYMBOLIC LOGIC 255

Module 23 Symbols 257

Module 24 Truth Tables 278

Module 25 Atom Smashing Molecular Arguments 297

Module 26 Polish and Other Systems of Symbolic Notation 302

Module 27 Semantic Trees 313

Module 28 Rules of Inference 331

PREFACE TO THE INSTRUCTOR

Logic for an Overcast Tuesday began as a byte in my Macintosh computer on a sunny fall day in a beach house on the coast of North Carolina and was finished many megabytes later in another sunny locale at the foot of mountain peaks in Southern California. Overcast Tuesdays are rare in Eden but, as Charles Sanders Peirce observed, they're the best days to do logic.

Writing my own book required me to rethink the overall goals of existing textbooks. I wanted an approach that retained rigor but spoke more directly to the students' interests. I thought we needed a book that attempted to provide continuity between the increasingly disparate subjects of Critical Thinking, Introduction to Logic, and Symbolic Logic, and I wanted adequate coverage of these topics. I also hoped to create a book that would explore new ways of organizing the material. My answer came in two parts: a modular system and a unifying theme. Finally, I wanted a book that addressed as well as exemplified creativity in logic.

COVERAGE

Logic for an Overcast Tuesday is divided into five units. Unit One, "Statements and Arguments," teaches logic students how to recognize when an argument is under way and how to use some rudimentary techniques for evaluating arguments and constructing good ones of their own. Unit Two, "Traditional Logic," introduces students to the techniques of Aristotelean logic in a way that pays attention to the origins of the techniques. Unit Three, "Symbolic Logic," brings into play the techniques of modern symbolic logic in a way that puts the student who is wary of symbolic language at ease. Unit Four, "Arguments and Fallacies," presents fallacies that illustrate how arguments can go astray. Unit Five, "Creative Thinking," introduces creative problem solving and offers a unique approach to logical induction.

In an important respect, *Logic for an Overcast Tuesday* is several logic textbooks in one; and it is unlikely that anyone will be able to cover all of the material in one semester. Each of the parts stands alone so instructors can use any combination of the five units in the book in the order they prefer.

MODULAR APPROACH

Logic for an Overcast Tuesday features a modular approach. Because many instructors do not assign all chapters or all sections of introductory texts, modules allow for greater flexibility: Instructors can pick and choose to suit their teaching style or syllabus. For example, a course in Critical Thinking can easily be constructed using selected units of this book. I recommend the following plan:

A Sample "Critical Thinking" Course

Key Module

Unit One: Statements and Arguments

Unit Four: Arguments and Fallacies

Unit Five: Creative Thinking

An Introduction to Logic course could be organized as follows:

A Sample "Introduction to Logic" Course

Key Module

Unit One: Statements and Arguments

Unit Four: Arguments and Fallacies

Unit Two: Traditional Logic

Unit Three: Symbolic Logic (select modules)

A more rigorous course in Symbolic Logic might be built around the entirety of Unit Three, "Symbolic Logic," and might skip Unit Four, "Arguments and Fallacies."

Notice that the Key Module is at the head of the organizational plan of any course based on this textbook. The Key Module, as the name suggests, holds the "key" to the organization of the modular system. This module, placed at the beginning of the book, gives definitions of crucial terms in an elementary fashion. Thus, an instructor wanting to teach a course with an emphasis on symbolic logic may omit much of the material from Unit One and concentrate on the more formal logic in Unit Three. Normally, the problem with selecting a later chapter in a book presupposes definitions of terms that are introduced earlier in the text, so the earlier material must be covered to understand the later material. The Key Module eliminates such worries since most of the terms are defined there.

UNIFYING THEME

I employ a *theme* to organize the material. Every module holds the key to unlocking a further component of the definition of validity. In Unit One students will come to know basically the following components of the technical definition of *validity*: argument, deductive argument, conclusion, and premises. The definition of validity appears as a footer on the first page of most modules. There, the definition of validity is reproduced, emphasizing one (sometimes more than one) component of the definition as it is discussed in that module. For example, in Unit One I give guidelines for distinguishing arguments from exposition. The component of the validity definition under discussion is arguments, so the footer on the first page of Module 4 is presented in the following way:

A deductive **argument** is valid, if and only if, if all the premises are true, then the conclusion must be true.

Also, in Unit One, where two components of the definition, "argument" and "premises," are discussed, the footer on the first page of Module 8 looks like this:

A deductive **argument** is valid, if and only if, if all the **premises** are true, then the conclusion must be true.

This footer may be used as a running index to the material in the text. It tells students where they are in relation to the definition of validity when deductive validity is relevant to the discussion. In Units Four and Five, where relevant, I use the definition of *soundness* or the definition of *good inductive arguments* as the footer.

CREATIVE THINKING

This is a textbook on *creative* as well as *critical* thinking. By this I mean (in part) that students, upon the conclusion of the course, will be able to generate good arguments of their own as well as criticize those of others. That is, this textbook affords both an active and a passive perspective to the student. For example, the book has students criticize the argument form of editorials and letters to the editor, and, in addition, it gives instruction, in a rudimentary way, on how they can write letters to an editor in a fashion that will bring their arguments respect while avoiding the editor's blue pencil. I have found this to be a very useful exercise.

Unit Five, "Creative Thinking," is the fullest expression of the book's emphasis on creativity. This unit is a new approach to induction that not only gives guidelines for recognizing good inductive arguments but also gives tips on how to generate good hypotheses, how to make accurate and comprehensive observations, and how to evaluate and create good inductive arguments.

SPECIAL FEATURES

- "Special Interest Boxes" include practical applications and topical examples in boxed features that appear throughout each of the five units of the book. In Module 29 I box material that illustrates how direct derivations mimic the way we think. In Module 23 I box instructions on how to symbolize IRS instructions. In Unit Four the boxes illustrate topical examples of informal arguments and fallacies.
- I introduce simplified statement diagrams. When statement diagrams were first introduced, they were intended to be a way to give a rough-and-ready idea of logical structure to novice logic students. What they have become instead is almost an alternate form of symbolic language. Such diagrams should not make too much of logical implication—not at this stage. They were originally intended to show how some asserted statements support other, more central statements. I've tried to get back to the original intent by reintroducing Monroe Beardsley's "Rules of Grouping and Direction." I've also retained the idea of diagramming *asserted* statements rather than just plain statements, since this makes the job more comprehensible.
- Statement diagrams shake the resolve of even the best students when the passage is more than five statements or so, and I judge they're right to be intimidated. As a remedy, I introduce *paragraph* diagrams, which allow us to keep the original intent of uncovering the rough-and-ready structure of arguments and allow more useful (and less painful) applications, even in extended arguments such as those by political columnists in newspapers. Paragraph diagrams are my own innovation and, I believe, a genuine improvement on diagramming techniques.
- In Modules 27 and 31 I introduce Semantic Trees for the Sentential Calculus and for the Quantificational Calculus. Few books utilize Semantic Trees, and those that do consign the technique to an appendix. I wanted access to Semantic Trees in my own teaching without having to escort my students through the narrow straits of direct deductions and difficult symbolic translation.
- In Unit Two, I give more detailed instructions on how to construct Venn diagrams than may be found in most logic texts. This unit not only includes

some tips on how to evaluate Venn diagrams but also includes an appendix of all possible Venn diagram forms.

- In Unit Three, the text takes the student through the construction of truth tables in much greater detail than other texts. Most problems in working truth tables are not, I find, problems in interpretation but rather problems in setting up the tables. When my students follow these "tips" few get the tables wrong, and the vast majority find interpretation to be very easy.

Finally, *Logic for an Overcast Tuesday* is a logic textbook with a difference. It has a quirky title, the writing style is informal and (I hope) sometimes humorous, my approach is often slightly irreverent or highly experimental, and I adopt a point-of-view about what logic is, where it came from, and where it is going.

ACKNOWLEDGMENTS

I want to thank a number of people for their assistance in the preparation of this manuscript.

First, and most importantly, I want to thank my logic students—especially those students at UNCW who compared my handouts to the textbooks we were using and who urged me to write a textbook of my own. I want to thank the subsequent logic students I taught who unselfishly contributed so much time, thought, and attention to helping me perfect the manuscript. This book is as much theirs as it is mine. I wish I could thank each one of them by name.

Next, my thanks to the editors at Wadsworth: Ken King and Donna Linden. *Periculum in mora,* but *per angusta ad augusta,* I hope.

I owe special thanks to Robert W. Burch of Texas A & M University and James C. Anderson of Colby College, who contributed so much to this project. I did my very best to make my revisions do justice to their suggestions. I'd also like to thank the publisher's other reviewers—Martin Dillon, SUNY–Binghampton; Jonathan Gold, Edward J. Kies, College of DuPage; Kevin C. Lavelle, University of New Mexico; David Paulsen, Evergreen State College; Phyllis Rooney, University of Iowa; James D. Stuart, Bowling Green State University; Ken Warmbrod, University of Manitoba; S. K. Wertz, Texas Christian University—all of whom reviewed parts of the manuscript. My thanks as well to Ferenc Altrichter, Jon Huer, Lloyd Rohler, and Steve Weiss of UNCW, who made many suggestions and contributions; to Norm Prigge, Jacquelyn Kegley, and Paul Newberry at CSUB, who helped see me through the finished product; and to Michael Wreen and Philip E. Devine, who read parts of the manuscript and gave helpful comments.

Without the help of the people mentioned, some of whom read draft after draft of this project from the formative stages of development to the present result and who generously offered comments, suggestions, and criticisms, this book would be greatly impoverished. Even so, Murphy's Law prompts us to expect that errors will inevitably get by even the closest scrutiny. Wherever the principle holds true, the fault is wholly my own.

LOGIC FOR AN OVERCAST TUESDAY

INTRODUCTION

This is a book mainly about the validity and invalidity of arguments.

The textbook emphasizes critical discrimination between good arguments and bad ones. More technically, and more precisely, we're mainly interested in that set of deductive arguments that are *valid*, and that set of inductive arguments that are *good*. We need to develop a sharp understanding of these concepts, and we need to unearth methods which will enable us to apply that concept in real-life settings.

As such, the long-term objective is two-part: to impart a skill and to foster a conceptual understanding. The one will serve to underscore the other on the theory that one's skill in logic increases insofar as his or her understanding of the concept of validity improves. However, we have a dialectical "give-and-take process:" when one's understanding of the concept increases, the best way to continue to improve that understanding is through practical application and the honing of skills.

We'll waste no time. Memorize the following definition, and memorize it with care. Get every word exactly right, repeat it, recite it, and know it inside and out:

> **A deductive argument is valid, if and only if, if all the premises are true, then the conclusion must be true.**

> *"It is easier to think well than it is to think badly."*
>
> —RENE DESCARTES

Don't be surprised that you don't yet understand this definition. Naturally, many of the terms will be unfamiliar to you and you shouldn't be intimidated by that. Think of this definition as the theme of the textbook.

Each word or component of this definition needs careful explanation and definition itself. The first thing we will study is the word 'argument'. As the course proceeds, you will learn what an argument is and how to distinguish it in real-life examples from samples of discourse that are *not* arguments. We will then take each component of the definition in turn.

Take this as a promissory note and cash it in. At the end of the reading of this textbook, you will understand the definition. If my textbook is well written, and you read it with care, study it with instruction from your teacher, and work the exercises diligently, then you will be able to take this note and cash it in: you will understand (with varying degrees of proficiency) what the concept of validity is about, and how to use it.

Students are understandably wary about they don't understand. Everyone is at least somewhat uneasy with the unfamiliar. But consider: you shouldn't be surprised that you will encounter concepts and ideas in college that you don't yet understand. The purpose of a college education is to instill those concepts and ideas. The key is to have confidence in your ability to grasp the unfamiliar, and to have confidence in the quality of your instruction at the college level. To that extent, you have something of a gauge to judge the success of your course work. If you apply yourself, you will come to know a good deal of what validity is about. If you understand that, then you've gone a long way toward understanding what logic is about and how to use it.

You should recite this definition until you have it memorized. You should know the definition as well as you know, say, the Pledge of Allegiance. As the book proceeds, you will learn about the individual components of the definition.

You need first of all to understand what an argument is, and how to distinguish it from non-arguments. To do that, you first need to know what statements are and asserted statements as well.

When you learn that, you need to know the special types of arguments—the inductive and the deductive—and how to recognize them.

The explanation of the definition of validity will require discussing *logical operators*. To some degree, you should already have a native speaker's rough-and-ready proficiency with some of them. You should have some idea of what the word 'and' means, for example. You will learn more about it later. You probably have a good idea what 'if . . . then' means in a sentence. You will learn even more in this course. You will also learn how to use concepts like 'all' with facility and precision.

The only operator that probably looks completely unfamiliar to you is the barbarism 'if and only if'. That has a specialized meaning for philosophers and other scholars. You will learn what that means also.

Finally, you will learn through practice and example what the necessity of 'must' consists of.

When you have done this, you will have a pretty good idea of what validity is about.

This definition of validity has a special kind of magic to help sort and arrange your thinking. I once read a science fiction novel that had an intriguing idea that will help you see just how special this definition is. In the novel, the protagonist was taught a new and powerful sort of artificial language. Just by learning the word for a thing, he instantly knew what the thing was and how it worked. In this fanciful language, let us suppose he learned the word for "television." Just by knowing the word, he was not only able to recognize it but also to say how it worked, what components were in it, and how to build one from scratch. Now this is, as I say, science fiction, and real life isn't that way. Such an artificial language is probably impossible. But, with this definition of validity, here we find something very much like it. The best way to learn how to understand the definition is to learn something about how to apply it. The more you learn how to apply it, the more you learn about what it means. The more you learn about what it means, the more you will learn how better to apply it still.

So, here's the definition once more. Memorize it, and when you've done that, we'll begin the task of learning what it means and how to use it, both in criticizing the arguments of others and in constructing good arguments of your own.

> **A deductive argument is valid, if and only if, if all the premises are true, then the conclusion must be true.**

KEY MODULE

LOGIC DEFINITIONS

(This module, the definitional module, is a necessary starting point for each of the units of the textbook. To master any unit of the book, some understanding of the definitions contained here must be first obtained. Refer to this Key Module as a valuable resource and as a reference for review.)

A **statement** is a sentence, or a sentential clause, that is either true or false. An **assertion** is a statement that positively declares the speaker's or writer's conviction that the statement is true.

How does a statement differ from a sentence or a proposition? Each of these terms has been employed as the unit of logical discourse by logicians from time to time. Some logicians prefer the term *sentence* to statement or proposition because it avoids some questionable assumptions that, arguably, words like 'statement' or 'proposition' entail.

We need not worry about these issues here, but it is important to see why 'statement' is preferable to 'sentence' as our unit of logical argument. The reason is that 'statement' more precisely describes conditions of truth or falsity than does the word 'sentence'. As you know from high school grammar classes, *declarative sentences* are sentences that are true or false. Interrogatives, exclamations, and so forth usually don't take truth values. So, why not stick to the declarative sentence terminology and forget about statements? One reason (as we shall see in the next section) is that sometimes, in unusual circumstances, compound interrogatives and compound exclamations can take values of true or false. But even that isn't the main reason. We use the terminology of statements here because the term 'sentence' fails to catch all of the cases of truth-value bearing sentences that we want. A sentence with a compound clause, for example, may actually contain two statements. For example:

EXAMPLE 1: "Cincinnatus put down his plough and picked up his sword" = two statements, *Cincinnatus put down his plough* and *Cincinnatus picked up his sword.*

Sometimes two or more sentences make up a single statement. For example:

EXAMPLE 2: "Is the city of Cincinnati named after the Roman citizen-soldier, Cincinnatus? You bet it is!" = just one statement, that the city of Cincinnati is named after the hero Cincinnatus. Here, "You bet it is" gives the answer to the question and asserts that it is true that Cincinnati is named after Cincinnatus.

Another difference between the two sentences is that example 2 looks like an *asserted* statement, whereas we cannot tell whether example 1 is asserted by the speaker or writer without further clue to its context. Someone may assert the whole sentence, "Either Cincinnatus put down his plough or picked up his sword." Notice that while the speaker utters "Cincinnatus put down his plough" she is not *asserting* it. Instead, she asserts the whole sentence, from the 'either' right on through the 'or'.

An **argument** is a series of asserted statements which contains *reasons* offered in support of a conclusion; **exposition** consists of any combination of sentences which do not invite belief in or assent to the truth of a conclusion on the basis of a set of reasons offered in support.

An argument contains (1) one assertion set forth as a conclusion, (2) one or more assertions set forth as reasons in support of the conclusion, and (3) a claim (often implicit) that the reasons support the conclusion. This claim is generally made by means of a **logical indicator,** but it may be merely suggested or implied, as may the conclusion itself.

An **argument chain** is a series of two or more arguments, each of which has a conclusion that is intended to serve as a **sub-conclusion** for a still more basic, overall conclusion. A sub-conclusion, then, serves as a reason for a more central conclusion, ultimately pointing the way to the main conclusion itself. The main conclusion can be recognized, therefore, by virtue of the fact that it is not intended to serve as a reason for anything else in the argument, though the sub-conclusions may serve as reasons for *it.*

An argument is **deductive** if it makes a claim that the conclusion follows from its **premises** with the force of *necessity;* a good deductive argument is said to be **valid.**

An argument is **inductive** if it claims at most probability or likelihood for the conclusion based on the evidence offered in its behalf. Logicians (usually) don't speak of valid or invalid *inductive* arguments; they may speak instead of *good* inductive arguments with conclusions that are either *warranted* or *corroborated* by the evidence.

Good deductive arguments conform to accepted and standard forms for an argument. If an argument commits an error in reasoning, then the argument commits a **fallacy.**

A **valid** argument is a deductive argument in which if the premises are true, then the conclusion cannot be false. Another way of putting this is to say it is a deductive argument which avoids the commission of fallacies. In a valid deductive argument, premises lend support to the conclusion in a way that makes the conclusion follow with the force of necessity. More formally, we say that *a deductive argument is **valid**, if and only if, if all the premises are* true, *then the conclusion* must *be true.*

A **sound** deductive argument is one that is both *valid* and has all true premises. More formally, we say that *a deductive argument is **sound**, if and only if, if all the premises are true, then the conclusion must be true, and in fact all the premises* are *true.*

When an argument faithfully follows only rules of inference, thus committing no fallacies whatever, but leads to apparently absurd conclusions, then we say that we have encountered a **paradox**. More formally, a logical paradox consists of two contrary or even contradictory statements to which we are led by apparently sound arguments.

TESTS FOR DEDUCTIVE AND INDUCTIVE ARGUMENTS

Definitions

A deductive argument is VALID, if and only if, if all the premises are true, then the conclusion must be true.

A deductive argument is SOUND, if and only if, if all the premises are true, then the conclusion must be true, and in fact all the premises are true.

Tests for the Validity of Aristotelean Syllogisms

An Aristotelean Syllogism is valid . . .

1. if the argument follows rules of inference without exception.
2. if the syllogism commits no formal fallacies.
3. if the syllogism is valid on the Method of the Three Rules.
4. if the syllogism is valid on the Venn Diagram Method.

Tests for the Validity of Molecular Arguments

A Molecular Argument in the Sentential Calculus is VALID . . .

1. if the argument follows only rules of inference.

2. if the argument commits no formal fallacies (not all formal fallacies have names).
3. if the argument is Truth-Table Valid.
4. if the argument meets the test of the Semantic Tree Method.

Tests for Good Inductive Arguments

1. The argument commits no inductive fallacies.
2. The argument accords with the scientific method.

UNIT ONE

STATEMENTS AND ARGUMENTS

MODULE 1

ARGUMENTS

1.1 ARGUMENT FOR ITS OWN SAKE

Nothing is more captivating than a good argument. Nothing challenges our abilities more, nothing else passes the time more readily, nothing piques our interest so much as a friendly disagreement backed up by good reasons.

Unfortunately, quarrels and brawls are also sometimes called arguments, and that gives the art of argumentation a bad name. A family quarrel, a heated dispute in a sports tavern, letting off steam about politics—each of these is called an argument. We've all heard the caveat: "Never argue about politics or religion," even if these are among the more interesting topics to argue. The presumption is that no good can come out of quarreling. But what *we* mean by an argument is a set of reasons asserted in support of a conclusion. A *good* argument is a compilation of *good* reasons offered in support of a conclusion. The fact that the word 'argument' is sometimes used as another name for a fight is a lamentable oddity of the language. Quarrels and logical arguments have little in common.

Granted: sometimes even the calmest and most rational arguments break down into quarrels, but when that happens the purposes of argumentation are lost. Logical arguments should resolve differences rationally, but we find nothing rational in a shouting match.

Rather than a *source* of disagreement, reasoned argument is what can *resolve* the disagreement.

Why do we argue? One reason is that argument is the way to knowledge. Knowledge doesn't expand by means of polite agreement; it grows and flourishes by rational criticism, by reasoned disagreement. The idea that science progresses by a process of harmony and agreement is a fantasy. In fact, the greatest gains in science are often made in the midst of heated dispute. Argument advances knowledge.

Undoubtedly, argumentation has utility, and the acquisition of knowledge is certainly a noble goal, but often, when we're good at it, the reason we argue is for its own sake—because it is a source of pure enjoyment. Every town has its redoubtable sports fan who is a source of statistics and traditions about sports. Such sports buffs are ever in search of equals in sports trivia who can give as good as they get. No one seriously expects to resolve the question, for example, of whether Babe Ruth deserves more credit for his 60 home runs in a season than Roger Maris does for his 61; the baseball fan will argue the point for the sheer pleasure of the competition.

When reading a newspaper, many people first open the comic section; for others it's the business section, the sports section, the lifestyles section, or the movie listings. For others still—those who have acquired a taste for argument—it's the editorial page where editors and columnists and letter writers square off with arguments. Readers with a background in logic take naturally to the challenge offered by professional masters of argument. The same holds true for those special television interview shows where skilled reporters take on professional politicians and other experts about issues of the day. *The McNeil/Lehrer Report, Firing Line, Nightline, Meet the Press,* and *Face the Nation* are examples of such shows. We develop a far greater understanding of the democratic process by exposure to such programs. In fact, we may even venture to say that argument is the backbone of democracy. The freedoms of expression guaranteed by democratic constitutions are precisely freedoms to argue and disagree.

This brings us to another point: arguments *persuade.* Sometimes they persuade us of conclusions that we shouldn't believe, just because the techniques employed don't warrant the conclusion. Other times they persuade us of conclusions whose truth we don't want to believe, but believe them we should because the arguments do warrant the beliefs. We may say of these cases that the arguments *rationally* persuade us.

1.2 ARGUMENT: LOGICAL VS. RHETORICAL ARGUMENT

We may discern two sorts of argument: rhetorical argument and logical argument.

> *"Histories make men wise; poets, witty; the mathematics, subtle; natural philosophy, deep; moral, grave; logic and rhetoric, able to contend."*
>
> —FRANCIS BACON

We know what a "rhetorical question" is: it is a question that is raised for effect or emphasis, without any serious expectation of a reply.

Rhetoric is "What *actually* persuades," whereas **logic** is "What *ought* to persuade."* Logic and rhetoric are separate, but not incompatible, forms of discourse. We can easily see why: logic, without rhetorical flair, may stand on its merits but fail to persuade the audience for whom it is intended. On the other hand, rhetoric, without serious regard for logical precision and accuracy, may persuade others to accept a belief which is spurious or prejudicial. Ideally, we should combine both, persuading our audience at the same time that we present them with arguments they *should* accept (because they are rational or worthy of belief).

Since the ideal of argument is both success in persuasion *and* truthfulness, we should strive to combine good rhetorical technique with good logical technique. Our arguments may be logical enough, but some skill in rhetoric and diplomacy is called for if we want to be successful in getting others to listen to us. Rhetoric and logic should be two sides to the same coin.

We'll see many examples of spurious arguments presented with rhetorical flourish when we consider the informal fallacies in Unit Four, but for now we need only look to television commercials to see the abuse of logic when powerful rhetorical style is present. Why should we believe that a product is worth purchasing, just because it's endorsed by a popular movie star? Obviously, that's a poor reason from a logical point of view to spend our hard-earned dollars. Nevertheless, many people buy products in the mistaken and illogical belief that if they own a pair of designer jeans, then they can be just like a movie starlet. Even a cursory examination of the logic involved shows how bad the persuasive content of such an argument must be. Nonetheless, people actually buy designer jeans just because they want to be like a movie starlet, even though such jeans may be made of inferior material and priced well beyond their actual value.

> *"Logic teaches us . . . not to undertake vain proofs; it teaches us not to attempt to disprove the beliefs of those who really differ from us in fundamental principles. We can successfully argue only with those who start from the same premises as we do. If our opponent starts from entirely different premises then there can be no question of conclusive proof, although we may be able to induce our neighbor to question his fundamental assumptions by pointing out to him the dubious character of the consequences which follow from his assumed principles. Tolerance; the avoidance of fanaticism; and, above all, a wider and clearer view of the nature of our beliefs and their necessary consequences is thus a goal or end which the development of logic serves. In this sense logic is a necessary element of any liberal civilization."*
>
> —M.R. COHEN, *A Preface to Logic*

*The explanation of this distinction is adapted from *Thinking Straight* by Monroe Beardsley, Englewood Cliffs, NJ: Prentice-Hall.

A similar point can be made about logic without regard to rhetorical persuasiveness. No matter how tight the argument, without a regard for the effect of the argument on its audience, an argument can fail to get the hearer to accept the beauty of its conclusion. Perhaps such a problem was evident in some manners of protest against the war in Vietnam during the '60s and early '70s. The anti-war protestors presented powerful arguments against further American participation in the war, but sometimes did so with more zeal than effect. Arguments in settings of confrontation are less persuasive than arguments which invite and engage dialogue.

Aim an argument to your audience, tailor it so that they will be inclined to regard it, then hit home with unyielding logical force, and you may find supporters where otherwise there were confused or misguided opponents. Notice that I'm not saying that one should resort to any rhetorical technique to win the day, but only those which are consistent with the rules of logical form and which appeal to the people spoken to.

MODULE 2

STATEMENTS

2.1 THE UNIT OF ARGUMENT

Within the unit of logic the basic molecule is the **argument.** Granting that, we may ask what is the basic unit of the argument itself. Some controversy exists on this point. Some logicians (such as Willard van Orman Quine) say it is the **sentence;** still others (such as Bertrand Russell) insist on the **proposition.** For our purposes, we shall settle on the **statement** as the basic unit of meaning in an argument.

What are the differences between the three and why does the controversy over them exist? Let us first compare the statement and the sentence as candidates for the argument's basic unit.

The virtue of advancing the sentence for an argument unit is its familiarity and economy: nearly everyone knows what a sentence is, nearly everyone can recognize a sentence, and commitment to the sentence as our argument unit precludes some thorny metaphysical problems. However, the *statement* addresses our needs more precisely in finding the unit of argument. We say that **a *statement* is a sentence, or sentential clause, that is either *true* or *false*.**

Now we see the reasons for preferring the statement over the sentence. First, not all sentences are true or false. Grammarians reserve mainly *declarative* sentences for determinations of truth or falsity. Simple questions, simple commands, and simple exclamations are neither true nor false.

A deductive **argument** is valid, if and only if, if all the premises are true, then the conclusion must be true.

Second, parts of sentences may be true or false. Compound clauses and other independent clauses may take truth values—for this reason, two or more statements may be contained in a single sentence. Thus, the statement nicely describes these units of truth or falsity which are so central to the art of argumentation.

Why prefer the term 'statement' to 'proposition' when searching for the unit of argument? The reasons here are somewhat more complex. Let us think of a proposition as as "the kernel of meaning" in a sentence. Some logicians even regard this kernel of meaning, the proposition, as language independent in certain respects. Thus, the English sentence 'It is raining' is said to contain the same proposition as the French 'Il pleut', and the German 'Es regnet'. Like the statement, propositions hold advantages over simple sentences since two or more propositions can be contained in the same sentence, and complex interrogatives, exclamations, and imperatives may sometimes be said to contain propositions which take truth values.

However, 'proposition' seems to involve odd claims or assumptions that 'statement' does not (or, at least, not so flagrantly). A rough analogy may be helpful here. Let's think of the sentence as the vehicle, the corpus or "body," which carries meaning. Analogously, the proposition is the "disembodied soul" of meaning in a sentence. By some accounts, propositions may be thought of as the *real* meaning of a sentence and the very same meaning can be expressed in other ways, by other sentences, even in different languages.

Continuing our analogy of the proposition to the disembodied soul, and the sentence as the body which carries the meaning, we may think of a statement as a middle ground between the two. Let's say that the statement is the body *and* soul of meaning. In other words, meaning cannot be made independent from the vehicle which carries it. On this account, we are "proposition atheists"—we deny the existence of propositions. True, commitment to statements seems to entail problems with translation. It may require us to deny that any two sentences can ever mean exactly the same thing, but nuances between sentences, tone, vocabulary, and even cultural and language differences make that translation gap a plausible outcome. Whatever the defects of deciding on statements as the unit of argument, they seem less severe than reliance on either sentences or propositions to do our work.

Even so, we need to be more precise. Let us say that **asserted statements** are the genuine units of arguments. **An asserted statement is a statement which someone is earnestly trying to convince us is true.** Statements can be uttered without asserting them. An actor, for example, can recite Shakespeare, "We are such stuff / As dreams are made on, and our little life / is rounded by a sleep . . . ," without having any personal convictions whatever about the illusion of life, or its dreamlike qualities. In fact, we find that most poetry and art show an absence of asserted statements, since their purpose is not argument but expression.

An argument, however, comprises asserted statements. The function of argument is to state not only truths, but to *assert* them, to convince us that they are true.

Below are some rules which will help us distinguish between types of sentences and their corresponding statements.

2.2 RECOGNIZING STATEMENTS

Logicians formally restrict the class of things we call "statements" to written or spoken sentences or parts of sentences that are true or false. Usually, these are declarative sentences (i.e., sentences which "declare" information, whether of fact, wish, intent or feeling).

Some declarative sentences contain independent clauses. You'll recall from your English grammar classes that **independent clauses** are clauses that can "stand by themselves." In other words, independent clauses are *statements.* Such clauses are introduced by conjunctions or by semi-colons. For example:

> **"We descended to one thousand feet and there we saw the runway at last."**

The clause "there we saw the runway at last," is an independent clause introduced by the conjunction *and.* It is a statement all by itself. So is the main clause of the sentence, "We descended to one thousand feet." Thus, we see that one sentence can comprise two or more distinct statements:

(1) **We descended to one thousand feet**

and

(2) **there we saw the runway at last.**

This is the meaning of our definition, that a statement is a sentence, or *part* of a sentence, which can be affirmed or denied. The *part of a sentence* which can constitute a sentence is the independent clause.

Contrast this with statements that have *dependent* clauses. A **dependent clause** is one that cannot stand by itself. Such dependent clauses are introduced by words like 'who', 'which', 'that', and by phrases beginning with (to name only a few): 'after', 'until', 'either', and 'if'. Dependent clauses, because they cannot stand by themselves, are *not* statements. They cannot be statements because they do not form a complete thought.

Each of the following examples contains dependent clauses. As such, each sentence contains only one statement (the dependent clause is underlined):

1. After Dirk Pitt raised the Titanic, he discovered the solution to a long-standing mystery.
2. Until you read a Stephen King novel, you don't know what a horror story is.
3. Sherlock Holmes, who smoked a pipe, was a very logical guy.
4. Mr. Spock is Hollywood's idea of a futuristic Sherlock Holmes, if you know what I mean.

Again, each of these sentences is to be treated as having one and only one asserted statement.

We considered cases where one sentence may yield two or more statements, but it's now time to consider that the reverse may be true: sometimes two or more sentences add up to only one statement. You have to use your judgment for this, but a case in point would be a complex question of the sort:

> **"Do I think that Anthony Burgess is a better writer than Stephen King? I sure do."**

Notice that the question is really a *rhetorical* question: it isn't asked because we lack information; it's asked for *emphasis.* In this case, the two declarative sentences add up to one statement (to rewrite it: "I sure do think that Anthony Burgess is a better writer than Stephen King").

When you're selecting out discrete statements from a passage, treat such self-answering questions, plus their answers, as one statement:

> (1) **"Do I think that Anthony Burgess is a better writer than Stephen King? I sure do."**

We said at the outset of this section that statements are usually declarative sentences. However, some other types of sentences (interrogatives, exclamations, etc.) can qualify as statements under certain circumstances. Here is a chart to help you:

These can be statements	These cannot
independent clauses	dependent clauses
complex exclamations	simple exclamations
complex interrogatives	simple interrogatives
negative interrogatives	simple imperatives

Simple exclamations are not statements.

EXAMPLE: "Good Golly, Miss Molly!"

Simple imperatives are not statements.

EXAMPLE: "Beam me up, Scotty!"

Simple interrogatives are not statements.

EXAMPLE: "What are the latest coordinates for Hurricane Earl, Meryl?"

However, some complex exclamations may qualify as statements.

EXAMPLE: "You're a fine spectacle, you miserable creature!"

Complex interrogatives may be statements.

EXAMPLE: "The question is, who is telling the truth?"

Negative interrogatives may be statements.

EXAMPLE: "It was you who clubbed him over the head with the candlestick, wasn't it, Miss Marple?"

A final case: sentences about fictional characters or places may fall short of being statements.

EXAMPLE: "Santa Claus likes Schnapps on his Wheaties."

This case shows why certain sentences about fictional characters can fail to come out to a statement which is true or false. We can say some "true" things about Santa Claus or Sherlock Holmes (that Santa Claus wears a red suit or that Sherlock Holmes smokes a pipe) because knowing these things is part of knowing the Santa Claus or Sherlock Holmes stories. However, we have no enduring myths about the breakfast habits of either, so the sentence in our example above is just nonsense.

2.3 OPINIONS, STATES OF MIND, AND ASSERTED STATEMENTS

"I believe that democracy is wonderful!"—Is this sentence an asserted statement, a bald opinion, a report of a state of mind, or possibly all three at once? These are three different but related notions, and we must get clear on the difference.

An argument is always an attempt to persuade or convince another that the conclusion is true on the basis of the reasons one presents. Because argument is a form of persuasion or discovery, the only statements that count are statements endorsed by the speaker or writer (i.e., *asserted* statements). An exception to this is the occasion when one reports an argument without endorsing it. For example, I may report to my class that the philosopher Michael Tooley argues that if abortion is justifiable, so are some forms of infanticide because no precise line can be drawn between infants and fetuses. In this case, I don't endorse or assert any of Tooley's statements (Tooley does!), but strictly speaking this is a report, not an argument. The important difference between a report and an argument is that one asserts statements in an argument, but one does not assert statements in a report. In other words, anyone who argues is in a real sense a partisan with opinions, and we should find nothing wrong with that.

Sometimes we hear the protest in a heated argument that "It's my opinion, and if it's my opinion, it can't be wrong." This statement is simply false (surely, some opinions are better than others!). This misconception probably arises from inexperience or timidity in argument, and also because of a natural confusion about what it is one is asserting when one states an opinion.

Sometimes, in stating an opinion, one is asserting a statement about a state of affairs and endorsing it. For example, a football fan may say, "The Denver Broncos will win the Superbowl, that's *my* opinion!" In this case, one is asserting something about the Denver Broncos, and the clause "that's *my* opinion!" is a way of emphasizing one's endorsement of the truth of the statement.

Other times, one is simply reporting a **state of mind** (called "a propositional attitude"). For example, when Sir Walter Raleigh wrote, "I wish I loved the human race . . . ," he was asserting a statement about his wish (a state of mind). It is, nonetheless, a statement.

"Belief" statements and "think" statements are somewhat more complicated. When we say we believe something, we mean that we *believe it to be true*. Thus, one is saying two things at once: one is a statement asserted about our state of mind (the belief), and the other is an assertion about a state of affairs (the content of the belief). Deciding which is more important in its context for the purposes of logic requires the exercise of careful judgment. For example, should a diplomat say, "I think that an outbreak of hostilities in the Middle East is inevitable in the next five years," the diplomat is asserting a statement about war in the Middle East, not about this thoughts (or thinking process). The "I think" is redundant here and can safely be omitted without loss, except possibly for loss of emphasis. However, when a philosopher of mind says, "I think, and I am no different from any higher mammal in this respect," she is saying something about having thoughts.

Let's return to the statement we considered earlier: "I believe that democracy is wonderful!" This is a report about one's beliefs; equally important, it is a statement about democracy. However, it is not the kind of statement one usually backs up with reasons. It is an asserted statement (certainly), but it is not the kind of statement one makes when one wants to engage in debate or discussion.

First person propositional attitudes may well be statements, but they are deceptive in that what's asserted is the fact that one *feels* a certain way, or that one has a belief (i.e., they report a state of mind). A true propositional attitude does not require reasons or evidence. If you *feel* that astrology is a legitimate science, you don't need to give evidence of your feelings any more than you need to justify the statement "I am in pain"—the report is enough. Consequently, no argument is under way.

A good rule to remember is never to say "I feel . . ." when you mean "I *think* . . ." Usually, when we say, "I *think* that astrology is a legitimate science," we are preparing the grounds for an *argument:* the hearer expects to be given *reasons* why you think this way or that. No such expectation is set up when one reports a feeling. "I don't doubt that you *feel* that way about astrology," you might reply, "but what do you *think* about it, and *why* do you think what you think?"

This is not mere quibbling. People often do report feelings about certain controversial issues without seeing any need to provide justification. I've often heard people say to me, "Don't ask me *why* I feel that way about abortion, I just do." Perhaps they have no reasons to offer!

Certainly, it is true that sometimes people say, "I *feel* that astrology is a true science" when they mean "I *think* or *believe* that science is a true science," and then they back it up with reasons. But notice that it is much more accurate to say, 'I *think* . . .' in such situations. We're not reporting on feelings, we're reporting on beliefs in such situations, and those beliefs need justification. For that reason, I think it is justifiable to legislate some usage: reserve the expression 'I *feel* . . .' for when you are reporting feelings, and 'I *think* . . .' when you are reporting the outcome of deliberations backed up by reasons.

Usually, anyone who says, "I *think* that abortion is immoral," is prepared to back up his contention with an argument. Here the speaker is not only saying something about his state of mind but is also saying something about the issue (in this case, the issue is abortion). By saying 'I think . . .' in such instances, he is *asserting* something about the *issue.* We naturally expect the person to say, "I think that way *because* . . . ," and then an argument is underway.

To see the difference, imagine a pair of hypothetical discussions, one in which someone says 'I feel . . .', and another in which someone says 'I think . . .'.

ABBOTT: "I feel that Joe Caligula would make a truly lousy mayor for Rome, N.Y."

COSTELLO: "But why ever do you feel that way? I know Joe; he's a capable guy."

ABBOTT: "I don't know why I feel that way. Maybe I got up on the wrong side of the bed, or maybe it's just my indigestion!"

Abbott is not offering an argument here; he's merely giving an explanation (in this case, a physical one) for his feelings. Compare that exchange with this one:

DRUSILLA: "I think Joe Caligula would make a superb mayor of Rome, N.Y."

MESSALINA: "Why do you say that?"

DRUSILLA: "Well, he's young and dynamic, he's tough on crime, and he's for the death penalty. Rome, N.Y. could use a man like that!"

This is clearly an argument. When Drusilla reports that she *thinks* Joe Caligula would make a good mayor, she is not saying merely that she has *thoughts;* she is asserting a conclusion which she can back up with reasons.

Notice an oddity about belief-statements: sometimes belief-statements work like 'I think'-statements, and other times they work like 'I feel'-statements. Consider two examples:

"I believe that Governor Axlerod should be empowered to review parole board decisions because she is given a public trust to protect the citizens of this state."

Belief-Statement 1

"I believe in Zeus!"

Belief-Statement 2

Belief-statement 1 is clearly an argument, but as belief-statement 2 stands, it is not. In the first example, the speaker is using the locution, 'I believe', to *assert* something about the Governor's review of parole board decisions, but the second example is merely a report of the speaker's belief (and, therefore, a true propositional attitude). One difference between them is the degree of commitment you might expect to find between the two speakers. The first speaker might well reject her conclusion in light of further argument, but the second belief is far more tenacious.

Sometimes we even hear the expression "I believe on the word of my friend!" This is an expression that has some currency in the South. That statement seems even less subject to possible revision than the second example. There's just no room for argument there!

Let's distinguish between the three uses of the word, 'believe':

believe that . . .

believe in . . .

believe on . . .

The first locution leaves lots of room for argument and resembles 'I think'-statements. The second belief-statement suggests that the speaker has a firm belief which may (or may not) be open to argument. The third belief-statement leaves no room for argument whatsoever (though the content of the belief may be true) and is best thought of as always expressing a propositional attitude.

2.4 CONDITIONALS AND LOGICAL CONNECTIVES

Here's an important rule about assertions: conditional ('if . . . then') statements, also called "hypotheticals," make up single assertions, and must be treated as such. A conditional is a set of statements linked by an 'if . . . then' but neither of the statements is asserted independently. The whole conditional may be asserted, but not any of its parts.

Consider, for example, that I say, "If I live until the year 2060, I'll get to see Halley's comet again." I am *not* asserting that I expect to live to the year 2060. Neither am I asserting that I expect to see Halley's comet again—in fact, I sincerely doubt that I'll get to live that long. However, I am asserting the *whole* conditional, that *if* I'm alive in 2060 A.D., *then* I'll be sure to look for Halley's comet—it's a great show! The 'if . . . then' fuses the whole sentence into a single assertion. Splitting up the statements in the conditional would be misleading; treating them as separate assertions would be a grave logical error.

Less obvious, but nonetheless true, is the fact that the rule holds for all compound sentences with one qualified case: conjunctions. Compound sentences (those which include the logical connectives *and, or,* and one special case—*if and only if*) make up *single assertions* which contain two or more *statements.* Suppose I say, "Either Pro-Lifers *or* Pro-Choicers are wrong about the moral status of abortion," I am then asserting the 'either . . . or' in the sentence—I am asserting the *disjunction* of the two statements.

Consider another example of the disjunction 'either . . . or'. Suppose you say in a moment of pique, "Look, either the movie *Witness* is the best movie of the '80s or my name is Dino De Laurentis!" You are not asserting that your name is "Dino De Laurentis." You are asserting the 'either . . . or'—the disjunction of the two statements.

Likewise, if I say, "It is not the case that Chicago is a city in North Carolina," then I am asserting the *negation* of the statement 'Chicago is a city in North Carolina'. (Compare 'Chicago *isn't* a city in North Carolina'—this is also an asserted negation.)

'If and only if' is a special logical connective used by logicians, especially for definitional or criterial purposes. If an architect says, "A building in Manhattan is the Empire State Building *if and only if* it is a building with a mooring tower for dirigibles on top of the eighty-sixth story," then the architect is asserting the whole compound joined by an 'if and only if'. We call this compound sentence a *biconditional* and the 'if and only if' is called *equivalence.* Thus, we say that the whole biconditional is asserted and not any of the statements that make up the compound.

Notice, however, that if I assert a *conjunction,* (e.g., "Both cable TV and network TV are getting more risqué in their programming,") then I am not only asserting the *conjunction* ('Both . . . and'), but also each of the *conjuncts* ("Cable TV is getting more risqué" and "Network TV is getting more risqué").

2.5 RULES FOR RECOGNIZING ASSERTIONS

1. When you **quote** someone else, the quotation itself is not a statement ("Yogi Berra said, 'It ain't over 'til it's over' ") unless you endorse the statement ("*As* Yogi Berra said, 'It ain't over 'til it's over' "). Notice, however, that while you are not asserting Yogi's quotation in the first example, the fact that Yogi said it is asserted in both examples.

2. Be careful of **hypothetical** (also called "conditional") statements. When you say, "If Bruce Springsteen ever does another album, it's sure to be a gold record," you are asserting the *whole* conditional, but not that Springsteen will ever do another album, nor that he will ever have another gold record.

3. Be careful of statements which report **states of mind** (called "propositional attitudes"). When you say, "I believe that they'll make another Rambo movie," or "I don't doubt that they'll do another

Rambo," you are asserting both that there will be another Rambo movie and that you believe it; but if you say, "I hope they don't make another Rambo movie," you are at most asserting your hopes—a state of mind.

4. When you are constructing or examining arguments, LOOK FOR THE STATEMENTS THAT ARE ASSERTED. These are the building blocks of arguments.

2.6 THE COMPOUND STATEMENT

Consider the following example:

> Some people think that a dog's behavior is learned. Either the owners have taught the dog what it knows or the dog has learned how to act like a dog from its mother, or by watching other dogs. This is wrong for a number of reasons. First of all, most dog owners have no idea, or only a vague idea, of how a dog is supposed to act. Nevertheless, almost all dogs all over the world have unmistakable similarities in their behavior. When was the last time you saw someone trying to teach his male dog to lift his leg when urinating? —Daniel Tortora, *Help! This Animal is Driving Me Crazy: Solutions to Your Dog's Behavior Problems*

What is the conclusion of this argument? In order to find this out, we have to learn something more about the nature of *statements* (as opposed to *sentences*).

In a previous section, we learned that philosophers discuss 'propositions', 'statements', and 'sentences', and we distinguished among the three of them. For the purposes of this book, we settled on the use of the expression 'asserted statement' to designate the unit (or "atom") of the argument. This was because we recognized some tricky metaphysical problems about 'propositions' and because we recognized that sometimes a single sentence can contain two or more distinct statements, each of which must be identified in order to attain the sort of precision essential for logical clarity.

In this example, we see the reverse: that two or more sentences can make up a single statement.

Let's lay out the argument sentence by sentence (but not "statement by statement"—at least, not yet) to see why this is so. Numbering each sentence in the order in which we find it, we have:

1. Some people think that a dog's behavior is learned.
2. Either the owners have taught the dog what it knows or the dog has learned how to act like a dog from its mother, or by watching other dogs.
3. This is wrong for a number of reasons.
4. First of all, most dog owners have no idea, or only a vague idea, of how a dog is supposed to act.

5. Nevertheless, almost all dogs all over the world have unmistakable similarities in their behavior.

6. When was the last time you saw someone trying to teach his male dog to lift his leg when urinating?

The author of this book is attempting to refute a commonly held view. He does so by briefly stating the view in the first two sentences, but in the third sentence he announces that the view is wrong. Sentence 3 makes no sense unless it is attached to the previous two sentences. Consequently, for our purposes, it is best to treat sentences 1 through 3 as a single asserted statement, and give all of that the status of the conclusion of the argument (for which reasons will follow).

If we were to underline the conclusion, the argument should look like this:

"Some people think that a dog's behavior is learned. Either the owners have taught the dog what it knows or the dog has learned how to act like a dog from its mother, or by watching other dogs. This is wrong for a number of reasons. First of all, most dog owners have no idea, or only a vague idea, of how a dog is supposed to act. Nevertheless, almost all dogs all over the world have unmistakable similarities in their behavior. When was the last time you saw someone trying to teach his male dog to lift his leg when urinating?"

If we were to number *asserted statements* in this argument, we should do so using circled numbers to indicate statements:

(1) Some people think that a dog's behavior is learned. Either the owners have taught the dog what it knows or the dog has learned how to act like a dog from its mother, or by watching other dogs. This is wrong for a number of reasons.

First of all.

(2) most dog owners have no idea, or only a vague idea, of how a dog is supposed to act.

Nevertheless.

(3) almost all dogs all over the world have unmistakable similarities in their behavior.

(4) When was the last time you saw someone trying to teach his male dog to lift his leg when urinating?

Should we diagram this argument, we find it's a straightforward argument chain:

(4)

. . . and

(3)

. . . and

(2)

. . . are reasons for

Notice also that sentence 6, while it is in the form of a question, is actually an asserted statement. It states a reason for believing sentence 5, that dogs all over the world exhibit similarities in their behavior. Why? Because no one actually teaches a male dog to lift its leg when urinating (that is the sort of thing which is genetically programmed in male dogs).

Let us look at one more example of a compound statement—one statement made up of two or more distinct sentences. Consider this passage:

> "Has anyone ever seen a housefly that died of natural causes? No, of course not! And do you know why not? Because they are naturally reckless. Flies are the aerial acrobats of the insect world, and they fly by the seat of their pants. When they want to fly away while they're clinging to the side of a wall or on the edge of your floor lamp, they drop off with reckless abandon. They free-fall for a breathless second, fully confident in their ability to get their wings working at full buzz. When airborne, they tantalize human pursuers who brandish deadly fly-swatters, daring them to swing wildly, cutting the air with savage swishes which set up turbulent air currents. Eventually, their recklessness catches up with them. Canny humans, who observe the fly's antics and see a likely victim hanging on a lamp, know enough to swing their swatters in a trajectory calculated to land full force a few inches beneath the spot where the fly is presently hanging. This catches the foolish daredevil in the midst of his death-defying leap, and squashes him against the lamp shade. The fly leads a precarious existence in exchange for the thrills of narrow escapes."

Once more, we see that the conclusion is best expressed as the compound of sentences 1 and 2. Thus, underscoring the conclusion, we have:

> <u>"Has anyone ever seen a housefly that died of natural causes? No, of course not!"</u>

The first sentence is a question, but the query is cashed in at the second sentence. The author is asserting that, indeed, no one has ever seen a fly that died of natural causes. He then goes on to explain why. Sentences 1 and 2 make up a single compound statement, one which here serves as the conclusion of the passage.

Summary:

1. Remember: one sentence can make up two or more statements.
2. Remember: two or more sentences can make up just one statement.
3. Remember: some questions, commands, and exclamations *can* be statements.

4. Watch the Logical Operators:
 'If . . . then' conditionals make up one asserted statement.
 'Either . . . or' disjunctions contain one asserted statement.
 'Both . . . and' statements are an exception since each statement between the *and* is an asserted statement.
 'It is not the case that . . .' makes up one asserted statement.

EXERCISES

Part I: Elementary Exercises

DIRECTIONS: *Read each of the following sentences and say whether it is an* **asserted** *statement or a sentence which includes one or more asserted statements. If the sentence is an asserted statement, then identify precisely what is asserted by the speaker. If the sentence is* **not** *a statement, say why.*

1. Mt. Whitney has an elevation of 14,494 feet.
2. Is Mt. Whitney the highest mountain in the world?
3. Mt. Washington, in New Hampshire, is the first place touched by sunlight in the United States each morning.
4. The Pacific Ocean is larger than the Atlantic Ocean.
5. What is the name of the ocean at Atlantic City?
6. Do wah diddy diddy dum diddy do!
7. Climb every mountain!
8. Mt. Washington is the highest mountain in North America.
9. The Bermuda Triangle extends from Bermuda to Cuba, from Cuba to Carolina Beach, North Carolina, and from there back to Bermuda again.
10. 'Denali' is the original name of Mt. McKinley.
11. In his book *Alaska* James Michener says, " 'Denali' is the original name of Mt. McKinley."
12. The Arctic Ocean is in the tropics.
13. Hubert said that the Arctic Ocean is in the tropics.
14. The Arctic Ocean is in the tropics—Hubert said so!
15. I suggest you learn more geography before you do these exercises.

Part II: Advanced Exercises

DIRECTIONS: *Say whether the following sentences contain* **asserted** *statements. If the sentence is an asserted statement, then identify precisely what is asserted by the speaker. If the sentence is* **not** *a statement, say why. These are "real life" quotations. Because they are taken from real life, they may prove to be more difficult than the "textbook" examples you worked in Part I. (Some of the examples may be open to discussion and legitimate disagreement.)*

1. Too bad that all the people who know how to run the country are busy driving taxicabs and cutting hair. —George Burns

2. I don't want to achieve immortality through my work, I want to achieve it through not dying. —Woody Allen

3. Say it ain't so, Joe. —A young boy to "Shoeless" Joe Jackson during the 1929 "Black-sox" Scandal

4. The only question to be settled now is, are women persons? —Susan B. Anthony

5. The race is not always to the swift, nor the battle to the strong, but that's the way to bet. —Damon Runyon

6. The Atomic Age is here to stay—but are we? —Bennett Cerf

7. When Congress passes a law, it's a joke; when Congress tells a joke, it's the law. —Will Rogers

8. Pity the meek, for they shall inherit the earth. —Don Marquis

9. I'll never make the mistake of being seventy again! —Casey Stengel

10. How much do you think I'll get for my autobiography? —Arthur Bremer

11. Remember, no one can make you feel inferior without your consent. —Eleanor Roosevelt

12. You can observe a lot just by watching. —Yogi Berra

13. Military justice is to justice what military music is to music. —Groucho Marx

14. Answer violence *with* violence. —Juan Peron

15. Get thee behind me, Satan! —*Matthew*, 16:23

16. I sometimes think that God in creating man somewhat overestimated His ability. —Oscar Wilde

17. If that's art, I'm a Hottentot! —Harry Truman

18. Wouldn't it be terrible if I quoted some statistics which prove that more people are driven insane through religious hysteria than by drinking alcohol? —W. C. Fields

19. History is bunk! —Henry Ford

20. Remember that God made your eyes,/ So don't shade your eyes,/ Plagiarize. —Tom Lehrer

21. I tremble for my country when I reflect that God is just. —Thomas Jefferson

MODULE 3

FINDING THE CONCLUSION

3.1 GUIDELINES FOR FINDING CONCLUSIONS

1. Read the passage quickly the first time. Is it an *argument*? If it is an argument, reasons will be given for a conclusion. If it is not an argument, then it may take the form of *description, narration,* or (most forms of) *poetry.* For example:

For the world, which seems
To lie before us like a land of dreams,
So various, so beautiful, so new,
Hath really neither joy, nor love, nor light,
Nor certitude, nor peace, nor help for pain;

—Matthew Arnold

This is an example of expository poetry (which is composed of assertions nonetheless). Or it may take the form of a *command:* ("Don't do what I do; do what I say!"). It may also be sheer, unsupported opinion: ("That man's a crook—that's my opinion!"). But an exposition *will not list REASONS.*

2. To pick out a conclusion, when reading the passage quickly for the first time, tentatively mark off the one, two, or more assertions you think *might* qualify for the main conclusion of the argument. Then, go back, re-read the argument, and carefully examine and eliminate some of the nominated 'conclusions' by asking yourself this question: "Is

---

A deductive argument is valid, if and only if, if all the premises are true, then the **conclusion** must be true.

this a reason for something *else* in the passage?" If it is, then your tentative conclusion is not the *central* conclusion in the argument (it may be a sub-conclusion which is used as a reason for a more basic conclusion in an argument chain). A central conclusion is not a reason for anything else in the argument.

3. Take heart, your work is almost over. When you "took nominations" for the main conclusion and came up with several candidates, you probably identified the major sub-conclusions. These form important secondary nodes in the limbs of the diagram tree. They, together with the main conclusion, compose the basic structure of the argument.

4. The conclusion may be either particular or general. One may reason from particulars to particulars, from particulars to generals, from generals to particulars, or from generals to generals (e.g., general-to-particular: All men are mortal. Socrates is a man. Therefore, Socrates is mortal).

5. Sometimes the conclusion will be left unstated, and sometimes the conclusion will be put obliquely. It is then up to you to supply the conclusion in your own words.

Consider the following argument (adapted from a vintage Carl T. Rowan editorial when Gerald Ford, Jimmy Carter, and Eugene McCarthy were all candidates for president of the United States):

1. Choosing between Ford and Carter for president is voting for the lesser of two evils.
2. Carter involves himself in controversies like the *Playboy* interview [where he said he was guilty of "lusting after women in his heart"].
3. Ford doesn't give a whit about the unemployed.
4. Despite Eugene McCarthy's laziness in the Senate, either waste a vote on McCarthy, or don't vote at all.
5. But as good Americans, you should vote.

Even though it is left unstated, the conclusion, clearly enough, is: you should vote for McCarthy.

6. Obviously, the most important part of this process is the identification of the main conclusion. *Remember these two rules:*

Rule 1. Ask yourself the question: "Is this statement a reason for anything else in the argument?" If so, it cannot be the main conclusion.

Rule 2. The main conclusion is never a reason for anything else in the argument. All the reasons offered should point to it.

3.2 LOGICAL INDICATORS

What is the main conclusion of the argument? What are the main supporting reasons in the argument? Most of the time, if the author of the argument is doing her job well, she'll come right out and tell us these things. For example, she may say, "The point I want you to see is that Dorothy Sayers is the most skilled of all mystery writers." She has just told us what the conclusion is. We should then look for her main reasons and she may well tell us what they are in this way: "My main reason for saying this is that Sayers' plots are the most technically developed plots of all!"

In other words, people who construct arguments will usually indicate to us what their conclusions and reasons are. We call these "logical indicators." The author may use an expression like 'therefore' or 'my point is that . . .' to signal a conclusion. We call such signals **conclusion-indicators.** The author may also signal reasons by saying something like 'because' or 'for the reason that . . .' and we call these **reason-indicators.**

Logical indicators (reason-indicators and conclusion-indicators) help to separate the premises from the conclusion. But sometimes there aren't any that are familiar, and sometimes (when logical indicators do appear), they're misleading; other times, no logical indicators may appear at all. For example, consider the argument:

> **"The appropriation for the B-1 bomber is extravagant. We need domestic reform."**

There are no reason- or conclusion-indicators in this argument (and it barely qualifies as an argument as it stands). However, it is clear that the first sentence is not a reason for the second, yet the second sentence can be construed as a reason for the first (obviously, the B-1 bomber's extravagance is not a reason for needing greater domestic reform. Does it work the other way around?).

The Author's Main Point

If you are in college, then you probably took a standardized aptitude test of one kind or another (such as the SAT or ACT) and displayed some success in attaining a reasonably high score. If you plan to attend graduate or professional school, you can expect to take other tests like them: the GREs, the LSAT, MCAT, and so on, through the alphabet soup provided us by ETS ("Educational Testing Service" of Princeton, New Jersey). If so, you can expect to find another section on "Verbal Ability." These usually consist of a wordy (often obscure) passage followed by multiple-choice questions. The question most asked is this: "What is the author's main point in this passage?"

We now can be more precise about what the test-givers are asking for: they want you to provide the **conclusion** of the argument. How do we pick out the conclusion? One way is to look for logical indicators in the argument. These are *signals* (but signals *only*) that an argument is underway. We discern two types of logical indications: (1) conclusion-indicators and (2) reason-indicators. These provide us with clues to the structure of an argument. However, they are *signals* only.

A signal is not an infallible guarantee. When we drive along a country highway, we may find a yellow traffic light and a sign that says: "Vehicle Approaching Intersection When Light Flashes." If you've driven on such a road and the yellow light was flashing, you know that sometimes you'll find a truck or a car and sometimes you won't. The signal is no guarantee—it is a guide.

In much the same way, conclusion-indicators are signals that a conclusion will follow, and reason-indicators are signals that the next statement is a reason for a conclusion. We can't guarantee the results. Sometimes conclusion-indicators don't always point out the main conclusion; sometimes reason-indicators don't always signal that a reason for a conclusion will be given. Why is this so? Probably because many people who use logical indicators don't know what they mean!

Below are lists of conclusion- and reason-indicators:

Conclusion-Indicators

(Words or phrases that usually show that the statement which follows is a *conclusion* or *sub-conclusion* of an argument):

*therefore . . . (This is the standard conclusion-indicator.)
*thus . . .
*so . . .
*hence . . .
*implies that . . .
*consequently . . .
proves that . . .
entails that . . .
as a result . . .
shows that . . .
indicates that . . .
suggests that . . .
permits us to conclude that . . .
we may deduce (or infer) that . . .
brings us to the conclusion that . . .
it must be the case that . . .
bears out the point that . . .
it follows that . . .
leads me to believe that . . .
accordingly . . .

Reason-Indicators

(Words or phrases that usually show that the statement which follows is a *reason* for the conclusion):

*These are probably the most commonly used logical indicators.

*since . . . (except when 'since' denotes time, as in: "I've been hungry *since* lunchtime.")

*because . . . (except when 'because' denotes a cause, as in: "The window shattered *because* the molecular structure of glass expands in cold weather.")

*for . . .

*for the reason that . . .

in view of the fact that . . .

as indicated by . . .

on the supposition that . . .

assuming that . . .

may be deduced from . . .

in that . . .

given that . . .

as shown by . . .

as indicated by . . .

owing to . . .

as is substantiated by . . .

inasmuch as . . .

Another Sort of Conclusion-Indicator

Sometimes an argument will have few, if any, reason-indicators, and perhaps no conclusion-indicators of the sort we considered. However, the direction of argument flow and the grouping of reasons displayed by the argument may help us to locate the conclusion, even if the passage is a lengthy one.

Remember that every writer seeks to communicate her main point and so she'll be likely to send up signals wherever she can that *this* is the conclusion of the argument. The conclusion-indicators we've already considered are only one way of doing this, but many times we can readily tell a conclusion merely by its context or location.

Remember this simple rule of thumb:

If a passage is well-written, then the conclusion should appear either in the first sentence or in the last (or both).

We might call this the **context conclusion-indicator.** It's a very reliable way of getting the point of the passage, though it's hardly infallible. Consider: if a writer truly wants to communicate simply and directly the main point of his argument, he'll most likely state the conclusion right up front, where everybody can see it. If the writer is particularly fastidious, he might also re-state it at the end of the passage, just to remind us. This is just good writing technique.

Sometimes, however, the writer might find it prudent to leave the conclusion unstated until the very end. Perhaps the writer is making a controversial point and wants us to wait until we read her reasons before baldly stating the conclusion. Imagine, for example, that our writer is attempting to persuade us that China has progressed far more under Communist rule than it would have under any other form of government. That would surely be controversial. If she had stated the conclusion in the first paragraph of her commentary, it might have turned off so many readers that they wouldn't bother to read her reasons. Consequently, she saves the surprise for last, hoping that we'll be carried along by the force of her reasons until we too can accept her conclusion. That's one reason for placing the conclusion at the end of an argument.

The 'context conclusion-indicator' is hardly a hard-and-fast rule—even a well-written passage may violate it for stylistic reasons, for example. But it is a very common technique of good writing, so it does bring us a measure of reliability.

Regrettably, many arguments are not well written. We might find the conclusion stated right in the middle of a long passage. Sometimes the author may not be clear himself about just exactly what his conclusion is—those of us who read "letters to the editor" know that such murkiness is not uncommon. The "Verbal Comprehension" section of the SATs will also contain such passages—we'll be asked to identify the "main point" of highly technical but often sloppy arguments as a qualification of our admission to college, inadvertently setting up such passages as models of writing we can expect to encounter in college (a lamentable thought!).

Naturally, we can't hope to apply our rule with much success in these cases. In that event, we need to muddle through as well as possible, and renew our pledge to ourselves and our readers that we will state conclusions to our arguments right up front whenever we can.

EXERCISES

DIRECTIONS: *Read the following passage carefully. Answer each of the multiple-choice questions that come after it. Choose the* **best** *answer. Mark the appropriate space on the answer sheet.*

It is now becoming apparent that overemphasis on the scientific method and on rational, analytic thinking has led to attitudes that are profoundly anti-ecological. In truth, the understanding of ecosystems is hindered by the very nature of the rational mind. Rational thinking is linear, whereas ecological awareness arises from an intuition of non-linear systems. One of the most difficult things for people in our culture to understand is the fact that if you do something that is good, then more of the same will not necessarily be better. This, to me, is the essence of ecological thinking. Ecosystems sustain themselves in a dynamic balance based on cycles and fluctuations, which are non-linear processes. Linear enterprises, such as indefinite economic and technological growth—or, to give a more specific example, the storage of radioactive waste over enormous time spans—will necessarily interfere with the natural balance and, sooner or later, will cause severe damage. Ecological awareness, then, will arise only when we combine our rational knowledge with an intuition for the non-linear nature of our environment. Such intuitive wisdom is characteristic of traditional, non-literate cultures, especially of American Indian cultures, in which life was organized around a highly refined awareness of the environment.

In the mainstream of our culture, on the other hand, the cultivation of intuitive wisdom has been neglected. This may be related to the fact that, in our evolution, there has been an increasing separation between the biological and cultural aspects of human nature. Biological evolution of the human species stopped some fifty thousand years ago. From then on, evolution proceeded no longer genetically but socially and culturally, while the human body and brain remained essentially the same in structure and size. In our civilization we have modified our environment to such an extent during this cultural evolution that we have lost touch with our biological and ecological base more than any other culture and any other civilization in the past. This separation manifests itself in a striking disparity between the development of intellectual power, scientific knowledge, and technological skills, on the one hand, and of wisdom, spirituality, and ethics on the other. Scientific and technological knowledge has grown enormously since the Greeks embarked on the scientific venture in the sixth century B.C. But during these twenty-five centuries there has been hardly any progress in the conduct of social affairs. The spirituality and moral standards of Lao Tzu and Buddha, who also lived in the sixth century B.C., were clearly not inferior to ours.

Our progress, then, has been largely a rational and intellectual affair, and this one-sided evolution has now reached a highly alarming stage, a situation so paradoxical that it borders on insanity. We can control the soft landings of spacecraft on distant planets, but we are unable to control the polluting fumes emanating from our cars and factories. We propose Utopian communities in gigantic space colonies, but cannot manage our cities. The business world makes us believe that huge industries producing pet foods and cosmetics are a sign of our high standard of living, while economists try to tell us that we cannot "afford" adequate health care, education, or public transit. Medical science and pharmacology are endangering our health, and the Defense Department has become the greatest threat to our national security. Those are the results of overemphasizing our yang, or masculine side—rational knowledge, analysis, expansion—and neglecting our yin, or feminine side—intuitive wisdom, synthesis, and ecological awareness.

______ 1. The central idea conveyed in the above passage is that

a. Rational thinking is linear, whereas ecological awareness is intuitive.
b. Our one-sided evolution through overemphasis on the scientific method has produced anti-ecological attitudes with alarming results.
c. Ecological awareness arises only through intuitive, non-linear systems.
d. We have overemphasized our yang, and neglected our yin.
e. All of the above.

______ 2. The author suggests that

a. We must abandon some Western attitudes and accommodate more Eastern outlooks in our world-view.

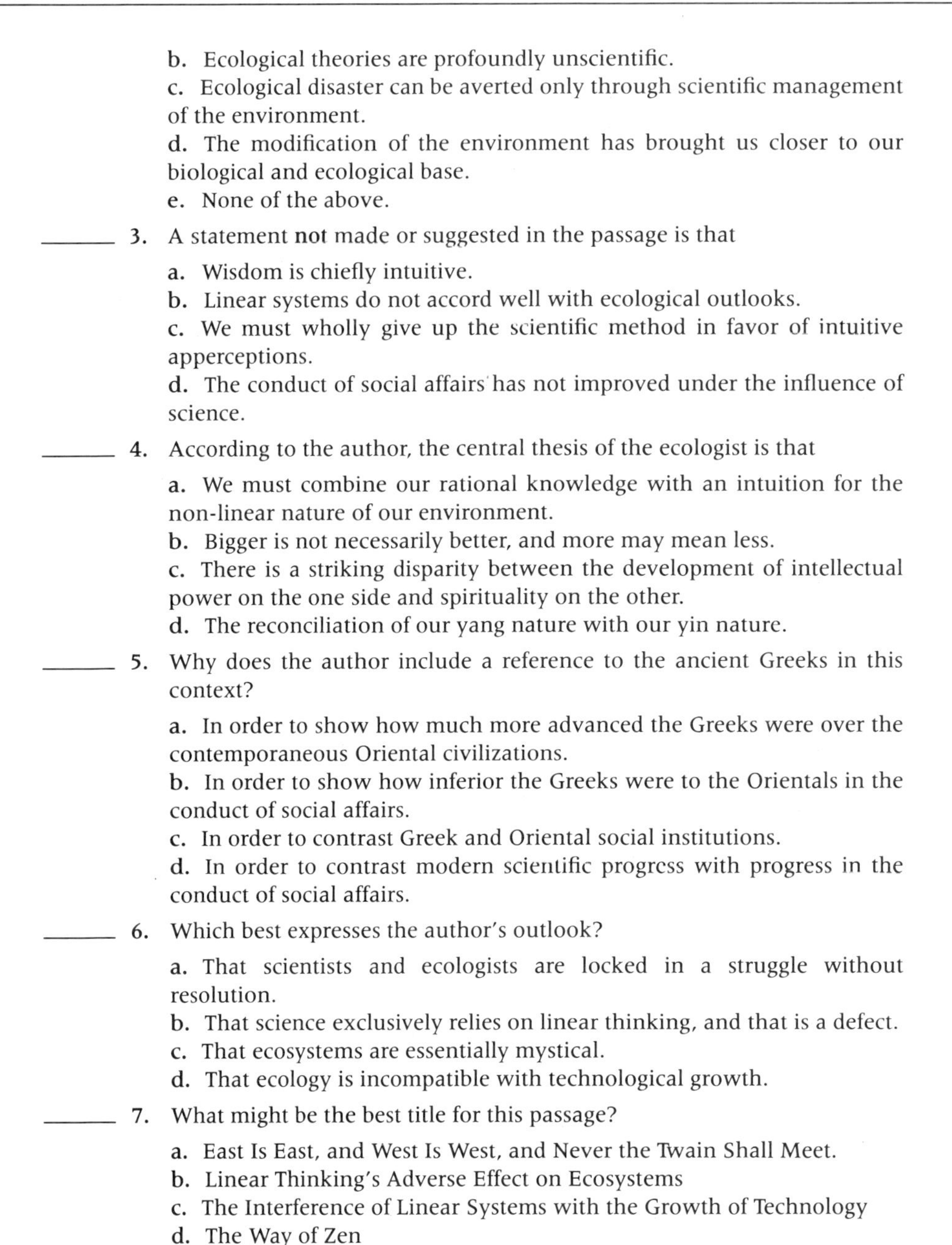

b. Ecological theories are profoundly unscientific.

c. Ecological disaster can be averted only through scientific management of the environment.

d. The modification of the environment has brought us closer to our biological and ecological base.

e. None of the above.

______ 3. A statement **not** made or suggested in the passage is that

a. Wisdom is chiefly intuitive.

b. Linear systems do not accord well with ecological outlooks.

c. We must wholly give up the scientific method in favor of intuitive apperceptions.

d. The conduct of social affairs has not improved under the influence of science.

______ 4. According to the author, the central thesis of the ecologist is that

a. We must combine our rational knowledge with an intuition for the non-linear nature of our environment.

b. Bigger is not necessarily better, and more may mean less.

c. There is a striking disparity between the development of intellectual power on the one side and spirituality on the other.

d. The reconciliation of our yang nature with our yin nature.

______ 5. Why does the author include a reference to the ancient Greeks in this context?

a. In order to show how much more advanced the Greeks were over the contemporaneous Oriental civilizations.

b. In order to show how inferior the Greeks were to the Orientals in the conduct of social affairs.

c. In order to contrast Greek and Oriental social institutions.

d. In order to contrast modern scientific progress with progress in the conduct of social affairs.

______ 6. Which best expresses the author's outlook?

a. That scientists and ecologists are locked in a struggle without resolution.

b. That science exclusively relies on linear thinking, and that is a defect.

c. That ecosystems are essentially mystical.

d. That ecology is incompatible with technological growth.

______ 7. What might be the best title for this passage?

a. East Is East, and West Is West, and Never the Twain Shall Meet.

b. Linear Thinking's Adverse Effect on Ecosystems

c. The Interference of Linear Systems with the Growth of Technology

d. The Way of Zen

MODULE 4

ARGUMENT VS. EXPOSITION

4.1 IN GENERAL

An argument is a set of reasons offered as support for the conclusion of the argument; exposition is all discourse other than argument.

Exposition, in a literary sense, is the laying out of intent or meaning, plot, or character in a written or spoken—as in storytelling—way. Our sense of exposition is a bit wider than that, however, since we mean to describe it as the exhaustive complement of argument in all spoken or written discourse. Anything spoken or written that is not argument (and not nonsense or gibberish) is exposition.

We may illustrate this distinction by means of a diagram:

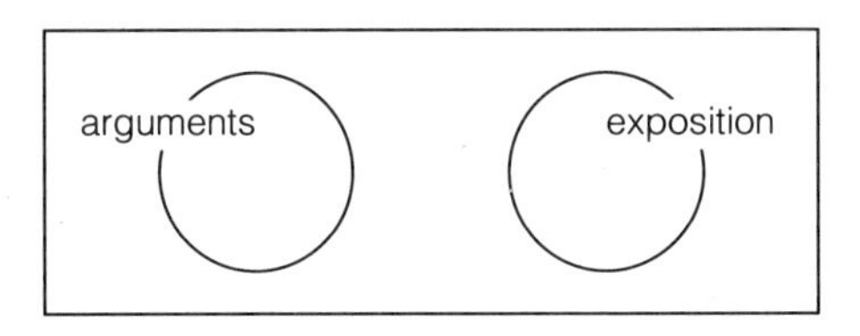

FIGURE 1

Suppose we draw two circles: one marked 'arguments'; the other marked 'exposition'. We may think of these circles as sets or more figuratively as "cor-

---

A deductive **argument** is valid, if and only if, if all the premises are true, then the conclusion must be true.

rals." In the first "corral," we place every argument ever written or spoken (we may therefore assume it is a very large corral). In the other, we place every piece of exposition ever written or spoken. Notice that the two corrals do not intersect with each other at any point, illustrating the fact that the domain of arguments and the domain of exposition are mutually exclusive. In other words, anything that is argument is excluded by the set of exposition, and vice versa.

This is a somewhat idealized way of thinking of the difference between argument and exposition, but in fact we shall find some borderline cases (especially when we consider 'explanation' at the end of this module). The problem is that sometimes the same passage can serve as argument on one occasion or as exposition on another, all depending on its context. Even so, we would do well to think of the two as exclusive—placed in a context, even the borderline cases cannot be both argument and exposition at the same time. We must decide which it is according to the way the passage is used.

A handy way of thinking about the distinction between argument and exposition is the difference between the first page and the editorial page of a typical newspaper. Consider the illustration below:

FIGURE 2

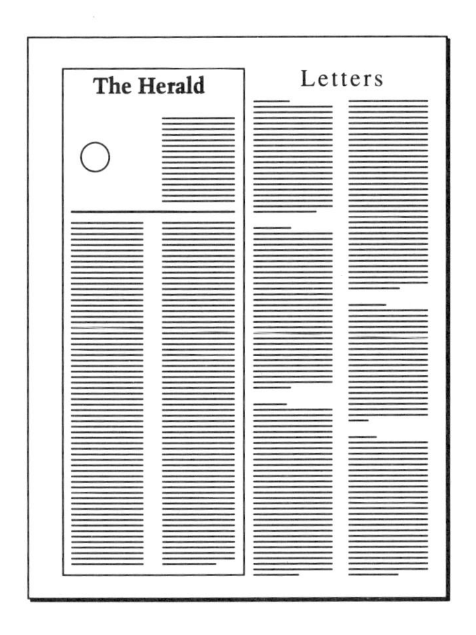

FIGURE 3

The first page (Figure 2) should be strictly exposition: uncritical reporting of events. The editorial page (Figure 3) should be where arguments are found, both in the editorials and (usually) among the letters to the editor. Notice that this is sometimes called "the opinion page." We might well take a moment to point out that opinions and arguments are hardly incompatible: the key thing to notice, however, is that argument is *supported* opinion—opinion backed up by *reasons*.

As an example of exposition, here is the lead story to the headline, "Men Walk on Moon," in the *New York Times*, July 21, 1969:

> **A Powdery Surface Is Closely Explored**
>
> *By John Noble Wilford*
> *Special to* The New York Times
>
> HOUSTON, Monday, July 21 —Men have landed and walked on the moon.
>
> Two Americans, astronauts of Apollo 11, steered their fragile four-legged lunar module safely and smoothly to the historic landing yesterday at 4:17:40 P.M., Eastern daylight time.
>
> Neil A. Armstrong, the 38-year-old civilian commander, radioed to earth and the mission control room here:
>
> "Houston, Tranquility Base here. The Eagle has landed."
>
> The first men to reach the moon—Mr. Armstrong and his co-pilot, Col. Edwin E. Aldrin Jr. of the Air Force—brought their ship to rest on a level, rock strewn plain near the southwestern shore of the arid Sea of Tranquility.
>
> About six and a half hours later, Mr. Armstrong opened the landing craft's hatch, stepped slowly down the ladder and declared as he planted the first human footprint on the lunar crust:
>
> "That's one small step for man, one giant leap for mankind."

We see why this is exposition rather than argument. Events are reported in a style that reveals them according to an order of importance. First of all—the most important point—men have landed and walked on the moon. Next, the time and place of the landing is recorded. The names of the astronauts are given. Then the fact of Neil Armstrong's famous walk and what he had to say as he stepped forth on the moon's surface is noted.

No argument is given here, only a sequence of events. This article is not designed to persuade or convince us of anything; it merely describes what happened. Granted, the reporter had to use his judgment as to what to report first—the most important item in any news story—but the reporter's interpretation of what is and what is not most important is not presented argumentatively and, in this case at least, is done cleanly and without controversy. To include an argumentative point of view in a news story is a serious journalistic flaw, one which is not made by John Noble Wilford in this place.

As an example of an argument, let's turn to page 16 of the same issue—the editorial page. Here is an excerpt from that editorial:

> **Ad Astra**
>
> In the long evolution of the human race up from the primeval ooze, no more significant step has ever been taken than yesterday's when man the worldling truly became but "little lower than the angels" and first set foot upon another planet. For thousands and thousands of years—all through

> the brief span of recorded history and through the dim ages of the darkening past—that puny creature endowed with the most powerful of all weapons, the human brain, has raised his eyes, his arms, his aspirations—first to the hills and the mountain tops, then to the skies and the stars.
>
> And now he has attained the unattainable; he has lifted himself from his little speck of matter to walk the surface and probe the depths of another world. True, it is an even tinier world than the one we know, presumably a lifeless world, a forbidding world—but another world.
>
> This is the year, this is the month, the week, the day when, so far as we can know, for the first time in all the eons of existence of this universe of ours, a sentient being has transported himself from his earthly habitat to a different sphere floating in the endless sea of space. The men that landed yesterday on the moon represented not any group or creed or nationality; they represented all humanity and they carried with them a little bit of all the hopes and struggles of mankind, to attain the heights throughout the ages.
>
> Yesterday, July 20, 1969, will be marked forever as the day man transcended the bonds of his nature and his environment, and the human race entered a new era leading to realms beyond comprehension and even beyond imagination. Man has realized the unrealizable because he dared to conceive the inconceivable; now one can believe that the limitations to the accomplishments of man are set only by the limitations of the human spirit. . . .

This is a long editorial, and I've reproduced only a fraction of it here. Nevertheless, we can see from the start that it is argumentative in tone. The editorial begins by insisting that this is the most important step ever taken by man in his entire history. Notice that such an assertion, even in the heady air of the triumph of the moonlanding, is not entirely without controversy. One can imagine, for example, that someone else might argue that even more important was the first step that Christopher Columbus took on the surface of the New World. We see that the author of this editorial would disagree, and he gives his reasons. That's the most important way to recognize that an argument is underway: that it is a passage which contains a conclusion and a set of reasons offered in justification.

When we read on, however, we learn that the editor, while he concedes that this is the most important *step* that man has ever taken, does not believe that continued exploration of space is the most important business that we should be engaged in at this time. The editorial concludes:

> When Vice President Agnew talks seriously of putting a man on Mars by the end of the century, however, it is time to take renewed thought of the nation's priorities. For a program leading to a manned journey to Mars would be of a staggeringly different financial dimension from that required to explore the moon. This country could undertake such a project by itself only at the cost of grossly neglecting its pressing social needs.
>
> After yesterday's dramatic event few will be rash enough to discount the possibility of an eventual landing on that planet, but if the objective is to be pursued at all in the foreseeable future, it should be pursued only in conjunction with other world powers—preferably with all of them under the auspices of the United Nations. Not only would costs for any one nation

then become endurable, but the goal itself would be a spur to that unity here on earth which is the best of all justifications for man's exploration of the universe.

This is what an editorial should be (even if you disagree with what it says) because it provokes one to consider new avenues of thought. This is an argument; it is designed to persuade or convince you, not merely to inform you. Perhaps the assertion in the beginning of the editorial that Armstrong's first step on the moon was symbolic of man's most important achievement in history is only mildly controversial. Much more so is the concluding paragraph which calls on a united worldwide effort to continue the exploration of space in order to make the costs of such exploration endurable and, even more importantly, would promote worldwide harmony and cooperation.

A basic rule of thumb: think of the front page of a newspaper as containing exposition and the editorial page as usually containing argumentation. This is a rule of thumb only. Not all editorials or letters to the editor are arguments. Sometimes they are merely informative (as when an editor informs us of a charity drive) or they may be otherwise expository (as when a letter writer thanks the editor for mentioning the charity drive in his column). Strictly speaking, however, the editorial page should be a forum for reasoned opinion. Because it usually is, we may find a helpful guide to the distinction between exposition and argument in the rough identification of the editorial with argument and the news story with exposition.

There are various types of exposition, some of which we'll consider here.

4.2 TYPES OF EXPOSITION

1. *Explanation:*

The "how to," "what is it?" and "how it works" of written or spoken discourse, explanation is a close relative of argument. Unlike argument, its main intent is to inform whereas the main emphasis of argument is on persuading or convincing someone of a truth which they do not yet know or believe.

An example of explanation is:

Egg Whites

The glory and lightness of a French soufflé are largely a matter of how voluminously stiff the egg whites have been beaten and how lightly they have been folded into the body of the soufflé. It is the air, beaten into the whites in the form of little bubbles, which expands as the soufflé is cooked and pushes it up into its magnificent puff. Correctly beaten egg whites mount to 7 or 8 times their original volume, are perfectly smooth and free from granules, and are firm enough to stand in upright peaks when lifted in the wires of the beater. . . . Another test of their perfect stiffness is that they will support the weight of a whole egg if one is placed on top of them. —Julia Child, Lousiette Bertholle, and Simone Beck, *Mastering the Art of French Cooking*, Vol. 1, p. 158

Notice that the authors of this cookbook are not arguing to any conclusion; instead, their point is that the test of a good soufflé is how well the egg whites are beaten. The authors are not trying to convince you of anything; they *know* what makes a good soufflé, and they're telling you how to make one too.

We will look at explanation in more detail at the end of this module.

2. *Poetry:*
Here is an amusing sample of poetry from *The New Oxford Book of English Light Verse* (Kingsley Amis, editor):

When quacks with pills political would dope us,
When politics absorbs the livelong day,
I like to think of the star Canopus,
So far, so very far away.

Greatest of visioned suns, they say who list 'em;
To weigh it science must always despair.
Its shell would hold our whole dinged solar system,
Nor even know 'twas there.

When temporary chairmen utter speeches,
And frenzied henchmen howl their battle hymns,
My thoughts float out across the cosmic reaches
To where Canopus swims.

When men are calling names and making faces,
And all the world's ajangle and ajar,
I meditate on interstellar spaces
And smoke a mild seegar.

For after one has had about a week of
The arguments of friends as well as foes,
A star that has no parallax to speak of
Conduces to repose.

—"Canopus," by B. L. Taylor

Even though this poem contains a series of asserted statements, no argument nor explanation is under way. The poem is to be appreciated in other terms—in this case, for the shared emotion we feel when we've been too long exposed to political arguments. Everyone can share that feeling at one time or another. The poet successfully conveys it, with a sense of humor.

The point is that poetry serves a specific expository function in the way it lays out meaning or emotion. Does this mean that poetry can never serve as argument? In fact, on rare occasions it does. Consider for example the soliloquy of *Hamlet* by Shakespeare. This is an argument and it is in poetic verse. However, the argument is not asserted by Shakespeare, the author, but by the character, Hamlet. Here, the exposition contains an argument, but we would say that the argument is not asserted—it serves the purpose of revealing part of the character of Hamlet. This complicates our distinction somewhat, but it is a necessary qualification.

3. *Description:*
Literary description is one of the more general categories of prose writing, but description is important to science and philosophy as well. S. E. Toulmin and K. Baier have put it well: "It is the purpose of such descriptions that explains why their typical merits and demerits are what they are—namely: exactness, minuteness, accuracy, detail, fullness, sketchiness, misleadingness." What's important to see here is that in a sense even explanation is a form of description, whereas argument is not. But description is even more general than explanation, especially when used in literary contexts. Explanatory description is the what, how, or how to of a thing, but prose description is intended to evoke a feeling or a setting, just as poetry does. Here is an example of general prose description:

> It was Christmas night, the eve of the Boxing Day Meet. You must remember that this was in the old Merry England of Gramarye, when the rosy barons ate with their fingers, and had peacocks served before them with all their tail feathers streaming, or boars' heads with the tusks stuck in again—when there was no unemployment because there were too few people to be unemployed—when the forests rang with knights walloping each other on the helm, and the unicorns in the wintery moonlight stamped with their silver feet and snorted their noble breaths of blue upon the frozen air. Such marvels were great and comfortable ones. But in the Old England there was a greater marvel still. The weather behaved itself. —T. H. White, *The Once and Future King,* p. 137

4. *Narration:*
Narration is a form of description which relates the particulars of an act or course of events. This is a form of storytelling. Consider the following example from a popular novel:

> "What's it going to be then, eh?"
>
> There was me, that is Alex, and my three droogs, that is Pete, Georgie, and Dim, being really dim. And we sat in the Korova Milkbar making up our rassoodocks what to do with the evening, a flip dark chill winter bastard though dry. The Korova Milkbar was a milk-plus mesto, and you may, oh my brothers, have forgotten what these mestos were like, things changing so skorry these days and everybody very quick to forget, newspapers not being read much neither. Well, what they sold was milk plus something else. They had no license for selling liquor, but there was no law yet against prodding some of the new vesches which they used to put into the old molocko, so you could peet it with vellocet or synthemesk or dremcorom or one of the two vesches which would

give you a nice quiet horrorshow fifteen minutes admiring Bog and all his holy angels and saints and your left shoe with lights bursting all over your mozg. Or you could peet milk with knives in it, as we used to say, and this would sharpen you up and make you ready for a bit of twenty-two-one, and that was what we were peeting this evening I'm starting the story with. —Anthony Burgess, *A Clockwork Orange*

5. *Journalism:*
Journalism is a formalized type of description. It is a report of events, but it is distinguished from narration by its particular purpose. Journalism requires its stories to be told in the order of the most important aspect of the events first, with details following. A well-written news story will follow this pattern. In the first paragraph, it will tell you what's most important or what is most likely to be of interest to you. Then it will go on to give you a fuller description or relate less important aspects of the story in the order of their importance. Consider this news report, taken from the sports page of a newspaper, on a typical baseball game:

Phillies 7, Reds 2

Cincinnati—Glen Wilson hit a two-run homer and Von Hayes had three hits, helping the Philadelphia Phillies to a 7–2 victory and a three-game sweep of the Cincinnati Reds.

The sweep was the Phillies' first at Riverfront Stadium since 1984. The Phillies have won just 29 of their last 79 games at Riverfront since mid-1974.

Philadelphia scored 26 runs on 43 hits with six homers in the three-game series.

EXERCISES

DIRECTIONS: *Say whether the following passages are* ***argument*** *or* ***exposition****. For those that are arguments, identify the conclusion; for those that are not, say whether it is explanation, poetry, narration, description, or journalism.*

______ 1. I've been reading *Backpacker* for several years now. And as much as I enjoy the variety, I don't agree with your attempt to make trail food into seven courses of haute cuisine ("Cooking Good," July). People are supposed to be roughing it. If they want superior cooking, they should come down out of the mountains and hit a French restaurant. —"Letters," *Backpacker Magazine*, March 1988

______ 2. Surgeon General C. Everett Koop Wednesday denounced doctors and other health workers who refuse to treat AIDS patients as a "fearful and irrational minority" who are guilty of "unprofessional conduct." —*Wilmington Morning Star*, 9/10/87

______ **3.** Most doctors believe that drinking and pregnancy don't mix. But according to a University of Helsinki study of 530 women, less than two drinks a week during pregnancy poses no significant risk to the fetus. Even those women who had three to four drinks a week during the first three months of pregnancy gave birth to normal children, but their alcohol intake was linked to a greater incidence of premature birth. Despite these findings, health officials still advise pregnant women to exercise restraint. "Since we don't know what level of alcohol is safe during pregnancy, the wisest course is to abstain completely," says Richard Morton, M.D., vice president for health services at the March of Dimes. "Nevertheless, the study provides reassurance for women who inadvertently may have had a few drinks before they realized they were pregnant." —*Glamour,* September 1987, p. 228

______ **4.** Life without cable? You bet! I signed up when it came to town, but found it often went on the fritz with no explanation (or refund), rendering the TV useless. The movies were lousy, MTV was tasteless as well as mindless, and most of the extra channels were stupefyingly dull. When they wanted a 30-percent increase, I got out my pliers and returned the box to them. Now my TV is in the land of the free. —"Letters," *TV Guide,* Vol. 35, No. 39, 9/26/87

______ **5.** The meteoric rise of Dr. Ruth Westheimer from local radio to national radio to local TV to cable TV to syndication has apparently ground to a halt. *Ask Dr. Ruth,* an advice show seen, starting last January, on 52 TV stations around the country has been attracting pint-size audiences . . . and insiders say the syndicator, King Features Entertainment, will cancel the program. No decision has been made, according to a King Features spokesman.

Lifetime, the service that kicked off Dr. Ruth's cable-TV career in 1984, plans to open negotiations in an attempt to lure her back to cable, possibly in a late-night slot with a revamped format. . . . —"News Update," *TV Guide,* Vol. 35, No. 31, 8/1/87

4.3 ARGUMENT VS. EXPLANATION

Most students find the difference between 'arguments' and 'explanations' hard to discern. Here are some tips on how to distinguish the two.

1. One key difference: argument proceeds in advance of discovery, or as a means to discovery; explanation states already-established knowledge in a matter-of-fact way. In this respect, an explanation (as the saying goes) "leaves no room for argument." Explanations present *facts;* arguments present *contentions* in the conclusion, though the reasons for the conclusion may be factual.

2. Another key, but related, difference: argument proceeds in order to *persuade* or *convince* the reasonable hearer or reader (or to make discoveries); since explanations are thought to be established knowledge, explanations are meant to *inform,* not convince.

3. A further clue: arguments give *reasons;* explanations may give *causes.*

4. Example: given different contexts, the same piece of discourse can be at one time an argument, but at another, an explanation. Consider:

"The sky of Mars seen from the surface of the planet would appear pink to the observer because Mars, the Red Planet, has a thin atmosphere which causes refraction of the sun's rays through the lower part of the visible spectrum, and because of a high content of rust particulates in the atmosphere which also contribute a pinkish tint to the sky."

In advance of the Viking Mars Lander probe of 1976, the color of the Martian sky could be speculation only. One can imagine scientists arguing over the appearance of the sky as Viking was setting up its equipment. After tests were made, however, and after the fact of the pink sky was established together with the confirmed reasons for it, this same passage might take on a new status: that of an explanation.

This is a complication to our basic rule, but real-life examples can be ambiguous in this way. Perhaps it will be useful to update the diagram showing the exclusivity of argument and exposition given in Figure 1 at the beginning of this module. In Figure 4, we include the relationship of explanation to argument and exposition:

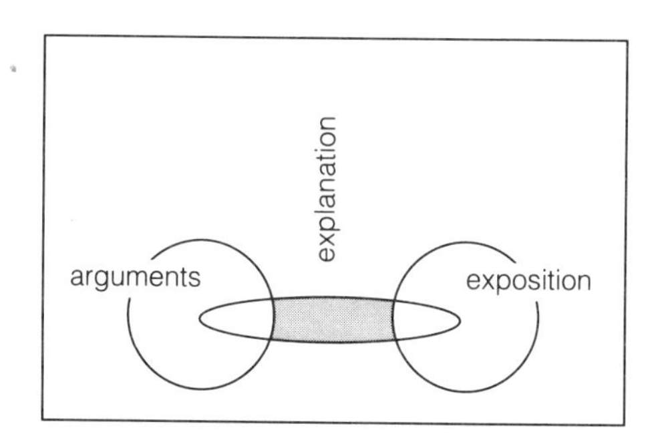

FIGURE 4

The oval in the middle represents the "corral" which holds explanations. Notice that some explanations may clearly serve as arguments whereas others clearly serve as exposition. However, we have a third region (placed between the circles for arguments and exposition) which is a "grey" area—because many examples of explanation do not clearly count as argument or as exposition except when they are placed in a wider context. This does not mean that such explanations are outside of the domains of argument or exposition entirely—it just means that we may be undecided into which corral to place them until we have more background.

You may therefore find it useful to think of explanation as a "third" category. When reading a passage, ask yourself whether it is argument, exposition, or explanation.

5. Another rule of thumb: sometimes 'thus' can be a conclusion-indicator, suggesting that an argument is underway, but 'thus' is the standard explanation-indicator. Every explanation has a "point" (not a 'conclusion') and we can find the point of an explanation by inserting an imaginary 'thus' until we see what statement in the "how to," "what it is," or "how it works" of an explanation makes the point. In our earlier example of whipping egg whites in a soufflé, we see that the point of the passage is the first sentence: "The glory and lightness of a French soufflé are largely a matter of how voluminously stiff the egg whites have been beaten and how lightly they have been folded into the body of the soufflé." Take this statement, read it at the end of the passage and place an imaginary 'thus' in front of it.

6. Even given all of these guidelines, some passages will still defy easy classification. Consider the following example:

"We're off to see the Wizard / the wonderful Wizard of Oz / Because, because, because / Of all the wonderful things he does. . . ."

In this case, we may be unable to determine whether the 'because' is a *reason-indicator* or an *explanation-indicator.* The fact of the matter is that some passages can, with equal justification, be regarded as arguments *or* as explanations, depending on your interpretation. Most passages are not so ambiguous, and the reader should have little difficulty in distinguishing the usual argument from the usual explanation. But keep in mind that according to some prominent logicians, such as Karl Hempel, that explanations can formally be treated as arguments for purposes of formal logic. This is surely true for scientific and mathematical explanations. However, many everyday explanations (e.g., how to repair appliances, what is the proper procedure for application for graduation), don't fall into the category of scientific or mathematical explanation, but they are explanations nonetheless. Thus, we can make a rough-and-ready distinction which will usually serve us, but no hard-and-fast rule exists for a sharp distinction in every case.

EXERCISES

Part I: Elementary Exercises

DIRECTIONS: *Say whether the following passages are better described as* ***explanations, arguments,*** *or* ***neither*** (i.e., *exposition). Here's a tip: compare the passage you are reading to the next several following it.*

_______ 1. The Central American nation has declared war because its neighbor has repeatedly refused to lift a trade embargo.

_______ 2. The neighboring country must be refusing to lift its trade embargo because the Central American nation has just declared war.

_______ 3. If this movie stars both Richard Dreyfuss and Meryl Streep, then it's sure to be a good movie.

_______ 4. Dress lengths will go up this summer. Short dresses reflect a period of prosperity, and dress manufacturers may want to influence the market.

_______ 5. Dress lengths will go up this summer. Necklines will go down.

_______ 6. Dress lengths will go up this summer. The midi-skirt and the maxi-skirt have been showing poor sales of late.

_______ 7. Magic has left the world since the Druids died off.

_______ 8. Magic has left the world since only the Druids knew the proper incantations.

_______ 9. Magic has left the world, my reason being that only the Druids, who died off long ago, knew the proper incantations.

_______ 10. I should be able to convince you since I have plenty of compelling reasons.

_______ 11. The Empire State Building is the most impressive building in the world. The World Trade Center is a pair of spikes and the Sears Tower is of bargain basement quality.

_______ 12. To change the oil, locate the bolt underneath the oil pan and remove it slowly using a socket wrench. Drain the oil in a pan. When all the oil has drained, replace the bolt and add the appropriate number of quarts.

_______ 13. Most books suggest that oil filters need to be replaced only after every other oil change; however, our recommendation is to change the oil filter every time you change the oil. Even though you may have changed the oil, a residue of grimy oil remains in the filter that you don't want to allow to contaminate your clean oil; besides, filters are very inexpensive and easy to replace.

_______ 14. New York is still the most populous city in the United States. Los Angeles has replaced Chicago as the second most populous city. Philadelphia is fourth.

_______ 15. If philosophy is not the most general of disciplines, then no field of study has taken its place. In recent times, philosophy has gone with the trends and required intensive specialization. Thus, no field of study exists which assumes the stance of the generalist in the study of the collective knowledge of humankind.

Part II: Advanced Exercises

DIRECTIONS: *Say whether the following passages are better described as* **explanations** *or as* **arguments**. *These are real-life examples taken from editorials, advertisements, and letters. Because they are taken from real life, they may prove to be more difficult than the textbook examples you worked in Part 1. Say why you identified it as one or the other. (Discussion will be helpful here.)*

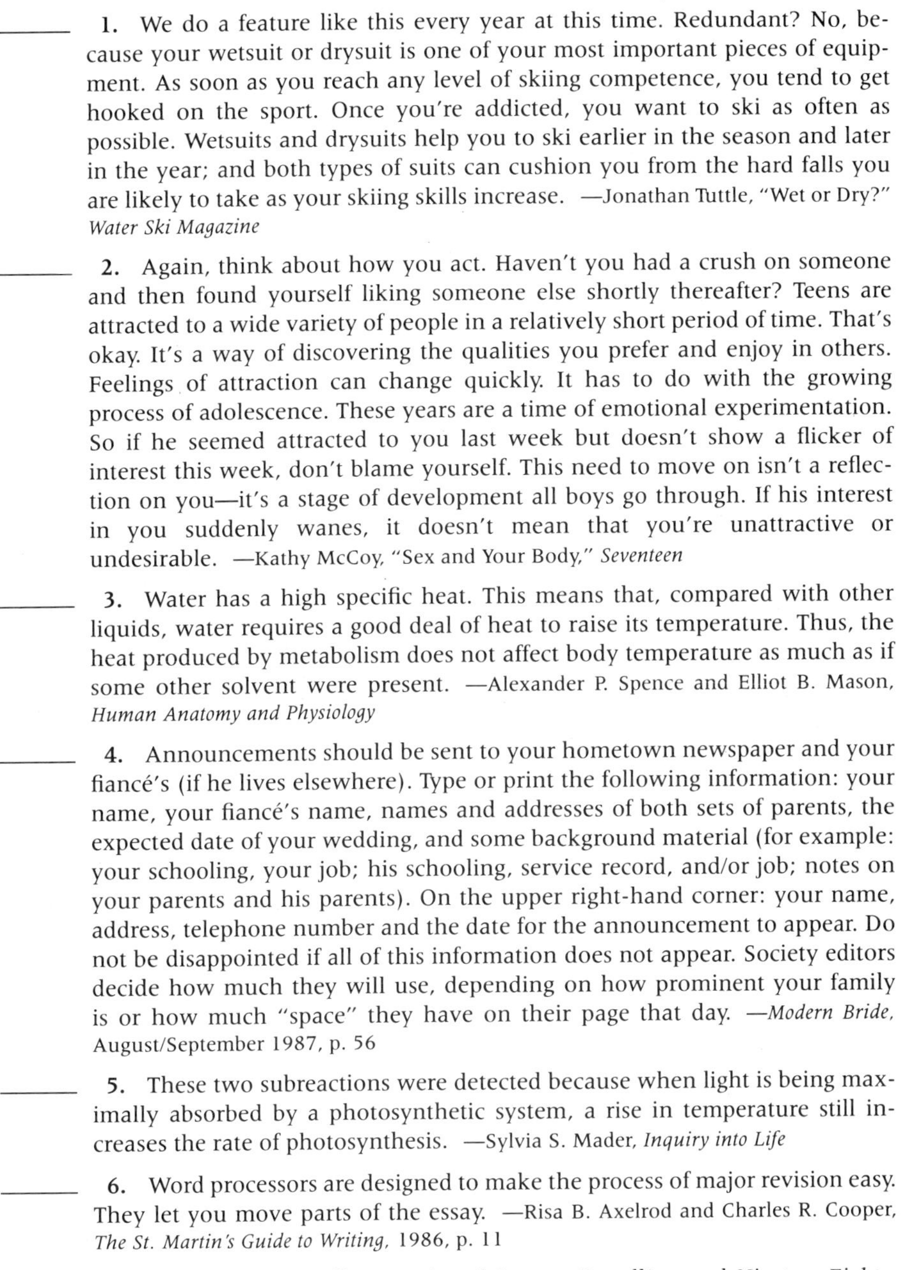

_______ 1. We do a feature like this every year at this time. Redundant? No, because your wetsuit or drysuit is one of your most important pieces of equipment. As soon as you reach any level of skiing competence, you tend to get hooked on the sport. Once you're addicted, you want to ski as often as possible. Wetsuits and drysuits help you to ski earlier in the season and later in the year; and both types of suits can cushion you from the hard falls you are likely to take as your skiing skills increase. —Jonathan Tuttle, "Wet or Dry?" *Water Ski Magazine*

_______ 2. Again, think about how you act. Haven't you had a crush on someone and then found yourself liking someone else shortly thereafter? Teens are attracted to a wide variety of people in a relatively short period of time. That's okay. It's a way of discovering the qualities you prefer and enjoy in others. Feelings of attraction can change quickly. It has to do with the growing process of adolescence. These years are a time of emotional experimentation. So if he seemed attracted to you last week but doesn't show a flicker of interest this week, don't blame yourself. This need to move on isn't a reflection on you—it's a stage of development all boys go through. If his interest in you suddenly wanes, it doesn't mean that you're unattractive or undesirable. —Kathy McCoy, "Sex and Your Body," *Seventeen*

_______ 3. Water has a high specific heat. This means that, compared with other liquids, water requires a good deal of heat to raise its temperature. Thus, the heat produced by metabolism does not affect body temperature as much as if some other solvent were present. —Alexander P. Spence and Elliot B. Mason, *Human Anatomy and Physiology*

_______ 4. Announcements should be sent to your hometown newspaper and your fiancé's (if he lives elsewhere). Type or print the following information: your name, your fiancé's name, names and addresses of both sets of parents, the expected date of your wedding, and some background material (for example: your schooling, your job; his schooling, service record, and/or job; notes on your parents and his parents). On the upper right-hand corner: your name, address, telephone number and the date for the announcement to appear. Do not be disappointed if all of this information does not appear. Society editors decide how much they will use, depending on how prominent your family is or how much "space" they have on their page that day. —*Modern Bride*, August/September 1987, p. 56

_______ 5. These two subreactions were detected because when light is being maximally absorbed by a photosynthetic system, a rise in temperature still increases the rate of photosynthesis. —Sylvia S. Mader, *Inquiry into Life*

_______ 6. Word processors are designed to make the process of major revision easy. They let you move parts of the essay. —Risa B. Axelrod and Charles R. Cooper, *The St. Martin's Guide to Writing*, 1986, p. 11

_______ 7. More than ten million copies of George Orwell's novel *Nineteen Eighty-Four* have been sold in the English language alone since it was first published in 1949, making it one of the best-selling serious novels of our age. It has been on so many required-reading lists in schools and colleges that it must also be one of the most widely read. Not surprisingly, therefore, the arrival of the *year* 1984 has brought with it an extraordinary upsurge of interest in the novel and its author.

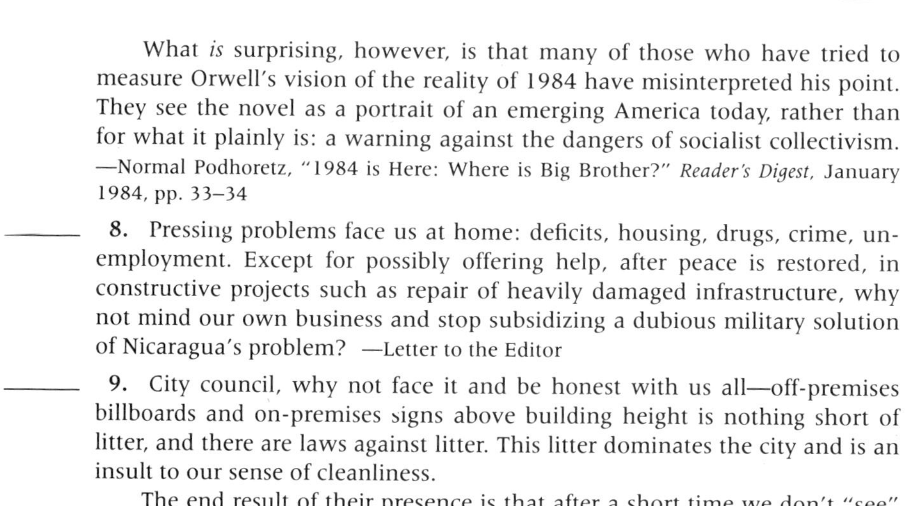

What *is* surprising, however, is that many of those who have tried to measure Orwell's vision of the reality of 1984 have misinterpreted his point. They see the novel as a portrait of an emerging America today, rather than for what it plainly is: a warning against the dangers of socialist collectivism. —Normal Podhoretz, "1984 is Here: Where is Big Brother?" *Reader's Digest*, January 1984, pp. 33–34

______ **8.** Pressing problems face us at home: deficits, housing, drugs, crime, unemployment. Except for possibly offering help, after peace is restored, in constructive projects such as repair of heavily damaged infrastructure, why not mind our own business and stop subsidizing a dubious military solution of Nicaragua's problem? —Letter to the Editor

______ **9.** City council, why not face it and be honest with us all—off-premises billboards and on-premises signs above building height is nothing short of litter, and there are laws against litter. This litter dominates the city and is an insult to our sense of cleanliness.

The end result of their presence is that after a short time we don't "see" the ugliness; we become "conditioned" to unsightliness, thereby not seeing other forms of pollution. —Letter to the Editor

______ **10.** It is always difficult to be meaningful and relevant, because there's just not enough time. Time to think seriously is hard to come by. I have been working all this past week in Los Angeles on a new television pilot for prime-time series. I left the taping at 4 o'clock this morning your time, chartered a plane and flew all morning to get here by 10. So I just want to tell you, if I fall asleep, don't worry, don't panic and don't disturb me. —An interview with talk show host Oprah Winfrey, *Time*, Vol. 129, No. 25, 6/22/87

______ **11.** As for gum disease, in a recent University of Texas experiment, most patients with advanced periodontal damage who had been unresponsive to traditional brushing and flossing showed major, rapid improvement when they combined the same regimen with a daily supplement of CoQ10. In some patients, symptoms of the disease disappeared completely in eight weeks. —*Ms.*, October 1987

______ **12.** The community lost a unique person when [Benjamin Peregrin] tragically drowned. He always greeted us with a smile when we brought our shoes in for repair at his shop on Wilshire Blvd. We are not sure he knew our names, but he could match shoes with a face.

His customers didn't need to present a ticket to pick up their shoes and this memory system seldom failed him. We will miss him. —Letter to the Editor

______ **13.** The current abundance of deer is evidenced in the increased sightings, deer/vehicle accidents and landowner complaints about crop damage. —*Deer Hunting Annual*, 1987 edition, 43920

MODULE 5

THE "THEREFORE" TEST

One helpful technique for selecting out the main conclusion of an argument is a handy and intuitive device we may call, the "therefore" test.

Suppose we have narrowed down the main conclusion of an argument to one of three important statements, ①, ③, or ⑤, but we don't know yet which of the three it is. We may find it helpful to test each against the others by reading one and then inserting a 'therefore' before another, to see if it makes sense. Then we should try it the other way. Chances are it will make clear sense one way, though not the other.

The operation looks like this:

Step 1: Insert a 'therefore' between ① and ③. Does it make sense that way? If "yes," skip to Step 3. If "no," go on to Step 2.

Read aloud:

①

'Therefore'

③

Step 2: Try it the other way. Insert a 'therefore' between ③ and ①. Does it make sense?

A deductive argument is valid, if and only if, if all the premises are true, then the **conclusion** must be true.

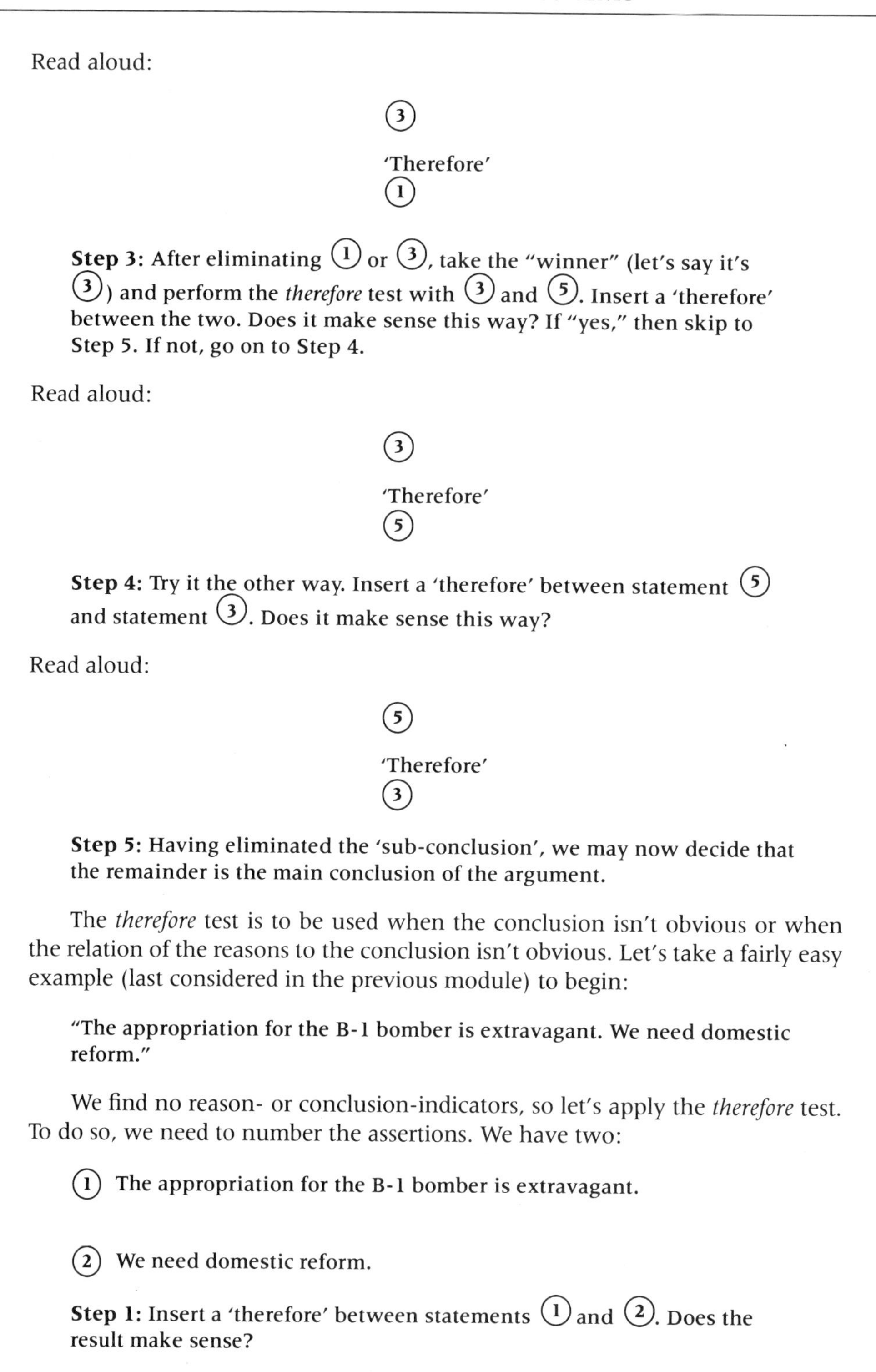

Read aloud:

③

'Therefore'

①

Step 3: After eliminating ① or ③, take the "winner" (let's say it's ③) and perform the *therefore* test with ③ and ⑤. Insert a 'therefore' between the two. Does it make sense this way? If "yes," then skip to Step 5. If not, go on to Step 4.

Read aloud:

③

'Therefore'

⑤

Step 4: Try it the other way. Insert a 'therefore' between statement ⑤ and statement ③. Does it make sense this way?

Read aloud:

⑤

'Therefore'

③

Step 5: Having eliminated the 'sub-conclusion', we may now decide that the remainder is the main conclusion of the argument.

The *therefore* test is to be used when the conclusion isn't obvious or when the relation of the reasons to the conclusion isn't obvious. Let's take a fairly easy example (last considered in the previous module) to begin:

"The appropriation for the B-1 bomber is extravagant. We need domestic reform."

We find no reason- or conclusion-indicators, so let's apply the *therefore* test. To do so, we need to number the assertions. We have two:

① The appropriation for the B-1 bomber is extravagant.

② We need domestic reform.

Step 1: Insert a 'therefore' between statements ① and ②. Does the result make sense?

(1) The appropriation for the B-1 bomber is extravagant.

therefore

(2) We need domestic reform.

We see that reading it this way doesn't quite work. We have no idea why the extravagance of the B-1 bomber's appropriation should ever lead us to the conclusion that we need domestic reform like health, education, and welfare.

Let's try the next step:

Step 2: Try it the other way around. Insert a 'therefore' between (2) and (1). Does the result make sense?

(2) We need domestic reform

therefore

(1) the appropriation for the B-1 bomber is extravagant.

This does indeed make sense. The money we need for domestic reform such as health, education, and welfare does support the conclusion that the expenditures on the B-1 bomber are extravagant. Our job is done, and the *therefore* test has helped us to see what the conclusion is.

Here's a much more difficult example. Let's see if the following passage is an argument and, if it is, what is the relation of the reasons to the conclusion. Notice that there are no clear conclusion- or reason-indicators in this example either:

> "The automatic camera being tested by the city of Pasadena, as well as its photo radar, would be considerably less useful if traffic lights were properly synchronized. I believe that the principle attraction of these devices is their revenue-generating potential. The problem for us drivers is that the city gets more return on its 'investment' by not synchronizing the lights."

The main point here is difficult to understand because the passage is poorly structured. Let's see if the *therefore* test will help us out.

When we number the statements, we have:

(1) The automatic camera being tested by the city of Pasadena, as well as its photo radar, would be considerably less useful if traffic lights were properly synchronized.

I believe that

(2) . . . the principle attraction of these devices is their revenue-generating potential.

(3) The problem for us drivers is that the city gets more return on its "investment" by not synchronizing the lights.

What is the conclusion to this argument? Notice that 'I believe that' is placed outside of statement ②. That's because it strictly speaking reports of state of mind but the author clearly wants to assert the *content* of the belief: "the principle attraction of these devices is their revenue-generating potential."

Could 'I believe that' be a non-standard conclusion-indicator here? Let's "nominate" ② for the conclusion and then use the *therefore* test to find out.

Step 1: Insert a 'therefore' between the first two statements. We have:

"The automatic camera being tested by the city of Pasadena, as well as its photo radar, would be considerably less useful if traffic lights were properly synchronized," *therefore,* "the principle attraction of these devices is their revenue-generating potential."

The passage makes somewhat more sense if we re-write it slightly—if traffic lights were synchronized, then automatic cameras and photo radar would be less useful, therefore, the principle attraction of these devices is their revenue-generating potential.

Notice that if we reversed the two statements, the 'therefore' in between would make no sense whatever. Still, this argument lacks something. Let's see if it makes more sense once we add statement ③.

Step 2:

① The automatic camera being tested by the city of Pasadena, as well as its photo radar, would be considerably less useful if traffic lights were properly synchronized. ③ The problem for us drivers is that the city gets more return on its "investment" by not synchronizing the lights.

Now, let's apply the 'therefore' and see what happens:

① The automatic camera being tested by the city of Pasadena, as well as its photo radar, would be considerably less useful if traffic lights were properly synchronized. ③ The problem for us drivers is that the city gets more return on its "investment" by not synchronizing the lights.

therefore

② . . . the principle attraction of these devices is their revenue-generating potential.

Now that makes much better sense. A little experimentation will show that we can't get the same success by placing the statements in any other order. For example, ② is not a reason for ①, nor is it a reason for ③.

Likewise, ① is not a reason for ③ nor is ③ a reason for ①.

However, ① + ③ taken together *are* reasons for ②. Thus, we have:

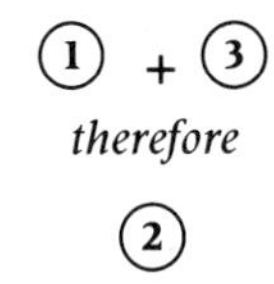

The conclusion of the argument is: the principle attraction of these devices is their revenue-generating potential.

Fortunately, few arguments we encounter will be so difficult as this to decipher. A good writer or thinker will make ample and clear use of conclusion-indicators and reason-indicators and will help us see the conclusion by placing it strategically at the very beginning or the very end of the passage. In this example, you see what happens when a writer neglects to do these things and puts the conclusion in the *middle* of the passage. When we encounter such arguments, that's the time to apply the *therefore* test.

MODULE 6

DIAGRAMMING ASSERTIONS

6.1 WORKING DIAGRAMS

What is the point of working diagrams for arguments? Basically, they show us the underlying structure of an argument. However, they have their limits, and we need to work within those limits.

A diagram should help us get the point of an argument. In the first instance, diagrams help to reveal the conclusion by pointing the way from supporting reasons to a relevant conclusion. Diagrams also help to reveal premises or evidence which stand as sub-conclusions. They help us organize our own arguments so we present them clearly and directly, without confusion. These are the advantages of diagrams.

Diagrams can serve as rough models of good argument form, but models can be difficult to apply to specific cases of arguments. Sometimes arguments are presented to us in a confused or sloppy manner. In these cases, diagrams can become exceedingly involved, and rather than serving as an aid to our understanding they become obstacles themselves. The idea is to get the basic overall structurc.

Another drawback to diagramming is that it is introduced at the beginning of a course in logic, yet one learns more about the structure of an argument as one learns more about validity, soundness, and techniques of formal reasoning. For this reason, we should look for only the broadest outlines of logical structure at this point in a course in logic.

A deductive **argument** is valid, if and only if, if all the premises are true, then the conclusion must be true.

Let us consider the main advantage of diagrams before proceeding with their explanation. Diagrams show how well we or others have followed two basic rules: **the Rule of Grouping** and **the Rule of Direction.**

The Rule of Grouping tells us to group like reasons for the same conclusion together. For example, if we have three good reasons for believing that life as we know it does not exist elsewhere in the solar system, we should list these three reasons together, and in order: (1) Mariner and Voyager satellites showed conditions on most of the other planets to be either too cold or too hot to support life; (2) the only reasonable conditions for life exist on Mars; but (3) the Viking explorer, which conducted tests for life on Mars, gave at best inconclusive results.

However, consider how this rendering of the argument violates the Rule of Grouping:

> **"The only reasonable conditions for life elsewhere exist on Mars, so life as we know it is unlikely anywhere else in the solar system because Mariner and Voyager satellites showed most of the other planets to be either too hot or too cold to support life. Moreover, the Viking explorer, which conducted tests for life on Mars, gave at best inconclusive results."**

Presented this way, the argument is sloppy. The reasons for the conclusion are all out of order, and the result may be confusion.

The Rule of Direction says that an argument should move in a single direction, one reason supporting another until the conclusion is reached. Let's consider the following argument:

> When I was a research assistant in physiology at the University of Iowa's School of Medicine in 1940, there were only half a dozen known vitamins.

Monroe Beardsley was among the first logicians to introduce the diagramming technique. Here is how he explains the Rules of Grouping and Direction:

"The writer's main problem, in formulating an argument, is to make the elements of the verbal texture—the syntax, the order of the words and topics, the connectives—bring into the open the logical relationships of the argument. Here are two fundamental rules to keep in mind. The *Rule of Grouping* is, briefly, that as far as possible, reasons for the same conclusion should be kept together, and their similar logical status called to the reader's attention. This can sometimes be done by parallel grammatical construction; in complex contexts, the reasons may have to be numbered and the system of numbering explicitly introduced. . . . The *Rule of Direction* is that when there is a series of assertions, each being a reason for the next one, the argument should move in a single direction, so the order of words helps to remind us of the order of the thought."

—MONROE BEARDSLEY, *Thinking Straight*

There was great concern at the time that we should all get our full quota of vitamins. Today we know that there are more than a dozen vitamins. It is not unlikely that in twenty years' time we'll have found that there are more than two dozen vitamins. What, then, is the point of taking pills in known quantity each day in order to be certain that you'll get your quota of vitamins? That suggests a certain overreliance on mankind's knowledge to date. There is only one way to be certain that you'll get all your vitamins, and that is to eat a wide variety of foods—especially those foods such as milk, eggs, grains and fish that we know to be highly nutritious. —Laurence E. Morehouse (with Leonard Gross), *Total Fitness in 30 Minutes a Week*

The argument follows this direction:

When I was a research assistant in physiology at the University of Iowa's School of Medicine in 1940, there were only half a dozen known vitamins. There was great concern at the time that we should all get our full quota of vitamins. Today we know that there are more than a dozen vitamins.

↓

It is not unlikely that in twenty years' time we'll have found that there are more than two dozen vitamins.

What, then, is the point of taking pills in known quantity each day in order to be certain that you'll get your quota of vitamins? That suggests a certain overreliance on mankind's knowledge to date.

There is only one way to be certain that you'll get all your vitamins, and that is to eat a wide variety of foods—especially those foods such as milk, eggs, grains and fish that we know to be highly nutritious.

We can see that whatever the special merits or defects of this argument may otherwise be, at least it displays the virtue of adherence to the Rule of Direction. The argument "flows" from beginning to end. The conclusion (the last box) is supported by a reason (the third box), which, in turn, is supported by another reason, and it, by still another.

Ideally, an argument which follows the Rule of Direction should look like this:

. . . where assertion 1 is a reason for 2 and 2 is a reason for assertion 3, and so on, until the conclusion, assertion 5, is reached. According to the Rule of Direction, good form demands that either the conclusion come up front as the first assertion in an argument, or at the end as it is diagrammed here, the last assertion in the argument.

6.2 A SIMPLE DIAGRAMMING TECHNIQUE

We can show these relations expressed by the Rules of Grouping and Direction by way of a diagramming technique to determine how well an argument follows them. Take the simple argument:

> "Terrorism will probably be on the increase because terrorists have been successful thus far in commanding a great deal of worldwide press attention."

We find this argument has two assertions and a reason indicator ('because'). Thus, we can show this by numbering the assertions:

(1) Terrorism will probably be on the increase

because

(2) terrorists have been successful thus far in commanding worldwide press attention.

By numbering the assertions and placing reason-indicators and conclusion-indicators outside the numbered, circled assertions, we then can show the relationship of the reasons to the conclusion by way of a diagram:

This simple diagram shows the two things we most want to find in a diagram: it clearly shows the conclusion, and it shows the relationship of the reason(s) to the conclusion. In this example, the second assertion supports the first (the conclusion).

6.3 DIAGRAMS FOR THE RULES OF GROUPING AND DIRECTION

There are two basic diagrams which show grouping and direction. The first, the most common, we call the **horizontal diagram.** The horizontal diagram looks like this:

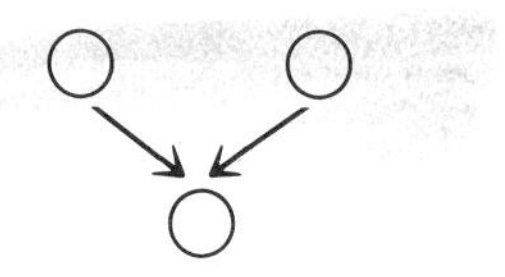

Blank Horizontal Diagram

We've deliberately left out the numbers of the assertions because horizontal diagrams (like chain diagrams) are indifferent to numerical order. We may think of the general form of the horizontal diagram as resembling a vendor holding a bunch of helium-filled balloons (as in the following example):

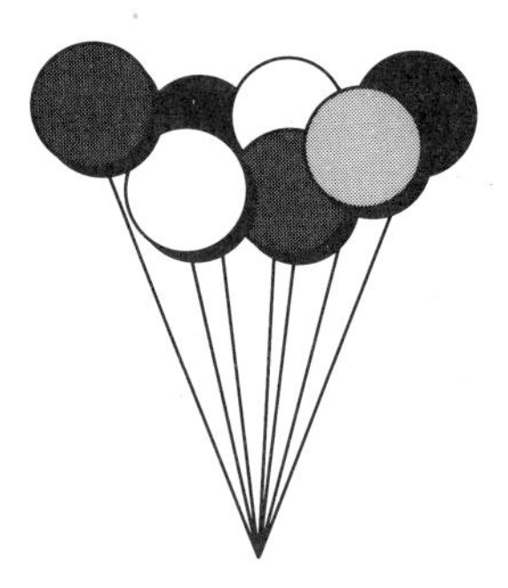

The numerals which "fill" the balloons aren't important to recognizing the shape that the "bunch" takes; what's important for recognizing the horizontal diagram is the way assertion-balloons more or less float together at the same level, all pointing down to the base (or 'conclusion') with equal, independent status.

The Rule of Grouping can best be illustrated by using a horizontal diagram, but now the numerals that fill the balloons are very important. Recall that we numbered assertions in the order in which they appeared in the passage we're planning to diagram. The Rule of Grouping says that like reasons for the same conclusion should hang together in numerical order. That means that when we read the numbers of the "balloons," those numbers should be in arithmetical sequence (for example: 1, 2, 3, 4, etc.).

Below is an example of a diagram which meets the Rule of Grouping:

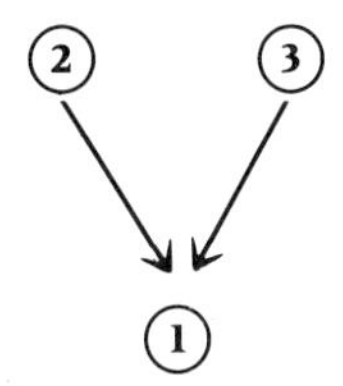

Horizontal Diagram

As it stands, this diagram faithfully represents the following argument:

(1) **For many people in the inner city, a full-time job may be hard to come by.**

(2) **Economic opportunities are few.**

(3) **Discrimination is still an obstacle for many.**

Even though there are no reason- or conclusion-indicators in the argument, we can see that assertions 2 and 3 each give separate reasons for believing the truth of the conclusion, assertion 1.

However, here is an example of a diagram which shows a *violation* of the Rule of Grouping:

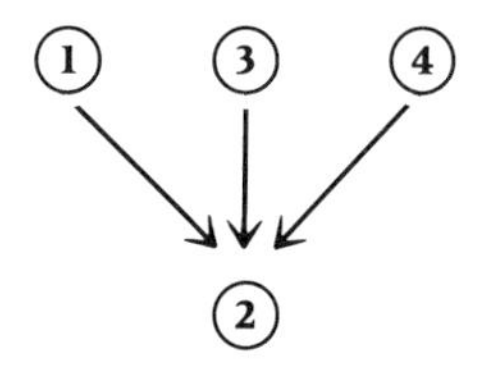

Violates Rule of Grouping

Here assertion 2 is the conclusion, which puts the order in which the argument is presented all out of kilter. For example, here is an argument which the diagram fits:

(1) Computerized cash registers in supermarkets keep detailed records of your purchases, especially if you pay by check where the computer records your driver's license number.

(2) That's why I say computers are an increasing threat to rights of privacy.

(3) Another reason is the detailed marketing surveys that new, more powerful computers allow.

(4) Finally, rapid electronic transmission of computer data allows terrible abuses and excesses by credit reporting agencies to become part of your credit record.

Do you see how confusing this gets? A simple reordering of the reasons and the conclusion, with a minimum of re-writing, would make the argument much easier to understand.

The second form of argument to consider is the **argument chain.** Once again, we can best illustrate the argument chain by means of a diagram with blank "balloons:"

Argument Chain

The analogy to a chain is obvious enough. Here, the top balloon is a reason for the middle balloon which, in turn, is a reason for the conclusion (the bottom balloon).

The Rule of Direction is best illustrated by a chain diagram in which the numerals "fill" the balloons of the chain, as follows:

Illustrates Rule of Direction

Recall the example we used earlier. You can see how the argument flows in a single direction:

> "Power breeds isolation. Isolation leads to the capricious use of power. In turn, the capricious use of power breaks down the normal channels of communication between the leader and the people whom he leads. This ultimately means the deterioration of power and with it the capacity to sustain unity in our society."

Now, let's consider a diagram which violates the Rule of Direction. Here is an example:

Violates Rule of Direction

This diagram shows that assertion 8 is a reason for 4, and 4 is a reason for 7. Any argument that was so out of sequence with regard to the Rule of Direction would be virtually impossible to follow. Yet, we find examples of this sort of bad argument form all the time (read the letters to the newspaper; you'll find plenty of examples).

The third form of argument we may consider is a derivative. We may call it the **linked horizontal diagram.** As the name suggests, this is a form of horizontal diagram, for the reasons still converge on the conclusion. Yet, we see something more in a linked horizontal. That something more is that the reasons are grouped together in a special relationship to each other. One is closely allied or *relevant* to the other in supporting the conclusion. This is not true in the example of the horizontal argument given earlier. Whereas few economic opportunities and the obstacle of discrimination each give *separate* support to the conclusion that jobs may be hard to come by in the inner city, these reasons are not (as the argument is stated) linked to each other in any special way.

Here is an example of a linked horizontal diagram:

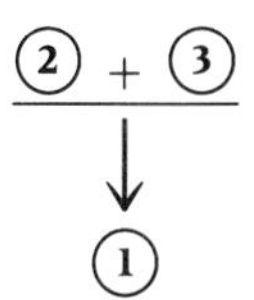

Linked Horizontal Diagram

(1) The Kansas City Royals may be contenders for the pennant this year

because

(2) Bo Jackson has joined the team

and

(3) he was the 1989 Most Valuable Player in the All Star Game.

The fact that slugger Bo Jackson has signed up with the Royals and the related, additional fact that he was the All Star Game's Most Valuable Player *together* support the conclusion that the Royals may be contenders during the season. Statements 2 and 3 are linked by the conjunction 'and' in this case. That's a clue—though by no means a guarantee—that the reasons are related. In fact, knowledge of baseball tells us that having Bo Jackson on your team when he was voted the top property in the All Star Game is one reason for believing that the team will be in the push for the pennant when September rolls around.

6.4 A CLUE FOR FINDING LINKED HORIZONTAL DIAGRAMS

The presence of conjunctions like 'and', 'also', and 'moreover' give us clues, not guarantees, that a group of reasons for a conclusion are related to each other in a special way. Conjunctions may also be special logical connectives, and these do guarantee a special relationship or "linkage" between reasons that converge on a conclusion. We will discuss these logical connectives later on in much more detail; for now, however, we must look at them briefly.

Logical connectives are a subset of the *logical operators* (briefly mentioned in Module 3 and discussed in more detail in Unit Three, Module 23, "Symbols"). The logical connectives are:

'both . . . and'	conjunction
'either . . . or'	disjunction
'if . . . then'	conditional
'if and only if'	biconditional
'it is not the case that . . .'	negation

The logical connectives have this special requirement: *the whole sentence must be treated as one assertion.*

Remember the difference between assertions and simple statements! The assertion is the unit of argument, not the statement. In a sentence containing a logical connective, remember that the logical connective is what is being asserted.

For example, consider this sentence:

"If Mozart were alive today, then he'd probably play lead guitar for the Boom Town Rats."

This *looks* like an argument with two statements; 'probably' is usually an inductive conclusion-indicator. In fact, it is *not* an argument; it is a single assertion. We must consider the truth of the sentence in full, not any part of it. (In fact, I rather doubt that, *if* Mozart were around, he'd want anything to do with that group at all.)

Let's take another example: "If the moon is blue, than we can expect snow in Hawaii." In this example, we have two *statements:* 'the moon is blue' and 'we can expect snow in Hawaii'. Taken by themselves, neither statement is true. But the statements are joined together by a logical connective, 'if . . . then'. Whereas each of the statements taken separately is false, certainly the assertion ("*If* the moon is blue, *then* we can expect snow in Hawaii") is true enough. (Can you imagine someone asserting "We get snow in Hawaii once in a blue moon?") This is because the *conditional*—the 'if . . . then'—is what is being asserted, not the individual statements.

All of the other logical connectives connect separate statements to each other, and they are to be numbered as separate assertions. For example, "Either the Grateful Dead is the greatest rock group in history or the Beatles is the name of the greatest rock group" contains two statements linked together by a logical connective 'either . . . or'. We represent that as *one* assertion.

We see that the logical connective 'either . . . or' is one way to make this special relationship between statements that establishes linkage. When two statements are linked together by a logical connective, we have just one assertion, and we represent it in a diagram with only one balloon.

However, suppose we argue as follows: "Either the Grateful Dead is the greatest rock group in history or else the Beatles hold that distinction. Surely it's not the Grateful Dead. It must be the Beatles!"

We diagram this argument as follows:

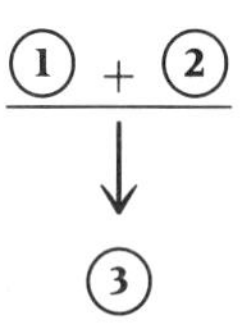

(1) Either the Grateful Dead is the greatest rock group in history or the Beatles hold that distinction.

(2) Surely, it's not the Grateful Dead.

(3) It must be the Beatles!

Notice that we underline the logical connectives 'either . . . or' and 'not'.

Now, consider this longer argument:

"If Anthony Burgess's book *A Clockwork Orange* is the best novel of the past few decades, then that argues for the conclusion that science fiction is the most important literary genre of the times. Either *A Clockwork Orange* is science fiction, or it is something very close to it. Moreover, only snobbery prevents us from calling it sci-fi because it makes use of time-honored sci-fi themes in a new and exciting way; critics are loathe to call new and exciting fiction "science fiction" no matter how much they mimic the genre. Therefore, I conclude that science fiction is indeed the most important literary genre of the times."

When we break this argument up into its constituent parts, the result is:

(1) If Anthony Burgess's book *A Clockwork Orange* is the best novel of the past few decades, then that argues for the conclusion that science fiction is the most important literary genre of the times.

(2) Either *A Clockwork Orange* is science fiction, or it is something very close to it.

Moreover,

(3) only snobbery prevents us from calling it sci-fi

because

(4) it makes use of time-honored sci-fi themes in a new and exciting way;

(5) critics are loathe to call new and exciting fiction "science fiction" no matter how much it mimics the genre.

Therefore, I conclude that

(6) science fiction is indeed the most important literary genre of the times.

This argument is more involved than anything else we've looked at so far. The first thing to notice is that assertion 1 is a conditional. We represent this by underscoring the 'if' and the 'then'. Because the 'if . . . then' is locked into the same assertion, we put the 'if' inside the statement which follows the numbered circle.

In assertion 4, we find a semicolon (;). The semicolon is a punctuation mark which serves the same purpose as a conjunction (this is true of the colon [:] as well). That signals another related but separate reason is to follow.

The final thing to notice is that this is an argument chain. An **argument chain** is a set of reasons which signals *directional links* that point to a final conclusion. That is to say, the "linkage" here is from top to bottom, following the Rule of Direction, one reason supporting another.

To diagram an argument chain, it is wise to take the argument in parts. Let's take the first two premises:

(1) If Anthony Burgess's book *A Clockwork Orange* is the best novel of the past few decades, then that argues for the conclusion that science fiction is the most important literary genre of the times.

② Either *A Clockwork Orange* is science fiction, or it is something very close to it.

We already know that assertion 2 contains two statements that are linked together by a logical connective 'either . . . or'. Therefore, we treat the whole 'either . . . or' sentence as one assertion. Moreover, we can see that they have some relevance to assertion 1, for assertion 2 fulfills the condition for the hypothetical in assertion one. So we can diagram them all as linked:

① + ②

Now, let's look at the remaining premises:

Moreover,
③ only snobbery prevents us from calling it sci-fi

because
④ it makes use of time-honored sci-fi themes in a new and exciting way;

⑤ critics are loathe to call new and exciting fiction "science fiction" no matter how much it mimics the genre.

Even though we have a conjunction 'Moreover', this is a new and separate line of argument. Taken as a whole, it too supports the conclusion, but from a different vantage point. This line of argument tells us that snobbery prevents us from calling *A Clockwork Orange* science fiction. But assertions 4 and 5 are related by the presence of the semicolon, by the presence of the reason-indicator 'because', and by their content as well. Thus, we see they are best represented as a linked horizontal diagram pointing to the sub-conclusion, assertion 3:

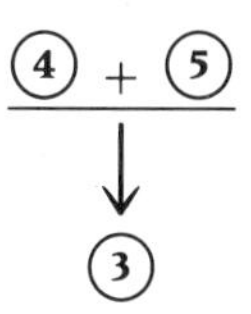

Taking both lines of argument together, they support the conclusion, assertion 6, clearly indicated by the conclusion-indicator 'Therefore'.

The argument in final diagram form looks like this:

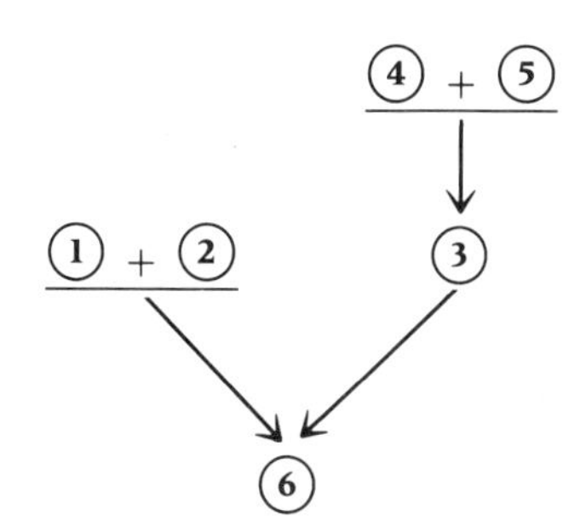

EXERCISES

Part I: Elementary Exercises

DIRECTIONS: *Diagram the following arguments. Two of the four problems already have the appropriate asserted statements numbered; two of them do not, and you must supply them. Place your diagrams in the boxes.*

1. (1) The automatic camera being tested by the city of Pasadena as well as its photo radar, would be considerably less useful if traffic lights were properly synchronized. (2) I believe that the principle attraction of these devices is their revenue-generating potential. (3) The problem for us drivers is that the city gets more return on its "investment" by not synchronizing the lights.

2. (1) Citizens so concerned about the acquisition of American real estate can take solace in a simple truth: (2) foreigners engaged in the acquisitions are paying for them. (3) That's a whole lot more civilized than was the previous acquisition of United States soil.

3. Select and number the assertions in the following passage. Select the reason-indicators and conclusion-indicators and underline them. Then diagram the passage.

> Nobody can be proud of the backhanded way Congress raises its members' salaries. But experience shows that it's either that way or no way, because few members will risk political oblivion by voting publicly to increase their own pay. —*The New York Times,* editorial, 2/1/89

4. Select and number the assertions in the following passage. Select the reason-indicators and conclusion-indicators and underline them. Then diagram the passage.

The outrage over the use of GOP polling "guards" in Orange County should serve as a reminder to all Latinos about the dangers of becoming Republican.

First, you get a full stomach, forget where you came from and act like any other Orange County conservative. Second, you'll become window dressing for Republican administrations which fight equal opportunity and civil rights for Latinos. Third, as long as you vote Republican, Bush will appear at your tortilla-making contests and President Reagan will assure you that Latinos and Republicans share many of the same values. —Letter to the Editor, the *Los Angeles Times*, 12/1/88

Part II: Advanced Exercises

Exercise 1

DIRECTIONS: *Take each of the following arguments, identify the assertions and number each, then construct assertion diagrams to identify the conclusion.*

1. How intimately sex and language are intertwined can be seen by reading pornography in a foreign language. When de Sade is read in the original no dictionary is needed. The most recondite expressions for the indecent, knowledge of which no school, no parental home, no literary experience transmits, are understood instinctively, just as in childhood the most tangential utterances and observations concerning the sexual crystallize into a true representation. It is as though the imprisoned passions, called by their name in these expressions, burst through the ramparts of blind language as through those of their own repression, and forced their way irresistibly into the innermost call of meaning, which resembles them. —Theodor Adorno, *Minima Moralia*

2. It seems that in Holy Scripture a word cannot have several senses, historical or literal, allegorical, tropological or moral, and analogical. For many different senses in one text produce confusion and deception and destroy all force of argument. Hence, no argument, but only fallacies, can be deduced from a multiplication of propositions. But Holy Scripture ought to be able to state the truth without any fallacy. Therefore in it there cannot be several senses to a word. —St. Thomas Aquinas, *Summa Theologica* [Aquinas is considering an objection to his own view in this place]

3. Now we of the Class of '29 were men of an entirely different stripe. We weren't picked for scholastic eminence or any other damn eminence. Come to think of it, we weren't picked at all. No school picked *us*, for Pete's sweet sake, *we* picked the school. We picked every school or college we ever darkened the door of. And, if you ask me, that's the main trouble with your schools and colleges today. They do the picking. And, of course, they're not up to it. If any one of those Deans of Admission ever sat on the Admissions Committee of a first-rate Club *anywhere*—either here or on the Continent—I'll eat two hats. —Cleveland Amory, *The Trouble With Nowadays*

4. Yet he had accomplished so much: the new hope for peace on earth, the elimination of nuclear testing in the atmosphere and the abolition of nuclear diplomacy, the new policies toward Latin America and the third world, the reordering of American defense, the emancipation of the American Negro, the revolution in national economic policy, the concern for poverty, the stimulus to the arts, the fight for reason against extremism and mythology. Lifting us beyond our capacities, he gave his country back to its best self, wiping away the world's impression of an old nation of old men, weary, played out, fearful of ideas, change and the future; he taught mankind that the process of rediscovering America was not over. He re-established the republic as the first generation of our leaders saw it—young, brave, civilized, rational, gay, tough, questing, exultant in the excitement and potentiality of history. He transformed the American spirit—and the response of his people to his murder, the absence of intolerance and hatred, was a monument to his memory. The energies he released, the standards he set, the purposes he inspired, the goals he established would guide the land he loved for years to come. Above all he gave the world for an imperishable moment the vision of a leader who greatly understood the terror and the hope, the diversity and the possibility, of life on this planet and who made people look beyond nation and race to the future of humanity. So the people of the world grieved as if they had terribly lost their own leader, friend, brother. —Arthur Schlesinger Jr., *A Thousand Days*

Exercise 2

DIRECTIONS: *Diagram the following argument; the assertions are numbered for you.*

(1) To be, or not to be—that is the question:
(2) Whether 'tis nobler in the mind to suffer
The slings and arrows of outrageous fortune,
Or to take arms against a sea of troubles,
And by opposing end them? (3) To die, to sleep—
No more—and by a sleep to say we end
The heartache, and the thousand natural shocks
That flesh is heir to. (4) 'Tis a consummation
Devoutly to be wished. (5) To die, to sleep—
To sleep—perchance to dream: ay, there's the rub;
(6) For in that sleep of death what dreams may come,
When we have shuffled off this mortal coil,
Must give us pause. (7) There's the respect
That makes calamity of so long life.
(8) For who would bear the whips and scorns of time,
Th' oppressor's wrong, the proud man's contumely
The pangs of despised love, the law's delay,
The insolence of office, and the spurns
That patient merit of the unworthy takes,
When he himself might his quietus make
With a bare bodkin?
(9) Who would fardels bear,
To grunt and sweat under a weary life,
But that the dread of something after death,
The undiscovered country, from whose bourn
No traveler returns, puzzles the will,

And makes us rather bear those ills we have
Than fly to others that we know not of?
(10) Thus conscience does make cowards of us all;
And thus the native hue of resolution
is sicklied o'er with the pale cast of thought;
And enterprises of great pith and moment,
With this regard their currents turn awry
And lose them in the name of action. —*Hamlet*, act 3, scene 1

Exercise 3

DIRECTIONS: *State, in your words, the conclusion of the argument in Exercise 2. Precisely what act is Hamlet contemplating?*

MODULE 7

DIAGRAMMING PARAGRAPHS

7.1 REVIEWING ASSERTION DIAGRAMS

In the previous section, we looked at the basic rules for diagramming assertions. We have a basic kind of assertion diagram, the horizontal diagram:

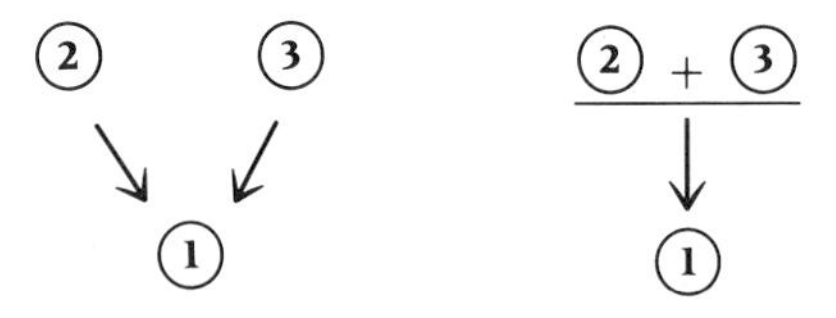

We considered that the clue to linked horizontal diagrams was the presence of grammatical conjunctions such as 'also', 'moreover', and 'and' or the presence of logical connectives: 'if . . . then', 'either . . . or', 'both . . . and', 'if and only if'. We noted that the horizontal forms of diagrams illustrate the Rule of Grouping, and that argument chains illustrate the Rule of Direction (as in the figure that follows).

A deductive **argument** is valid, if and only if, if all the premises are true, then the conclusion must be true.

Chain Diagram

Finally, we saw an example of a complex chain-horizontal–horizontal link diagram:

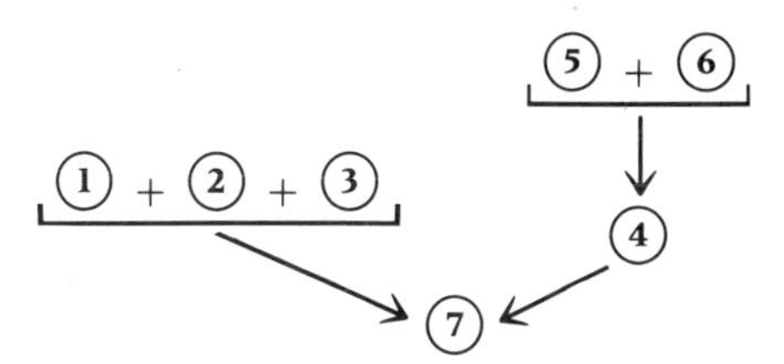

Chain-Horizontal–Horizontal Link Diagram

However, even the complex diagram shown here hardly does justice to the complexity and length of arguments we encounter in ordinary discourse. The diagram in the figure above has only seven assertions. A typical newspaper editorial may contain one hundred or more assertions. Short of monastic devotion to crossing all the T's and dotting all the I's that would be required to lay out an assertion diagram for the argument presented in an editorial, what can we do to uncover the basic structure of such an argument?

To answer this question, first of all we need to ask still another question: what is it we hope to learn from the diagramming method?

The main thing we want to learn is the *main point* of an argument. In other words, we want to know the conclusion. A diagram can help us decipher the conclusion in a poorly structured or difficult passage.

Sometimes we encounter a difficult argument about a subject in an arcane or unfamiliar discipline, and we may not be able to discern it's point. Knowing something about the structure of arguments, then, can be an invaluable aid to interpreting such passages and that may have the further benefit of opening up to us a subject with which we've had little or no prior acquaintance.

The first thing we do is separate the paragraphs and number them. Assuming the argument is a long one, we can't reasonably lay out each of the assertions (although we'll keep the assertion-diagram method in reserve, as I'll explain later).

The second thing we should do is select "candidates" for the paragraph which most likely contains the main conclusion. Think of this as a "nominating" process. Here's a hint: the Rule of Direction tells us that the main conclusion of an argument should be either at the very beginning or at the very end of the passage. Good form dictates this, but the rule is sometimes violated for considerations of style or simply because the writer just doesn't know the rules. Most published writing presupposes this kind of good form, however, so the very first and the very last paragraphs of any argument are always good candidates to nominate for the conclusion-bearing paragraph. Select one or two more that might contain the main point.

Having done this, you're now entitled to a small surprise: you've probably already unlocked the main structure of the argument! If you're any good at this at all (and knowledge of reason- and conclusion-indicators certainly suggests you are), then one of these will be your main conclusion, whereas the others are very likely key reasons or sub-conclusions in the argument.

Finally, determine the relationship of the reasons to the conclusion. Ask yourself, of the candidates nominated, is this one a reason for that, or is that one a reason for believing this? Eventually, the main conclusion will emerge, and with it the broad outline of the argument structure.

Let it be understood: in a long passage, we are mainly interested in getting to the broad outlines of the argument. We can always fill in the fine details later on. That's why we're introducing the paragraph-diagramming method. Obsessive concern with delineating every assertion and every relation between one assertion to another will not get us to the general understanding of what's going on, which is a necessary preliminary to any further research.

Let's draw an analogy: if we are trying to recognize the main structure of a tree it is necessary only to get a general idea of its shape. If we encounter an unfamiliar tree in a forest, and we want to know what kind it is, one way of checking it out is to consult a good illustrated book of trees. As these things go, we're not likely to have such a book with us on a hike through the woods, so we have to memorize the general shape of the tree,* its base, its trunk, and the major limbs. Then, when we have access to our book, we'll compare (as in the illustration on page 75).

Immediately, we can rule out the yellow birch—it's too small and spare, let's say. The tree we saw looks weightier, and its shape is broader. The tree we saw is more like the live oak or the Spanish chestnut, judging by the broad outlines. Since the tree we saw in the woods was an evergreen, we might conclude that we saw a live oak.

Naturally, no tree will look exactly like the illustration—that's beside the point. The point is that we can recognize a tree by its overall shape. That's why

*. . . or the shape of its leaves, or the color and texture of its bark.

Live Oak

Spanish Chestnut

Yellow Birch

we look at the base, the trunk, and the major limbs. Attempting to memorize every twig and branch would be foolish and time wasting, and not at all helpful to our task.

Much the same sort of thing applies to diagramming paragraphs in arguments. It's important to examine the general shape and the context of the paragraphs in the argument. Two individuals might draw diagrams that differ in their details, but if they understand the argument, they should be able to agree on the broad outlines: the base (the conclusion), the trunk (the supporting subconclusions), and the major limbs (the main supporting reasons in the argument). That's all we need. We'll fill in the twigs and the branches later.

Excessive attention to detail in the organization of each and every assertion might prevent us (so to speak) from seeing the forest for the trees.

7.2 THE PARAGRAPH DIAGRAMS

I contend that the paragraph, and not the sentence or the statement or the individual word, is of fundamental importance to understanding the meaning of the argument. This point is controversial among philosophers just as the basic unit of matter is controversial among physicists (whether it's the pi meson, the neutrino, or the quark). Sentences, assertions, statements, or words themselves *look* more basic, until we realize that we can't fully understand them outside of their context. (The word 'scale' for example has fully forty or more separate meanings in a good dictionary.) We trust the writer to understand what a paragraph is when we undertake this method; a paragraph, grammarians teach us, is a division of a written work that contains a complete thought. If true, we are better served by diagramming whole paragraphs in a written work than we would be by diagramming separate assertions.

Paragraph diagrams are very similar to assertion diagrams except that we will place their respective numbers of occurrence in *boxes* instead of in circles. Thus, for a simple two-paragraph argument, we may have:

. . . where box 1 is the first paragraph we encounter in the argument, and box 2 (here, the conclusion) is the second.

Box diagrams are otherwise analogous to assertion diagrams. We have the familiar forms: the horizontal box-diagram, the linked-horizontal box-diagram, and the box-diagram chain:

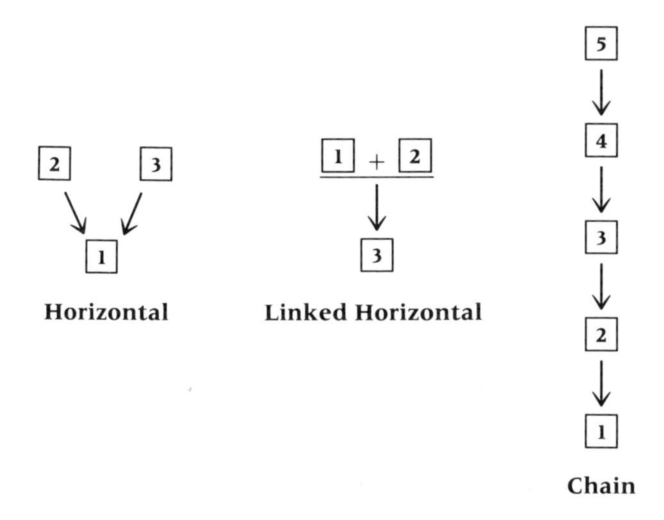

In identifying the order of the paragraphs, we certainly don't want to have to write out every sentence, so all that suffices is to indicate the very first word of the paragraph and the last, and connect them by three dots. For example, consider the following excerpt from a newspaper editorial:

> The maneuvering continues over whether volunteer firefighters in the Chadbourn area will let a black person join their not-so-merry crowd.
>
> The firefighters don't want to. They don't even bother to claim that everybody is welcome if he's qualified and that skin color doesn't matter. It's perfectly obvious that it does.

We number the paragraphs in the order we find them, place the numbers in a box, and then pick out the first one or two words of the paragraph and the last, connecting them by three periods (. . .). Thus, we have:

[1] The maneuvering . . . crowd.

[2] The firefighters . . . does.

We then proceed to number the rest of the paragraphs in the same manner:

Apparently prodded by the possibility of legal action—certainly not by embarrassment—the firefighters and the town have set up a grievance board to hear complaints from people who are rejected for membership. It seems a pointless gesture, considering that firefighters get two members to the town council's one.	[3] Apparently . . . one.
On the other hand, people who aren't satisfied by the decisions of the grievance board can take them to Town Council, which has the ultimate responsibility anyway.	[4] On the other . . . anyway.
The firefighters work for the town. The town has no business discriminating on the basis of race. The sooner this silly color line is broken, the better for everybody concerned.	[5] The firefighters . . . concerned.

Following the guidelines for diagramming (just as we did for assertion diagrams), we have this result:

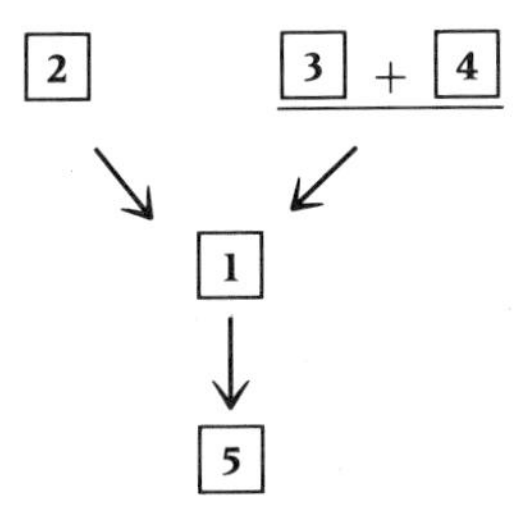

Taking "nominations" for the main conclusion as we've already learned to do, paragraphs 1 and 5 turn out to be likely candidates. Number 5 tells us a little more than paragraph 1 does, and 1 clearly supports 5 (and not the other way around), so 5 is our main conclusion.

Clearly, the argument is not well-organized according to the Rule of Direction. Paragraphs 3 and 4 have an apparent relevance to each other—the lead clause "On the other hand, . . ." reveals the connection. Therefore, we'll link them. Paragraph 2 supports 1, and so do 3 and 4 (jointly), thus we discern the structure.

7.3 JOINT PARAGRAPH-ASSERTION DIAGRAMS

We can get even more precise as to what the conclusion is by adding an assertion diagram to the conclusion paragraph (in this case, it's paragraph 5). We reproduce the paragraph in full, and number the assertions, circling the numbers as we go, and underlining the reason- and conclusion-indicators, and the logical connectives (if any). Thus, we have:

(1) **The firefighters work for the town.**

(2) **The town has no business discriminating on the basis of race.**

(3) **The sooner this silly color line is broken, the better for everybody concerned.**

Clearly, assertions 1 and 2 have a special relationship. The fact that the firefighters work for the town is important when we realize that the town has no business discriminating on the basis of race. Thus, we'll link the two. Notice, however, that these two assertions aren't especially relevant to assertion 3. Nevertheless, 3 must be the conclusion, especially given the overall context of the whole argument. Thus, we may assume that the editorialist *meant* assertions 1 and 2 to support assertion 3. We'll construct the diagram that way:

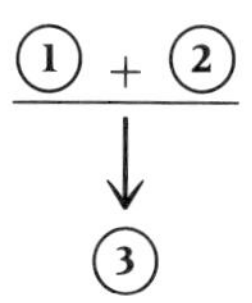

It would be worthwhile to notice that, without extracting the general shape of the argument from its paragraph form, unlocking the precise conclusion in the last paragraph might have been an impossible job—the argument is so sloppily presented! We can now assert with confidence, however, that the editorialist wants to argue that the sooner the color line is broken in this town, the better.

7.4 DON'T DIAGRAM EVERYTHING IN SIGHT!

Lengthy passages may not be well written. Other times, they may give a great deal of explanation or background. We don't need to diagram everything in order to see the main outlines (or, as we put it in our analogy, the base, the trunk, and the major limbs of the tree).

The following "Letter to the Editor" is an example of a sophisticated argument, well-structured, but with subtle humor. It's subject is obscure. In order to understand what the letter writer is getting at, we might want to put together a rough paragraph diagram. Here is the letter:

Editor: Perusing, as is my usual pleasant pursuit of an early Sunday morning, the *Sunday Star News* of Jan. 25, I encountered within an otherwise harmless film review a statement so unexpectedly extraordinary as to cause me nearly to choke on a blueberry muffin. Having recovered my composure, I feel compelled to comment.

"This movie is so boring" (reads the review in question), "that copies of the script should be sold to monks as Gregorian chants. They would never know the difference." A simple statement in and of itself, yet its implications raise several interesting points.

Firstly, while my own experience with the monastic routine is admittedly not vast, I feel safe in asserting that Gregorian chants are, as a general rule, not hawked door-to-door on a commercial basis.

Secondly, even if they were (and keeping in mind that most monasteries do not still practice the Gregorian music tradition), I feel safe in asserting that even the most disenchanted (!) young novice would not at all be likely to buy a boring film script while under the impression that he was purchasing a Gregorian chant.

Lastly, and most disturbing aesthetically, is the reviewer's equation of boring film scripts with Gregorian chants, a subtle but nonetheless odious comparison which more than implies that the two entities share a common quality: i.e., boringness.

I feel safe in asserting that I am not alone in finding Gregorian chant a unique quality which, while not perhaps possessing the melodious excitement of Prince or the intellectual vigor of the Boom Town Rats, does manage nevertheless to transcend the mundane with such an emotional intensity that, yes, even a film review which causes one to choke on a blueberry muffin seems bland, even tame in comparison. . . .

May I suggest that . . . all future reviews containing such aesthetic confusion be sold to monks as an unpleasant example of the ill-considered metaphor. They would never know the difference. —Norman Bemelmans, Wilmington, N.C.

This argument deals with an arcane subject—Gregorian chant.* Notice also that the editor took the liberty of editing the text—we can tell that because we detect three dots (. . .) in several places at the end. This is probably because the letter was so long, space considerations required that it be condensed. We must trust the editor to have done a good job of bringing out the point.

The letter writer's style is difficult and his vocabulary is impressive, but the argument is actually very well structured. That structure helps us to understand his point. Let's see what it is.

*Pope Gregory I in the sixth century A.D. was the first to devise a musical notation called "Gregorian chant." This is a monadic chant, called "plainsong," which was adopted by the Roman Catholic Church. An example is "Dies Irae," a haunting melody which has been used by some composers (and horror movie directors!) as "demon music."

First, we need to number the paragraphs, as given below:

	1	Editor: Perusing . . . comment
	2	"This movie . . . points.
FIRSTLY:	3	while my own . . . basis.
SECONDLY:	4	even if . . . chant.
LASTLY:	5	and most disturbing . . . boringness.
	6	I feel . . . comparison . . .
	7	May I suggest . . . difference."

Paragraphs 3 through 5 are prefaced by 'Firstly', 'Secondly', and 'Lastly'. These are reason-indicators, and they help us see the structure.

Paragraphs 1 and 2 are lead-ins to the argument. Strictly speaking, they are not part of the reasons-to-conclusion process; they are *explanations.* They tell us why the letter writer is taking the trouble to write his letter, and what he's responding to (a movie review). As such, we don't need to diagram them. Paragraphs 6 and 7 are then the only candidates for the conclusion, but since 6 supports 7, we know that we have an argument chain. The result is:

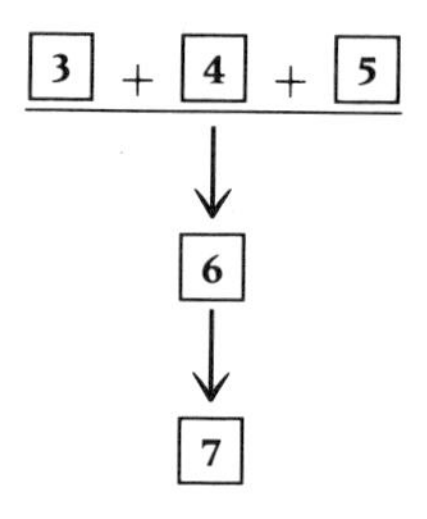

Notice that the argument faithfully follows the Rules of Grouping and Direction.

Now the point of the argument is plain enough: the writer is insisting that the movie reviewer used an "ill-considered metaphor" when he compared a boring film script to Gregorian chant. The structure of the argument helped us to understand his point.

EXERCISES

Exercise 1

DIRECTIONS: *Here is a short argument in paragraph form. Work a paragraph diagram and determine the conclusion of the article.*

Editor: I am a 20-year-old college student who had thought that the change in the legal drinking age would not affect my lifestyle, because I have never indulged in drinking and do not frequent local clubs. Due to a recent event, however, I have realized that even I have to pay a social price.

On Sept. 26 a group of my friends and I want to a small and well-behaved party at Wrightsville Beach. Around 10 o'clock two sheriff's deputies arrived and placed themselves directly at the foot of the staircase leading up to the porch. When a person descended the stairs, he was greeted with a flashlight full in the face and, if evidence of drinking was apparent, a demand for an I.D.

After a few offenders were given court orders, one officer mounted the stairs uninvited and held a private discussion with the hostess. When I later asked her what he had said, she could give no sure answer and knew only that everyone had to go inside and all people under 21 should leave.

I am convinced that hounding party goers to the extent of posting guards at the front door oversteps the bounds of enforcing a law and becomes harassment. The officers did catch a few offenders, but the main effect was to needlessly disrupt a quiet party.
—Cathy Butler, Wilmington, N.C.

Exercise 2

DIRECTIONS: *Below is a rather lengthy argument. Work a paragraph diagram and determine the conclusion of the article.*

Is Another 1929 Possible?

Could America's financial structure come crashing down in the late '80s much as it did in the waning days of the '20s?

Economy watchers have been asking themselves this question increasingly since mid-1982 when Mexico inaugurated a new era in international finance by threatening to default on its $80 billion foreign debt.

The answer, regardless of what economists preached in the '50s and '60s, is decidedly yes. The more people consider the question, the more they can't help being struck by the Age of Reagan's uncanny resemblance to that earlier period of free-market Republicanism, the Age of Hoover, Coolidge, and Harding.

The similarities are striking, not just in economics, but in politics and culture as well. In both the '20s and '80s, people were fascinated with illicit drugs—alcohol on the one hand, cocaine on the other—and the organized crime that inevitably goes with them. Both decades had their symbol of young urban affluence—the flapper and the yuppie. In both periods, Christian fundamentalism and "creationism" were rising forces out in the hinterlands. In 1928 there was a strong drive among both Republicans and Democrats to draft Henry Ford for president; in the '80s there was a somewhat shorter-lived movement among Democrats to draft Lee Iacocca.

But it is in economics where the real similarity lies. Both the '80s and the '20s were boom periods, in which the gap between stock market speculation and actual production grew wider and wider.

Each decade opened with a burst of inflation and a subsequent but short-lived recession. Recovery followed, although certain sectors (heavy industry in the '80s, agriculture in both the '80s and the '20s) continued to limp behind. From 1919 to 1921, the price of what fell by half and the price of corn by two-thirds. Labor also fared badly. Union membership declined by 28 percent from 1920 to 1930, virtually the same rate as the 15.4 percent decline between 1980 and 1985. While manufacturing output increased 29 percent from 1923 to 1929, average weekly wages for manufacturing employees rose just 5 percent. Like the '80s, the '20s were also a golden age of union-busting, in which a new breed of "specialists" pioneered in the use of yellow-dog contracts (requiring workers to pledge not to join unions) and informers to weed out troublemakers.

As the bull market gathered steam in both decades, an unprecedentedly large portion of the middle class stopped what it was doing to devote itself to stocks. Treasury Secretary Andrew Mellon persuaded Congress to slash income taxes in 1921, 1923, 1926, and 1928; for the very wealthiest, the income tax rate dropped from 73 percent in 1921 to 24 percent in 1929. Federal expenditures fell 40 percent over the same period.

The '80s have seen more or less the same combination of rising speculation, easy credit, and tax cuts. In August 1982, when the current bull market was just getting underway, the Dow Jones stood at just 777; in May, 1987 it was nearly 2,400. Over the same period, volume on the New York Stock Exchange more than doubled.

But this is only the tip of the iceberg. Other markets have surged forward more dramatically. On the government securities market, volume has quadrupled since 1980. Trading on government-securities futures tripled in 1984 alone. Meanwhile, high-flying savings and loan associations have sprung up like mushrooms after a spring rain. In states with lax regulations, like California, a savings and loan requires just $3 in capital to generate $100 worth of assets, usually in highly unstable real estate loans. Paradoxically, New Deal reforms such as federal banking insurance have encouraged bankers to fly higher and faster. In the old days, skittish depositors kept their bankers on a short leash and would jerk them back at the first sign of trouble. These days, depositors are less vigilant, thanks to federal insurance. The leash is longer and wonderfully elastic and as a result the potential for disaster is probably greater.

Where will it end? Will the '80s emulate the '20s right through to the bitter end? Is Reaganomics laying the groundwork for the Crash of '89 just as Andrew Mellon's economics did in the '20s? Will Wall Street lay another egg?

Unfortunately, the future is a lot more difficult to read than the past. But for those hooked on a cyclical view of history, the parallels go on. In both decades protectionism was on the rise, although free-market Republicans in the White House opposed it. As governor of Massachusetts, Calvin Coolidge first gained fame by breaking the Boston police strike in 1919; as president, Ronald Reagan's popularity soared when he smashed the air traffic controllers' strike in 1981. In 1983 banks began to offer interest on checking accounts; the last time they did that was in 1923. Beginning in 1927, 2,000 U.S. Marines found themselves locked in combat with a Nicaraguan nationalist leader named Cesar Augusto Sandino. . . . —Daniel Lazare, *In These Times*

Exercise 3

DIRECTIONS: *Here is another lengthy argument. Notice that there are about forty-six separate assertions in it. A paragraph diagram will do the job with only fifteen paragraphs.*

Work a paragraph diagram and then determine which paragraph contains the conclusion. When you've done that, then work an assertion diagram on the paragraph containing the conclusion. Identify the assertion which is the conclusion of the argument.

The Quiet Assault on America's Constitution

The American Constitution, the world's oldest written national charter, has been seriously threatened only twice: First by the secession of the slave states in the 1860s and, a century later, by former President Richard M. Nixon's attempt to corrupt it.

We are now witnessing a third threat, far less overt than the two earlier, but more insidious because it masquerades under the guise of legality and conservatism but is, in fact, subversion—if not of the letter of the Constitution, then of its spirit.

First, there is the attempt to write into the Constitution not fundamental principles of government, but comprehensive controls of private morality. Our "conservatives" have forgotten what the greatest of conservatives, John Marshall, said: "A Constitution is intended to endure for ages to come, and consequently to be adapted to the various crises of human affairs." It should, therefore, in all but specific structural areas, confine itself to fundamental principles such as those set forth in the preamble, leaving to successive generations of voters and of judges the right to interpret these in light of changing social, economic and political circumstances.

Second, there is overt departure from constitutional principles. The Constitution, for example, provides that "all bills for raising revenues shall originate in the House of Representatives." Now, however, the budget is drafted in the executive office, and the President claims the right to dictate its terms to the Congress; now, too, the Senate not only claims, but exercises, control over the budget: The bill passed last month was in a sense fraudulent. It originated not in the House but the Senate, and was added as a "rider" to a House bill on a wholly different subject, which was then discarded.

And, under pressure from the military or out of sheer paranoia, the Congress has acquiesced in another departure from the constitutional control of appropriations. Article 1, Section 9, provides that "A regular statement and account of the receipts and expenditures of all public money shall be published from time to time." Since its founding in 1947, there has never been a statement, regular or irregular, of any of the receipts or expenditures of the Central Intelligence Agency.

Perhaps the most dangerous transformation in our constitutional system is the evaporation of the great principle of the supremacy of the civilian to the military authority. That was very much an American principle, and it flourished down to almost our own day; after all, it was not so long ago that President Harry S. Truman fired Gen. Douglas MacArthur for trying to take conduct of the Korean War into his own hands.

But with the creation of the atomic bomb and the deepening of the Cold War, there has been a profound shift in the center of gravity in the American government. Ours has become a "security" state and, as a result, military considerations influence every aspect of our life. Thus, most recently, the claims of the CIA for "security" have taken precedence over the claims of freedom of the press, freedom of assembly, freedom of scientific research, freedom of travel and due process of law.

Another serious challenge to the integrity of our constitutional system is the erosion of the principle of separation of powers as if affects the judiciary.

The Constitution imposes few requirements upon the President; perhaps the most important is that, "He shall take care that the laws be faithfully executed." Under the Judiciary Act of 1789, this authority is exercised chiefly through the Department of Justice, headed by the attorney general.

We are now confronted with an astonishing spectacle. Instead of "faithfully executing" the laws as interpreted by the high court, the President and his appointed officials in the Department of Justice are busy circumventing and frustrating them. Indeed, the Justice Department has adopted a policy of ignoring Supreme Court decisions on such matters as school segregation, busing, redistricting on racial grounds, affirmative action in employment and tax exemptions for private schools that engage in discrimination.

With respect to busing, the assistant attorney general stated that "we have concluded that involuntary busing has failed." Perhaps it has, but that is not a matter for the Justice Department to conclude, but for the courts to adjudicate. The Justice Department concluded, too, that it would not represent the United States in the denial of tax exemption of private schools, thus presenting us with the spectacle of the Supreme Court forced to appoint an outside attorney to argue the position of the Treasury Department and of an earlier court.

So, too, in another case—one of classic importance—that challenged the obligation of Texas to provide public education to the children of illegal immigrants. An earlier Justice Department had argued (quite rightly) that the term "persons" in the 14th Amendment covered all children, whatever their parentage. The current Justice Department, however, announced that it no longer had any interest in the question and withdrew from the case. Fortunately, the Supreme Court struck down the Texas law.

No less disturbing is the record of the Justice Department in a series of redistricting cases. There, under improper pressure from congressmen representing the states involved, it reversed its earlier commitment to upholding the election laws of the Johnson Administration.

More serious, in the long run, than these particular incidents, is the philosophy that animated them. Thus Attorney General William French Smith has called upon the courts to respect "the groundswell of conservatism evidenced in the 1980 election." And, Smith added, if the courts persisted in their independence, we will witness "attacks by persons who see majority rule thwarted by the legal system itself." Needless to say, such "thwarting" is what the Constitution itself requires, if and when a majority rule flouts the Bill of Rights.

When we reflect on this process of politicizing the Department of Justice and the courts, we should keep in mind the words of Abraham Lincoln's attorney general, Edward Bates, that "The Office of the attorney general is not properly political, but strictly legal." More than a century ago, James Russell Lowell reminded us that "new occasions teach new duties." True enough: But it is new duties they teach, not a repudiation of the old and familiar. The most fundamental duty of Americans, as citizens and officeholders, is that of respect for a Constitution and a Bill of Rights that have served us well for two centuries. —Henry Steele Commager, *Los Angeles Times*, October 17, 1982

MODULE 8

DEDUCTIVE VS. INDUCTIVE ARGUMENTS

8.1 DISTINGUISHING DEDUCTIVE FROM INDUCTIVE ARGUMENTS

We have learned to distinguish arguments from exposition. We know that arguments always involve an attempt to give a set of reasons in support of a conclusion, whereas exposition (which includes poetry, narration, description, explanation, and so forth) does not. Arguments always take this sort of form:

reason 1.
reason 2.
.
.
.
reason n.
conclusion.

Whereas there may be as many as ten, twenty, or one hundred reasons or as few as one, there can be no more than one main conclusion. If one or more reasons point to two or more conclusions, we will say that two or more arguments are underway with the same premises.

We may now distinguish between two different types of argument.

A deductive **argument** is valid, if and only if, if all the **premises** are true, then the conclusion must be true.

A **deductive** argument is an argument which makes the claim that the conclusion follows from the reasons (called "premises") with the force of *necessity.*

An **inductive** argument is an argument which makes the claim that the reasons offered in support of the conclusion (here the premises are called "the evidence") do so with the force of *likelihood* or *probability.*

We may notice the first distinguishing feature of deductive and inductive arguments. All arguments have **premises** as the reasons offered in support of the conclusion, whereas inductive arguments have **evidence** which serve as premises. In other words, an argument looks like this:

premise 1.
premise 2.
.
.
.
premise n.
conclusion

. . . where n = any number.

However, an inductive argument will look like this:

evidence 1.
evidence 2.
.
.
.
evidence n.
conclusion

Naturally, as before, there may be one, two, ten, or one hundred instances of evidence offered in support of the conclusion.

The fact that inductive arguments offer only *evidence* in support of conclusions is the first and most striking way of distinguishing inductive from deductive arguments. When Sherlock Holmes first meets Doctor Watson (in *A Study in Scarlet*), Holmes concludes right away that Watson was in medical service in Afghanistan. What is the reason for coming to such a startling and revealing conclusion? The evidence, of course. Watson has a deep tan—he didn't get it in England! He looks like a medical type, but has the bearing of a military man. His arm is still—probably an injury. His face shows hardship and the effects of a disease. Where in a hot climate could Watson have seen such hardship? Probably in a conflict. Britain was at war in Afghanistan at the time. Therefore, it is likely he was in military medical service in the war, and just returned.

Notice the detailed evidence that Holmes produces. This is a clue that the argument underway is inductive, and not deductive. Even though Holmes characteristically states his conclusion with utmost confidence, this is at best a good inductive argument.

Deductive premises, on the other hand, *may* often be recognized by the frequency of **logical connectives** (or variants of logical connectives) in the course of an argument. (This is a rule of thumb only.) The basic logical connectives are:

'both . . . and'

'either . . . or'

'if . . . then'

'if and only if'

'it is not the case that . . .'

An argument that uses logical operators in this way might be the following (the logical connectives are boldfaced): "**If** temperature and light remain constant, **then** existing weather conditions can be reliably predicted to continue. **Either** a cold front is moving in **or** prevailing winds might cause the front to shift to the north. In fact, **both** the cold front is expected to be diverted to the north **and** temperature and light will remain constant. Therefore, it must be the case that existing weather conditions will continue to prevail."

Notice that this argument relies on the interworking of the logical connectives and—given the accuracy of the data—the conclusion must certainly follow. This is a deductive argument.

Deductive arguments attempt to make good on a claim that the conclusion *necessarily follows* from the premises. The reason that they can make this claim—if we have a good deductive argument; there are bad ones which falsely make the claim—is that they rely on certain *associations of ideas* which bind the conclusion to its premises.

The clearest way of understanding what a deductive argument is may be through considering the process of elimination. We know that something's wrong with our antique Studebaker; it won't start. We read the manual, and it says: "if your Studebaker won't start, then one of the following must be faulty: your battery, your terminal connections, your lead wires, your solenoid, or your generator. There are no other possibilities." We've already replaced the battery, the terminal connections are new, the lead wires are new, and the generator was fixed just last week. Thus, we argue: "Assuming it's not the battery, the terminal connections, the lead wires, nor the generator, then the solenoid must be faulty—there are no other possibilities." If you're sure about the other parts, and the manual is accurate, you can be certain that the problem is with the solenoid. Replace it.

We see that this argument uses the process of elimination—in fact, all deductive logic can be expressed as process of elimination* though that may not

*To understand why this is so, consider that a celebrated proof in mathematical induction has shown that all of the first order predicate calculus can be expressed with only two logical operators: negation ($\sim$) and disjunction ($\vee$).

always be obvious. Sometimes, that's one way of recognizing that a deductive argument is under way. Let's consider the thinking that went into repairing our Studebaker. While a great deal of evidence, experiment, and trial and error is going on (for example, testing the faulty generator with a voltmeter), nevertheless the main technique of argument here is deductive process of elimination. By testing the battery, terminal connections, and others, we are ruling out possibilities in order to arrive at the answer. As Arthur Conan Doyle once had his character Sherlock Holmes say to Dr. Watson: "When you've eliminated the impossible, whatever remains, no matter how improbable, must be true." Granted, you just had your mechanic replace the generator only last month. You want to say it can't be faulty. However, you have tested every other possibility, and eliminated them. My advice to you is: replace your generator, and get a new mechanic.

This is a deductive argument. The premises confer certainty on the conclusion. Given the truth of the premises, there is no other possibility.

Sometimes the conclusion-indicators give us a clue as to whether the argument is inductive or deductive. However, this is not a sure thing—be careful.

Some standard deductive conclusion-indicators are:

'it follows that . . .'

'necessarily . . .'

'it must be the case that . . .'

'we may deduce that . . .'

'implies that . . .'

Some standard inductive conclusion-indicators are:

'it is likely that . . .'

'it is probable that . . .'

'we can be assured that . . .'

These are guides or signals to deductive or inductive arguments. Most conclusion-indicators, however, are not so decisive. Consider, for example:

'therefore . . .'

'thus . . .'

'hence . . .'

'so . . .'

'we may infer that . . .'

> *"When you have eliminated the impossible, whatever remains, however improbable,* must be the truth."
>
> —Sherlock Holmes in *The Sign of the Four* (ARTHUR CONAN DOYLE)

None of these conclusion-indicators gives us a clue as to which s argument is underway. The *best* way to tell whether an argument is deductive or inductive, however, is by assessing the *force* with which the conclusion is asserted. **Conclusion force** may be utterly decisive (as in 'it is certain that . . .') or it may bring us up short of certainty (as in, 'it is very, very likely that . . .').

Here, then, is an overview of how to distinguish deductive from inductive arguments:

Deductive	Inductive
uses *premises* for reasons	uses *evidence* for reasons
deductive conclusion-indicators	inductive conclusion-indicators
"necessity" for conclusion force	"likelihood" for conclusion force

EXERCISES

Part I: Elementary Exercises

DIRECTIONS: *Say whether the following passages are better described as* ***inductive*** *or* ***deductive*** *arguments:*

______ **1.** If you shot at me, you must have a gun. You don't have a gun. It couldn't have been you.

______ **2.** Mozart died at age thirty-five. Since he was so prolific a composer, he must have done much of his work at an early age.

______ **3.** Lee Harvey Oswald acted alone in the assassination of President Kennedy. That was the conclusion of the Warren Report, which conducted an extensive investigation. Nothing I have seen since causes me to revise that judgment.

______ **4.** George Washington was born February 22, 1732. It's now March of 1988. If, by some miracle, Washington were still alive, he'd be 258 years old now.

______ **5.** We've received radio signals from the planet Oklafar, the second planet around the star Betelgeuse. We think we may find intelligent life there.

______ **6.** University Park is located in the geographic center of Pennsylvania. Pennsylvania is a large state. Altoona is less than fifty miles away from University Park. Altoona must be in Pennsylvania too.

______ **7.** The formatting is clean and crisp. The layout is filled with graphics. The whole job took less than a week to complete. This report looks like it was done on a French *Pomme de Terre* computer.

______ **8.** We found a killer whale washed up on the beach; it had a huge gash in its side. Swimmers are missing. A derelict rowboat was found drifting with no one on board. "Jaws" is back!

______ **9.** The state of California divides its universities between the California State University system, which is mainly undergraduate schools, and the University of California system, which includes graduate schools. Cal State/Bakersfield must be mainly an undergraduate university.

______ **10.** A "Yuppie" is a "young urban professional." At seventy-eight years of age, Ronald Reagan is no Yuppie.

Part II: Advanced Exercises

DIRECTIONS: *Say whether the following passages are better described as inductive or deductive arguments. These are real-life examples as opposed to textbook examples—they're taken from editorials, advertisements, and letters. Because they are taken from real life, they may prove to be more difficult than the textbook examples you worked in Part I.*

______ **1.** As an army officer during World War II, I wore my trusty Colt 45 pistol religiously in combat. To my knowledge, few officers fired their handguns once during those days, let alone 8,000 rounds. It is ridiculous to use this capacity as our criterion for selecting a pistol. —"Letters," *Time,* 128 (22) 1987, p. 10

______ **2.** The longitude of New York is 74° 0′ 3″ West; of Philadelphia 75° 10′ West. What is their difference in longitude? Explanation: Since New York and Philadelphia are both west of the prime meridian of Greenwich, their difference in longitude must be the distance by which Philadelphia is farther west than New York. This distance is found by subtracting the lesser longitude of New York from the greater longitude of Philadelphia. —Abraham Sperling, *Mathematics,* 1957

______ **3.** Harmless error doctrine requires reviewing courts, upon finding an error in the proceedings below, to determine what the sentencer would have done if there had been no error. The appellate court must decide what the result of the trial would have been if the sentencer had heard a different bundle of evidence. This process is ill-suited to capital sentencing review. —James C. Scoville, "Deadly Mistakes: Harmless Error in Capital Sentencing," *University of Chicago Law Review,* Vol. 54, p. 755

______ **4.** In demonstrating that Reaganism has yet to run its course, Irving Kristol seems to take particular glee from the fact Reaganomics by running up a huge Federal debt, has foreclosed the liberal agenda for the foreseeable future. Has it occurred to him that Mr. Reagan's debt has also foreclosed any Government agenda? For a few years of "morning in America," this administration has squandered our Government's wealth, its economic power and its ability to respond with flexibility and ingenuity to future economic crisis. —Letter to the Editor, the *New York Times Magazine,* July 12, 1987, p. 70

______ **5.** The current abundance of deer is evidenced in the increased sightings, deer-vehicle accidents and landowner complaints about crop damage. Expanding deer herds and the problems they create generally result in longer hunting seasons and more liberal bag limits for deer hunters. —*Deer Hunting Annual*

______ **6.** ENGINEER SCOTT: "But, Mr. Spock, why'd ye ever do such a fool thing as riskin' yer life by jumping into the matter transformer?"

CAPTAIN KIRK: "Yes, Spock. I'd like to know the answer to that myself."

MR. SPOCK: "Well, Captain. I knew that if I remained standing where I was, the Klingons could focus the mind probe on me, which is fatal to Vulcans. And I knew that if I moved anywhere else on board, I would be exposed to lethal Zebulon radiation. Logic dictates that it would be irrational to leap into a faulty mechanism, but, as paradoxical as it sounds, Captain, sometimes it is only logical to do the irrational."

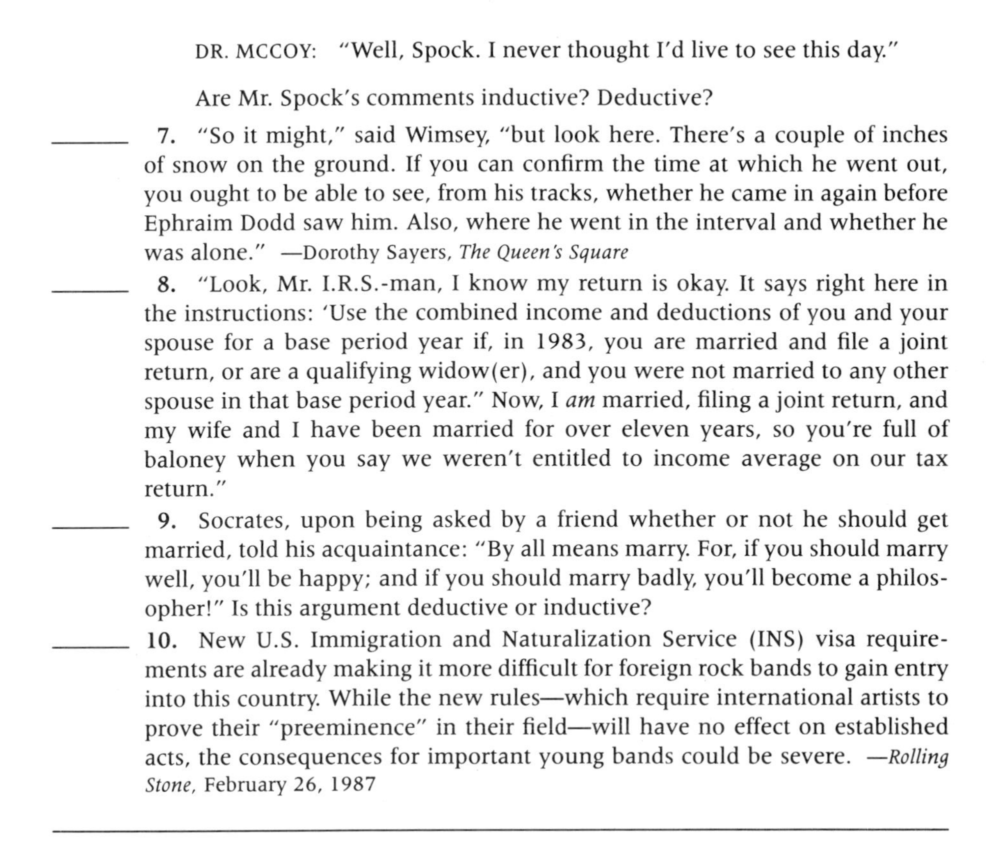

DR. MCCOY: "Well, Spock. I never thought I'd live to see this day."

Are Mr. Spock's comments inductive? Deductive?

______ **7.** "So it might," said Wimsey, "but look here. There's a couple of inches of snow on the ground. If you can confirm the time at which he went out, you ought to be able to see, from his tracks, whether he came in again before Ephraim Dodd saw him. Also, where he went in the interval and whether he was alone." —Dorothy Sayers, *The Queen's Square*

______ **8.** "Look, Mr. I.R.S.-man, I know my return is okay. It says right here in the instructions: 'Use the combined income and deductions of you and your spouse for a base period year if, in 1983, you are married and file a joint return, or are a qualifying widow(er), and you were not married to any other spouse in that base period year." Now, I *am* married, filing a joint return, and my wife and I have been married for over eleven years, so you're full of baloney when you say we weren't entitled to income average on our tax return."

______ **9.** Socrates, upon being asked by a friend whether or not he should get married, told his acquaintance: "By all means marry. For, if you should marry well, you'll be happy; and if you should marry badly, you'll become a philosopher!" Is this argument deductive or inductive?

______ **10.** New U.S. Immigration and Naturalization Service (INS) visa requirements are already making it more difficult for foreign rock bands to gain entry into this country. While the new rules—which require international artists to prove their "preeminence" in their field—will have no effect on established acts, the consequences for important young bands could be severe. —*Rolling Stone,* February 26, 1987

8.2 A COMPLICATION: INAPPROPRIATE CONCLUSION FORCE

Unfortunately, we find a complication. Sometimes an argument proceeds along swimmingly, listing data after test results and other pieces of evidence, and then all of a sudden the arguer announces: "Therefore, it is absolutely certain that my conclusion follows." All along, we were under the impression that the argument was inductive, but the arguer pulls the rug from under us and claims the force of a deductive argument for her conclusion.

Alternately, someone argues: "Both Venus flytraps and pitcher plants are found in southeastern North Carolina. Therefore, it is probable that Venus flytraps are found there." This is exceedingly odd because the arguer doesn't exert the warranted amount of force for his conclusion. Surely, given the truth of that premise, Venus flytraps *must* be found there. This is a deductive argument, but the arguer obviously doesn't know enough about logic to know that he can assert his conclusion for certain, and so he hedges.

If people were more logical, then we'd never have to worry about inappropriate conclusion force. Unfortunately, that's not so. So, let us introduce two new categories:

1. Deductive arguments mistakenly passed off as inductive

and

2. Inductive arguments mistakenly passed off as deductive.

Detecting these is a matter of judgment. Sometimes an argument makes a pretty good deductive argument, but a lousy inductive one, yet it is passed off as inductive. And sometimes an argument makes a pretty good inductive argument, but a poor deductive argument, yet it is passed off as deductive by the arguer.

The issue is one of *appropriate* vs. *inappropriate* conclusion force.

Deciding between them is as much a judgment call as if we were "Monday Morning Quarterbacks" trying to decide whether a referee's penalty call was warranted, given the instant replays.

Try your luck on the following exercises.

EXERCISES

DIRECTIONS: *Decide whether the following arguments show appropriate or inappropriate conclusion force. If you encounter a deductive conclusion-indicator but the argument would make a better inductive argument, then the conclusion force is inappropriate. Likewise, if you encounter an inductive conclusion-indicator, but the argument warrants a more certain conclusion, the conclusion force is inappropriate.*

These exercises will be more productive if they engender discussion.

______ 1. "These Christmas Tree lightbulbs don't work. They're wired in series. I know that the red bulb is burned out (it's burned black). Therefore, I'll replace the red bulb. Then they'll all light up."

______ 2. "If we get six inches of snow, the Chancellor always cancels classes. We got at least six inches of the white stuff this morning, so it's very likely that we won't have class today."

______ 3. "They say that if you have a red dawn, figure on rain; but if you have a red sunset, the next day is sure to be sunny. We had a red sunset yesterday afternoon. We're sure to have a sunny day."

______ 4. "Nobody else could have done it; let's assume that's true. Then, it's probable that Ngaio Marsh is the killer."

______ 5. ". . . You appeared to be surprised when I told you, on our first meeting, that you had come from Afghanistan."

"You were told, no doubt."

"Nothing of the sort. I *knew* you came from Afghanistan. . . . The train of reasoning ran, 'Here is a gentleman of a medical type, but with the air of a military man. Clearly a army doctor, then. He has just come from the tropics, for his face is dark, and that is not the natural tint of his skin, for his wrists are fair. He has undergone hardship and sickness, as his haggard face says clearly. His left arm has been injured. He holds it in a stiff and unnatural manner. Where in the tropics could an English doctor have seen much hardship and got his arm wounded? Clearly in Afghanistan.' The whole train of thought did not occupy a second. I then remarked that you came from Afghanistan, and you were astonished."

"It's simple enough as you explain it," I said, smiling. —Arthur Conan Doyle, *A Study in Scarlet*

8.3 TEN LITTLE INDIANS

We rarely encounter a sustained argument that is exclusively deductive or exclusively inductive. Political speeches, scientific research, scholarly monographs—each of these generally uses a blend of both inductive and deductive reasoning, though the overall argument will usually emphasize a deductive or an inductive mode of reasoning (for example, scientific research, which relies heavily on evidence, is usually largely inductive).

Detective mysteries provide a useful example of how the two forms of reasoning work together. My guess is that most detective stories are primarily exercises in deductive reasoning, although every mystery relies to some degree on the evidence that the detective uncovers. In the end, however, the novel is—more often than not—conclusively resolved on or near the last page (that's why the mystery novel with a missing last page is a source of such frustration and consternation). How can that be, unless the central argument of the mystery attains the certainty conferred only by deductive reasoning?

The model strategy of the detective novel is *process of elimination,* the very soul of deductive reasoning. To make this point, let's look at the broad outlines of Agatha Christie's novel, *Ten Little Indians,* which takes its title from the familiar nursery rhyme.

The story is set on an isolated island—aptly called, "Indian Island"—off the coast of Britain. Ten guests have been invited to stay on the island by a mysterious host. Once there, they have no way of leaving the island, and one-by-one the guests are murdered in much the manner described by the nursery rhyme.

Here is the "cast of characters" (ten in all):

The Guests

Mr. Justice Wargrave	the "hanging judge"
Vera Claythorne	the ex-governess
Capt. Philip Lombard	the soldier of fortune
Emily Brent	the elderly spinster
Gen. MacArthur	veteran of "the Great War"
Dr. Armstrong	the physician
Anthony Marston	the playboy
Mr. Blore	the police inspector
Mr. and Mrs. Rogers	the butler and the cook

As the story proceeds, one guest after another is murdered. A careful check by the living guests shows no one else could be on the island—a storm comes up, making passage by boat impossible, and the island is small enough to search for hidden fiends. Since no one else could be there, one of the guests must be the murderer.

The problem is that the number of guests dwindle due to "nefarious acts" until, as the nursery rhyme has it, "then there were none."

How could that be? Let's consider each of the premises, one by one.*

Premise One: Anthony Marston choked to death on poison—the first to die.
'Therefore', Anthony Marston can't be the murderer.

Premise Two: Mrs. Rogers was given an overdose of sleeping pills.
'Therefore', Mrs. Rogers can't be the murderer.

Premise Three: General MacArthur was killed by a blunt instrument.
'Therefore', Gen. MacArthur can't be the murderer.

Premise Four: Emily Brent was injected with a hypodermic filled with poison.
'Therefore', Emily Brent can't be the murderer.

Premise Five: Mr. Rogers was murdered with an axe.
'Therefore', Mr. Rogers can't be the murderer.

Premise Six: Justice Wargrave was murdered with the gun.
'Therefore', Justice Wargrave can't be the murderer.

Premise Seven: Dr. Armstrong was drowned by the murderer.
'Therefore', Dr. Armstrong can't be the murderer.

Premise Eight: Mr. Blore was thrown to his death from a window.
'Therefore', Mr. Blore can't be the murderer.

. . . And then there were two.

Premise Nine: Philip Lombard was shot to death by Vera Claythorne.
'Therefore', Philip Lombard can't be the murderer.

Premise Ten: Vera Claythorne hanged herself.

The obvious conclusion? Vera Claythorne was the murderer! By process of elimination, since no one else could be on the island, this is the only conclusion possible.

Unfortunately, although the deduction *looks* good, the conclusion is false. Vera Claythorne was *not* the murderer of all the victims. As you expect, a good murder mystery will have an unexpected twist—and here it is: *one of the premises must be false.*

We know that, given a valid deductive argument, if all the premises are true, then the conclusion *must* be true. Since we take it as given that Vera Claythorne was not the murderer of all the other guests, then we must have been mistaken in assuming that each of the others were murdered.

*My apolgies to those who would like to read the novel, but haven't. Read it anyway.

The overall form of the argument is deductive and if we assume that each of the premises is true, then it *looks like* it follows that Vera Claythorne was the murderer. In fact, she *did* shoot Philip Lombard, the soldier of fortune. But she shot him only because she assumed that *he* had to be the murderer because she knew that she didn't do it. Emotional problems and guilt drove her to commit suicide, but the opportunity was carefully arranged. She found a hang-noose already prepared for her. If she didn't kill the other guests, who did?

The only conclusion is that one (or more) of the other premises is false. It's possible, for example, that two or more of the other guests committed the murders (but, in fact, the murders were committed by one person). How can that be?

Here's where *inductive* considerations of faulty evidence enter in. How do we conclude that someone is dead? Most often, he (or she) is *pronounced* dead by a qualified physician. Dr. Armstrong lived long enough to pronounce most of the others dead. Was his diagnosis of one of the victims incorrect? Is it possible he lied to the others that someone was dead when he or she really wasn't? Was *he* the murderer after all?

In fact, as we learn at the end of the novel, Justice Wargrave was incorrectly pronounced dead (he died later in the novel). Premise Six is false.

RECOMMENDED READING

A useful and enjoyable exercise to illustrate the common linkage between inductive and deductive argument forms in lengthy argument chains is to read some good mysteries.

Here are a few suggestions:

Agatha Christie, *Ten Little Indians,* New York: Pocket Books, 1983.

———, *Murder on the Orient Express,* New York: Dodd, Mead, Inc., 1985.

Arthur Conan Doyle, *A Study in Scarlet,* New York: Readers' Digest Associates, Inc., 1986.

Umberto Eco, *The Name of the Rose,* New York: Warner Books, Inc., 1986.

Stephen Greenleaf, *Death Bed,* New York: Ballantine Books, Inc., 1987.

Dorothy Sayers, *The Nine Tailors,* New York: Harcourt, Brace, Jovanovich, Inc., 1966.

Julian Symons, *The Thirty-First of February,* New York: Carroll and Graf, Publishers, 1987.

MODULE 9

SOME USES OF LANGUAGE

9.1 EMOTIVE LANGUAGE

We may usefully distinguish between two broad uses of language. Sometimes we use language in an *informative* way and other times we use language *emotively.* **Informative language** (sometimes called cognitive language) provides information which can give us the basis for argument. **Emotive language** is language which merely expresses emotion.

Emotive language consists of words or statements which possess a *tone* or *force* that evokes emotion. Emotive tone reveals the feelings of a speaker (". . . a cancer in the White House"), whereas emotive force can arouse feelings in an audience ("Communist!"). *Euphemisms* and *buzzwords* are categories we reserve for language which contains emotive tone or force (or both).

Euphemism is the substitution of inoffensive language for language which is explicitly offensive, and so euphemisms are mainly expressive of emotive *tone*—though they may sometimes also have emotive force, especially when they

> *"Contrariwise," continued Tweedledee, "if it was so, it might be; and if it were so, it would be; but as it isn't, it ain't. That's logic."*
>
> —LEWIS CARROLL, *Through the Looking Glass*

A deductive **argument** is valid, if and only if, if all the premises are true, then the conclusion must be true.

work their way into everyday language. The term 'buzzword' is slang, and so it is somewhat harder to define formally, but it has a nice ring about it since we may take **buzzwords** to be terms which primarily arouse emotions in an audience (i.e., an audience "buzzes" upon hearing the word—think of a conservative political audience murmuring with agreement when their candidate describes his opponent as a "liberal" or some other sort of "disagreeable" term). Thus, we may say that a buzzword has mainly emotive force, though it may also sometimes express the feelings of the speaker.

Euphemisms and buzzwords tend to distract us from rational argument.

Let's consider a few examples. On October 19, 1987, the Dow Jones Industrials plunged a record 508 points—this was a stock market crash. 'Crash' accurately describes what happened to the market even though the word does have some emotive impact (few words, if any, have *no* emotive tone or force whatsoever). As a result, many experts who were advocates of the market took pains *not* to describe the drop in the market as a plunge. Some analysts took to calling the crash a "correction in the stock market." This is a euphemism (if anything is!). These analysts hoped to quiet down fears of investors by substituting the technical-sounding term, "correction" for the more accurate word, "crash."

On the other hand, some journalists and critics of the market began using buzzwords like "market meltdown!" and "Bloody Monday" to describe the crash. These buzzwords served only to heighten fears unjustifiably.

Cool and rational analysts tried to steer a middle course. They recognized that what occurred was a bona fide crash, and so they used the term. But by rejecting euphemisms like "a correction in the stock market," they avoided deceiving investors into thinking that everything was running smoothly on Wall Street. On the other hand, they also rejected using buzzwords like "meltdown" because they didn't want to fuel a panic. Instead, they carefully analyzed the causes of the market crash and proposed effective remedies.

Another example: I once asked a native Californian if he wasn't concerned about the many predictions that a major earthquake would occur there and he replied, "Oh, we just think of them as a kind of weather."

The Californian was *euphemizing* the very real dangers posed by earthquakes to life and property. By describing earthquakes as "weather," he was comparing them to ordinary, everyday phenomena. Such a euphemism might cause the Californian to understate those dangers to himself and his family. This is one way in which euphemisms can serve as a vehicle of dangerous self-deception.

Here's a short list of some euphemisms and buzzwords for your enjoyment.

Euphemism or Buzzword	Translation
"slight mechanical adjustment"	no landing gear (euphemism)
"terrorist"	freedom fighter (buzzword)
"freedom fighter"	terrorist (euphemism)
"energetic disassembly"	explosion (euphemism)

Euphemism or Buzzword	Translation
"deferred maintenance"	overgrown landscaping (euphemism)
"Rocky Mountain Oysters"	bull testicles (euphemism)
"The Public Safety Unit"	Idi Amin's Death Squad (euphemism)
"card-carrying liberal"	a liberal (buzzword)
"right-wing conservative"	a conservative (buzzword)
"roach motel"	roach trap (euphemism)
"Long Island Iced Tea"	the name of an alcoholic beverage (euphemism)
"Society for the Propagation of the Faith"	The Inquisition (euphemism)

There's a place for emotive language, especially when we need to express emotions in contexts such as poetry, but emotive language can derail an argument and send it out of control or in the wrong direction. Let's consider one more example which shows the degrees of emotionalism we may find in the simple reporting of a news story. Suppose that the mayor of our city has just passed an unpopular real estate tax through the city council. Let's suppose that our city has three newspapers and each of them editorially regards the tax as unnecessary or burdensome.

Here's how one of the newspapers might report the story:

The Morning News

Mayor's Fat Tax Hike Bleeds Property Owners

FIGURE 1

The Morning News is generally a reliable newspaper but the editor has slipped into emotive language in an inappropriate place and manner in reporting this story. He describes the tax as a "fat" one; he suggests that the mayor has "hiked" the tax and that it will "bleed" property owners. Such buzzwords are rarely appropriate, but they are never acceptable in a front-page headline. No doubt, the publisher will sell more newspapers than usual because the tax is bound to be an unpopular one, but we may justifiably regard this headline as a lapse in editorial judgment.

Here's another version of the story as headlined in a crosstown daily. This newspaper is given to sensationalism as part of its journalistic daily fare. Notice how *The Daily Dagger* abandons all pretense to objective reporting:

The Daily Dagger

TAX, TAX!
SPEND, SPEND!

"Hizzoner" jacks up taxes again!

FIGURE 2

The editor of this paper resorts to slogans and epithets. He describes the mayor with the derogatory epithet "hizzoner" (a mockery of "his honor"). The information content of the news story is very low indeed! Almost no reporting is given on the front page—the headline says it all!

Perhaps the editor will also run an unflattering photograph of the mayor on the front page as well. The editor will dig through his photo archives until he finds one that will make the mayor look as sinister as possible. We find no merit in this newspaper's presentation of what may be an issue with two arguable sides to it. Raising taxes is rarely popular; perhaps the mayor showed some conviction and courage in proposing this new tax.

Finally, let's look at our city's newspaper of record, *The Herald*. This is a fine, old newspaper with a tradition of excellence, concerned only with reporting the news as objectively as possible (it doesn't even have a comic page!). This is how the editor headlines her story.

Notice the restraint, the cool presence of mind, expressed in the headline of the newspaper? Why, the headline isn't even in bold print! We don't find any euphemisms or buzzwords either; the editor is concerned with relating the story

Academic Buzzwords

—An Expose of Higher Ed Jargon
by Philip E. Devine and Robert Hauptman

. . . The following is a collection of jargon which is currently used in higher education by both professors and administrators. These terms reflect the money, status, power and sex struggles that protect academics from an unhealthy preoccupation with teaching and scholarship. Our definitions are meant to be humorous, sarcastic, ironic or, on occasion, preposterously all too true. [What follows are selected examples:]

Adjunct Professor: Slave.

Bombinate: Buzz: the prototypical buzzword.

Brainstorm: To waste time.

Chair: Head of a department. You need four of these and a table for a dining room set.

Curriculum Vitae (also VITA): Glorified resume.

Disallocate: Rip-off.

Excellence: A totally meaningless expression used in job descriptions, college catalogues and letters of reference.

Fullbright Scholar: CIA agent.

Grant, Fellowship, Scholarship: Money.

Hidden Agenda: Motives with non-explicit purpose. (If you understand this, you should join the CIA.)

Meaningful: Meaningless.

Objective: Biased in accordance with my values.

Overqualified: What they use to nullify you when all else fails.

Plagiarism: Copying. If more than one source is used, this is called research.

Publish or Perish: Publish and perish anyway.

Rights: Power.

Rigorous: Trivial.

Scholarship: Footnotes (lots of them).

Strategic Plan: A faculty purge.

Tenure: Early retirement at full pay.

Test: Chance to even the score with your students.

University: (1) An organization for creating obstacles to teaching or research, (2) a college, (3) a Vo-tech Institute.

Value-Judgment: Your opponents' unsupported opinions.

Verbalize: Talk.

Warm Boot: Dismissal with excellent references.

[From *The Journal of Educational Public Relations,* Vol. 10, No. 1]

The Herald 25¢

Property Tax Increases to 3%

FIGURE 3

as objectively, *factually,* as possible. Look at the length of the story columns on the front page: we find a lot of *information* there. Even the headline is informative (the tax is raised to 3%!). This is the ideal use of informative—as opposed to *emotive*—language.

Let's say that the editor of *The Herald* also thinks that the tax is ill-advised—perhaps even irresponsible. Well, she'll keep her opinion to herself in the news story, on the *front* page. She'll lay out her opinions and her arguments against the tax on the *editorial* page.

What do we find there? Again, we find a careful analysis of the impact of the tax, but still no euphemisms or buzzwords. This is the ideal basis for argument. The editor even gives the mayor some grudging credit for courage and conviction in proposing the tax, even though she thinks he's wrong! She relates her argument with objectivity and restraint even there, and, as a result, the argument has far more impact on thoughtful people than it would otherwise. (See "Academic Buzzwords" for selected examples in higher education.)

9.2 METAPHORS AND SIMILES

Metaphors are close relatives of euphemisms and buzzwords; in fact, euphemisms and buzzwords *are* metaphors of a certain sort. A metaphor is a kind of comparison, but one in which the comparison is taken almost exactly. Thus, a **metaphor** is a figure of speech which takes a word or phrase that has a standard use and uses it for purposes of comparison in a non-standard (i.e., non-literal) way. For example, a traffic "bottleneck" is a metaphor since it is used to describe a type of traffic jam where the traffic bunches up at an exit ramp or merged lane. The result looks very much like the point at which a bottle narrows near the top—the "neck" of the bottle (itself a metaphor).

We can better understand what a metaphor is if we first of all define a simile—a close relative of metaphor. A **simile** is a comparison or analogy between two different things where the comparison is made explicit by the words 'like' or 'as'. For example, the popular mystery writer Stephen Greenleaf describes a view of San Francisco Bay with this rather wild simile: "The whitecaps frolicked in the bay like frisky Easter bunnies." Here, Greenleaf is comparing two things which are so essentially different that the result is humorous.

Likewise, metaphors are comparisons but the analogy between the two things is putatively so tight that the 'like' or 'as' which designates the comparison is omitted—thus, the comparison is *implicit* rather than *explicit.* "Now is the winter of our discontent / Made glorious summer by this sun of York," says Richard III in Shakespeare's play. Notice that Shakespeare does not say, "Now is *like* the winter of our discontent." He omits the 'like', rendering a metaphor instead of a simile. By employing the metaphor, Shakespeare is suggesting that Richard's discontent is almost an exact correspondence to winter. The figure of speech here calls to mind the chill, the gloom, and the darkness of winter and suggests likewise that Richard's discontent is chilling, gloomy, and dark. Fortunately, thanks to this "sun" of York, winter is made "glorious summer," and thus the gloom is lifted by the light of day.

Metaphors operate at more than one level of meaning. Taken at its surface meaning, a statement which contains a metaphor is literally absurd. Taken at a deeper level, however, the metaphor makes sense in the context of the statement. For example, if we say, "This set of exercises is a nightmare," we can't literally mean that a set of logic exercises can be any sort of dream, bad or good. But if we look at the figurative meaning (i.e., the meaning *beneath* the surface), it makes perfectly good sense. Like a nightmare, the exercises keep us awake nights and cause us anxiety. In fact, the comparison is so tight that we can dispense with the 'like' which characterizes a simile, and assert (figuratively, of course) that the exercises *are* nightmares.

Metaphors are as close to poetry, perhaps, as the average writer ever gets. Thus, they have some distinct advantages. Metaphors are an outlet for creativity, for one thing. They afford us the means to make sharp comparisons between certain properties of otherwise unlike things. They also point the way to further thought because they enable us to reflect even more on previously unsuspected ways in which the two things share properties in common.

A very famous metaphor was used by the political philosopher John Locke when he said: "In the beginning, all the world was America." Locke, who wrote at the beginning of the eighteenth century, was writing about the imagined condition of the world before governments were instituted. He probably meant that the state of nature was much like America was then: vast, rich, mostly unexplored, and free. We can imagine that Locke was comparing America to the Garden of Eden. (Notice that Locke wrote this sentence long before the Declaration of Independence.) Nowadays, the metaphor brings to mind an even greater basis for comparison. The United States remains free and prosperous after the institution of national government, and the metaphor seems to have vindicated Locke's political views to some extent. Much of that is coincidence, but the metaphor is obviously a very rich one, which is probably why it still excites us in the twentieth century.

Thus, we detect two beneficial features of the use of the metaphor: (1) it suggests a tight comparison; and (2) it exhibits creativity. But metaphors also have some potential drawbacks, and we must consider these as well.

If metaphors can be an aid to explorative thinking, they can also run away with us, get us off the subject, or act as a substitute for careful and critical thinking. This is especially true with "mixed metaphors." Senator Jesse Helms of North Carolina, for example, once complained that the U.S. State Department was trying to discredit him for his strong stand against communism. He said, "We can't continue down the slippery slope of kicking our friends in the teeth." This mixed metaphor confuses the audience. A **mixed metaphor** is a figure of speech that jumbles together two or more distinct avenues of thought or comparison. They are a sign of sloppy thinking. Helms mixed his metaphors when he compared our foreign policy to a downhill slide on the one hand, and our foreign policy to a "kick in the teeth" on the other. The image that this mixed metaphor conveys to the hearer is an absurd one: can you imagine kicking someone in the teeth while rolling downhill? One can only imagine that our foreign policy is exceptionally dexterous, which is surely *not* what Helms is trying to say!

Mixed metaphors, then, can be a source of unintended humor, making the speaker or writer look ridiculous. *The New Yorker Magazine* frequently collects mixed metaphors and runs them in print. Like its cartoons (for which *The New Yorker* is famous), the mixed metaphors add comic relief to the otherwise serious and dignified tone of the magazine. Here is a sample:

Block That Metaphor!

> Mobile, Ala.—In the dwindling twilight of a storm-tossed Thursday, Charlie Graddick grabbed the burnished levers of political demagoguery to whip up a hometown crowd and breathe life into a bid for governor that has seen more switchbacks than a snaky mountain road. From the Orlando (Fla.) Sentinel

I count at least nine separate metaphors in this passage. The point is not only that it's badly written, but it displays bad thinking as well. Upon reading this paragraph, the reader has to be confused.

Mixed metaphors can often signal whether or not a book is worth reading. This is especially true about books in the area of pop psychology and popular metaphysics. If these books have something worth saying, they should present their ideas clearly, in an organized fashion and with a minimum of metaphors. Open the book to any page and count the metaphors; if you find a great number of them, you may have a basis for suspecting the value of the book's advice. Can you spot the metaphors in the following excerpt?

Block That Metaphor!

The most common, and perhaps the most malignant, form of anger is resentment. Anger by itself is not destructive. We can channel this valid feeling into appropriate action and resolution. But when anger is allowed to fester, the residue of resentment eventually hardens into an all-consuming bitterness. Rather than endure the pain of verbally expressing angry feelings as they occur, most of us choose the repressive route. We become pros at throwing our angry thoughts and feelings down into the subconscious. We develop a pressure-cooker lifestyle, a lifestyle boiling with resentment and capable of blowing up at any time. We become apathetic because we are afraid that if someone gets to know us intimately, the cover will be blown and the pot of destructive anger discovered. Rather than shoot randomly without a target, as in the case of fear, we wonder why we should shoot at all. In fear of losing control, we avoid encounters that could add to the anger and put us over the brink into a world of total lack of control and unrelenting pain. When the pain becomes too great and care from others becomes too little, we ask, "Why relate at all? Why build more expectations that will go unmet?" These questions are the questions of a stuck person who needs to start all over. From the book, *Hooked on Life*, by Tim Timmons and Stephen Arterburn

To avoid mixed metaphors, sustain the basis for the analogy throughout. If we begin by comparing a philosopher to a watchdog, for example (as Plato once did in *The Republic*), then maintain the comparison. "Why is a philosopher like a watchdog? Because he will lick his master's hand, and bark at a stranger, whether or not the stranger ever did it any harm, or the master any good. Therefore, a philosopher is like a watchdog in that he knows what he knows, and he knows what he doesn't know." Admittedly, this is more simile than metaphor, but it makes the point. Plato is using a riddle to reveal something about the relation of wisdom to philosophy and, in doing so, he retains the comparison to a watchdog the whole time.

Another problem with metaphors is that they can be either too rigid or too loose. When they are too rigid, they may serve to block new and more fruitful avenues of thought. When they are too loose, they have the tendency to allow the reader to run away with them down unintended paths.

The expression "the Iron Curtain" is a good example of a metaphor which is too rigid. The term was applied to the nations which fell under the domination of the Soviet Union after World War II. The metaphor suggests an impenetrable barrier. One of the results of this unfortunate metaphor is that the United States may have been deceived by the connotations of the term into taking too rigid a negotiating stand with the nations of Eastern Europe for many years, treating them all as equally inaccessible when in fact some (perhaps Poland, Hungary, and Yugoslavia) would have been receptive to forming greater ties with the United States. Thus, we see that metaphors can sometimes "color our thinking" too restrictively.

When metaphors are too loose, they may get us into the opposite sort of trouble. For example, when Plato compared a philosopher to a watchdog, he may not have mixed his metaphors, but the comparison could bring with it too many other (and unwanted) associations for the reader. For example, imagine this hypothetical student essay: "A philosopher is a watchdog, not only because he knows what he knows and he knows what he doesn't know, but also because, like the watchdog, he is the guardian of ideas. When he encounters a new idea that's strange to him, he attacks it and brings it down. He scowls at the unfamiliar and the new, and he protects the public from intruders and original thinkers. The philosopher and the watchdog alike are narrow-minded."

9.3 EVALUATIVE AND "VALUE-NEUTRAL" LANGUAGE

Evaluative language is language which expresses *value judgments.* Value judgments are assertions of worth. Whenever we employ expressions such as "good," "just," and "beautiful," we are making value judgments. Evaluative language is central to a wide range of everyday and professional discourse—from ethics to politics, from art to the social sciences.

Evaluative language presents special problems for logicians, and the means that philosophers have employed to deal with the expression of value in logic remain a largely unresolved issue. These issues are much too complex to explain in detail, but perhaps an example will suffice to show the sorts of difficulties involved.

Suppose that a popular opinion poll shows that a majority of voters prefer (i.e., "value") Proposition 103 over Proposition 100, and the same poll shows that voters prefer Proposition 100 to Proposition 104. Paradoxically, it doesn't follow that a majority of voters would prefer Proposition 103 to Proposition 104! The reason may be that Proposition 103 has provisions that would cancel out all the provisions of Proposition 104 (whereas Proposition 100 would not). The point is that a good rationale may exist for refusing to draw the conclusion that voters would prefer Proposition 103 to 104; namely, that the electorate would not be irrational in voting the other way.

We have a formal name for this paradox; it's called "the voter's paradox." As with all paradoxes, we retain the faith that a resolution is possible. The resolution here has to do with, perhaps, the inappropriateness of cross-comparisons in complicated pieces of legislation or with reasonable shifts in standards. The reason that ordinary logic cannot adequately encompass a logic of value is precisely that value judgments are based on underlying standards, and those implicit standards need to be made explicit in order for logic to deal with evaluative language.

Nonetheless, evaluative language is so much a part of ordinary discourse that any introduction to logic needs to address some of the concerns. Those concerns are similar to many of the concerns already expressed with regard to propositional attitudes, emotive language, and metaphors, so we will examine some of them in that light.

> In their book, *Introduction to Logic,* authors Charles W. Kegley and Jacquelyn Ann Kegley give us these tips on what to look for in assessing evaluative language logically:
>
> We shall assume that in arguments we need (1) to spell out and specify any value judgments being made, (2) to indicate whether any reasons are given for these value judgments, and (3) to show what role they play in the argument or in the decision-making process.

The key to identifying value judgments is a basic and underlying distinction we need to make between *facts* and *values. Facts* are the backbone of informative language. We may say that a fact is a statement which in some sense corresponds to external, measurable, and verifiable phenomena, or that it coheres with the existing body of knowledge. But *values* on the other hand relate to standards, ideology, criteria, or rules.

When we make a value judgment, we need to unpack the system of standards which underlie it (for example, we can point to the Bible or to Christian morality as our standard for appeals), but we can also fruitfully assess the information which forms our basis for making the judgment. Nearly every value judgment is predicated on the basis of some factual state of affairs. For example, we may argue that capital punishment is wrong because it fails to act as any greater deterrent to crime than does life imprisonment. In this example, we can assess the "factual" side of the argument, and try to ascertain whether or not *in fact* capital punishment is no more of a deterrent than imprisonment using the sorts of statistical analysis familiar to students of sociology. Of course, given the nature of statistical analysis, we can only hope to arrive at probabilistic conclusions in such a dispute (and statistics, as the same students will note, is a discipline rife with paradoxes of its own).

That brings us to yet another topic. Because evaluative language is so difficult to treat from a logical point of view, many philosophers in the recent past have advocated using discourse that is entirely value free. We call this "value neutral" language. The hope is that we can analyze the factual components of discourse that are used in value judgments, and determine whether those statements are true or false. The lauded ideal of value neutrality is to restrict our discourse only to verifiable factual statements without ever making reference to language which is "value laden."

However, many philosophers are coming to the conclusion that the ideal of value neutrality in our language is an impossible hope. Some go so far as to say that standards and ideologies "color" the very way we ascertain what is factual, or that they shape the very assumptions behind methods of verification. This is reasonable, given the fact that values make up such a large part of both our everyday and our professional language.

Perhaps the only samples of discourse that are wholly value neutral (as a political scientist once quipped to me) are environmental impact statements and municipal planning reports! Certainly, bureaucratese is *so* bland that it could never serve as a model for good writing, but its blandness probably conceals some value commitments anyway, however much they appear to be toned down.

What we can watch out for, however, is language that is "value loaded." We may say that some discourse is **value loaded** on the condition that it employees language which pre-judges the worth of the topic under discussion and that it does so with little or no argument in support. To recall our earlier discussion, buzzwords and euphemisms are "loaded" in this way. They may carry an emotive content, but they may also carry an implicit evaluation as well. Whether we call the Contras in Nicaragua "terrorists" or "freedom fighters" amply illustrates our point. Suppose someone says, "We need another billion dollars to aid the freedom fighters in Nicaragua." No argument is given, but the speaker has already committed herself to a favorable evaluation of the Contras. "Freedom fighters" here is an example of a value loaded term.

On the other hand, if someone else says, "We must cut off any further aid to the terrorists in Nicaragua," once more no argument is given, though an implicit value judgment is made.

The rule is this: whenever we have a choice between a *value loaded* term and one which is more *descriptive,* use the *descriptive* term. For example, use the term "Contras" instead of "freedom fighters" or "terrorists." Now, when our speaker says, "We need another billion dollars to aid the Contras in Nicaragua," we've prepared the grounds for a more reasonable discussion of the *merits* of the issue. We now have the basis for a reasoned argument instead of a heated quarrel.

A great deal of time and energy is wasted on debates over terms that are implicitly loaded when a little effort to find more descriptive terms might place the argument in better position for resolution. Such terms abound in our daily discourse. To take a familiar example, consider the term 'affirmative action'. The term *Affirmative Action* was coined by the U.S. Supreme Court to title a temporary program of compensatory hiring and admissions for minorities. No doubt, the term was devised because it had "positive-sounding" connotations (what could be more positive-sounding than "affirmative" and "action?"). Critics of the program have countered that affirmative action is really 'preferential hiring'. But, after this discussion, we are now in a good position to see that the term *preferential hiring* is equally loaded for the *negative* connotations of the term. If we want to place the argument on a more objective basis, perhaps we should search for still another term that more accurately describes the goals of the program. None exists but we might suggest something like, "compensatory hiring and admissions." That term has certain positive connotations as well, no doubt, but it places the program in a more descriptive stance since it makes clear that the standard appealed to in the program is one of compensatory justice, and many ethicians and political philosophers have elaborated the criteria for a just act of compensation.

Racial and sexist epithets are likewise value loaded terms. Fortunately, we've made a great deal of progress in dispensing with racial epithets. Many have settled on the term *black* after a good deal of reflection to designate people of Negroid races. Only two decades ago, the terms 'Negro' and 'colored' were commonly used to describe people of that race, but a debate surfaced in which blacks objected to these terms, suggesting that they had negative connotations. The terms 'African-American' and 'Afro-American' still have some advocates, so the discussion continues.

A similar debate is now being carried on about sexist epithets and sexist language. Some terms and some uses of language are either explicitly or implicitly demeaning or offensive to women. They reflect old and exploitative stereotypes of women. In a more enlightened age, we need first of all to recognize when the language we use is demeaning or offensive, and then we need to "repair" the language. If necessary, we need to coin new terms that are not loaded.

The fact that the debate is still ongoing is reflected in the indecision we face, for example, whenever we write a standard business letter. The old guide books used to instruct us to use "Dear Sir" as a standard salutation. That is obviously unacceptable because it carries with it the implicit suggestion that the business person to whom you are writing is a man. Before the '60s, few women were permitted important roles in business in America, so the salutation reflects a bias against women in business. On this point there is wide agreement, but the question remains: what do we write in its place?

If you know the name of the business person to whom you're writing—no problem, just address the letter to her. But often we don't know whether the recipient of the letter will be a man or woman. What do we do in those cases? Some people have suggested we use "Dear Sir or Madam" as the opening salutation, but others have pointed out that it too is unacceptable (consider some of the connotations of "madam!"). We may be reminded of the debate over 'Afro-American' and 'black' in the civil rights movement to see how far we have yet to go in eliminating sexist language. For my own part, when I write a business letter to (say) "Acme Widgets and Gadgets Co.," I compromise by using this opening salutation: "Dear Acme." It's impersonal, but at least it's not sexist.

You may have noticed that I alternate between 'he' and 'she' whenever I devise examples in this book. For a very long time, 'he' was the standard unmarked term for gender. That was supposed to be indifferent between masculine and feminine in designating gender, but of course it wasn't. Many women's rights advocates recommended the use of 'he or she' in place of 'he' as the pronoun of unmarked gender. That does the job all right, but it is often unwieldy (maybe we'll have to use some clumsy terminology to divest language of sexist connotations). Sometimes this is abbreviated to 'he/she', but this has shortcomings as well. One women's rights advocate joked that reading this expression aloud ("he *slash* she") sounded like a sex crime was underway! In their place, perhaps alternating 'she' and 'he' in about equal numbers is the best way of dealing with this problem while preserving some of the familiar meter of the English language.

Like all disputes about evaluative language, other standards and criteria intrude when looking for a more descriptive term. Some of these are aesthetic concerns, like meter and rhythm. The traditional marriage ceremony, for example, concludes with these words: "I now pronounce you man and wife." Why *man and wife*? Why not, *husband and wife*? Perhaps one of the reasons was the presumption of male dominance in the marriage, as some suggest, but notice that the language is in iambic meter. At least another reason for the choice of terms was that only "man" preserved the poetry of the marriage ceremony which is basically in iambic pentameter verse. What more appropriate place for poetry is there than a marriage ceremony? Of course, changing the language to *husband and wife,* while it sacrifices some of the poetic meter, may be more advisable because it deletes the presumption of inequality of sex roles.

While we still debate many of these replacement terms, a large consensus has been reached on very many others. What follows is a list of some of these, together with a list of those still under scrutiny:

Loaded terms	**Descriptive replacements**
"businessman"	"business person" or "executive"
"man power"	"labor power"
"mankind"	"humanity"
"man" (in general)	"person"
"the girls in the office"	"the secretaries"
"mailman"	"mail carrier"
"chairman"	"head of the department"
"congressman"	"member of Congress"

But what do we do with these . . .?

"manhole cover"	???
"Bachelor of Arts"	???
"Master of Arts"	???
"freshman"	???

As you can see, some of these are very problematic. Perhaps we can dispense with "manhole cover" in favor of "cover" but "freshperson" has very unfavorable connotations when put in place of "freshman." The same problem arises with the obvious replacements for "*Bachelor* of Arts" and "*Master* of Arts." Perhaps we can say that the "man" in "freshman" no more connotes the status of a male than the "van" in "caravan" suggests the presence of a moving van (or the "cat" in "catamaran" denotes a feline), but if we do that, we can make the same objection to replacing "business person" for "businessman," or "chairperson" for "chairman," even though these terms have come into wide acceptance.

A decade or so ago, *Ms.* magazine proposed substituting "pe-man" for man on the argument that "woman" unfairly connoted the presence of a womb. That stratagem has thankfully fallen to the wayside.

What is obvious is that we have yet a long way to go in the task of deleting sexist connotations from language. One hopes that we can do that task with ingenuity and some regard for the poetry of language. That task falls to you just as much as it does to professional pundits of logic and language. This project can be as much fun as it is important. Let's hear your suggestions.

EXERCISES

Exercise 1

DIRECTIONS: *Devise a non-sexist, standard, opening salutation for a business letter.*

Exercise 2

DIRECTIONS: *Here is a list of terms selected at random. Imagine that they are euphemisms or buzzwords which disguise implicit meanings. Devise a definition for each of them, showing how they can be used in a sinister or deceitful fashion.*

Make your own Metaphor!

Term	Your suggested meaning	Euphemism or Buzzword?
Example:		
clemency list	"enemies list"	euphemism
rose garden		
Disneyland		
health club		
surprise package		
pugilist		
butcher shop		
shooting star		
mother's milk		
cotton candy		

Exercise 3

DIRECTIONS: *The following passages are taken from books on psychology and popular psychology. To some extent, they each employ metaphors, propositional attitudes, and evaluative and emotive language. Take note of each when you find it. For example, count the metaphors and assess their role in making their respective points clear. Are the terms "loaded," or is the language as descriptive as it can be? Are the authors talking about their own states of mind, or are they making descriptive assertions about issues? Do they unacceptably employ euphemisms or buzzwords? From the samples offered here, say whether or not you think the book deserves further reading.*

1. The F.O.O.L.

You are behaving as a fool if you look outside of you for an explanation of how you should feel or what you should do. Taking credit as well as responsibility for yourself is the first step to eliminating this erroneous zone. Be your own hero. When you get out of the blaming and hero worship behavior you'll be moving over from the external to the internal side of the ledger. And on the internal side there are no universal shoulds, either for yourself or for others. —Dr. Wayne W. Dyer, *Your Erroneous Zones*

2. If you will now bring together the means we possess for uncovering what is concealed, forgotten and repressed in the mind (the study of the ideas occurring to patients under free association, or of their dreams and of their faulty and symptomatic actions), and if you will add to these the exploitation of certain other phenomena which occur during psycho-analytic treatment and on which I shall have a few other remarks under the heading of "transference"—if you bear all these in mind, you will agree with me in concluding that our technique is already efficient enough to fulfill its task, to bring the pathogenic psychical material into consciousness and so to get rid of the ailments that have been brought about by the formation of substitutive symptoms. And, if in the course of our therapeutic endeavors, we extend and deepen our knowledge of the human mind both in health and sickness, that can, of course, only be regarded as a peculiar attraction in our work. —Sigmund Freud, Third Lecture, *Five Lectures on Psycho-Analysis*

3. Self is the starting point, not the finality. Those who remain fixed within themselves, close doors against the world and attempt to live by the false and fantasied dictates of "autonomy" are fine candidates for emotional starvation. Self-imposed solitary confinement, in terms of life-style, is the cut-off of inspiration, creativity and constructiveness. Counter to the widespread belief that the lone wolf is strong and admirable for his supposed fount of inner resources, he is in fact the victim of Negative Self-Image. What appears to be strength is, at bottom, weakness: "If I work with, love with, live with Others, I stand the chance of failing with them. If I fail, I'll be disliked." —Harold M. Newburger and Marjorie Lee, *Winners and Losers: The Art of Self-Image Modification*

MODULE 10

CREATING ARGUMENTS

10.1 A SAMPLE LETTER TO THE EDITOR

In previous sections, we have learned that the difference between exposition and argument can be illustrated by comparing the typical lead story of a newspaper and a typical letter to the editor. One expects a letter to the editor to be argumentative in tone, whereas one expects a news story to be informative.

The letter-to-the-editor format is a useful one for instruction in logic: the good letter is concise, clear, and to the point. The best letters are well-argued, well-organized, dispassionate, and forceful. In other words, the best letters to the editor should exemplify the lessons we've learned. However, we've applied these lessons only passively thus far, in the mode of criticism of arguments which others have written. Now we will consider a new role, a more active role, and attempt to apply the lessons we've learned by constructing good, well-organized arguments of our own.

Perhaps the best way of doing this is by writing a "letter to the editor." This need not be a letter we actually expect to publish in a newspaper; we're doing this merely as an instructive exercise. However, in this assignment, the idea is to write about something we genuinely care about. Logic has the virtue of being "topic neutral," so you can properly write about anything you wish in order to complete the assignment. However, in order to do the assignment well, imagine you are actually writing this letter for publication in the opinion page of your local newspaper. That means that you have to keep an audience in mind—a

---

A deductive **argument** is valid, if and only if, if all the premises are true, then the conclusion must be true.

useful objective for any writer. It also means that you need to beware of the "editor's blue pencil."

10.2 THE EDITOR'S BLUE PENCIL

The editor of the opinion page of the local newspaper shares some concerns in common with the author of a section of a logic textbook that deals with recognizing arguments. Both have criteria or standards for good argument form; both look for precision and clarity in the arguments they receive (whether from correspondents in the case of the editor or logic students in the case of the logician). Both the logic teacher and the editor should be somewhat indifferent to the subject matter of the arguments they receive. The logician just wants to see good arguments, the editor wants to allow readers to express their opinions on a wide range of subjects.

Both the editor and the logic teacher will want arguments that get right to the point and express their conclusions clearly. The editor needs to develop skills, just as the logician does, in identifying conclusions. This enables her to "caption" the point of the letter in a headline.

Granted, the editor has some concerns that differ from those of the logic teacher. She needs to keep in mind the fact that the newspaper needs to sell, so an editor might want to publish letters that arouse controversy or otherwise provoke interest from the readership. But that's not the only difference. Typically, the editor is faced with severe constraints on space, so she will want letters that are concise as well as clear and precise. When the editor is faced with a letter that is interesting but much too long for publication, she pulls out the infamous "blue pencil" (blue because blue marks will not show up in photo reproduction) and begins slashing text.

The editor's job is often a thankless one. In order to fit a lengthy argument into a small space, much of the text will have to be cut. The editor needs to exercise careful judgment, cutting and trimming in a way that brings out the main point of a correspondent's letter, while preserving the meaning of the text and retaining its basic structure. Here's an example of a letter which ran too long, forcing the editor to cut it.

> The board's silence will be seen as recognition of petty, political, vindictive and/or self-seeking motives which are too shameful to admit to. . . .

The editor makes use of the convention of three dots called "ellipsis" to show that text was deleted. A letter writer should do his best to avoid the sinister three dots that will be inserted if the argument runs too long because, no matter how good the editor is and how well-intentioned, she still sees your argument through her own eyes, and some things in the text which you would prefer to emphasize might get deleted by editorial incisions.

Logicians may not share an editor's concerns for space, but they too value conciseness of expression. When an argument displays the virtue of conciseness (i.e., using the fewest number of premises necessary to derive the conclusion), logicians may describe the argument as "elegant." Still, compactness of argument is less a concern for logicians than it is for space-conscious editors.

We may detect another reason for editorial use of the blue pencil. Sometimes an editor will delete text when the argument *rambles*. This is a logical fault as well as a fault in rhetoric and letter writing in general. When an editor is forced to do this, her job is to attempt to uncover a structure that the original may lack or display poorly. Rambling is a serious fault.

10.3 GETTING YOUR LETTER PRINTED

Newspapers try to publish a broad spectrum of opinion. A newspaper in a small city may have an editorial policy to publish most of the letters submitted; the editor can usually do so because he receives proportionally fewer letters than a big city newspaper does. But sometimes even the small city newspaper gets swamped, especially on a heated topic, and the editor has to select letters on much the same criteria as the metropolitan newspapers. What are those criteria? We can now make some rough generalizations about what an opinion-page editor will look for.

One of the first criteria used by editors is **topicality.** A newspaper is mainly chartered to cover news and issues on a daily basis. When the paper has just run a story on the news of the day or an editorial has just argued a current issue, the editor will naturally favor letters which address that news or those issues.

If the topic is controversial enough, the editor can expect to be inundated with letters on the subject. If he cannot run them all over the next few days, he must favor certain letters for print over others. One thing he will look for is **originality.** If the letter writer takes a new slant on the issue or has a special insight on the news, that letter will get priority for publication.

Space considerations always intrude on an editor's selection of letters to run, so he may also want to select letters which are **concise** and to the point. Some editors recommend that letters should be approximately 250 words in length, or about the length of a single page of typewritten text, double-spaced. If the letter is longer and space isn't too great a consideration, perhaps the editor will run the letter in full, but, more often, he takes out his blue pencil and trims the letter to fit. Most editors will try to do this fairly and with an eye to the letter writer's style. The point, however, is that you probably don't want the letter cut by someone else, so the rule is to do your own cutting beforehand. Keep it short. Knowing how to cut the length of your argument is a key feature of good writing ability, but cutting is a painful process and it requires self-discipline.

Next, an editor will look for **clarity** in the argument. We already know what that means: you can usually make your point clear by putting it right up front. It should be the first sentence of your argument unless it's a controversial or surprising contention, in which case it might be better to state your reasons first. This has the effect of persuading the reader before revealing the surprising conclusion. In this case, the conclusion of your argument should be the last sentence of your letter.

Furthermore, you should liberally employ reason- and conclusion-indicators to make the structure of your argument clear. "I want to argue that the school funding proposals now on the table should not be passed by the school board because they are badly conceived and come at an inappropriate time," a letter to a newspaper might begin.

Your letter should be **coherent.** This means that the letter should be consistent in its point of view and the argument should hang together without confusion. In one recent letter to the editor of a local newspaper, a writer argued: "So Quayle is no Jack Kennedy! I say 'thank Heaven'. The idolization of the Kennedys is something I don't understand."

The writer's point could be clearer, but so far we can get the gist of it. She is saying that no harm is done if vice-presidential candidate Dan Quayle doesn't compare well to the Kennedys. We should now expect the writer to give reasons why the comparison wouldn't be favorable to Quayle, and that prompts us to expect some reservations about the character, abilities, or political outlook of the Kennedys (some of which she gives us). However, in the very next sentence, she says:

> **"Robert Kennedy was deserving of more credit than his brother. Certainly his character was deserving of more admiration, and Rose Kennedy has had more sorrow than should be the lot of any one person. The assassination of two sons and the loss of another in World War II were terrible crosses to bear."**

This part of the argument actually somewhat *undercuts* her contention that the Kennedys are not worthy of "idolization." We can see that the writer is confused. We thought she was going to find fault with the Kennedys, but instead she extols their virtues! She provides no motivation for or transition to the very next paragraph where she writes, "The fact remains that Jack Kennedy was no angel, nor is Ted." Moreover, she offers no reasons for this. She is somewhat inconsistent in her list of reasons, and that produces confusion in the reader's mind. The letter becomes *incoherent.*

To make sure your letter is coherent, keep your main conclusion in mind the whole time you're writing and find relevant and compelling reasons to back it up.

That brings us to the last point: the editor will look for **quality in argument.** Do the reasons actually support the conclusion? Are the factual assertions true? Do they rationally persuade us?

This is actually about the last point an editor will consider. His job is merely to display a representative range of opinions in the letters column and let the reader make up her own mind as to the quality of the reasons offered and the compellingness of the argument form. But, when pressed for space, it is a factor he'll include in his selection process. Given a choice between a very well-argued letter and a poorly-argued one, when confronted with printing only one or the other, the editor will surely print the better one.

Thus, we may list the criteria important to the opinion-page editor in approximately this order:

Criteria for Getting Your Letter Printed

1. Topicality
2. Originality
3. Conciseness
4. Clarity
5. Coherence
6. Quality of Argument (Do the reasons support the conclusion?)

Notice that the logician looks for many of the same criteria, but in very much the reverse order. For the logician, *quality of argument* comes first, closely followed by *coherence* and *clarity. Conciseness* is important to the logician, but since the logician does not have the constraints on space that face the newspaper editor, she does not require as great of a reliance on this trait. *Originality* may be a consideration for the logician too, but strictly speaking she is more concerned about getting the mechanics of the argument right. Least important is *topicality.*

Logicians' Criteria for a Good Argument

1. Quality of Argument (Do the reasons support the conclusion?)
2. Coherence
3. Clarity
4. Conciseness
5. Originality
6. (Topicality?)

As we can see, these are quite the reverse of the opinion-page editor's criteria, but the important point is that they share most of the same criteria even if it is in reverse order.

10.4 EMOTIVE LANGUAGE AGAIN

A good thought to keep in mind when you are writing a letter to the editor is that you want to persuade, in a cogent fashion, people who are either undecided or uninformed about the issue you address or people who are inclined to disagree with you on the issue. When you write such a letter, you shouldn't think of yourself as a cheerleader for your team, who presumably shares your point of view. If you do that, you persuade nobody but perhaps the people who already agree with you. That doesn't take much talent.

However, persuading the undecided, the uninformed, the indifferent, even your opponents—now that takes talent! Let's face it: you won't succeed if you insult them, engage in name calling, mischaracterize the issue, ridicule them, patronize them, or overemotionalize.

Here are some ideas on letter form that might just win these people over. First of all, exercise restraint. A cool head makes clearer sense. That's not easy to accomplish; if you didn't already feel strongly about your issue, you probably wouldn't bother to take the time to write a letter to the editor in the first place. But, if that's the case, write a first draft, then wait until the next morning before you decide to send it. (For some reason, highly emotive prose never looks quite as good in dawn's early light.) Then re-write it.

Sometimes (perhaps very often!) people use the letters column in the newspaper as an outlet for letting off steam. If you agree with the letter writer, then you can let off steam too. That's well and good, but if you don't share the writer's enthusiasm for his side of the issue, you will likely be offended or made even more adamantly opposed to the letter writer's point of view. Restraint is the valve that shuts down such excessive enthusiasm.

For this reason, keep your letters free of obvious euphemisms or buzzwords, avoid emotionally loaded language, and keep your analogies and metaphors tight.

Nowadays the preferred style is *understatement*. Perhaps that wasn't always the fashion (compare a typically understated contemporary documentary to the zeal of a 1930s newsreel!) but it is now and, from a rhetorical if not a logical point of view, the degree to which your letter is understated may have an effect on how well-received it is by its intended audience.

Next, let the facts do the talking. Inexperienced letter writers load up their letters with adjectives and superlatives, but these add nothing to the basic premises of the argument. They're merely adornments and, as such, they're distracting. A good letter lays out many specific facts, each of which is open to some sort of corroboration or verification. Unsupported generalizations should be avoided.

Finally, keep an eye on the structure of your letter. A carefully planned letter makes its conclusion clear by placing it prominently and labeling it with conclusion-indicators. The major sub-conclusions should also be given prominent positioning. Follow the direction and grouping rules. Let your argument flow in a single direction, and group like reasons for the same conclusion or sub-conclusion in a tight cluster.

As a teacher of logic, I read a lot of letters to the editor. They're extremely useful as a source for class (as well as for this book). I usually ask my students to bring the daily local newspaper into class so we can examine the letters for logical structure and fallacies. Many of these letters are so badly written and argued that many students are discouraged from writing letters of their own (Ralph Nader once commented that every city has three types of outcast: the town drunk, the village idiot, and the town citizen—presumably, you can identify the last type of outcast by reading the opinion page of the newspaper).

In a recent logic class, we read a series of letters about the state-to-state uniform drinking age which Congress passed into law. The letters were of special interest to the students, but the letters tended to be highly emotional, full of bad grammar and fallacies, and obviously self-serving. One day, we opened the newspaper and read one letter on the subject aloud. I praised the letter for its restraint, understatement, and use of specifics. Even though there was a contained anger under the surface, the writer let the facts do the talking. I suggested that this was exactly the sort of letter I wanted my students to emulate because it would succeed in persuading reasonable people who were interested in learning the truth. It was only then that I discovered that one of my own students had written the letter and she was present in class.

The student wrote the letter without any prompting from me. I reprint the letter below with her permission:

> Editor: I am a 20-year-old college student who had thought that the change in the legal drinking age would not affect my lifestyle, because I have never indulged in drinking and do not frequent local clubs. Due to a recent event, however, I have realized that even I have to pay a social price.
>
> On Sept. 26 a group of my friends and I went to a small and well-behaved party at Wrightsville Beach. Around 10 o'clock two sheriff's deputies arrived and placed themselves directly at the foot of the staircase leading up to the porch. When a person descended the stairs, he was greeted with a flashlight full in the face and, if evidence of drinking was apparent, a demand for an I.D.
>
> After a few offenders were given court orders, one officer mounted the stairs uninvited and held a private discussion with the hostess. When I later asked her what he had said, she could give no sure answer and knew only that everyone had to go inside and all people under 21 should leave.
>
> I am convinced that hounding party goers to the extent of posting guards at the front door oversteps the bounds of enforcing a law and becomes harassment. The officers did catch a few offenders, but the main effect was to needlessly disrupt a quiet party. —Cathy Butler, Wilmington, N.C.

10.5 A FEW LAST TIPS

Here are a few other tips. Use the *active* voice as the rule. For some reason, composition students, whether they are in high school English class or in graduate school in philosophy, believe they must use the passive voice in order to write well or to write objectively. Some scholars think that use of the personal pronoun, 'I' in a serious work makes the writing appear subjective and parochial; they teach us to twist sentences into passive voice contortions, if necessary, to avoid ever using the personal pronoun 'I'. But ask yourself which of the following sentences is clearer and more informative:

PASSIVE VOICE: Often the passive voice is emphasized in scholarly writing to avoid subjectivity.

ACTIVE VOICE: Some scholars emphasize the passive voice to avoid subjectivity in scholarly writing.

Notice that the statement in the active voice identifies the source; that's one reason for using it.

In a very useful reference source entitled *The Careful Writer* [New York: Atheneum, 1973], Theodore M. Bernstein discusses the use of the active and passive voice. Bernstein was a managing editor of *The New York Times,* and anyone who wants to write a successful letter to the editor should consult this definitive work on good writing. Bernstein emphasizes that use of the passive voice has its place, but on the whole he prefers the active voice for most purposes in writing:

The passive voice, used without cause, tends to weaken writing. It also usually requires the use of more words. Which may be another way of saying the same thing. Compare "A good time was had by all" with "Everyone had a good time," or compare "Our seas have been plundered by him" with "He has plundered our seas."

Bernstein recommends using the passive voice when the subject of the action is less important than the action, when the subject is "indefinite or unknown," when the intention is to emphasize the actor by identifying her at the end of the sentence (a strategic position), or when you wish to be diplomatic and avoid strong language.

The custom of writing in the passive voice has fallen out of favor, even in scholarly writing. You'll notice that I usually employ the active voice in this book: I believe that makes it easier to read and understand.

Another rough-and-ready tip: write short sentences. That helps you avoid grammatical tangles and makes your writing easier to read. Grammar and logic are close relatives; make sure you write in complete sentences or you may fail to clearly communicate a single thought. Also, for the same reason, if you have a choice between a short word and a longer word (as in "Lincoln *freed* the slaves" vs. "Lincoln *emancipated* the slaves"), use the short word as long as it accurately says what you want.

Finally, you might want to preview Unit Four, "Arguments and Fallacies," in this book before you send off the letter and make sure you haven't committed any of the fallacies listed there.

Having done all this, you have the makings of a good letter.

Writing a letter to the editor is one of the best exercises you can do in this stage of a logic course because it combines most of the features of good argument we've learned throughout the rest of this unit on reasons and arguments. A letter to the editor should be an argument. It should be well-structured. It should avoid emotional appeals and let the facts support the conclusion.

Most of what we did and will do in a logic book is spectator sport. You are given sets of exercises already made for you and you pick out the right answers. That can't be avoided for most of our purposes. However, writing a letter to the editor requires you to use what you've learned actively. It's a creative process.

EXERCISES

DIRECTIONS: *Write a letter to the editor.*

MODULE 11

TAUTOLOGIES, CONTRADICTIONS, AND CONTINGENT STATEMENTS

11.1 TAUTOLOGIES

Informally speaking, a **tautology** is a statement which is true no matter what. Tautologies are closely related to the concept of a logical truth, and a **logical truth** is any expression whose denial must be false. For example, 'All Joshua trees are Joshua trees' is a statement which is undeniably true and, therefore, it is a logical truth since its negation, '*Not* all Joshua trees are Joshua trees', is a statement which must be false.

Notice that 'All Joshua trees are Joshua trees' is clearly true whether or not any of us know what a Joshua tree is. This is to say that logical truths are statements which are true independent of their *semantical content;* we don't need to know what the term 'Joshua tree' means in order to understand that the statement is true.

Tautologies* differ from logical truths in that they are "ordinary language" statements that are true as a matter of the meaning of their component expres-

---

A deductive argument is valid, if and only if, if all the premises are **true**, then the conclusion must be **true**.

*Strictly speaking, in formal logic we define a tautology as any statement of the sentential calculus that is true in every row of its Truth Table (see Unit Three, "Symbolic Logic") and a logical truth is a proper subset of those statements in the quantificational calculus which are true under all interpretations. In saying that "logical truths are statements which are true independent of their semantical content," we set aside for now more complex issues regarding the meanings of the quantifiers (e.g., 'all') and the copula of the predicates ('is' and 'are').

sions. Another way to say this is that tautologies are statements that are in some sense always true just by virtue of their semantic content.

We rarely find authentic tautologies in practice; usually, we have to manufacture them. Take a famous example by philosopher Williard Van Orman Quine. He defines the word 'bachelor' as "a never-married man;" thus, 'All bachelors are never-married men' can be rendered a tautology, given our usage of the word here.

However, even this is not without some problems. In ordinary usage, sometimes divorced men call themselves bachelors. And a Bachelor of Arts is, in a sense, a bachelor, though many who hold such degrees are women, or married.

The point is that because tautologies rely on the semantics of the terms in the statement, and that words often have many different senses and usages, we may find difficulty in discovering candidates for tautologies that are always true without controversy. Nevertheless, the concept of tautology is a valuable one, even if it need be somewhat idealized and abstracted from actual practice.

One instance in which it's valuable to be able to recognize tautologies (or *near*-tautologies) is when advertisers, con-artists, or others sometimes use such expressions in a way that seems to convey some profound meaning which isn't really there.

For example, we occasionally see bumper stickers which read: "If guns are outlawed, then only outlaws will have guns!" The platitude may strike fear in the hearts of some motorists as they read the slogan and envision a society made lawless by the passage of gun control laws. Some people may even be persuaded to oppose gun control laws because of the slogan. However, this is faulty thinking, and the error is made clear when we recognize the tautological element in the statement.

If *all* guns are outlawed, then surely it is true that only outlaws will have guns. If it's true, then it's *tautologically* true. In such a society, even if you're a law-abiding citizen in other respects, if you were to own a gun, you would be breaking the law. In that sense, you *would* be an outlaw, but in that respect only. Even police would be outlaws in that one respect if they carried guns in a society in which *all* guns were outlawed.

The slogan *seems* to say more than it really does. Philosophers sometimes say it's *trivially* true—trivial because it doesn't really tell us much we didn't already know. The trick of proponents of such slogans is that they dupe us into thinking that the slogan says something profound, when it doesn't really say much at all.

Here is an example of a tautology:

"The weather for catching fish is that weather, and no other, in which fish are caught."

—W. H. BLAKE

Here's another example. If you ever drive along U.S. Route 301 in Maryland near Glen Burnie, you might see these roadside signs advertising a housing development:

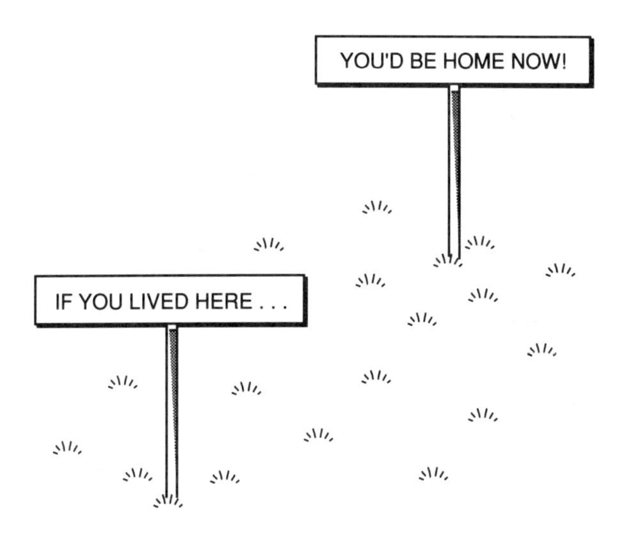

The first time we see such signs, we might credit the advertiser with some cleverness, but this gimmick has been used by realtors since the '50s (at least!) and by now we're inclined to be annoyed whenever we see the slogan. Why are we annoyed? Because, if the statement is true, it's *trivially* true; and, if it's not a tautology, then it's false.

Why do I say it may be false? The reason helps us to understand the special difficulty about ordinary language tautologies: they do not admit of nuances of meaning. I can make the statement a false statement simply by insisting that "I might live here, but I can never call anywhere but the hills of Pennsylvania home." This is a slightly different sense of home—home is "where the heart lies." However, if I define 'home' simply as "the place where I live," then it is indeed true, but *trivially* so, that if I lived here, I'd be home now.

More reasons why tautologies may be valuable should be mentioned here, though they are a bit technical. Philosophers argue whether there are any undisputed tautologies, but if there were, when we'd have a basis for a greater understanding of the concept of synonymy. Being able to clearly determine synonyms may be of great value in translation from one language to another, or in the development of artificial intelligence in computer science. Moreover, the idea of tautologies when they are truth-functionally defined (for example, in Truth Tables) can assist formal derivations in strict deductive logic.*

---

*See Unit Three, "Symbolic Logic."

11.2 CONTRADICTIONS

A **contradiction** is a statement which is *false* no matter what. As such, a natural language contradiction is just the opposite of a tautology (which is always true).

We probably encounter contradictions most often as the distinguishing mark of two (or more) statements which are inconsistent. We say of such statements that they *contradict* each other. For example, if at nine o'clock this morning I claim that "Arizona is the most beautiful state in the nation," and then at nine o'clock this evening I make yet another claim, that "California is the country's most beautiful state," I've contradicted myself. Of course, you may well point out that I've simply changed my mind, and then the inconsistency may be resolved.

Suppose, however, I say the following in one breath: "Arizona is the most beautiful state and California is too!" Now I have conjoined the two claims into one statement, and that statement is about as clearly a contradiction as we can get in ordinary language.

If I were to make such a claim, you'd have a right to be confused. "They can't *both* be the most beautiful state," you will surely insist. One or the other must be. Explain yourself!" The statement as it stands must be false, and natural discourse abhors a contradiction. No one can hope to utter a contradiction without qualification and at the same time strive to tell the truth.

It's easy to frame formal contradictions in a symbolic language but they're as difficult to pin down in ordinary discourse as tautologies are. Let's introduce some simple symbols and see why. If I come across the statement 'P and *not*-P must be true' (where 'P' stands for some determinate statement), I have a formal contradiction. But let's say that 'P' stands for the statement 'It is raining'. In this case, I am claiming both that it is raining and not raining. How is that possible? Well, you might say, it may be raining in Terre Haute, Indiana but not in Truth or Consequences, New Mexico. Fair enough, but suppose I insist: "It is both raining and not raining at the same place at the same time:" The novice logic student replies: "Maybe it's drizzling!"

Bravo for you! You're demonstrating just how impossible it is to accept a contradiction without looking for some way to qualify—to *resolve*—the apparent contradiction. Yet, drizzling is a form of rainfall, isn't it? Whether you answer "Yes" or "No" to that question, the contradiction stands, and a contradiction has a way of making us feel uneasy.

People do sometimes slip into contradictions, whether by design or by just plain bad thinking. The results are sometimes amusing and sometimes annoying. The great British logician and philosopher Bertrand Russell once told a story about how he and another person went to see a yacht together. Obviously unimpressed, the other person remarked, "Oh, I thought it was bigger than it is!"

Russell replied, "No, it's no bigger than it is." (If you don't know what a contradiction is, you won't get the joke.)

Consider the lyrics to this popular song:

It never rains
In southern California,
But it pours,
Oh man, it pours!

Now that you know what a contradiction is, don't you find that song annoying? What can the songwriter possibly mean? Perhaps he is being ironic ('irony' is a figure of speech one uses to say one thing, but mean another). Perhaps the Chamber of Commerce likes to advertise the absence of rain in California, but the songwriter hints that the truth is somewhat different. Perhaps the song is suggesting something even more, and at another level of meaning—like the aphorism, "It never rains, but it pours" (suggesting that troubles never come when you're prepared for them).

You see? We can't accept contradictions at face value. Taken literally, 'It never rains, but it pours' is a contradiction. Thus, we look for a second level of meaning to help us resolve the contradiction as apparent only.

Why do we go to such lengths to attempt to find explanations or qualifications when we are presented with an apparent contradiction? That raises a particularly interesting logical point. Formally speaking, we can validly derive anything at all (or nothing at all!) from a logical contradiction. Everything follows! Nothing follows! Contradictions make our thinking utterly incoherent, and both language and thought break down as a result. Contradictions render communication impossible.

Perhaps that's why cynical people sometimes employ them in their speech or writings. Suppose a politician confronts a politically divisive issue, like abortion, and she tells a conservative audience in Wilkes-Barre, "I'm against abortion," but then turns around and tells a liberal audience in Ann Arbor, "I'm for freedom of choice!" We may well insist that she isn't telling us the truth. Perhaps she is confident that the inconsistency won't be found out. If she *is* found out, she may gain political mileage by qualifying the apparent contradiction, saying something confusing like: "Oh, I'm for freedom of choice all right, except in some cases when abortion is wrong." Don't be fooled, you just don't know where she stands on the issue.

Demagogues and pseudoscholars also sometimes appeal to contradictions to make themselves seem profound. We can't understand them on a literal level, so we look for some deeper meaning (where none may exist). It's in our nature to expect people to tell the truth, and con-artists like to take advantage of our good natures. But if you know what a contradiction is, stick to your guns, and insist that the contradiction is real. If the reply doesn't satisfy you, then vote for somebody else or throw the book away.

One last word about usage: contradiction is an attribute of statements only (and only a special class of statements at that). It doesn't apply to practices for the simple reason that practices cannot be true or false. If a high school principal allowed last year's senior class to go on a class trip, but denies this year's class permission to do the same, she isn't contradicting herself. She may be unfair, or maybe budget cuts prevent her from authorizing the practice, but practices have to be judged on ethical grounds instead of logical ones.

11.3 CONTINGENT STATEMENTS

A **contingent statement** is a statement which is "contingently" true or false; its truth or falsity depends on conditions or events. Most of the statements ever uttered or written are contingent in this way. You don't know whether a contingent statement is true or false until you check it out.

Tautologies and contradictions, on the other hand, are only a small set of the statements we encounter; they're special cases. They are "necessarily" true or false, and their truth or falsity depends on the meaning or form of the sentence. But contingent statements must be *verified* or in other ways established.

Checking out contingent statements for truth can be as sophisticated as conducting scientific laboratory experiments or as simple as just going out and looking for yourself. Consider the following example:

The author of your textbook is 6 feet 5 inches tall.

Is this statement true? Chances are, you don't know unless you've met me. This is a contingent statement because you have to check it out in order to know whether it's true or false. You may ask someone who knows me, or you can come see for yourself (I'm in Room 104-F, Faculty Towers).

Many times we encounter a contingent statement and we already know that it's true or false without having to check it. But the point is that it's contingent because it is the sort of statement that you *can* check out. Here are a few examples:

Wayne Gretzky now plays for the Los Angeles Kings.

Jessica is married to Roger Rabbit.

Seoul, Korea was the site of the 1988 Summer Olympics.

The capital of Alaska is Boston.

Tom Cruise plays a logic professor in "Top Gun."

The Pacific is the ocean at Miami Beach.

Most of you know right away that the first three statements are true and the second three are false, but they're all contingent statements because they're the sort of thing you can check out. You can check that Seoul was the site of the 1988 Summer Olympics by consulting an almanac, and any world atlas will tell you that Juneau is the capital of Alaska (not Boston). You don't know anything about Jessica and Roger Rabbit? Go see the movie. None of these statements are in any sense *necessarily* true or false. A director with a sense of humor could have cast Tom Cruise as a logic professor in "Top Gun," Wayne Gretzky could have elected to stay in Edmonton, and a good-sized earthquake might have moved

Miami Beach to the Pacific Ocean. Thus, these statements are *contingently* true or false.

Summary:

1. **A tautology is a statement that is *true* no matter what.**
2. **A contradiction is a statement that is *false* no matter what.**
3. **A contingent statement is a statement which may be either true or false, depending on external conditions or events. (Most statements ordinarily encountered are contingent.)**

EXERCISES

DIRECTIONS: *Say whether the following statements are apparent tautologies, apparent contradictions, or contingent statements. If a statement appears to be a tautology, suggest an interpretation that may make it false (or contingent). Likewise, if a statement appears to be a contradiction, suggest an interpretation in which it may be true (or contingent).* **Note:** These exercises are open to discussion.

Con **1.** Togo is a nation in Africa.
Con. **2.** Bernaise sauce is made with egg yokes.
Taut. **3.** "A man's gotta do what a man's gotta do."
______ **4.** Television is a "vast wasteland."
______ **5.** The ozone layer is disintegrating.
Tauto **6.** Every event has a cause.
______ **7.** Every effect has a cause.
______ **8.** Kangaroos are found in Australia.
______ **9.** Ash trees are the best trees for landscaping, but elms are even better.
______ **10.** All humans are mortal.
______ **11.** All humans are rational animals.
______ **12.** The universe tends ultimately to disorder.
Contradiction **13.** No Americans are residents of the United States.
______ **14.** Every action has an equal and opposite reaction.
______ **15.** A clean desk is the sign of a sick mind.

MODULE 12

TRUTH AND VALIDITY

12.1 IN GENERAL

Perhaps no word is misused more frequently in everyday conversation than the logical term 'validity'. In proposing a new marketing technique, an executive might remark, "That's a *valid* idea," but this is a mistaken use of the term. Ideas are not valid or invalid; *arguments* are.

People commonly confuse *truth* and *validity.* The two terms are distinct. Validity says something about the proper *form* of an argument; truth concerns the *content.*

Informally, we may say that a **valid argument** is one that is of such a form that the conclusion *necessarily follows* from the reasons offered in its support. The key word here is 'necessarily'. If the premises (i.e., the *reasons* in a deductive argument) are true, then the conclusion *must* be true. Notice that validity is a phenomenon of deductive arguments only.* Deductive arguments are either *valid* or *invalid;* inductive arguments are either *justified* or *unjustified.*

A deductive argument is **valid,** if and only if, if all the premises are true, then the conclusion must be true.

*Some logicians like to say that inductive arguments can be valid, but they use the notion of validity with regard to inductive arguments in such a way that the term takes on entirely new meaning. To say that an inductive argument is "valid" in this way tells us little more than it's a *good* argument. For this reason, I prefer the traditional insistence that the term 'validity' be restricted to deductive arguments only.

The concern of valid arguments is *form,* but inductive arguments are chiefly justified on the basis of their *content.* A deductive argument bears close analogy to a calculation: if a teenager has been married three years, we can validly deduce that she married quite young. That's an obvious calculation to make. On the other hand, an inductive argument relies on the evidence presented—for example, if a patient has all the symptoms of malaria, we can say that whether or not the patient has contracted malaria there is a justified probability of it (though a good physician may well want to know still more about the patient before she administers treatment).

Truth is important to deductive arguments but a valid deductive argument with true premises is not called a true, but *sound,* argument. We can reduce this concept to a formula:

soundness = validity + all true premises

A sound inductive argument is a justified argument with true premises.

Let's see how validity differs from truth. As far as validity is concerned, it doesn't really matter if the premises of a deductive argument are false: you may still have a valid argument. If that strikes you as peculiar, consider that we often argue: "Sure, I may be mistaken but, for the sake of argument, let's say I'm right." We argue this way all the time when we have incomplete or uncertain information. We say, "Assume I'm right; if I am, it necessarily follows that such-and-such or so-and-so is true."

Let's consider this matter schematically. Assume that we have deductive arguments limited to two premises and one conclusion. This is true of Aristotelean syllogisms, but other forms of deductive arguments may have as many premises as you wish, or as few as only one; they must, however, have one conclusion. Given certain rules, called **rules of inference,** and the relevance of the premises to the conclusion, any one of the following arguments can be valid.

Consider a simple example of a valid argument that corresponds to argument form 7 in Table 1 (p. 130). We see that this is an argument with false premises and a false conclusion. Such an argument may be valid or invalid: the point to remember is that *it can be valid.* Here's an example:

PREMISE 1: **Rome is in Germany and London is in France.**
PREMISE 2: **New York is in California.**
CONCLUSION: **Therefore, Rome is in Germany.**

We see that both of the premises and the conclusion are false (premise 2 is unnecessary to the argument, but that doesn't affect its validity). How can this be valid? Imagine that our geography pupil is a good reasoner but has bad information. We know that both premises are false, but assume for the sake of argument they were true (let's say, unknown to us, Germany just changed the name of one of its cities to 'Rome' and France changed a name of a city to 'London'). In that case, you see that the deduction would yield a true conclusion—it would *have* to be true, given the information in the premises.

These Argument Forms *Can* Be Valid

	1	2	3	4	5	6	7
Premise 1	true	true	false	false	false	true	false
Premise 2	true	false	true	false	true	false	false
Conclusion	true	true	true	true	false	false	false

TABLE 1

Let's take another example. Argument form 6 in Table 1 has a true first premise, a false second premise, and a false conclusion. Such arguments may or may not be valid—the point to remember is that such an argument *can* be valid depending on whether or not it follows proper rules of inference. Here's an example:

PREMISE 1: **Either Lee Harvey Oswald acted alone, or there was a conspiracy to assassinate President Kennedy.**

PREMISE 2: **It is not the case that Lee Harvey Oswald shot Kennedy.**

CONCLUSION: **Therefore, there was a conspiracy to kill the president.**

The truth of premise 1 is obvious enough but, if we believe the Warren Commission Report, both premise 2 and the conclusion are false. Nevertheless, we still argue like this all the time. Certainly, our information on the assassination of President Kennedy is incomplete, however careful the Warren Commission was in its investigation. This is an entirely reasonable argument. We say: "I *know* what the Warren Commission Report says but go along with me on this: for the sake of argument, let's say that Lee Harvey Oswald did not act alone. In that case, there had to be a conspiracy. New evidence suggests that Oswald wasn't the lone gunman." Even though it may turn out that we're wrong, and that Oswald was a lone gunman after all, such arguments are still the grist of good investigations. This is good reasoning. In other words, it's a valid argument *even if* it has some false premises.

A simple calculation shows us that any argument with two premises and one conclusion has eight possible combinations. We see that seven of the possibilities *can* be valid arguments. That includes the case where both premises and the conclusion are true as well as the case where both premises and the conclusion are false. In fact, we have only one case ruled out, and that's the case where both premises are true and the conclusion is false. In fact, that's the meaning of our formal definition of validity:

A deductive argument is valid, if and only if, if all the premises are true, then the conclusion must be true.

Here, then, is the one arrangement of truth and falsity among premises and conclusion that must be an *invalid* argument:

The Invalid Schema	
Premise 1	true
Premise 2	true
Conclusion	false

TABLE 2

We already said that a **sound** argument is a *valid* deductive argument with all true premises. Since the argument is valid, and in fact all the premises are true, then the conclusion must also be true. Here is the schema of a *sound* deductive argument:

The Schema of a Sound Argument		
Premise 1	true	+ Validity
Premise 2	true	
Conclusion	true	

TABLE 3

12.2 RULES OF INFERENCE

Once more, we emphasize that validity is a matter of form and truth is a matter of content. An analogy may be helpful here: when we say that validity is *formal,* we mean 'formal' in much the sense that we use the term with regard to a "formal" dinner dance. When an invitation specifies "formal attire," we know that the hosts require a certain sort of dress—for a man, a tuxedo; for a woman, an evening gown. Analogously, formal logic also specifies a certain sort of attire. Once that "attire" is assumed, we formally guarantee the truth of the conclusion.

We give a name to these "forms"—**rules of inference.**

Not all possible rules of inference have formal names; the number and variety of valid argument forms constitutes an infinitely large set. However, a small subset of these valid argument forms can generate, through repeated applications of a combination of them, *all* possible valid arguments. We will consider several of them here in order to get a deeper understanding of what validity means.

Two of these rules of inference have Latin names: ***Modus Ponens*** and ***Modus Tollens***. We retain the Latin names out of tradition, possibly because the English translation is clumsy—the former literally means "the mode of putting forth" and the latter translates as "the mode of taking away." Consider an example of *Modus Ponens:*

PREMISE 1: If wishes were horses, then beggars would ride.

PREMISE 2: Wishes are indeed horses.

CONCLUSION: Therefore, beggars will ride.

Of course, we know right away that premise 2 is not true, but the argument form is valid. Assuming, for the sake of argument, that the premises were true, then we could be assured of the truth of the conclusion without any further appeal to evidence. So, let's do that! You see, then, that the conclusion follows after all.

Since validity is a matter of form (and not of truth content), we can extract the form from the arguments by inserting "placeholders" for the statements. We call these placeholders **atomic statement letters.**

The atomic statement letters are capital letters of the alphabet. Let's extract the discrete statements from our example and substitute the statement letters 'P' and 'Q'. We then have the following result:

Let 'P' stand for: "wishes are horses"
Let 'Q' stand for: "beggars will ride"

Notice that at this stage logic is indifferent to tense or grammatical mood. We have deleted the subjunctive 'would' in "beggars would ride" and substituted the verb 'will'.

Utilizing the logical operators, which in this case is just 'If . . . then', we have the following:

PREMISE 1: If P, then Q.

PREMISE 2: P.

CONCLUSION: Therefore, Q.

We call this form *Modus Ponens*—a rule of inference.

The same "suit of clothes" will also fit many other arguments. Let's use two new statements and substitute them for the statement letters 'P' and 'Q' respectively. We have:

P: "the engine stalls"

Q: "we need to drop the nose of the aircraft slightly"

PREMISE 1: If the engine stalls, then we need to drop the nose of the aircraft slightly.

PREMISE 2: The engine has stalled.

CONCLUSION: Therefore, we need to drop the nose of the aircraft slightly.

We can see that the argument is clearly valid since it fits the *Modus Ponens* form. Moreover, the premises in this argument are also *true,* which guarantees the truth of the conclusion. Since the argument is both valid and it has all true premises, we know it to be sound as well. Our earlier example about "wishes are horses" was valid but not sound.

Next, let's consider an example of *Modus Tollens.* It also employs the logical operator 'If . . . then'.

PREMISE 1: **If this is Tuesday, then this must be Belgium.**
PREMISE 2: **This is decidedly not Belgium.**
CONCLUSION: **Therefore, it must not be Tuesday.**

Substituting the statement letters 'R' and 'S' for the two statements and removing the logical operators 'If . . . then' and 'not', we have:

R: **"this is Tuesday"**
S: **"this is Belgium"**
PREMISE 1: **If R, then S.**
PREMISE 2: **It is not the case that S.**
CONCLUSION: **Therefore, it is not the case that R.**

This is the rule of inference called *Modus Tollens.*

We'll conclude this section with one more rule of inference called **Disjunctive Syllogism.** This is "process of elimination" rule of inference.

'Disjunction' refers to statements with the logical operator 'Either . . . or'. A "syllogism" is a kind of argument. (Strictly speaking, this is not a true syllogism—see Unit Three, "Traditional Logic"—but we retain the name for the sake of tradition.)

Here is an example of the rule of inference called Disjunctive Syllogism:

PREMISE 1: **Either my name is Rumpelstiltskin or you've got toys in the attic.**
PREMISE 2: **My name is not Rumpelstiltskin.**
CONCLUSION: **You've got toys in the attic!**

Selecting out the logical operators 'either . . . or' and 'not' and replacing statements with statement letters, we have:

P: **"My name is Rumpelstiltskin"**
Q: **"You've got toys in the attic"**
PREMISE 1: **Either P or Q.**
PREMISE 2: **It is not the case that P.**
CONCLUSION: **Therefore, Q.**

This is Disjunctive Syllogism. Any uniform substitution of statements for statement letters here will result in a valid argument. It doesn't matter whether the statements are *true*—that is a consideration of *soundness*, not of *validity.* Nonetheless, we'll replace statement letters with new statements which *are* true so we can have a sound (as well as valid) argument.

Let 'P' stand for: "water freezes at 100° centigrade"
Let 'Q' stand for: "water boils at 100° centigrade"

"Either water freezes at 100° centigrade or it boils at that temperature at sea level, I don't know which. However, I've just been assured that it doesn't freeze at 100° centigrade, so it must boil at that temperature."

This argument is both valid and sound. To say it's valid is to say that the truth of the premises assures the truth of the conclusion without any further appeal to evidence. Soundness requires more than that—it requires a valid form *and* all true premises. In order to ascertain the truth of the premises, we need more evidence. In this case, we need scientific evidence from experiments which disconfirms the hypothesis that water freezes at 100° centigrade.

If we can only be assured that the first premise is true (either water boils at 100° centigrade or else it freezes) then we need to conduct just one set of experiments. Once we satisfy ourselves that water never freezes at that temperature, then we know, without needing any more experiments, that the conclusion must be true. This is the meaning of validity: assuming the truth of the premises, the truth of the conclusion is guaranteed. Valid forms of thinking exempt us from having to conduct any additional tests on the boiling point of water; we now know what it is without further testing. In this sense, good thinking is a shortcut to knowledge.

12.3 FALLACIES

There are infinitely many ways that arguments can go wrong. When an argument goes wrong, we say that a *fallacy* has been committed.

More formally, we say that a **fallacy** is the violation of a rule of inference.

We've considered three rules of inference thus far; let's see what a fallacy looks like.

Recall that both *Modus Ponens* and *Modus Tollens* are arguments which properly use the logical operator 'if . . . then'. Here is an improper use of the 'if . . . then' operator in an argument which results in a fallacy:

PREMISE 1: If creationism is true, then Darwin was wrong about a great deal about the origin of species.

PREMISE 2: In fact, the latest scientific evidence reveals that Darwin was very often wrong about a great deal.

CONCLUSION. Therefore, creationism is true.

This is a fallacious argument. Even if the premises are true, the conclusion can be false. Given that Darwin was very often wrong about many things, it still doesn't follow that creationism is true. In fact, many scientists (e.g., Stephen Jay Gould) believe that Darwin was wrong about many important features of his theory, but these scientists do not conclude that this counts as decisive evidence against the theory of evolution (Gould himself is an "evolutionist").

Formally, the fallacy looks like this:

PREMISE 1: If P, then Q.

PREMISE 2: Q.

CONCLUSION: Therefore, P.

This is called "the fallacy of affirming the consequent." We say that it is a fallacious or invalid form because at least one example of a uniform substitution of statements for statement letters will yield an invalid argument. For example:

Let 'P' stand for: "You married well"
Let 'Q' stand for: "You'll be happy"

"If you married well, then you'll be happy. You are indeed happy. Therefore, you married well."

Obviously, this is invalid; you may be happy and single!

A fallacy based on the improper inference from the 'either . . . or' statement together with the affirmation of one of the disjuncts is called "the fallacy of affirming a disjunct," and it looks like this:

PREMISE 1: Either P or Q.

PREMISE 2: P.

CONCLUSION: Therefore, it is not the case that Q.

To take an example, let's substitute the statement 'there's a loose connection somewhere' for P, and 'the vacuum cleaner's motor is burned out' for Q. We have this result:

PREMISE 1: Either there's a loose connection somewhere or the vacuum cleaner's motor is burned out.

PREMISE 2: There is a loose connection.

CONCLUSION: Therefore, it is not the case that the motor is burned out.

This *looks* like it might be a good argument, but anyone who has ever fixed an appliance can see the fallacy right away. Upon repairing the loose connection, you'd be advised to test the motor as well—it might be burned out too.

12.4 PARADOXES

Apparently good arguments can sometimes lead us to absurd conclusions. When this happens, we are said to have encountered a **paradox.**

Paradoxes can be a special source of delight and diversion for the sophisticated student of logic, but they serve a special purpose for logicians and philosophers because they show us the limits to which we can take formal logic. They are a signal that something has gone wrong, but paradoxes aren't *just* bad arguments. Their resolutions are often elusive and paradoxes therefore cause us to inquire even deeper into the philosophy of logic.

Perhaps the most famous paradox of all time is the "Liar's Paradox." This paradox was well known to the ancient Greeks. Traditionally, the paradox is given as follows:

All Cretans are liars.

I am a Cretan.

Therefore . . .

Therefore, what? If I say I am a Cretan, then I must be a liar; but if I am a liar, then what I say must be false, in which case I cannot be telling the truth when I say I am a Cretan. . . . In other words, if I'm telling the truth, then I must be lying; and if I'm lying, then I must be telling the truth!

Figure 1 and Figure 2 show a more modern version of the Liar's Paradox. If you find yourself shaking your head in confusion upon reading these examples, you're in good company. Logicians and philosophers have spent a great deal of time and effort in attempting to resolve such paradoxes because paradoxes, unchallenged, represent a wholesale threat to the enterprise of logical reasoning.

When we encounter a paradox, we should immediately suspect that the argument really isn't valid or sound. In other words, we should look for hidden fallacies.

Consider the following classic paradox:

"There is a barber in a village who shaves everyone who does not shave himself. Does the barber shave himself? He does if he doesn't; and if he doesn't, he does."

Confused? Notice that it won't do to attempt to resolve the paradox by declaring the barber is a woman (the example refers to the barber as "*him*self") but if that thought crossed your mind, you're on the right track. One way to approach a paradox is to look for hidden and unjustified assumptions.

Actually, this is one paradox that's fairly easy to resolve. Since the conditions listed result in an impossible state of affairs, we can safely say that no such village can possibly exist. There is no such barber.

The great logician and philosopher Bertrand Russell once encountered a paradox very much like the Barber's Paradox just described. Russell was engaged in a major project that wasn't going well, and he introduced set theory into his program in an attempt to save it. Many later philosophers have disputed the legitimacy of that move, but even more telling was the fact that the introduction of set theory resulted in a major paradox, the "paradox of non-self-membered sets" (also called "Russell's Paradox"). The paradox was a close

The statement on the next page is false.

FIGURE 1

relative of the Barber's Paradox, and Russell describes in great detail what consternation this paradox caused him. In his autobiography, he tells how he kept a pencil and a sheet of paper on his kitchen table. He sat down at the table each morning, pondering the problem. When he left the table at the end of the day, the sheet of paper remained blank.

The resolution of paradoxes is often a very complicated matter, but we persist in the faith that some resolution is indeed possible, whatever the paradox. Ironically, one of the oldest paradoxes, the Liar's Paradox, requires some of the most complicated of explanations to resolve it, but every paradox (we should suspect) when placed under scrutiny will show some element that reveals that the argument really isn't valid or sound after all. In other words, we should suspect that a fallacy is present.

Sometimes, however, paradoxes can be useful as well as revealing. One such is the "paradox of strict implication." This paradox shows us that from contradictory premises, any conclusion whatsoever can be validly deduced. For example, the following is a valid argument:

PREMISE 1: P.
PREMISE 2: It is not the case that P.
CONCLUSION: Therefore, Z (or anything at all).

This looks startling, but consider what it *really* means: it means that such an argument must be *unsound.* Of course it does! The premises can't both be true—they contradict each other. The fact that the argument is formally valid doesn't mean it's a good argument; it describes an impossible state of affairs. So, naturally enough, the argument is absurd.

By the way, the philosopher and logician Willard Van Orman Quine believes he has detected the Liar's Paradox in the epistles of St. Paul. Judge for yourself:

> One of themselves, *even* a prophet of their own, said, The Cretans *are* always liars, evil beasts, slow bellies.
>
> This witness is true. Wherefore rebuke them sharply, that they may be sound in the faith. —Titus 1:12–13

Quine draws the (unjustified) conclusion that St. Paul heard the Liar's Paradox but did not understand its logical properties and so he used it to drive home a righteous message. However, another conclusion is possible that allows for Paul's complete understanding of the paradox:

St. Paul had a sense of humor.

The statement on the previous page is true.

FIGURE 2

EXERCISES FOR UNIT ONE

Exercise 1

DIRECTIONS: *Determine which of the following are arguments and which are not. Place your answers ('argument' or 'not an argument') in the space to the left of the passage.*

1. My point is simply that on such occasions liberty does not mean very much since the mentally ill person is being "controlled" by distorted thought processes.

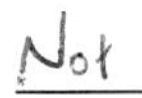

2. A section of crane being trucked to the State Ports Authority Monday afternoon caught a power line across Shipyard boulevard, snapping a utility pole and causing two transformers to break open and leak coolant.

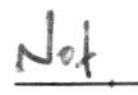

3. Maternity leave has been on the minds and tongues of many. Talk shows have discussed it, women's groups have debated it, and recently the U.S. Supreme Court has had its say.

On January 13, the court ruled that states may require employers to give pregnant workers job protection not available previously.

4. Nothing can have value without being an object of utility. If it be useless, the labor contained in it is useless, cannot be reckoned as labor, and cannot therefore create value. —Karl Marx, *CAPITAL*

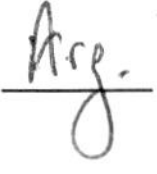

5. If the Old North State can control, sell, and profit from the sale of booze, it can certainly have a lottery.

If lotteries were good enough for our first president, George Washington, they are good enough for me.

6. For reasons best understood by the company's promotions department, Timex has commissioned a poll of 1,000 Americans to determine the country's "attitude toward adventure." Among the findings: Twenty-one percent have participated in adventuresome sports activities in the last five years; men are more adventuresome than women; the West is slightly more adventuresome that other parts of the world. The final 48-page report, prepared by Research & Forecasts Inc., concludes that "Americans believe being adventurous means relying on one's own instincts and surviving natural elements." Well, not *always* natural. While only 5% of "adventuresome people" have tried hang gliding, only 3% have ridden camels and only 1% have gone on jungle treks, a full 7% of all respondents claim personal adventures with the occult. —*Sports Illustrated,* October 20, 1986

7. Between any two different real numbers x and y, there is another real number. In particular, the number $z = (x + y)/z$ is a number midway between x and y. Since there is also a number s between x and z and another number t between z and y and since this . . . can be repeated ad infinitum, we say . . . that between any two different numbers (no matter how close together), there are infinitely many other real numbers. —Edwin J. Purcell and Dale Varberg, *Calculus With Analytic Geometry,* Prentice-Hall, p. 10

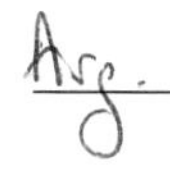

8. Not only did you fail to print any portions of the final environmental-impact statement on proposed drilling at the Arctic National Wildlife Refuge, you mentioned that there is "at least a 5 percent chance of recovering 9.5 billion barrels [of oil] or more." This means that there is possibly a 95 percent chance that less will be recovered. I don't believe that one can logically justify jeopardizing the well-being of this critical breeding area under such uncertain conditions. —Letter to the Editor, *U.S. News and World Report,* December 28, 1986

9. The Florida Breeding Bird Atlas is a major undertaking that will require the assistance of hundreds of volunteers throughout the state. Participants need not be expert birds. —Florida Naturalist, cited in *The New Yorker,* November 10, 1986

_______ 10. By the end of this decade the United States will have sent spacecraft past all the planets of the solar system from Mercury to Neptune. Four spacecraft are leaving the solar system, venturing toward the stars beyond. It is now time to start thinking of exploring the stars—of sending probes to the pinpoints of light dotting the night sky. —*Astronomy,* January 1987

Exercise 2

DIRECTIONS: *Read the passages below and determine whether each is an* **argument** *or an* ***explanation****. Put your answer in the space to the left.*

_______ 1. Now there is a way to wear certain prescription drugs instead of having to remember to swallow them.

Transdermal (through the skin) medication comes measured in a waterproof patch that sticks to skin. Medicine in a patch works fast because it doesn't have to pass through the digestive system first. Instead, the patch releases a continuous, even flow of medicine that's absorbed directly into the blood. Though the adhesive sometimes causes a rash, patients report fewer side effects than with oral medication. —*Woman's Day,* December 2, 1986

_______ 2. In another major decision last week, the Supreme Court directed a chilly breeze down Madison Avenue's way. By a vote of 5 to 4, the court ruled that even truthful ads for lawful goods and services may be restricted by the state to protect "the health, safety and welfare" of its citizens.

The court's pronouncement grew out of a case from Puerto Rico, where the former owners of the Condado Plaza Hotel and Casino had invoked the First Amendment guarantee of free speech to challenge a local ban on advertising by gambling casinos. Because of the crime and corruption that gambling could attract, Puerto Rico could have outlawed betting altogether, stated Justice William Rehnquist for the court, and the "greater power to completely ban casino gambling necessarily includes the lesser power to ban advertising of casino gambling." More ominously for other advertisers, Rehnquist suggested that the same logic could be used to justify governmental restrictions on other "products or activities deemed harmful, such as cigarettes, alcoholic beverages, and prostitution." —*Time,* July 14, 1986

_______ **3.** It's so fast and simple to use the AT&T Card, you're going to get spoiled.

To use it, just dial the number you want to call, and enter your AT&T Card number. That's it. It's the easiest way to plug into the dependable AT&T Long Distance Network from virtually any phone.

In fact, you can make calls even without carrying your AT&T Card. Just remember your AT&T Card number. . . .

So, if you travel frequently, get the AT&T Card. It's as simple as dialing. —Advertisement for AT&T

_______ **4.** He, to get the cold side outside, put the warm side fur inside. That's why he put the fur side inside, why he put the skin side outside, why he turned them inside outside. —Anonymous, *The Modern Hiawatha*

_______ **5.** Officials have told high school students they cannot stage the musical farce *A Funny Thing Happened on the Way to the Forum* because it is morally unacceptable and degrading to women.

_______ **6.** Children are taught to think of themselves as members of a group. The school is a society with its own way of doing things. And the most important goal for teachers is to give children the confidence of membership by teaching them, for instance, how you place your outdoor shoes in the cubbies and put on your indoor shoes. Children learn about social relationships: the way you behave with your teacher, with your elder brother, with your mom. In the first grade, before doing anything academic, the teacher spends the first part of the year getting the children socialized to the ways of the school and the habits of working together in groups. Children are even responsible for cleaning the school. —*U.S. News and World Report,* January 1987

_______ **7.** Calcium goes to work your whole life to build and maintain strong, healthy bones. Yet, as you get older your bones may begin losing calcium. This can lead to "osteoporosis" causing severe problems later in life. And while it can be caused by a lack of calcium, getting enough calcium in your diet is very easy to do.

One of the most natural ways to replenish your calcium is with dairy foods like milk, yogurt, cottage cheese and cheese. All you need are three servings a day to meet your recommended dietary calcium allowance of 800 milligrams. And dairy foods offer you a wide variety of delicious and nutritious ways to get a lot of your other essential nutrients. —Advertisement in *Health* Magazine

Exercise 3

DIRECTIONS: *Determine whether the following passages are* ***inductive*** *or* ***deductive*** *or* ***no argument at all****. Place your answers in the appropriate space.*

_______ **1.** Aquino's tumultuous rise to power sounds a note of warning to all dictators who have chosen to ignore the pleas of their citizens for freedom and democracy. People power has finally arrived, and I hope it is here to stay. —Letters, *Time,* 129(4), 1987: p. 6

_______ **2.** Opponents of the amendment are raising a more serious and thoughtful argument against the plan, focusing on three main considerations: 1) what effect change would have on the two party system, 2) whether it would favor geographical or popular representation, and 3) whether it would give the edge to a liberal or conservative candidate. —*Time,* May 4, 1970

_______ **3.** Sherlock Holmes: "Like all other arts, the Science of Deduction and Analysis is one which can be only acquired by long and patient study, nor is life long enough to allow any mortal to attain the highest possible perfection in it."

_______ **4.** Slowly everyone shook his head. There could be no other explanation. The drinks themselves were untampered with. They had all seen Anthony Marston go across and help himself. If followed therefore that any cyanide in the drink must have been put there by Anthony Marston himself. And yet—why should Anthony Marston commit suicide? —Agatha Christie, *Ten Little Indians*

Exercise 4

DIRECTIONS: *You are given several short arguments below. Using the* **Assertion Diagram** *Method, select the assertions and identify each of them in the order in which you encounter them by selecting the first word of the assertion and the last, and by connecting them with three periods (i.e., . . .). Then diagram the argument.* **Be sure to get the conclusion right.**

(I've already set out the assertions in problem 1 for you. Follow this example for the others.)

1. Nothing can have value without being an object of utility. If it be useless, the labor contained in it is useless, cannot be reckoned as labor, and cannot therefore create value. —Karl Marx, *CAPITAL*

Assertions

(1) Nothing . . . utility.

(2) If . . . value.

2. My point is that on such occasions liberty doesn't mean very much, since the mentally ill person is being "controlled" by distorted thought processes. . . .

Assertions

①

②

3. Gary Wills contends in his article "A Fantasy Breeds a Scandal" (*National Affairs,* Dec. 29) that the underlying cause of the Iran-Contra crisis is President Reagan's reliance on the moral certainties of his old B movies. Quite the contrary. Although Brass Bancroft, the Secret Service character that Reagan played, might very well be involved with the midair interception of a jet carrying terrorists to safety or an air strike on a state that supported terrorism, he would never trade arms with an enemy for hostages. In fact, the president became embroiled in a scandal and lost public support when he failed to follow those old verities that Mr. Wills finds so ridiculous.
—The Mail, *Time,* January 5, 1987

Assertions

①

②

③

4. Editor: I oppose the new mandatory seat belt law. It's just more government regulations on its people, of which there are already too many.

Seat belts may save lives in one situation while causing others to lose theirs in another. As an adult, it should be my option whether to wear them, not the government['s]. It concerns me the United States is slowly becoming a police state because of so many such laws.

The mandatory seat belt law should be repealed so its citizens can enjoy the freedom of choice under the Constitution. —Letter to the Editor, *The Wilmington Morning Star*

Assertions

①

②

③

④

⑤

⑥

UNIT TWO

TRADITIONAL LOGIC: THE LOGIC OF CATEGORICAL SYLLOGISMS

MODULE 13

ARISTOTELEAN LOGIC: THE SYLLOGISM

13.1 RECOGNIZING SYLLOGISMS

The development of logic did not begin with Aristotle, nor did it end with his death. Certainly, Socrates and Plato also made important logical discoveries. Socrates, for example, helped to perfect the logic of the dialectic, and Plato advanced the art of logical definition. Nevertheless, Aristotle has made such an important contribution to logic that we name a whole branch of the field in his honor.

Aristotle is credited with perfecting the categorical syllogism. A **syllogism** is an argument which employs three terms, including a middle-term and two end-terms, a major premise, a minor premise, and a conclusion. Syllogisms are readily identified by their "operators" (which we will later call "quantifiers"*): 'All', 'No', 'Some', and 'Not all'. Each of the premises employs one operator, an end-term, and the middle-term. The conclusion employs an operator and both of the end-terms.

'All' is the universal affirmative operator.
'No' is the universal negative operator.
'Some' is the particular affirmative operator.
'Not all' is the particular negative operator.

A **deductive argument** is valid, if and only if, if all the premises are true, then the conclusion must be true.

*See Unit Three, "Symbolic Logic."

Here is a straightforward example of a syllogism:

MAJOR PREMISE: **All tigers are panthers.**
MINOR PREMISE: **All Siberian tigers are tigers.**

CONCLUSION: **Therefore, all Siberian tigers are panthers.**

A fast way to decipher the structure of the syllogism here is to pick out the two **end-terms.** The end terms are the terms which appear exclusively in the conclusion. By convention, we designate one of the two end-terms as the **minor-term** (or "S"-term); it is the *first* term you encounter in the conclusion. Thus, 'Siberian tigers' is the minor-term.

By the same convention, we designate the other end-term, **major-term** (or "P"-term), as the *last* term you encounter in the conclusion. Here, 'panthers' is the major-term (see Figure 1).

That leaves only 'tigers' as the **middle-term** (or "M"-term); this is the term which occurs twice in the premises (see Figure 2).

Since 'panthers' is the major-term, the premise that contains it must be the major premise: "All tigers are panthers." Notice that tigers is the middle-term here. By convention, we usually write the major premise first. (Some traditional reasons for this will be considered shortly.)

Since 'Siberian tigers' is the minor-term (and notice that 'Siberian tigers' are a subset of 'tigers' and not to be confused with each other since there are other sorts of tigers, like "Bengal" tigers) then the premise which contains it is the minor premise: "All Siberian tigers are tigers." The term 'tigers' occurs here as well because it is the middle-term.

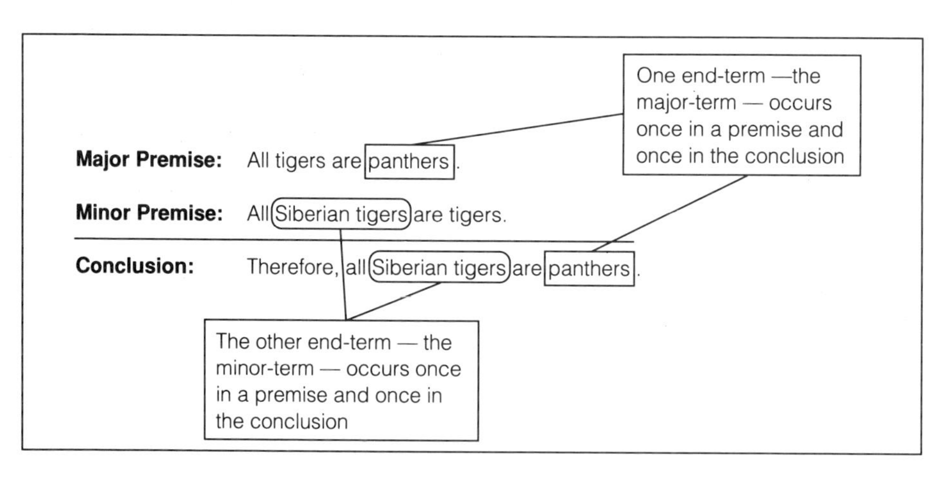

FIGURE 1

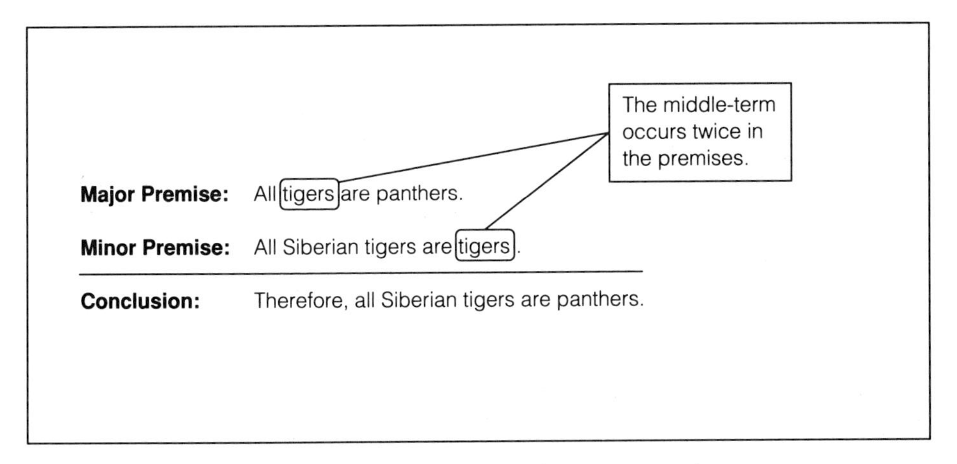

FIGURE 2

Keep these reminders handy:

The middle-term is the term which occurs twice in the premises.
The **end-terms** are the terms which appear in the conclusion.

Each of the premises and the conclusion has an operator (or 'quantifier') which ranges over the quantities expressed by the class terms 'Siberian tigers', 'tigers', and 'panthers'. In each case the quantifier is 'all' so both of the premises and the conclusion are *universal affirmative* statements. This is indeed a syllogism according to Aristotle's (and our) definition of 'syllogism'.

Here is another, less obvious, example of a syllogism. In fact, this is the classical example of a syllogism:

MAJOR PREMISE: **All Greeks are mortal.**
MINOR PREMISE: **Socrates is a Greek.**

CONCLUSION: **Socrates is mortal.**

Now, you may look at this example and wonder how it meets the definition of a syllogism. You can see readily enough that 'Greek' or 'Greeks' must be the middle-term, since it occurs twice in the premises (we will use the plural, 'Greeks'). You can also see that 'Socrates' and 'mortal [things]' must be the end-terms, since they each show up in the conclusion. And the operator 'All' is obvious in the first premise. But, you may ask, where are the operators for the minor premise and for the conclusion?

Perhaps because it anticipates this very question, this longstanding example is a classic. The name 'Socrates' should be treated as: "All things identical with

Socrates." (You should render all proper names in this way.) Thus, both premises and the conclusion have universal affirmative operators.

Here is another example of a syllogism:

Sheboygan is not Schenectady.
Schenectady is not Scranton.
Scranton is not Sheboygan.*

At first, you may think that the operators (if any) must be particular negative, since these are the names of particular cities. However, since you have seen the 'Socrates is mortal' example, you know that each of the proper names take *universal* operators. 'Schenectady' is the middle-term; 'Sheboygan' and 'Scranton' are the end-terms. Thus, in standard syllogistic form, this example should read:

All things identical with Sheboygan are not things identical with Schenectady.
All things identical with Schenectady are not things identical with Scranton.
All things identical with Scranton are not things identical with Sheboygan.

Notice that I've underscored each of the occurrences of the word 'not'. This is important, since their appearances change the status of each of the operators to **universal negative** (or "No"). We will discuss this in more detail in the next section, but even now it is important to remember to pick out negations in a sentence to determine the value of the operator.

Now, let's identify the terms again. We know that 'Schenectady' is the middle-term. Let's designate it as 'M'. Let us designate the term 'Sheboygan' as 'P' (for the Latin word, *primus,* or "first"). Next, we designate 'Scranton' as 'S' (for the Latin, *secundus*). Thus, we can construct the schematic:

All P are not M. = No P are M.
All M are not S. = No M are S.
All S are not P. = No S are P.

As the equal signs suggest, the schematic on the left is **equivalent** to the schematic on the right (equivalent, that is, in preserving "truth values," (i.e., if 'All P are not M' is **true,** then so is 'No P are M, and vice versa'. If 'All P are not M' is **false,** then so is 'No P are M', and vice versa).

Finally, for convenience we designate the logical operators with discrete terms of their own. This is done as shown in the following table:

*Three American cities of about the same population. Sheboygan is in Wisconsin, Schenectady is in New York, and Scranton is in Pennsylvania.

Operation	Logical Operator	Term
Universal Affirmative	All	A
Universal Negative	No	E
Particular Affirmative	Some	I
Particular Negative	Not all	O

Our final schematic, substituting logical term-letters for the logical operators, is as follows:

E. P are M.
E. M are S.

E. S are P.

13.2 MAJOR AND MINOR PREMISES

A word about the Aristotelean tradition of categorical syllogisms: Aristotle identified a **major** and a **minor premise** in every syllogism. The **major premise** was the premise which contained the major-term; the **minor premise** contained the minor-term.

The **major-term** is originally the one which was designated *primus,* or the P-term. It occurs as the *predicate* of the conclusion. The **minor term** is the one originally designated as *secundus,* or the S-term; it occurs as the subject of the conclusion.

Consider this syllogism:

"Dogs are noisy critters, because some dogs howl at the moon and all noisy critters are howlers."

'Dogs are noisy critters' is the conclusion, so 'dogs' is the minor-term (because it is the subject of the conclusion), and 'noisy critters' is the major-term (because it is the predicate of the conclusion). This becomes:

MAJOR PREMISE: **All noisy critters are howlers.**
MINOR PREMISE: **Some dogs are howlers.**

CONCLUSION: **Therefore, all dogs are noisy critters.**

'Howlers' is the middle-term, 'noisy critters' is the P-term, and 'dogs' is the S-term. Thus, the final schematic is:

A. P are M.
I. S are M.

A. S are P.

Using the methods we'll introduce in Unit Two, we will be able to determine that this is an *invalid* argument form.

MODULE 14

A DATE WITH BARBARA*

14.1 THE SCHOLASTIC NOTATION

Recall again the table for logical operators for Aristotelean logic:.

All S are P	A	Universal Affirmative
No S are P	E	Universal Negative
Some S are P	I	Particular Affirmative
Not all S are P (Some S are not P)	O	Particular Negative

TABLE 1

The Scholastic notations **A, E, I,** and **O** provide the basics for a simple mnemonic device for remembering valid syllogistic forms. Notice that each of the notations uses a vowel for a symbol. This is handy because Latin is a lan-

A **deductive argument** is valid, if and only if, if all the premises are true, then the conclusion must be true.

*The scholastic mnemonic device for determining validity of categorical (Aristotelean) syllogisms.

guage rich in vowels (like all Romance languages). The upshot is that we can now use these vowels to select names for each of the valid syllogistic forms.

However, this method has severe limitations which will soon become obvious. First, this is a mnemonic device; consequently, it requires rote memorization. (If your memory is faulty, the drawbacks to this method become at once obvious.) You will undoubtedly come to prefer (as I do) the mechanical rules and Venn diagrams for determining valid syllogistic forms. But note that these mechanical rules and diagrams were not devised, in the latter case, until the middle of the nineteenth century by the logician John Venn. Some of the rules we shall apply were not devised until the twentieth century.

There is another, and more severe, drawback to the Scholastic Method. It requires us to distinguish between *strongly (unconditionally) valid* forms and *weakly (conditionally) valid* forms of syllogisms. The rules and diagrams do not have such liabilities. (We will consider what "strongly" and "weakly" valid forms mean for us in a moment.)

Let's consider a classical example of a strongly valid syllogism:

All Greeks are mortal.
Socrates is a Greek.

Therefore (∴) Socrates is mortal.

Putting this in proper form, let 'Greeks' = M (the middle-term), 'mortal [things]' = P, and 'things identical with Socrates' = S (the two end-terms), and we have:

Example 1

All M are P
All S are M

∴ All S are P*

Whenever we have a syllogism in this form, even when substituting terms for term-letters, the syllogism remains valid. For example, let's substitute 'whales' for M, 'mammals' for P, and 'humpback whales' for S. This yields the equally valid syllogism:

All whales are mammals.
All humpback whales are whales.

∴ All humpback whales are mammals.

*Notice that this is a *universal affirmative,* or an *A*-form statement, designating the class of *all* things identical with Socrates (the class has only *one* member, i.e., Socrates himself. It remains, nonetheless, universal affirmative and not particular).

You may suspect that any consistent substitution of terms for term-letters in Example 1 will yield a strongly valid syllogism, and you are correct. We can now generalize to the rule:

All Syllogisms of the Form

All M are P

All S are M

∴ All S are P
are strongly valid

We can make the relationship even more schematic, and pick out the Scholastic notation, together with the appropriate terms (*in their proper order*—this is important), and we have:

Example 2

A. MP

A. SM

A. SP

Since every properly formed syllogism will have three Scholastic operators, each valid form will have three vowels. Example 2 has the three vowels A, A, and A. So, let's give it a name, one which has all and only three vowels. The Scholastics did this, and they selected the name **BARBARA** for the valid form which appears in example 2. Let's do the same:

A. MP

A. SM

A. SP

"BARBARA"

Now, you'll be delighted to know that there are only a very few (fifteen, to be exact) strongly valid syllogistic forms (and only a few more "weakly" valid forms), so that we can give proper names to each of them. It will come as no surprise to learn that's just what the old monks did. However, they selected Latin names. *'Barbara'* is itself a Latin *and* an English name (etymologically, it means "red beard" or "foreigner"*). Unfortunately, most of the rest of the names they selected do not share this happy coincidence. In fact, names like *'Ferison'* and *'Festino'* sound distinctly alien to our American ears. Nevertheless, here is the complete table of strongly valid syllogistic forms, together with the names given them by the old monks:

*'Barbarosa' was the name given by the Romans to the barbarians from the North. It was a descriptive name. The Romans, being an Italian people, were dark-haired and dark-complected. The Germanic peoples, however, had the distinction of being light-skinned, and, sometimes, redheaded. Thus, the Romans began referring to foreigners derogatorily as "red beards."

Barbara	Celarent	Darii	Ferio
A. MP	E. MP	A. MP	E. MP
A. SM	A. SM	I. SM	I. SM
A. SP	E. SP	I. SP	O. SP
Cesare	**Camestres**	**Festino**	**Baroco**
E. PM	A. PM	E. PM	A. PM
A. SM	E. SM	I. SM	O. SM
E. SP	E. SP	O. SP	O. SP
Disamis	**Datisi**	**Bocardo**	**Ferison**
I. MP	A. MP	O. MP	E. MP
A. MS	I. MS	A. MS	I. MS
I. SP	I. SP	O. SP	O. SP
Camenes	**Dimaris**	**Fresison**	
A. PM	I. PM	E. PM	
E. MS	A. MS	I. MS	
E. SP	I. SP	O. SP	

TABLE 2

Notice that the order of the term-letters is identical for *'Barbara', 'Celarent', 'Darii',* and *'Ferio':*

MP
SM
SP

We designate this particular order as a **figure** or "overall shape." We reserve the name *figure* to designate the arrangements of the terms in a syllogism.

If we look at each of the other rows, we find that the 'P' and the 'M' terms occur each time in the first row, though the order in which they occur may change according to the syllogism. Likewise, the 'S and 'M' terms always appear in the minor premise row. However, by a convention adopted by the old monks, the conclusion is always arranged with the 'S' (or minor-term) first, and the 'P' (or major-term) second. We can take advantage of this observation and designate four figures:

Figure 1	Figure 2	Figure 3	Figure 4
MP	PM	MP	PM
SM	SM	MS	MS
SP	SP	SP	SP

TABLE 3

Now we can group the named syllogisms by figures:

Figure 1	Figure 2	Figure 3	Figure 4
Barbara	Cesare	Disamis	Camenes
Celarent	Camestres	Datisi	Dimaris
Darii	Festino	Bocardo	Fresison
Ferio	Baroco	Ferison	

TABLE 4

These syllogistic schemata will always yield strongly valid syllogisms with appropriate substitution of terms. Take '*Datisi*', for example. '*Datisi*' is Figure 3. Consequently, the proper schema is:

A. MP
I. MS
I. SP

Wherever we have an assignment of the Latin vowels ('A', 'E', 'I', or 'O') in a complete syllogism consisting of a major and a minor premise and a conclusion, we call the result the **mood** of the categorical syllogism. Thus, **AAA, AII,** and **EIO** are all "moods" of various syllogisms. The syllogism is complete when we've also determined its **figure.** Therefore, in the traditional schema, a categorical syllogism is symbolized by both its mood and its figure.

Let's make the following substitutions of terms for term-letters for *Datisi* (mood **AII,** Figure 3). We have:

M = rugby players
P = rugged things
S = ballet dancers

Thus, the result is the valid syllogism:

All rugby players are rugged.
Some rugby players are ballet dancers.
∴ Some ballet dancers are rugged.

14.2 STRONGLY VS. WEAKLY VALID SCHEMATA

Now, let's consider the complicated distinction between strongly (unconditionally) valid and weakly (conditionally) valid syllogistic schemata. Strongly valid syllogisms are always valid, but weakly valid syllogistic schemata are valid if and only if we make certain assumptions about the existence of the entities designated by key terms.

The names that the old monks gave to the weakly valid syllogisms are *'Darapti'*, *'Falepton'*, *'Fesapo'*, and *'Bramantip'*: four names in all, although some of them are repeated for different figures given certain assumptions which we shall consider in a moment.

Consider the weakly valid syllogistic schema *'Darapti'* (see Table 5). *'Darapti'* is a Figure 1, so it is rendered as:

A. MP

A. SM

I. SP

But *'Darapti'* is valid if and only if we make the explicit assumption that the term which 'S' designates *exists*. Let's put this into English and see what this means. Let 'M' = Venus flytraps; let 'P' = carnivorous things and let 'S' = plants from outer space:

All Venus flytraps are carnivorous things.

All plants from outer space are Venus flytraps.

∴ Some plants from outer space are carnivorous.*

This syllogism is valid if and only if we assume that the entity designated by the S-term actually exists (i.e., that there are indeed "plants from outer space").

We can now add the weakly valid syllogistic schemata to our table, listing the existence assumptions necessary to their actual validity. They are:

Figure 1	Figure 2	Figure 3	Figure 4	Existence Assumptions
Darapti	Falepton		Falepton	**'S' exists***
Fesapo	Fesapo			**'S' exists**
		Darapti	Fesapo	**'M' exists[†]**
		Fesapo		**'M' exists**
			Bramantip	**'P' exists[‡]**

*I.e., "there really are S's."
[†]I.e., "there really are M's."
[‡]I.e., "there really are P's."

TABLE 5

Notice that the names *'Darapti'*, *Falepton'*, and *'Fesapo'* are repeated many times, but each time under a different figure (the order of the term-letters reverses), or given a different existence assumption.

*According to one North Carolina biologist, Venus flytraps came to earth from spores from outer space. He contends that since nothing like Venus flytraps exists in any other part of the earth other than southeastern North Carolina, they must have an extraterrestrial origin. (I merely report this; I do not endorse it.)

In modern times, we tend to prefer to spell out these relationships, but how did students of logic learn them in the Middle Ages? Why, they learned a poem, of course (that's how these things were done back then). And, as you might expect, the poem was in Latin. For your further interest, I reproduce the poem below (no, you needn't memorize it).

"Barbara, Celarent, Darii, Ferioque prioris;
Cesare, Camestres, Festino, Baroco secundae;
Tertia, Darapti, Disamis, Datisi, Falepton,
Bocardo, Ferison habet: quarta insuper addit
Bramantip, Camenes, Dimaris, Fesapo, Fresison."

14.3 A NEW MNEMONIC DEVICE

T. H. White opens his novel *The Once and Future King* (the basis for the play and motion picture "Camelot") by recounting young King Arthur's education: "On Mondays, Wednesdays and Fridays it was Court Hand and Summulae Logicales, while the rest of the week it was the Organon, Repetition and Astrology."

Arthur draws on his early education throughout his reign as king to aid him in his unprecedented tasks of the unification of England and the imposition of Rule by Law. When Merlyn instructs the king and Sir Kay of the historical causes for an impending war, Kay and Arthur have this exchange:

> Kay said definitely: "I can't stand any more history. After all, we are supposed to be grown up. If we go on, we shall be doing dictation."
> Arthur grinned and began in the well-remembered sing-song voice: Barbara, Celarent, Darii, Ferioque prioris, while Kay sang the next four lines with him antiphonetically.

Here, Arthur mocks his early rote learning, though T. H. White emphasizes the importance of the future king's early education by devoting a large part of the novel to Arthur's training. Anyone well-educated in Europe in the Middle Ages had to learn Latin, so the Latin names would not have been as unfamiliar to Arthur as they may well be to the modern student.

Can we breathe life into the old lyrics? Let's consider once more the various moods and figures of the strongly valid syllogistic forms. We have a total of fifteen:

	Moods			
Figure 1				
MP	AAA	EAE	AII	EIO
SM				
SP				

Figure 2				
PM	EAE	AEE	EIO	AOO
SM				
SP				
Figure 3				
MP	IAI	AII	OAO	EIO
MS				
SP				
Figure 4				
PM	AEE	IAI	EIO	
MS				
SP				

TABLE 6

Now, let's update the Scholastic's old memory aid with more familiar names for each of the strongly valid syllogistic forms. I've chosen geographical names:

Figure 1	Figure 2	Figure 3	Figure 4
Canada	Seattle	Britain	Valverde
New Haven	Manchester	Pacific	Miami
Tahiti	Leighton	Monaco	Levittown
Mexico	Rangoon	Wellington	

TABLE 7

Memorizing this short list of place-names ought to be quite easy for you compared to the horrors of the Latinized '*Darii*' or '*Camestres*'. Nevertheless, such memorization is no longer necessary.*

This list of geographical names may well represent the first time the old list has been updated in as much as a thousand years. Surprised? The reason this revision has been neglected may be in part due to the fact that easier and more graphic tests of the validity of syllogistic arguments have been devised—in particular "Venn" diagrams. We will look at this method shortly.

*Nevertheless, all of the later developments in logic are built on the foundations constructed by the old monks. If you were to take other courses in logic, even symbolic logic, you'd encounter terms like '*Barbara*' and 'middle-term', and wonder where they came from. You no longer need to rely on rote memory, as the medievals did, to do such work in logic, but the terminology and the traditions have outlived the technique.

EXERCISES

Exercise 1

DIRECTIONS: *Take each of the* **strongly** *valid syllogistic schemata, substitute English terms for term-letters, and write out the syllogisms in English. Do you see why they're valid? (Recall, such an argument is valid on the condition that if all the premises are true, then the conclusion must be true.)*

Exercise 2

DIRECTIONS: *Take each of the* **weakly** *valid syllogistic schemata, substitute English terms for term-letters, supply the necessary existence assumptions, and write out the syllogisms in English. Do you see that given the existence assumptions they are also valid?*

Exercise 3

DIRECTIONS: *Take each of the* **strongly** *valid syllogistic schemata, remove the vowels, and assign* **English** *names for each of the vowel groups (other than the geographical names I devised in this section). You may use proper names or you may choose names of things instead.*

Exercise 4

DIRECTIONS: *Complete the project in Exercise 3 by adding new English names for the* weakly *valid schemata (e.g.,* 'Falepton' *and* 'Bramantip'*).*

MODULE 15

THE ARISTOTELEAN SQUARE OF OPPOSITION

15.1 THE SQUARE

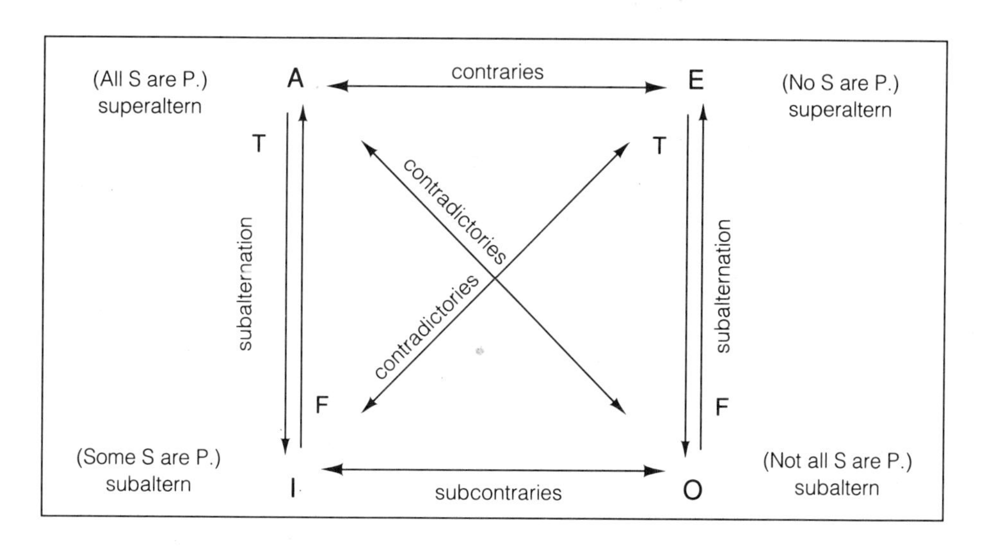

When two statements are incompatible, and their denials are incompatible, they are contradictories of each other:

A deductive argument is valid, if and only if, if **all** the premises are true, then the conclusion must be true.

"All chameleons are lizards" vs. "Not all chameleons are lizards."

If the first is *true,* then the second must be *false;* if the second is *true,* then the first must be *false.*

When two statements are logically incompatible in such a way that they cannot both be true, though both can be false, they are contraries of each other:

"The pelican is brown" vs. "The pelican is white."

Both cannot be true, but both can be false (i.e., the pelican might have been tie-dyed purple).

When two statements are logically incompatible in such a way that they cannot both be false, though both can be true, they are sub-contraries of each other:

"Some baseball players went to Yale" vs. "Some baseball players never went to Yale."

Both cannot be false, but both can be true: consider Ron Darling of the N.Y. Mets, a Yale graduate, and Casey Stengel, *not* a Yale graduate.

EXERCISES

DIRECTIONS: *Say whether the following pairs of statements are contradictories, contraries, subcontraries or none of these.*

1. Men are mortal.
   Women are mortal.
2. The moon is blue.
   The moon is harvest gold.
3. All earthquakes are dangerous.
   Not all earthquakes are dangerous.
4. A few comedians are funny.
   Not all are.
5. All raisins come from California.
   No raisins come from California.
6. Some coal is anthracite.
   No coal is anthracite.

MODULE 16

GUIDELINES FOR DETERMINING EQUIVALENCE IN CATEGORICAL SYLLOGISMS

16.1 EQUIVALENT STATEMENT FORMS

The main difficulty you may face in translating ordinary English statements into standard categorical form will probably be with the occurrence of negated terms in a sentence. For example, the British logician Bertrand Russell tells a story about a remark his cleaning lady once made: "I ain't never done no harm to no one."*

Obviously, we're not meant to take *all* the negations in this sentence literally (what's why it's amusing—because the speaker is an unpolished user of the language). However, sometimes very skilled users of the language will implant a group of negations and negated terms in a sentence in order to obscure intelligibility (have you ever read the fine print in an insurance contract?) In such cases, we either rely on professionals to interpret the language for us, or we must interpret the obscure language ourselves.

Not only insurance providers make use of negations in this seemingly obstructive way. So may accountants, lawyers, and diplomats, each for a variety of possible reasons, perhaps to keep their contracts and agreements intelligible only to other skilled professionals or to tone down the apparent harshness of what

---

A deductive argument is valid, if and only if, if **all** the premises are true, then the conclusion must be true.

*Bertrand Russell, *My Philosophical Development.* I am indebted to Ken Warmbrod of the University of Manitoba for this example.

they are saying. However, even reading IRS instructions meant for the general public, we may encounter sentences with such negations and negated terms. Thus, it's wise to pinpoint what was really meant underneath the obscure language.

We do this by learning the **converses, obverses,** and **contrapositives** of statements. Here is a list of them:

Conversion

Statement		*Converse*
E: No S are P.	::*	No P are S.
I: Some S are P.	::	Some P are S.

Obversion

Statement		*Obverse*
A: All S are P.	::	No S are non-P.
E: No S are P.	::	All S are non-P.
I: Some S are P.	::	Some S are not non-P.
O: Not all S are P.	::	Some S are non-P.

Contraposition

Statement		*Contrapositive*
A: All S are P.	::	All non-P are non-S.
O: Not all S are P.	::	Some non-P are non-S.

The one caution I impress on you is to distinguish between *negations* (e.g., the occurrence of the word 'not' or the contraction *n't* [as in *don't* or *can't*] in a sentence, and the occurrence of *negated terms* ('non') in a sentence).

For example, "I don't think you're right" contains a *negation* (the *n't* in *don't*). We may translate this sentence as "It is not the case that I think you're right." The fragment "I think you're right" is obviously affirmative, but in logic we use the standard expression, 'it is not the case that . . .' to change the value of the sentence from affirmative (in this case) to negative.

Sometimes we double negations in a sentence for emphasis. Suppose we replied, "Well, it's not the case that I don't think *you're* right!" Here we mean nothing more than *you're right!*—an affirmative assertion. Such double negations are sometimes made inadvertently by unskilled speakers of the language (such as Russell's cleaning lady). Other times we use them for emphasis (as in this paragraph); then we intend the double negation to be an ironic affirmation.

*The double colon '::' means that the expression on the left can be replaced by the expression on the right without loss of truth value.

To sort this out, apply the following "rule of thumb:"

A double negation is an affirmative: count the number of negations in the assertion. If the result is an odd number, the assertion is negative; otherwise it is affirmative.*

"I haven't got nothing" really means that I have *something:* 'haven't' = "have *not*" and 'nothing' = "*no* thing." The total number of negations in the assertion = 2, an even number; therefore, the assertion is affirmative. Compare this to mathematics, where the product of two negative numbers is always positive: $-2 \times -2 = +4$.

However, the assertion 'All worms are non-snakes' contains a *negated term* ('snakes' is negated by 'non'). Be careful of these. 'Non-snakes' may well be a negated term, but the value of the whole assertion is *affirmative.* We identify it as a universal affirmative statement.

When you want to determine the equivalent expression in Aristotelean logic, you may consult the table for converses, obverses, and contrapositives, OR you may do *most* of what you need to find out an *easier* way. The following is a shortcut that works most of the time. You may wish to adopt these suggestions as rules of thumb when interpreting statements of quantity that contain a confusing number of negations.

16.2 A SHORTCUT

1. Identify the discrete terms and assign appropriate term-letters to them.

Example: "Some of the signatories of the treaty are not non-aggressors."

Let 'S' = "signatories of the treaty"
Let 'P' = "aggressors"

The sentence converts to:

"Some S are not non-P."

Be careful to identify and preserve the negations in the sentence.

*Note well: this is a rule of thumb only. When the rule works, it works only with complete assertions, not with sentences or statements. Negations embedded in *'if . . . then'* assertions seem to violate the rule, as in "If Tom doesn't come to the party, then he won't sing for us." Notice however that there are two statements here, but only one *assertion.*

2. Ask yourself if the expression is either **universal** or **particular.**

EXAMPLE: **"Some S are not non-P" is particular.**

You can tell because 'Some' is a particular-type operator. So is 'Not all' (or 'Some are not'), whereas 'All' and 'No' are universal operators.

3. Ask yourself if the expression is either *affirmative* or *negative.*

EXAMPLE: **"Some S are not non-P" is negative.**

This looks like a double negation, but be careful; it isn't. We have one negation ('not') and one *negated term* ('non'). Thus, following our rule of thumb, the presence of one negation makes this an odd number of negations, so the statement is negative.

4. Apply the following table:

Universal Affirmative	A.	All S are P.
Universal Negative	E.	No S are P.
Particular Affirmative	I.	Some S are P.
Particular Negative	O.	Not all S are P. (Some S are not P.)

We now know that our sentence, 'Some of the signatories of the treaty are not non-aggressors', is *particular negative,* or an O-sentence. Therefore, it should be translated as 'Not all S are non-P', or the obverse, 'Some signatories of the treaty are non-aggressors'.

NOTE: The only equivalences this procedure will not work for are the two CONVERSION rules because these turn around the relationship of the term-letters.

No S are P. ≡ No P are S.

Some S are P. ≡ Some P are S.

EXERCISES

DIRECTIONS: *Find the equivalent expression and say whether it is A (universal affirmative), E (universal negative), I (particular affirmative), or O (particular negative). (Note that some of the examples do not translate well using the shortcut just described. Identify these and translate them according to the rules of equivalence listed in the table at the beginning of this module.)*

1. No seahawks are not tarheels.

2. All telephone companies are not the same.
3. Some of the patrons of Djangos are not non-dancers.
4. All unjust men are nonchalant.
5. Some soft drinks are uncolas.
6. All non-aggressors are non-combatants.
7. Some non-partisan supporters of the president are not team players.
8. It is not the case that all dolphins are playful.
9. Some dolphins are not even-tempered.
10. It is not the case that all husbands say "I'm not here" at "He's Not Here," a tavern in Chapel Hill, when their wives call.

MODULE 17

ALL BANKS ARE NOT THE SAME

17.1 MISUSING OPERATORS: A PET PEEVE

In a television commercial for Manufacturer's Hanover Bank in New York City, comedian Tim Conway portrays a bumbler who comes into the bank with a knapsack full of pennies. The inevitable happens, and Conway spills all of his pennies onto the bank lobby floor. Immediately, the security guard pitches in and helps pick up the pennies. Tim Conway is impressed with the friendly service the bank offers. Then, an announcer with a rich baritone voice proclaims this slogan: "All banks are not the same."

Put in its best light, the statement is ambiguous; in its worst light, it's false and misleading. This is so because the statement is badly constructed both on grammatical *and* logical grounds.

Let's consider how grammar can sometimes affect logical structure. A good example of how a term's placement in a sentence can change its meaning is illustrated with the word 'only'. Notice the difference in meaning between these two sentences:

1. "He gave his car *only* to his son."
2. "He *only* gave his car to his son."

Do you see the difference? In the first statement, permission to use the family station wagon was granted to the son and *only to the son,* so that when

---

A deductive argument is valid, if and only if, if **all** the premises are true, then the conclusion must be true.

Sonny is indicted for reckless driving in a sixteen-car pileup, we have little doubt that Sonny's the culprit, for the car was his responsibility alone.

On the other hand, the second statement should be restricted to other contexts. In defense of the boy's father and his decision to loan his reckless son the car, an observer might placate the victims and relatives of the victims in the accident by saying, "He did nothing that you wouldn't have done. He *only* gave his car to his son."

We see then that the placement of certain terms in a sentence—terms like 'only' or 'all'—can materially affect the meaning of the statement. We face this problem when we hear the Manufacturer's Hanover ad. Consider the difference between:

1. "All banks are *not* the same."

   and

2. "*Not* all banks are the same."

In everyday usage, statement 1 may be ambiguous. Surely what Manufacturer's Hanover means to say is "*Not* all banks are the same." As commonly used, that interpretation is a possibility—and surely that's the message that the advertisers really want to get across. But statement 1 allows another interpretation, one that is false and misleading: "All banks differ."

This is presumably *not* the message that the advertisers want to convey to their public. Logic strives for precision. This means that logic should try to legislate against ambiguity. Think of the equivalent of 'All banks are not the same' as the universal negative statement, 'No banks are the same' (whatever the common usage on Madison Avenue) and that may be interpreted to mean *that every bank is different.* But this gives us no particular reason to open an account with Manufacturer's Hanover because the slogan doesn't tell us that there's anything particularly special about them. Consider the diagram in Figure 1.

Consider this discussion of the word 'only' by a former editor of the *New York Times:*

"ONLY"

Normally, the proper positioning of *only* requires no more than asking yourself, "What does it actually modify?" Thus a headline that says, "$35,000 Bond Thief Only Nets Paper," does not conform to the normal order; the *only* patently modifies "paper" not "nets," and so should adjoin it. An interesting exercise for developing *only* awareness was cited in the publication *Word Study,* distributed by G. & C. Merriam Company, as follows: "Eight different meanings result from placing *only* in the eight possible positions in this sentence: 'I hit him in the eye yesterday.'" Try it. —THEODORE M. BERNSTEIN, *The Careful Writer*

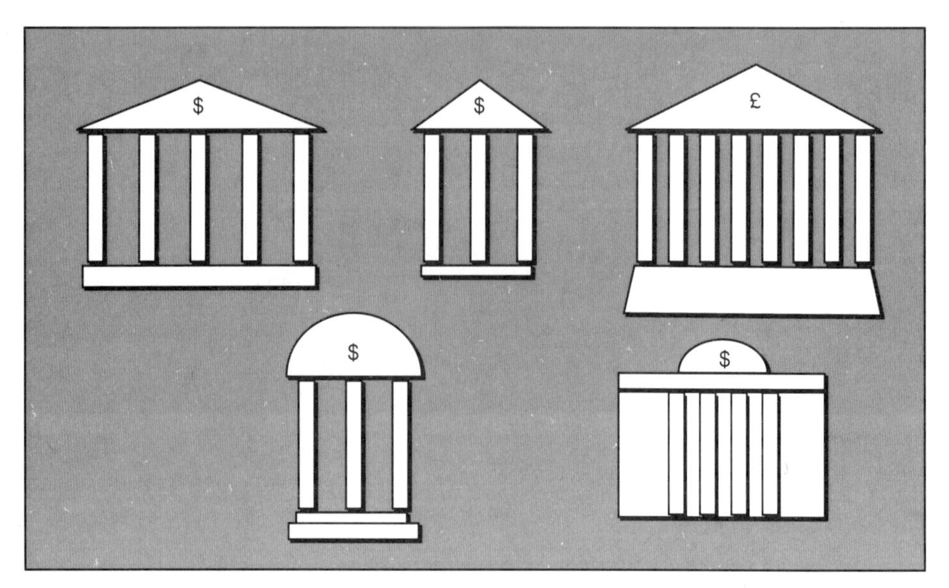

FIGURE 1
"All banks are not the same."

Figure 1 shows us the logical meaning of 'All banks are not the same'. As you can see from the diagram, no one bank has anything particularly special to offer. We are left without a good reason to open an account with any one of them.

Now, consider Figure 2. This is a diagram which illustrates the meaning of the particular negative statement, 'Not all banks are the same'. The suggestion

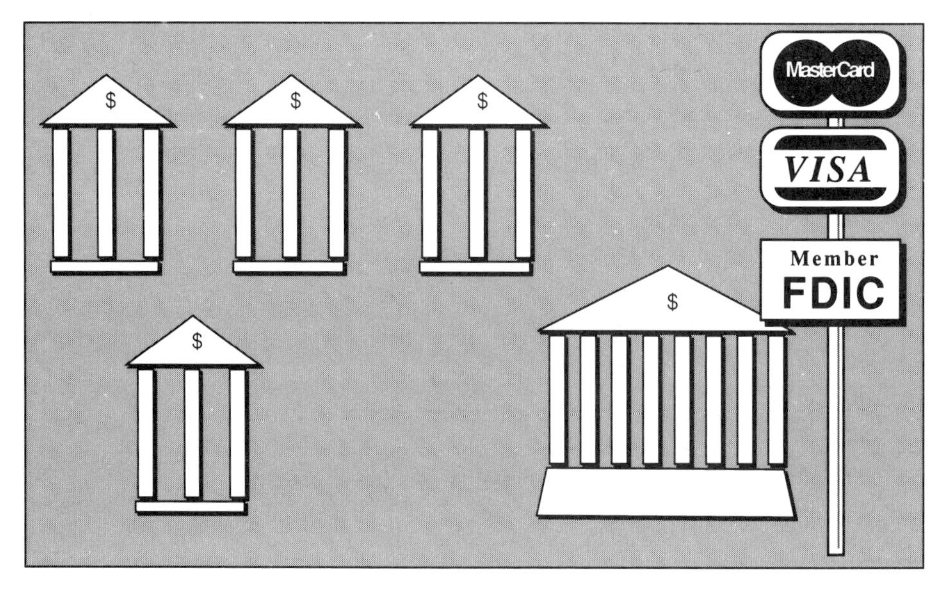

FIGURE 2
"Not all banks are the same."

here is that one (or possibly more) bank(s) is (are) different, that a bank offers special services, and that this is a good reason to open an account with them.

This diagram shows us the message that Manufacturer's Hanover surely wants to convey.

Notice, 'All banks are not the same' is logically equivalent to 'No banks are the same', whereas the separate statement, 'Not all banks are the same', is equivalent to neither of them. The relationship can be spelled out:

"All banks are not the same." ::* E. "No banks are the same."

However, it is not the case that 'Not all banks are the same' is equivalent to either 'All banks are not the same' or 'No banks are the same'. That is because 'Not all banks are the same' is an O-statement (particular negative) and the others are E-statements (universal negative).†

17.2 MISUSING ARISTOTELEAN OPERATORS

Editor: I would like to express my opinion on the issue of the drinking age being raised to 21. I don't believe that some people realize that the young people are mostly employed by places which sell beer. We need these jobs to live—to pay our way through college and to support our families.

All teens don't live to get drunk. Some of us are responsible adults, and we need everyone's help to fight this new highway bill. Letter to the Editor, *Wilmington Morning Star*

The letter writer undercuts an otherwise persuasive argument by misusing the operator 'all' in the sentence 'All teens don't live to get drunk'. The very next sentence makes clear that the writer means "Not all teens live to get drunk." Expressed this way, the meaning is clear and unambiguous. It also puts more polish on the argument.

Logical and grammatical conventions demand that, given a choice between an ambiguous and a precise formulation of a statement, we should use the more precise statement. Treat all sentences of the form 'All X are not-Y' as expressing universal negation (equivalent to 'No X are Y'). When one employs the operator in the way it's used in the letter above, we should consider the usage misleading and incorrect.

*Once more, the double colon '::' means that the one expression can be replaced by the other without loss of truth value.

†For a detailed and technical discussion of the ambiguities here, see Module 32 "Advanced Quantificational Logic."

EXERCISES

Exercise 1

DIRECTIONS: *Select the term(s) which act(s) as Aristotelean operator(s) in the sentences below. Consider whether the following statements make* **ambiguous** *or* **unambiguous** *use of the Aristotelean operators. Then read the story which follows and place the statements in context. Where ambiguous, give an unambiguous interpretation using the appropriate operators carefully.*

1. Some backpackers didn't have blisters on their feet.
2. No long distance calls from this phone, credit cards only.
3. Only the brave deserve the fair.
4. Not all chow chows are as unruly as mine.
5. The heroine got only her due.
6. The heroine only got her due.
7. All people named 'Gwen' are not alike.

Exercise 2

DIRECTIONS: *Read the story that follows. How would you re-tell the story to eliminate the ambiguity? When is Gwen unjustifiably confused (because she doesn't know the conventions of logic and grammar)? When is she justifiably confused (even when the logic and grammar are flawless)? Pay special attention to the placement of the logical operators in the sentences.*

Packer's Progress

Gwen was frantic to escape her screaming dogs and barking infants. She howled, "Enough!" Just then she noticed a notice outside her Logic class:

> "Weekend backpacking:
> *Gwen's need only apply!"*

Gwen was puzzled. Did the sign mean that since she was named 'Gwen', she had no worries about getting accepted (perhaps because backpacking Gwen's were in such short supply), or did the sign mean that applications would be accepted from people named 'Gwen' alone? Gwen never found out but her application was accepted anyway, so she went backpacking the very next weekend.

After walking ten miles with the other twenty or so grubby Gwen's, they stopped for a rest. Everyone took off smoldering hiking boots. Gwen noted that some backpackers didn't have blisters on their feet. Later, while climbing a mountain and eating a box lunch, Gwen decided she should call her small ones. Luckily, on the side of an oak tree, she saw a pay phone. But when she walked up to the phone, she spotted a small sign:

No long distance calls from this phone.
Credit cards only.

This confused Gwen but she never made the call because her attention was diverted by a black dog chewing a hole in her foot. She yelped and was rescued by a red-faced man who said, "Not all chow chows are as unruly as mine."

Gwen smiled faintly. "Is this your dog?" she asked.

"That's no chow chow," he replied. "It's a labrador retriever." Gwen smiled her most enchanting smile. The man grinned.

"By gosh, this just goes to show that all Gwen's are not alike."

Gwen blinked and shook her head. "Does that mean I'm special?" she wondered.

Gwen was entranced by the burly man but her rapture was interrupted by a sharp pain in her foot. "Dog!" the man yelled. "Leave her alone." The retriever hopped away and the man introduced himself as Gregory.

"That was very brave, sir," she said.

"And only the brave deserve the fair," he answered.

"But do the brave *only* deserve the fair," she asked, "for I am no beauty?"

Gregory smirked. "You're right," he chortled. Gwen smacked him on the face, then whispered: "The heroine got only her due."

He smacked her back. "The heroine only got her due," said Gregory. Gwen picked up her pack, continued her hike and wondered the whole time if she and Gregory had begun a budding romance or a small-scale skirmish.

MODULE 18

THREE RULES FOR DETECTING VALID SYLLOGISMS

18.1 CHECKING SYLLOGISMS FOR VALIDITY

A very simple method exists for determining if a syllogism is valid—the method of the **Three Rules**. A syllogism is valid if and only if:

1. **The middle-term is distributed exactly once.**
2. **No end-term is distributed exactly once.**
3. **The number of negative premises must equal the number of negative conclusions.***

Apply each of these rules to a syllogism: if the syllogism meets all three rules, it is strongly valid; if it violates even only one of them, it is invalid.

In order to properly apply these rules, you need to know if the terms are *distributed* or *undistributed.* Consult Table 1 in 18.2, "Distribution Patterns," for further explanation of these concepts.

Notice: to say that a syllogism is valid *if and only if* it meets these three rules means it is valid just in the case where it meets *all three* of these conditions; if it

---

A deductive argument is **valid,** if and only if, if all the premises are true, then the conclusion must be true.

*From *Logic* by Wesley Salmon, Prentice-Hall, 2nd ed., p. 53.

fails on *any one,* it is invalid. In other words, if a syllogism meets rules 1 + 2 + 3, it is valid; but if a syllogism violates *any* of 1 or 2 or 3 (or any combination), then it is *invalid.*

However, before we can work any examples, we need to know the distribution rules. Notice that we don't have to understand the concept of distribution in order to apply the rules (I explain what distribution means in the next module for those who wish to understand the concept better). All we need to do is apply the rules and we'll be able to say without fail whether or not any syllogism is valid.

Examine and memorize the rules, then we'll work some examples to demonstrate them. First, however, we need to learn the table for distribution patterns.

18.2 DISTRIBUTION PATTERNS

Let 'UA' = Universal Affirmative
Let 'PA' = Particular Affirmative
Let 'UN' = Universal Negative
Let 'PN' = Particular Negative

UA Statement	subject distributed	predicate undistributed
UN Statement	subject distributed	predicate distributed
PA Statement	subject undistributed	predicate undistributed
PN Statement	subject undistributed	predicate distributed

TABLE 1*

Or in schematic form:

Mood	Subject	Predicate
UA	D	U
UN	D	D
PA	U	U
PN	U	D

TABLE 2

One of my students suggested a mnemonic device for remembering these distribution patterns. You may find such a memory aid helpful so I'm reproducing one example here:

*See Module 19.1 for an explanation of terms.

"**All** have DUties," (UA)
"**No,** Don't Deny it!" (UN)
"**Some** of Us Understand it," (PA)
"**Not all** of Us Do!" (PN)

Notice that the operators are in boldface and the distribution patterns (according to subject and predicate) are underscored. Thus, we have the pattern:

All	D	U
No	D	D
Some	U	U
Not all	U	D

TABLE 3

18.3 DETERMINING VALIDITY WITH THE THREE RULES METHOD

To work a solution to any given syllogism, we need to apply and re-apply a two-step procedure to each of the premises and the conclusion.

First, we must determine the subject and predicate of each of the premises and the conclusion, then assign the *distribution pattern* to each according to its mood and Table 2.

Second, we then scan the "checkoff list" of three conditions (1 through 3) and see if the syllogism meets each of the rules. If it does, it's *valid;* but if it fails on any one of them, it's *invalid.*

To assign the *distribution pattern,* let's consider an example. The following statement doesn't yield to any obvious standard subject-predicate form, but a moment's reflection will bring us the desired result:

PREMISE: **"All God's critters have a place in the choir."**

What's the subject here, and what's the predicate? The syllogistic subject/predicate statement is always of the form:

[Operator] *subject . . . predicate.*

Clearly, the operator is the Universal Affirmative 'All'. The subject is also easy enough to detect: 'God's critters.' However, the predicate is not in standard form (i.e., of the form: *is* or *are* [*a certain thing*]). To obtain the standard form, we need to introduce the copula (i.e., a word such as 'is' or 'are') into the second term. The second term is '[*things that*] have a place in the choir'. Introducing the copula, we have: *are things that have a place in the choir,* and that's our predicate.

Now, let's take a whole categorical argument and convert it into standard syllogistic form so we can apply the Three Rules:

"All God's critters have a place in the choir, but some things that have a place in the choir do not have hooves, so not all God's critters have hooves."

Converting the categorical argument into standard form, we have:

All God's critters are things that have a place in the choir.
Not all things that have a place in the choir are things that have hooves.

Therefore, Not all God's critters are things that have hooves.

Notice that we converted the second premise to standard syllogistic form, first, by recognizing the negation ('not') and converting the particular negative operator to the standard 'Not all', and, second, by transforming the second term into the standard predicate form, 'things that have hooves'.

Now let's apply our two guidelines.

First, determine the subject and predicate of each of the premises and the conclusion, then assign the *distribution pattern* to each according to its mood and Table 2.

Here is the chart from Table 2 once more:

Mood	Subject	Predicate
UA	D	U
UN	D	D
PA	U	U
PN	U	D

TABLE 4

For example, look at the first premise: The operator is universal affirmative. The subject term is 'God's critters' and from the table we see it's *distributed* ("D"). The predicate term in the first premise is *'are things that have a place in the choir'* and applying the table we find it is *undistributed* ("U").

Here is the completed distribution pattern:

	D	U
[All]	*God's critters*	*are things that have a place in the choir.*
	U	D
[Not all]	*things that have a place in the choir*	*are things that have hooves.*
	U	D
Therefore, [Not all]	*God's critters*	*are things that have hooves.*

Thus, the distribution pattern for *this* argument is:

UA **D, U**
PN **U, D**

PN **U, D**

Now we can follow the second guideline (applying the Three Rules):

Second, scan the "checkoff list" of three conditions (1 through 3) and see if the syllogism meets each of the rules. If it does, it's *valid,* but if it fails on any one of them, it's *invalid.*

Let's put our example (*All God's critters . . .*) to the test.

Rule 1 says that "the middle-term must be distributed exactly once." Here the middle-term is: *'things that have a place in the choir'*. We see that the middle-term is undistributed (**U**) in the major premise and undistributed (**U**) in the minor premise as well. Since the result is that the middle-term *is not distributed at all,* this syllogism violates Rule 1. Therefore, the syllogism is *invalid.*

We could stop here, for we need only one violation of a rule to determine invalidity. However, for the sake of illustration we will continue working through the other two rules.

Rule 2 says that "no-end-term can be distributed exactly once." This rule requires two steps: we must check one end-term, and then the other. Let's start with the S-term (the minor-term): *'things that have hooves'*. We see that the S-term is distributed (**D**) in the minor premise and distributed (**D**) in the conclusion as well. That means that the S-term is distributed *twice.* That accords with the rule which prohibits end-terms from being distributed only once. However, we must now proceed with the second end-term.

The P-term (the major-term) is 'God's critters'. We see that 'God's critters' is distributed (**D**) in the major premise and undistributed (**U**) in the conclusion—that means it is distributed *once* overall in the syllogism, so rule 2 is likewise violated and the syllogism is also invalid for this reason.

Rule 3 says that "the number of negative premises must equal the number of negative conclusions." This is the easiest rule to check. Our example is

UA
PN
PN

The first premise is affirmative, the second is negative, and the conclusion is negative. Since the rule says that the number of negative premises must equal the number of negative conclusions, we count the negatives. We have one negative premise, and we have one negative conclusion. Since 1 = 1, the syllogism meets rule 3.

Even though the syllogism accords with rule 3, notice that it failed to accord with the other two other rules so the syllogism overall is *invalid.* (We emphasize once more that in order to be valid, the syllogism must meet all three rules without fail.)

EXERCISES

DIRECTIONS: *Determine whether the following syllogisms are valid or invalid by means of the Three Rules test.*

1. All surfers are sharkbait.
Some sharkbait are seals.
∴ All surfers are seals.

2. Some backpackers are bearbait.
Not all bearbait are berries.
∴ Not all backpackers are berries.

3. Some software is copy-protected.
No science is copy-protected.
∴ Software is no science.

4. Some meanders are frets.
All frets are ornamental borders.
∴ All meanders are ornamental borders.

5. All accents are mellifluous.
Some mellifluous sounds are heard in Brooklyn.
∴ All accents are heard in Brooklyn.

6. All droves are herds.
Not all herds are packs.
∴ Not all droves are packs.

7. Some old movies are colorized.
No colorized movies are any good.
∴ Some old movies are not any good.

8. All scorpions are arachnids.
Some arachnids have venomous stings.
∴ Not all scorpions have venomous stings.

9. No shrewdness is a skulk.
Some skulks are sounders.
∴ Not all shrewdness are sounders.

10. All wisps are snipes.
Some bevies are snipes.
∴ Some wisps are bevies.

11. A kindle is a brood.
Some broods are dogs.
∴ No dogs are kindles.

12. All gams are schools.
Some gams are shoals.
∴ Some schools are shoals.

13. No swarms are coveys.
Some coveys are bevies.
∴ Not all bevies are swarms.

14. All drifts are droves.
Some drifts are hogs.
∴ Some droves are hogs.

15. All falls are coveys.
No coveys are nides.
∴ No nides are falls.

MODULE 19

DISTRIBUTION IN CATEGORICAL SYLLOGISMS

We say that a term in a categorical syllogism is **distributed** if a statement makes the claim about *every* entity to which the term applies.

Recall that every categorical statement has exactly *two* terms (and every properly formed syllogism has exactly *three* terms).

When you are considering why a term is or is not distributed, keep this rule in mind:

> **In a universal categorical statement, the *subject* is distributed; in a negative categorical statement, the *predicate* is distributed.**

19.1 UNIVERSAL AFFIRMATIVE STATEMENTS

Consider: for universal affirmative (A) statements, such as 'All labrador retrievers are excellent swimmers' ('All A are B'). In this example, 'labrador retrievers' is the **subject term,** and 'are excellent swimmers' is the **predicate.** By this statement, the *claim* is made that the predicate 'is an excellent swimmer' applies to each and every member of the class of labrador retrievers; the subject is **distributed.**

A deductive argument is valid, if and only if, if **all** the premises are true, then the conclusion must be true.

Even if such a claim turns out to be false, as in 'All labrador retrievers are cats' the subject is *still* distributed. What is important for our purposes is that the *claim* is made.

On the other hand, it doesn't follow that every member of the class of excellent swimmers is a member of the class of labrador retrievers. The statement is *silent* on the question of whether some, any, or all excellent swimmers are labrador retrievers. In fact, many excellent swimmers (e.g., Mark Spitz, an Olympic champion) are not labrador retrievers. So, the predicate is **undistributed.**

We may represent this relationship in the form of a diagram (called "Euler Circles" after the eighteenth century logician Leonhard Euler). Think of the following circles as containers (or "corrals"). The outer circle represents the enclosure of 'B'-things; the inner, the enclosure of 'A'-things. Whether or not there are actually any B-things or A-things, we can still stipulate that, if there were, all A-things would be enclosed in the corral of B-things. Thus, we have:

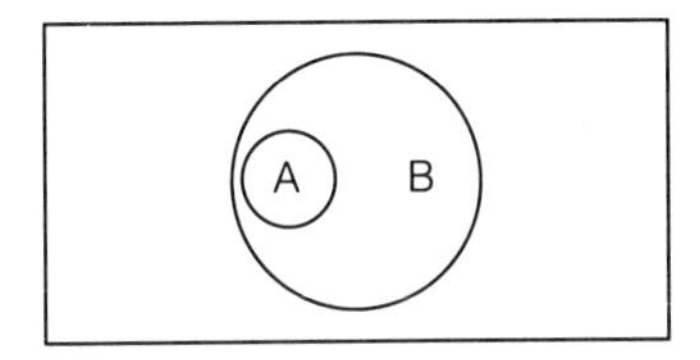

"All A are B."

We see that every member of the class 'A' is a member of the class 'B', but it is not the case that every member of the class of B-things is distributed in the A-corral. There could be B-things outside the A-enclosure.

19.2 UNIVERSAL NEGATIVE STATEMENTS

Let's consider the example that 'No mutts are pedigrees'. As you can readily see, both the subject and the predicate terms are **distributed.** This sentence makes a statement about every mutt and about every pedigree.

Consider the following diagram:

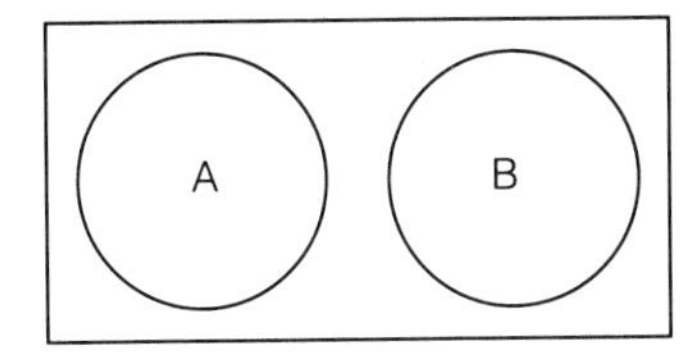

"No A are B."

What this diagram shows is that every A-thing is separate from every B-thing, and that every B-thing is separate from every A-thing. It is in this sense that both terms are distributed. Consider: every mutt is separate from every pedigree, and every pedigree from every mutt.

19.3 PARTICULAR AFFIRMATIVE STATEMENTS

Consider the sentence 'Some rottweilers* are rabid'. We see that both terms are undistributed. Certainly, we are not making a claim about all rottweilers. Neither are we committed to claiming anything about all rabid creatures. Let's represent this relationship with Euler Circles.

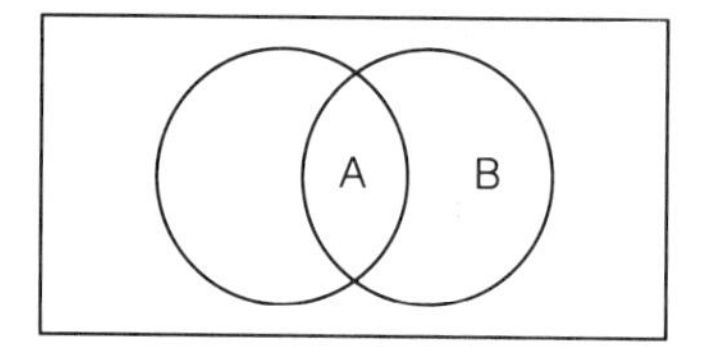

"Some A are B."

We see that the statement 'Some A are B' says something definite about the existence of an A-thing, and includes the A-thing in the B-corral. This means that at least one A-thing is a B-thing. Note, however, that while we say something about *some* A-things, we are not talking about *all* A-things. Correlatively, we say something about at least one B-thing (the ones that are also A-things), but not about all B-things. So neither term is distributed.

To return to our example, we see that while we are talking about at least one rottweiler, we make no claim about the whole breed. Similarly, while we are talking about at least one rabid creature (the one or ones that are also rottweilers), we make no claims about every rabid creature. Fortunately, whatever the whereabouts of the other rabid creatures, we have kept the rabid rottweilers securely enclosed in our corral.

19.4 PARTICULAR NEGATIVE STATEMENTS

Let's take the statement 'Not all dogs are named after Woodrow Wilson'. This statement is equivalent in its truth-value to the statement 'Some dogs are *not* named after Woodrow Wilson'. However, even though the form 'Some A are

*A rottweiler is a large breed of dog with a stocky body, short hair, and a large head. The dog is an old German breed, and has a chancy disposition. Doberman pinschers are downbred from rottweilers.

not B' is usually the preferred expression for particular negative statements, we will employ the operator 'Not all' as standard.

Notice that while we have made a claim about only one or a few dogs (that they are named after Woodrow Wilson), we have said something about every other creature named after Woodrow Wilson (there could be possums, ferrets, and even people so named): the entire class of B-things is separated from the one or more A-things that are not so-named. Thus, the term 'dogs' here is **undistributed,** while the term 'things named after Woodrow Wilson' is **distributed.**

Puzzled? It may help to consult a diagram. Consider the "Euler" representation of 'Not all A are B':

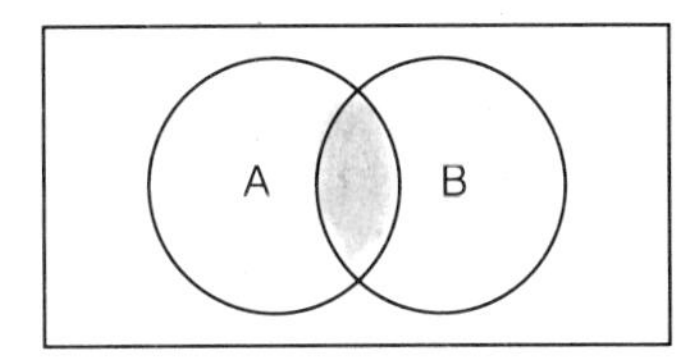

"Not all A are B."

Notice how this differs from the diagram for 'Some A are B'. Here, the A-thing is well outside the B-corral. Thus, we can see that when we say, "Not all A are B," we are talking about one or a few A-things. Yet, remember the diagram for universal negative statements, and you will see that here all the B-things are separated from the one or more A-things. That meets our definition for distribution. Thus, the A-term is undistributed, but the B-term is distributed.

MODULE 20

CONSTRUCTING VENN DIAGRAMS

Until about 1860, students of logic had to memorize the weakly and strongly valid syllogistic forms in the fashion of the medieval Scholastics in order to learn logic. Up to this point, most of logic was learned by rote memory. So routine was this method of instruction in logic that in the year 1781 the great philosopher Immanuel Kant was moved to say that all of logic is "complete."

Kant would have been in for a surprise. The most important work in logic done in a thousand years was to take place over the next century.

Perhaps the most important of that work was done by the logicians and mathematicians: George Boole, Charles Sanders Peirce, and John Venn. We will learn the full impact that Boole and Peirce had on logic in Unit Three when we discuss "sentential" logic. John Venn, however, was able to take the categorical syllogism of Aristotle and the scholastics and find a method (as opposed to a mnemonic device) with which to determine validity. We call that method the **Venn diagram.**

Venn's method was sheer genius. He had a fundamental grasp of what came easy to logic students. He had an intuition that mnemonic devices, however ingenious, were not the stuff of thinking. Rather, he believed (in a flash of insight that is widely shared among computer scientists nowadays) that what humans absorb best is *pattern recognition.* Consequently, he constructed a method that took advantage of the human proclivity for recognizing patterns.

A deductive argument is **valid**, if and only if, if all the premises are true, then the conclusion must be true.

20.1 PRELIMINARY REMARKS

Here is a good, intuitive way of grasping what Venn's method was about. Let's imagine a big brown box that contains everything in the universe:

. . . where the 'U' in the left-hand corner designates "the universe." We must imagine this to be a very big brown box indeed. Yet, we won't be interested in *everything* in the universe. Instead, we'll restrict our attention to just those things under discussion.

Consider, for example, the statement 'All palominos are horses'. We can designate 'palominos' with the term-letter 'P' and 'horses' with the term-letter 'H'. Now, imagine two 'corrals'—one contains all palominos and the other all horses. Imagine those corrals overlapping and focus on the part of the universe which contains both. We have the following diagram:

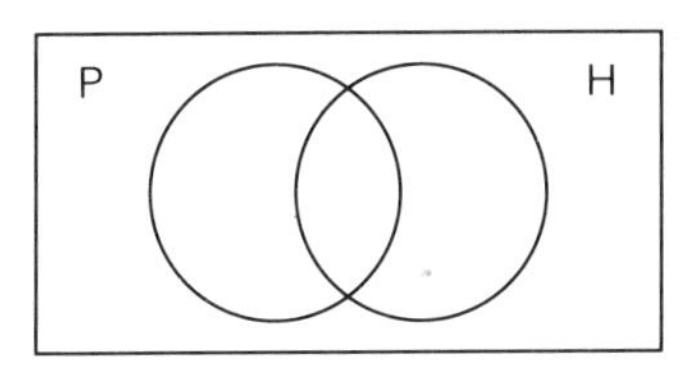

. . . where the 'P'-corral holds palominos, and the 'H'-corral holds horses. Pay attention to the "fences." We see that due to the overlapping of the corrals, the fences run right through the middle forming separate compartments. We may number those compartments as follows:

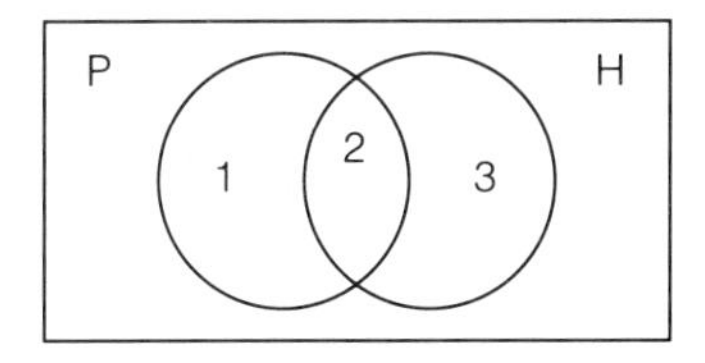

Thus, we have a total of *three* compartments;* compartment 2 represents the overlap between palominos and horses. Since 'palominos' is our subject, we would expect to find any existing palominos in compartment 2.

**Four,* if we include the space in the box outside the circles (which designates the rest of the universe!).

Furthermore, we can expect that compartment 1 (the set of all palominos which are *not* horses) to be *empty.* We can represent this as well by the simple device of *shading it.* Thus, we have:

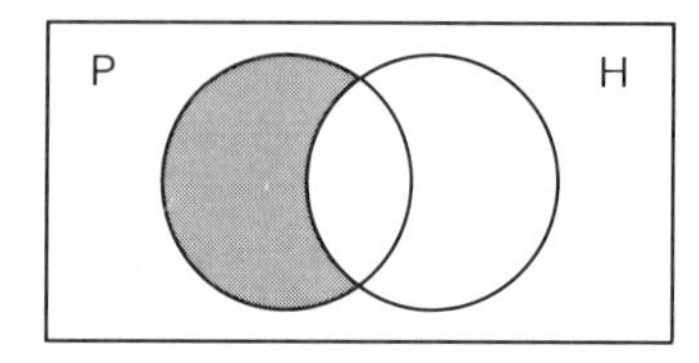

The result bears an uncanny resemblance to the emblem on a well-known credit card. (In fact, it looks somewhat like a "MasterCard.") The point to notice however is that we have just successfully represented the **universal affirmative** statement 'All P are H' by means of this pattern of "corrals" and the device of shading. Thus, if anything is a palomino, it must also be a horse and located in compartment 2.

We can represent all of the Aristotelean logical operators using Venn diagrams. Consider the **universal negative** statement 'No horses are zebras'. Let 'H' = horses and 'Z' = zebras. We can see that the intersection of the two corrals must be empty. Representing this by shading, we have:

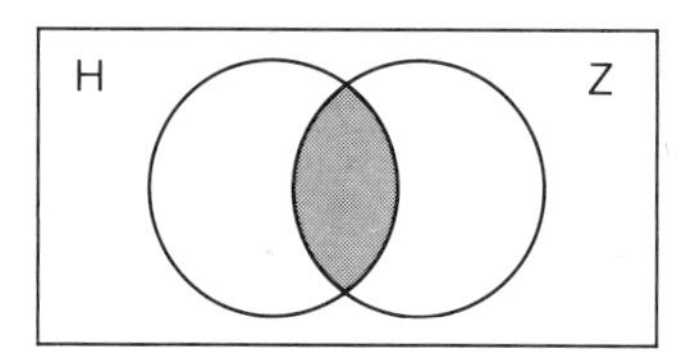

Since compartment 2 is shaded we know it is *empty:* no horses are zebras. If anything at all is a horse, it cannot be found in compartment 2.

How do we represent the **particular affirmative** and the **particular negative** operators? In order to show this, we introduce a further symbol: the asterisk (*). Think of the asterisk as representing the existence of some "critter". There may be one or more, but there is *at least* one. For example, take the **particular affirmative** statement 'Some horses are palominos'. That tells us that there is at least one horse which is a tan-colored animal we call a palomino. We represent it by showing that it must exist in the part of the corral between horses and palominos which stands as the intersection between the two. Thus, we have:

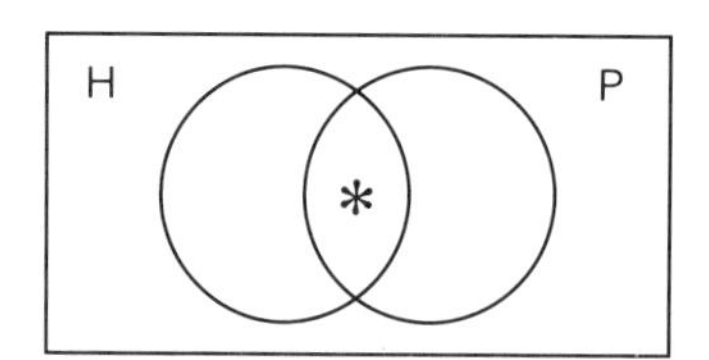

. . . and we see that compartment 2 is actually *occupied*. There really *are* horses that are palominos.

Of course, not all horses are palominos: some of them are albinos or mustangs, and some mustangs are calico in color. Let 'H' = horses and 'P' = palominos, then we can represent 'Not all horses are palominos' by showing that such critters exist *outside of* the intersection of the two corrals, as follows:

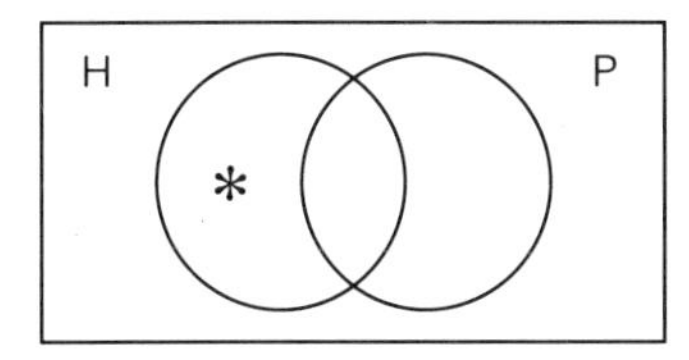

We see that compartment 1 is occupied, so we know that some horses are *not* palominos (whether or not any other horses are!). Important: Don't disregard the blank spaces in compartments 2 and 3. They also tell us something. The blanks tell us that we *don't know* whether those spaces are occupied or empty.

Thus, we have three features to read in any diagram:

1. shading (which means the compartment is empty)
2. asterisk (which means the compartment is occupied)
3. blanks (which means that we *don't know* whether the compartments are occupied *or* empty on the basis of the information given us)

20.2 THE FOUR OPERATIONS

We can now show all four operations on any two statements beginning with a subject 'S' and a predicate 'P' as follows:

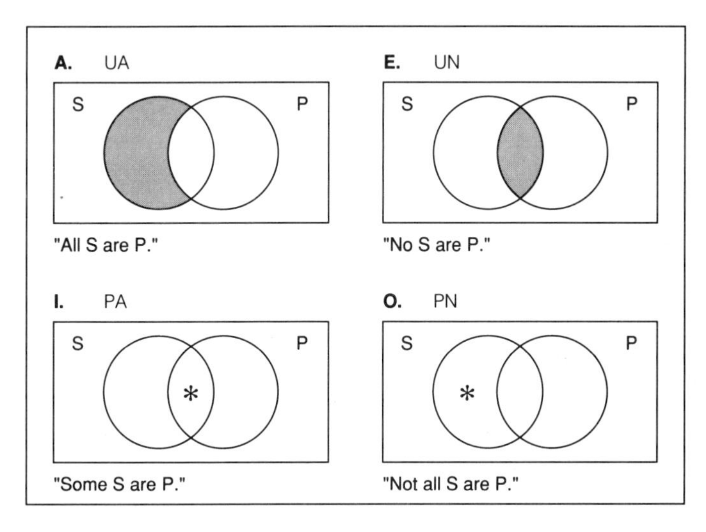

The Four "MasterCards"

20.3 THE COMPLETED VENN DIAGRAM

Notice in the diagram above that we are employing only *two* terms—the S- and P-terms of the Aristotelean syllogism. However, we remember that a syllogism is a complete argument only when it employs *three* terms: the S-, P-, and M-terms. So, how do we show the relationship of the three? What follows is the completed Venn diagram:

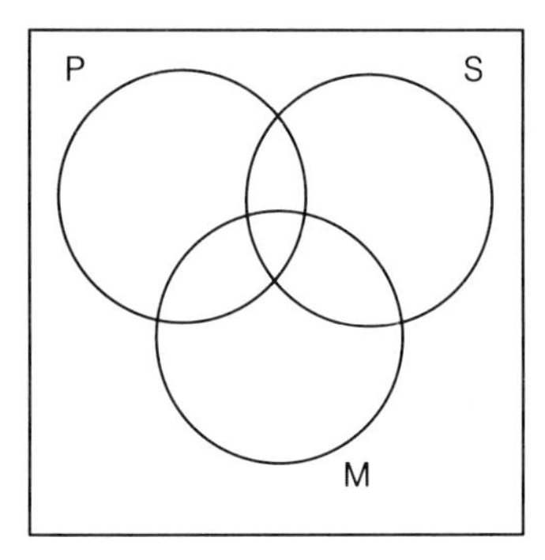

We represent the M-term by a third circle that cuts through the other two. Notice that we have increased the number of separate compartments by adding the third circle. In fact, we now have *seven* compartments* to consider, as shown below:

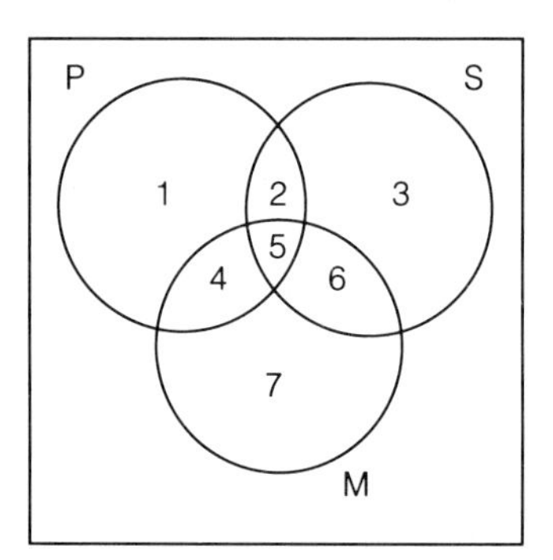

Think of the circles as corrals once again where each line represents a fence. The intersection of the P- and S-circles (formerly just compartment 2) now has a fence running through the middle of it, splitting it in two. If anything were in the intersection, it would be on one side of the fence or the other (i.e., in compartment 2 or in compartment 5). Keep this in mind.

The same holds true for the other intersections. The intersection of circles S and M has a fence down the middle, creating two compartments numbered 4

**Eight,* if we include the space outside of the circles (which designates the rest of the universe).

and 5. The circles P and M likewise have a middle fence, creating compartments 5 and 6. Once you think of our "critters" (whatever they are) as being on one side of the fence or the other, we're then ready to solve a syllogism using the Venn Diagram Method.

20.4 EMPLOYING VENN DIAGRAMS

Remember: Diagram *only* the premises. You will not diagram the conclusion. The Venn Diagram Method takes advantage of the special human capacity to recognize patterns as a means to solve problems in logic, a capacity that many philosophers argue is what makes the difference between human intelligence and computer (or "artificial") intelligence. Think of the Venn diagram as a way of representing your assumption that the premises are all true. You want to see if the conclusion follows. Remember:

> **A deductive argument is valid, if and only if, if *all* the premises are true, then the conclusion *must* be true.**

You're about to see what validity means in practice, and you will do so in the form of a pictorial representation—a Venn diagram.

Follow these guidelines to find solutions to Venn diagram problems, then we will work some examples.

1. **Direct your attention to two circles at a time.** Diagram one premise at a time. Since each premise contains two term-letters, you will confine your attention to the two circles which represent these term-letters.

   For example, see how the P-term and the M-term relate. We can use cross-hatching (this is not the *shading* operation) to show the S-circle as a "phantom" so we can confine our attention to the P- and M-circles.

   Direct your attention to these two circles, as depicted here:

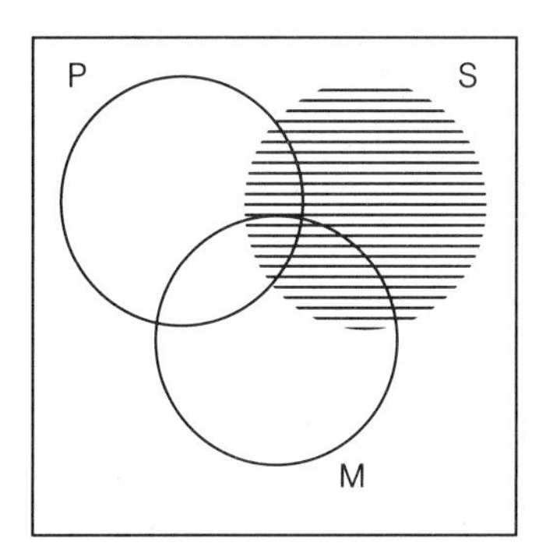

Notice that we're not paying any attention to the S-circle here (which is depicted, for now, as a "phantom" circle). You will still take cognizance of *all* the fences, however, when you interpret this diagram, but for the first premise, confine your attention to the first two circles (in the major premise, that includes the P- and M-circles).

In the second premise (the minor premise), confine your attention to the S- and M-circles, as depicted here:

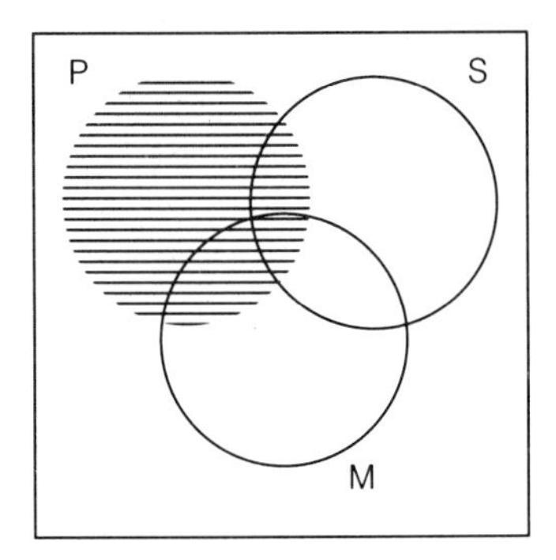

As in the previous example, we are disregarding the "phantom" circle in working the second premise. The "phantom" circle in this case is the P-circle; we confine our attention to the S- and M-circles for now. Notice the *outline* of the intersection of the S- and M-circles on which the phantom P-circle faintly and only partially intrudes. (Later we will take cognizance of *all* the fences in the relevant intersections.)

2. Identify the intersections. When you are diagramming a premise, select the intersection described by the appropriate term-circles. For example, if your premise is 'All P are M', identify the P-circle and the M-circle, then pick out the intersection of the two circles. In the following diagram, the compartments which make up the intersection of the P- and M-circles are 4 and 5:

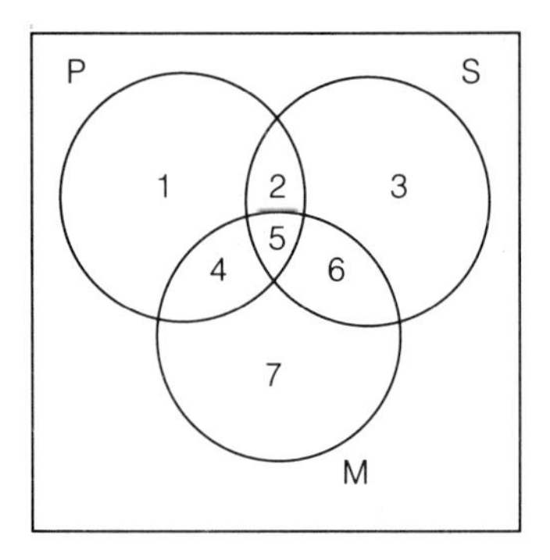

A good technique to make your work clear is to mark off the intersection with "tick" marks after you've identified it. The intersection of the P- and M-circles, compartments 4 and 5, is highlighted in the figure given below:

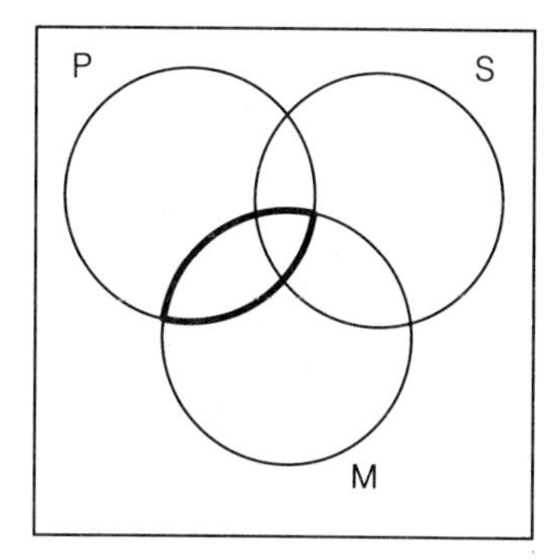

In the case of 'All S are M', compartments 5 and 6 are highlighted, as follows:

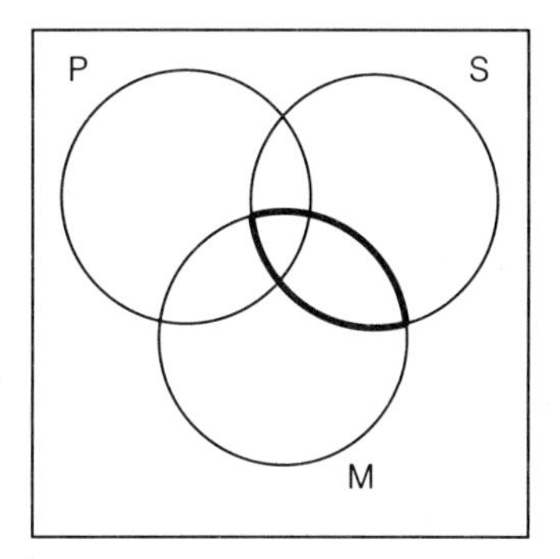

3. Do all your work in the *subject* circle. For example, if your premise reads, "All S are M," the operation (shading or asterisk—in this case, shading) goes in the subject circle, which in this case is the S-circle.

To determine what operation to take and where to locate it, consult this table:

Operator	Symbol	Operation	Intersection Location
UA—'All'	A	shading	outside the intersection
UN—'No'	E	shading	inside the intersection
PA—'Some'	I	asterisk	inside the intersection
PN—'Not all'	O	asterisk	outside the intersection

Thus, for 'All S are M', the operation is shading, and the shading goes *outside* the intersection—*but all your work is confined to the subject circle, here, the S-circle.* Thus, your diagram should look like this:

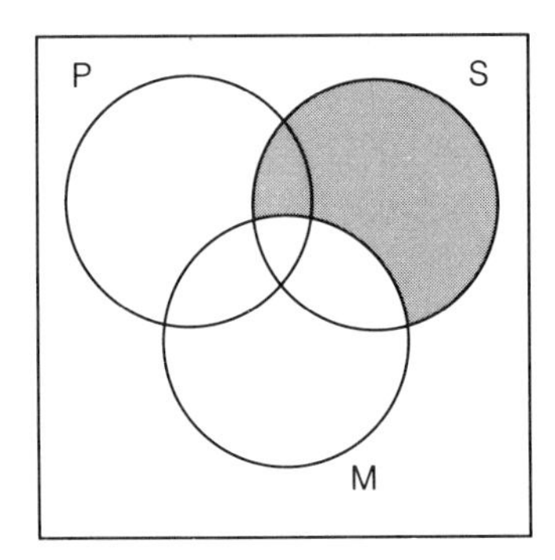

Notice that *two* compartments (2 and 3), where S is not M, have been shaded. *Every operation is done in two compartments.*

4. ***Never* diagram the conclusion.** Since the diagram serves as your assumption that the *premises* are true, diagramming the conclusion would hopelessly confuse your diagram, defeating its purpose in the first place.

5. **Test for *invalidity* by asking the question: "Given the diagram, is it possible for the conclusion to be *false*?"** Remember, you never test for validity. Your job is to try to find the counterexample, to prove *invalidity.* If you cannot, then you know the argument is valid.

Another way of putting this is to remember that your diagram serves in place of the assumption that all your premises are true. Assume the premises are true: is it *possible* for the conclusion to be false?

Here are the five guidelines once more:

---

1. **Direct your attention to two circles at a time.**
2. **Identify the intersections.**
3. **Do all your work in the subject circle.**
4. **Never diagram the conclusion.**
5. **Test for invalidity by asking the question: "Given the diagram, is it possible for the conclusion to be *false*?"**

20.5 APPLYING THE FIVE GUIDELINES FOR DIAGRAMMING: AN EASY EXAMPLE

Here is a simple syllogism:

All pizzas are cheesy.
No ice cream is cheesy.

No ice cream is pizza.

Here the P-term is 'pizzas', the S-term is 'ice cream', and the M-term is 'cheesy things'. Thus, replacing the terms with the appropriate term-letters, we have:

All P are M.
No S are M.

No S are P.

We follow guideline 1, and diagram two circles at a time. In the major premise, the circles we diagram are the P-circle and the M-circle.

In the minor premise, we diagram the S- and the M-circles.

According to guideline 2, we highlight the intersections. In the major premise, we have:

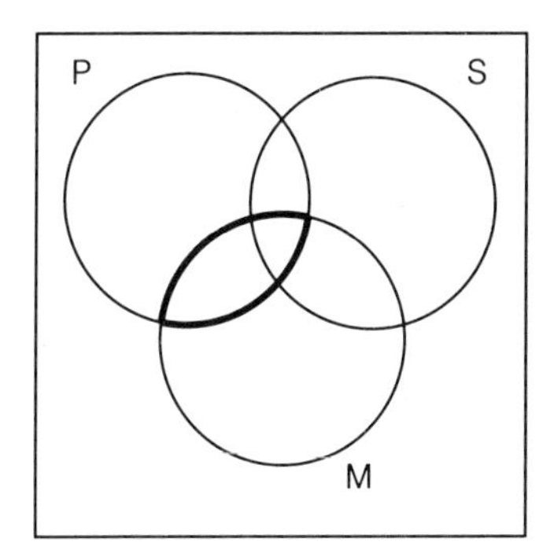

. . . and in the minor premise, we get:

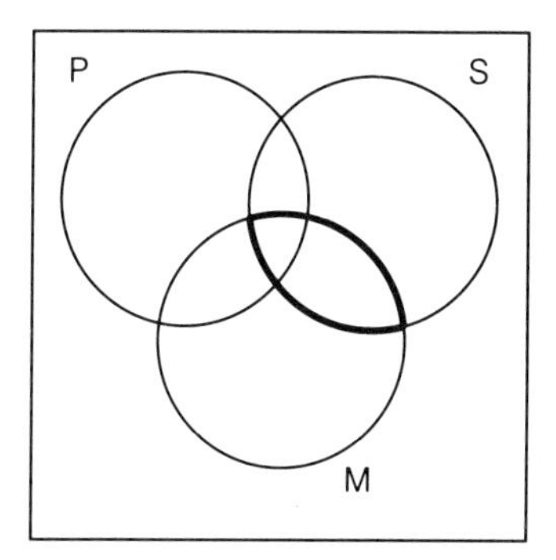

We do all our work in the subject circle, according to guideline 3. So, the subject circle in the major premise is the P-circle. Consulting our table, we know this is a "UA" operator, '[All] P are M'. So, this is a shading operation, and the shading goes *outside* the intersection, or in the two outside compartments of the P-circle: compartments 4 and 5 (as shown below):

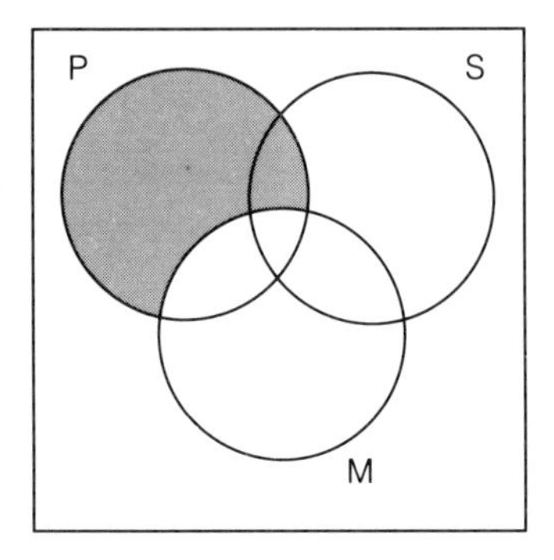

In the minor premise 'No S are M', the subject circle is the S-circle because 'S' is the subject of the sentence 'No S are M.' If the example read, "No M are S," then the subject would be the M-circle. 'No S are M' is a "UN" operator. Thus, it takes a *shading* operation, and the shading goes *inside* the intersection (as follows):

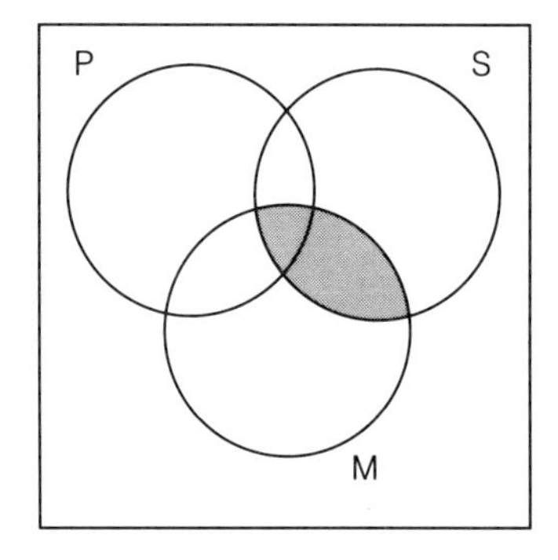

Taking the two premises together, we have the following *combined* diagram:

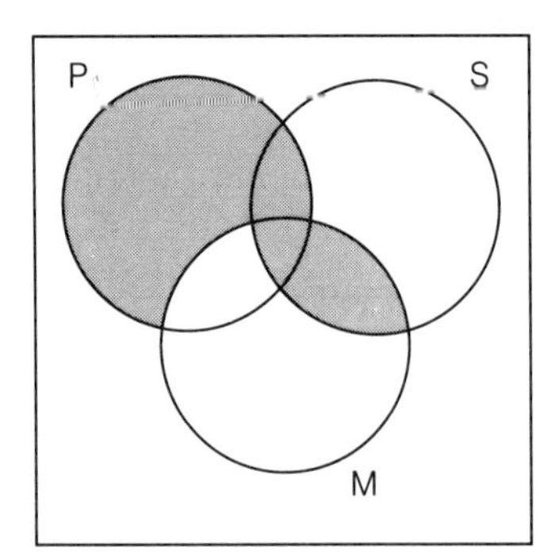

Notice that your Venn diagram is complete. You have diagrammed both of the premises, so you must now stop diagramming. Guideline 4 says, "Never diagram the conclusion."

We are now ready to go on to guideline 5: Test for invalidity by asking the question: "Given the diagram, is it possible for the conclusion to be *false*?"

The conclusion reads: "No S are P." Your question is, given the diagram, is it possible for 'No S are P' to be false? We see that wherever we have an intersection of the S- and P-circle, we find shading. Remember what we learned earlier in this section:

We have three features to read in any diagram:

1. shading (which means the compartment is empty)
2. asterisk (which means the compartment is occupied)
3. blanks (which mean we don't know whether the compartment is empty or occupied)

Shading means "empty." Therefore, wherever the S- and P-circles intersect, the compartments are empty. Therefore, it is impossible for the conclusion to be false, since it is a fact that 'No S *are* P'.

Therefore, this is a *valid* argument.

20.6 AN EXAMPLE OF AN INVALID ARGUMENT, PROVEN BY THE VENN DIAGRAM METHOD

Here is a plainly invalid argument:

No coffee mugs are Mercator projections.
Some coffee mugs are printed objects.
____
Therefore, no printed objects are Mercator projections.

The middle-term here is 'coffee mugs'; the P-term is 'printed objects', and the S-term is 'Mercator projections'. Putting the term-letters in place of the terms themselves, we get this result:

No M are P.
Some M are S.
____
No S are P.

Note: although we need to identify the P-, S-, and M-terms, we don't really need to decide what is the major premise or the minor premise using the Venn diagram test. This is because those procedures place the syllogism in a standard form which makes them recognizable on the Scholastic's rote memory test. As it turns out, the first premise here is the minor premise, and the second is the major, but we need only take the premises in the order in which we encounter them for our purposes in a Venn diagram (which is what we've done here).

Our first step is to follow guidelines 1 and 2. Guideline 1 tells us to confine our attention to two circles at a time when constructing the diagram. In the first premise, we have 'No M are P'. Thus, we work the part of the diagram concerning the M-circle and the P-circle. The next step, guideline 2, instructs us to identify the intersections. (We will look at the first premise now, and come back to the second premise later.) Here is the intersection of the M-circle and the P-circle (we've highlighted it):

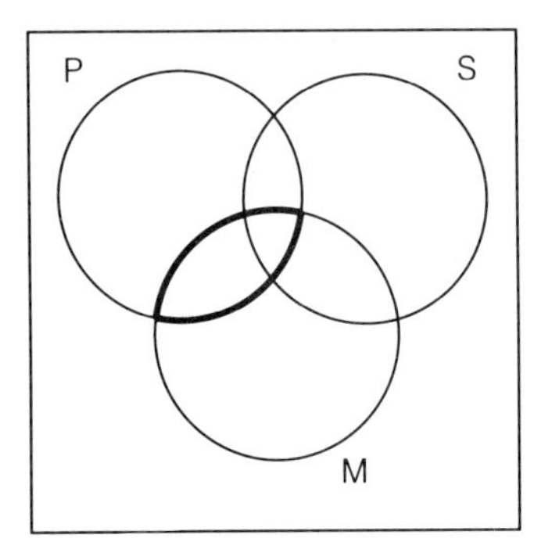

Guideline 3 tells us to do all our work in the subject circle. The subject of the sentence 'No M are P' is the term-letter M. Thus, we perform our work in the corresponding M-circle. Now, we have to decide the operation and the location. The table for determining locations and operations follows:

Operator	Symbol	Operation	Intersection Location
UA—'All'	A	shading	outside the intersection
UN—'No'	E	shading	inside the intersection
PA—'Some'	I	asterisk	inside the intersection
PN—'Not all'	O	asterisk	outside the intersection

Once again, we see that this is a *shading* operation ("UN"), and that the shading is located *inside* the intersection of the M- and the P-circles. Thus, we have:

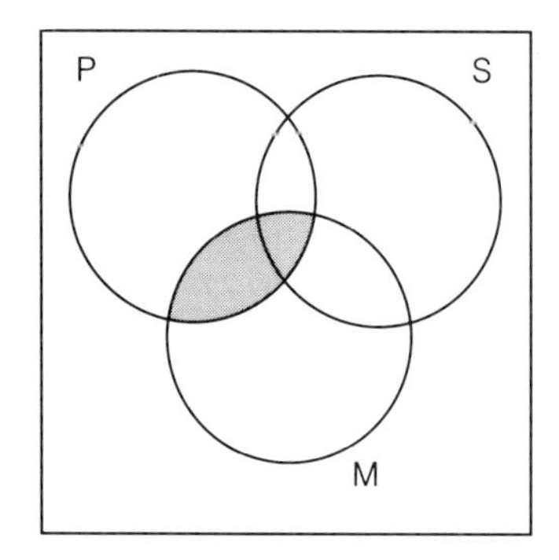

Having completed the first premise, we now turn to the second, 'Some S are M'. Repeat guidelines 1 and 2, confining our attention to the compartments that make up the S- and M-circles. Highlighting the intersection, we get:

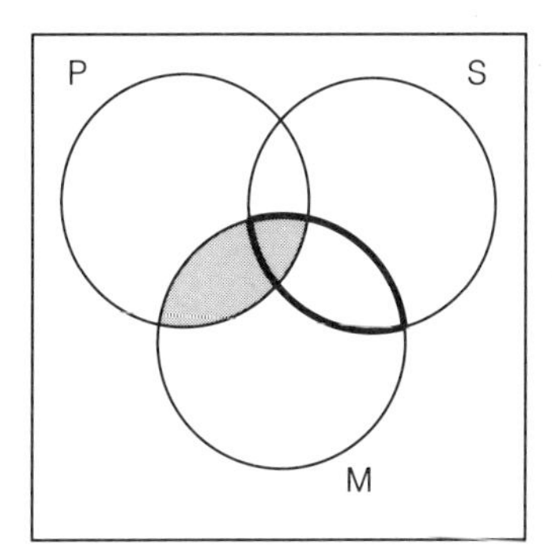

Notice that the intersection of M and P is already shaded, and that part of this shading is in the intersection of M and S as highlighted.

Next, locate the operation in the table. The premise 'Some S are M' is particular affirmative ("PA"). Thus, we know that this is an *asterisk* operation and that the asterisk goes *inside* the intersection of S and M. Compartment 5 is already shaded, meaning it's empty. An asterisk means *occupancy.* Normally, we would have to worry about whether the asterisk would go in compartment 5 or 6, or both. But since we know that 5 is empty, we place the asterisk in compartment 6. That compartment must be occupied.

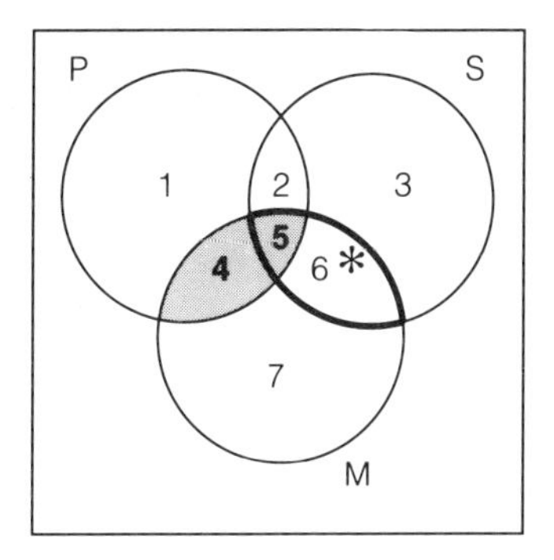

Notice that we have also followed guideline 4, and located our work in the subject circles. The subject of the second premise, 'Some S are M', is the term-letter S. Thus, our work must be in the S-circle. We've done that: compartment 6 is within the S-circle (in fact, it's within the intersection of S and M).

Thus, the completed diagram is as follows:

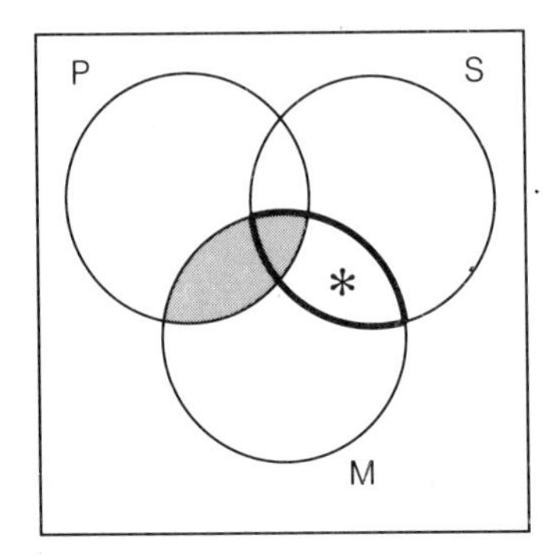

Now, we are left with the interpretation of the diagram. Guideline 5 says: Test for invalidity by asking the question: "Given the diagram, is it possible for the conclusion to be *false*?"

We see that it is indeed possible for the conclusion to be false, given the diagram. The conclusion reads, "No S are P." However, we see that compartment 2 is blank: it could be occupied, or it could be empty. This means that compartment S *could be occupied.* If that were so, then the conclusion would be false. Thus, the syllogism is *invalid.*

The next drawing shows an arrow pointing to a compartment which serves as a counterexample to the supposition that our syllogism is valid:

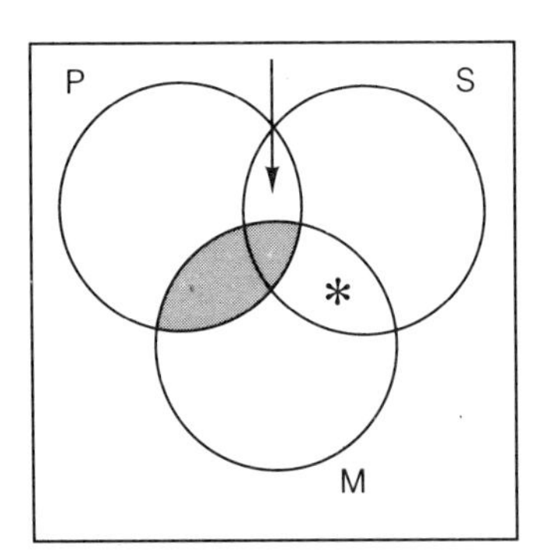

Remember: Your job is to attempt to prove *invalidity.* You shouldn't try to prove the syllogism valid—that would be a mistake. All you need to show invalidity is the merest *possibility* that the conclusion *could* be false given the premises.

20.7 BRIDGES: A MORE DIFFICULT EXAMPLE USING PARTICULAR AFFIRMATIVE AND PARTICULAR NEGATIVE PREMISES

Here is a new syllogism employing PA and PN premises:

Not all gizmos are doodads.
Some gizmos are gadgets.

Therefore, not all gadgets are doodads.

'Gizmos' is the M-term, 'gadgets' is the S-term, and 'doodads' is the P-term. Thus, we have:

Not all M are P.
Some M are S.

Not all S are P.

The diagram for 'Some M are S' is shown below:

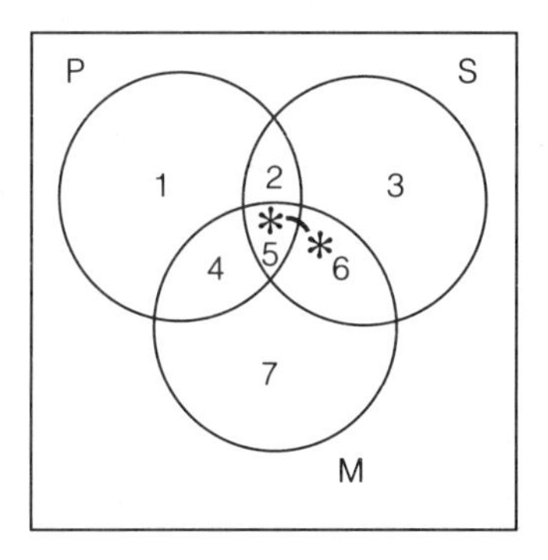

'Some' is a PA operator, thus we know that it is an *asterisk* operation and it goes *inside* the intersection of M and S. Now, however, we see we have a problem. There's a *fence* between the two intersections. But then, where does the asterisk go? Maybe it's in compartment 5, maybe it's in compartment 6, *maybe it's in both*! How do we represent that? We represent it by means of a bridge, a straight line which connects asterisks on both sides of a fence.

Remember: A bridge means: "this compartment is occupied, or that one, or both."

Thus, we have a *bridged* asterisk to show our first premise.

Let's look at the next premise: "Not all M are P." We do our work in the subject circle, the M-circle. This is a particular negative (PN) premise. That means it is an *asterisk* operation and that the asterisk is located *outside the intersection.* The intersection of the M- and the P-circles is compartments 4 and 5. *Since we do all our work in the subject circle, we confine our attention to the M-circle.* But here's a familiar problem. Once we locate the part of the circle that's *outside* the intersection, we again have two compartments to worry about: 6 and 7. There's another fence between them. Once again, we need a bridge to jump the gap between the two compartments. Thus, we have:

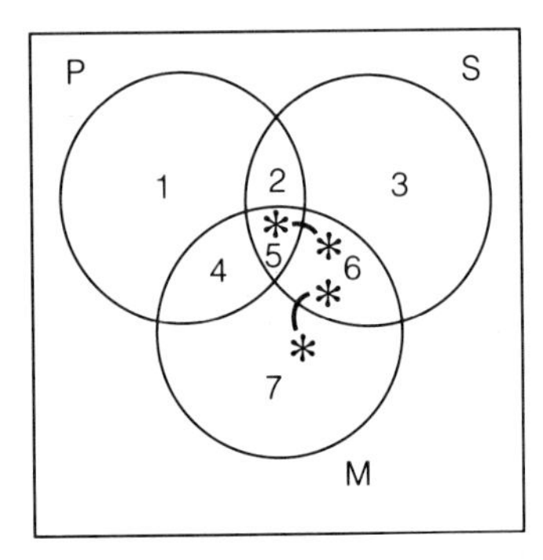

This diagram is quite a bit more difficult to interpret than the previous ones we looked at since we have significantly increased the number of possibilities. However, guideline 5 is quite straightforward: Test for invalidity by asking the question: "Given the diagram, is it possible for the conclusion to be *false*?"

The conclusion of the argument is 'Not all S are P'. Is it possible, given the diagram, to interpret the conclusion as false? Yes, it is; the argument is *invalid.*

How did we arrive at this conclusion? Maybe if we re-name the circles, that will help to explain the process. Once again, let's remember that the P-term stands for 'doodads', the S-term stands for "gadgets', and the M-term stands for 'gizmos'. We substitute the names for the terms in this version of the completed diagram:

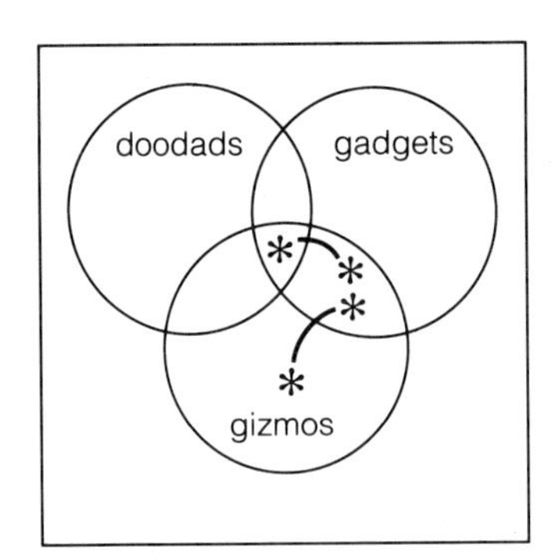

We have different sorts of compartment occupants here, but let's suppose these inhabitants can cross bridges if one is provided. The conclusion says that 'Not all gadgets are doodads'. Let's suppose we caught these inhabitants and put them in compartments of a corral connected by bridges. For our purposes, let's suppose that we surprised our inhabitants on a day that had all of the gadgets in the compartment with the doodads (compartment 5) and all of the gizmos in compartment 7 (they crossed the bridge, you see). Thus, we see that it is indeed possible, given the argument, that 'Not all gadgets are doodads' is false. The gadgets are completely inside the compartment that identifies them as doodads. Maybe gadgets are doodads, maybe they aren't; the point is that the possibility that they are doodads is not precluded by the information given in the premises. That's why the argument is invalid. Given the premises, it's possible for the conclusion to be false. Yet, we know that a deductive argument is valid, if and only if, if all the premises are true, then the conclusion must be true. This is not the case here. Therefore, the argument must be invalid.

I purposely used non-descript terms like 'gadgets' and 'doodads' for this example because nothing in common sense rules out imagining that 'Some doodads are gadgets', or that 'No doodads are gadgets' for that matter. The point is that where bridges are built, whatever the terms in the syllogism are, we must imagine that it's *possible* to have cross-overs. Sometimes imagining cross-overs defies common sense. If we were to say, "Some men are jellyfish," it's hard to imagine human beings occupying jellyfish compartments. Nevertheless, and even though we know the statement to be false, the operation on Venn diagrams *requires* us to assume the possibility of cross-overs as a condition of invalidating the argument. Remember: validity is distinct from truth. Our object is to imagine the set of circumstances that violates the conclusion, as in this obviously invalid syllogism.

Not all men are sharks.

Some men are jellyfish.

Therefore, not all jellyfish are sharks.

And here is the completed diagram:

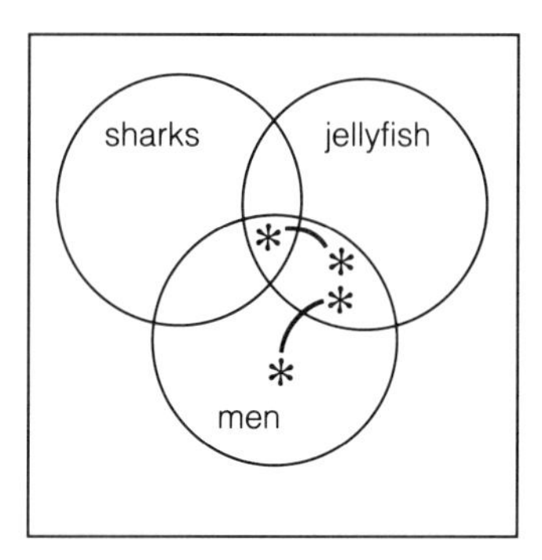

Notice that this Venn diagram is identical (except for substitution of terms) to the "gadget-doodad" diagram given. It is likewise *invalid.*

EXERCISES

Operator	Symbol	Operation	Intersection Location
UA—'All'	A	shading	outside the intersection
UN—'No'	E	shading	inside the intersection
PA—'Some'	I	asterisk	inside the intersection
PN—'Not all'	O	asterisk	outside the intersection

Exercise 1

DIRECTIONS: *Use the table above to determine how to fill in the circles in the cards in each of the figures below.*

1.

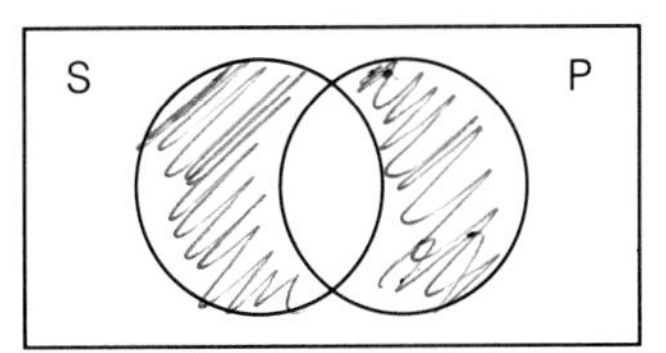

"All S are P."

2.

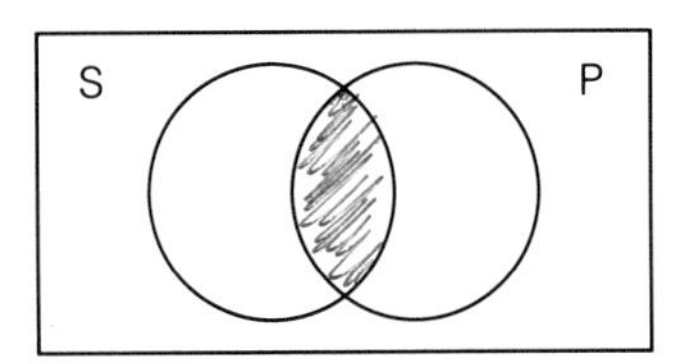

"No S are P."

3.

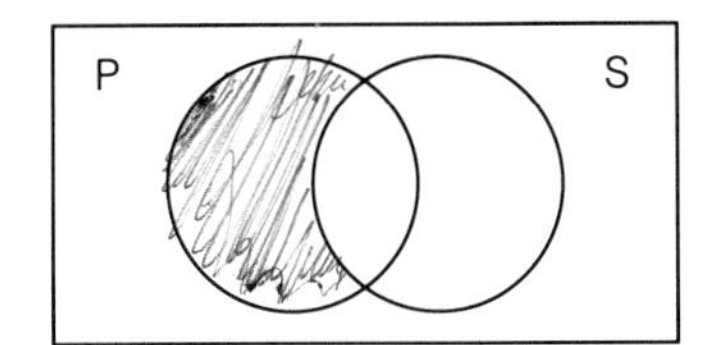

"Some S are P."

4.

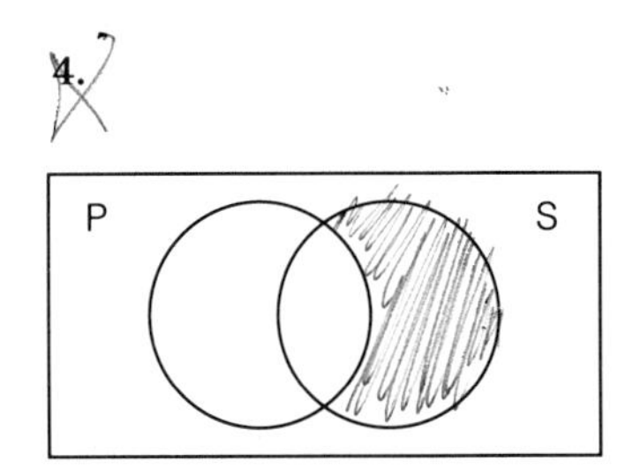

"Not all P are S."

Exercise 2

DIRECTIONS: *The following diagrams have three terms. Be careful to use bridges when necessary.*

1.

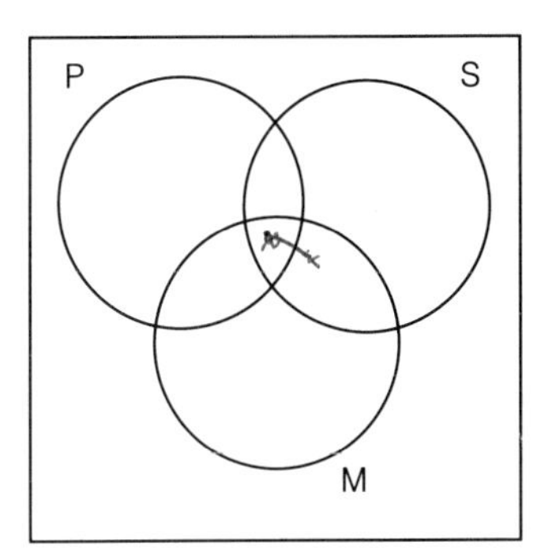

"Some M are S."

2.

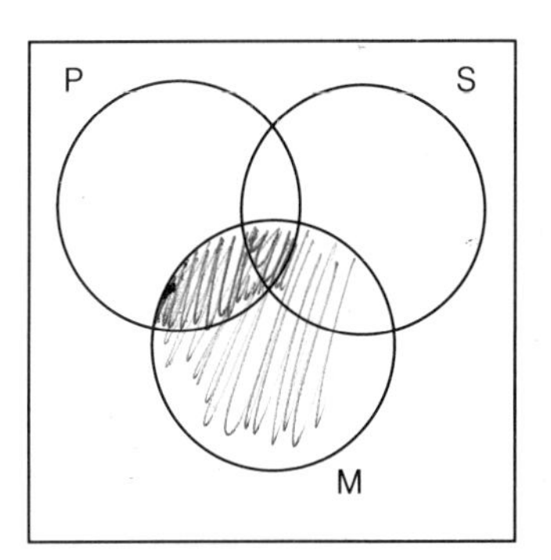

"No P are M."

3.

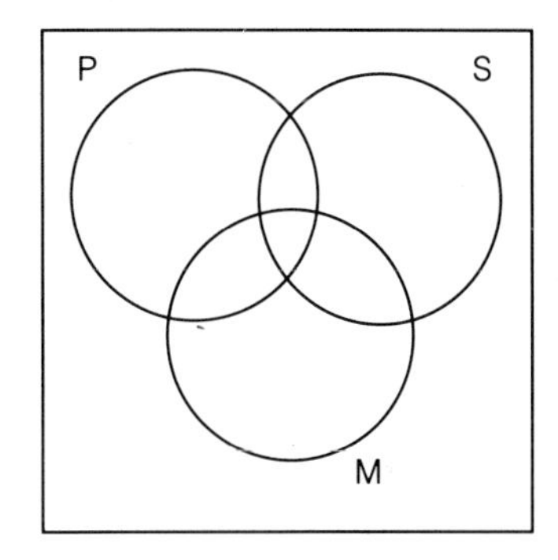

"Not all P are S."

4.

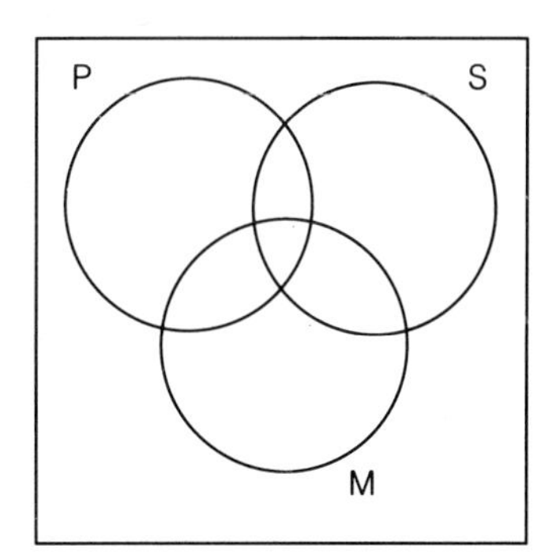

"All M are S."

MODULE 21

TIPS FOR TESTING VENN DIAGRAMS

21.1 INTRODUCTION

Many students find the Venn diagram more difficult to use as a test for validity than the Three Rules Method. That's to be expected. The real value of Venn diagrams is that they force us to *understand* validity, whereas the Three Rules merely *demonstrate* the validity or invalidity of syllogistic arguments.

The difference is analogous to working out the right answer to a math problem as opposed to consulting the answer key in the back of the book. The answer key will give you the correct answer every time, but you won't know *why* it's the right answer until you work it out yourself. Similarly, unless you already understand the difficult and non-intuitive concept of distribution in categorical statements, you will only be able to *apply* the Three Rules without understanding *why* they work. Interestingly, Venn diagrams are derived from a method of depicting categorical distribution patterns known as "Euler" circles. They show you precisely why the Three Rules get the results they do.

Knowing the Three Rules is handy; you can use them to check your results with the Venn diagram. But there is no substitute for interpreting Venn diagrams when it comes to understanding what validity is all about.

What follows are three tips to help you interpret Venn diagrams. The first requires you to get clear on the existence commitments made by categorical statements; the second is based on the time-honored Aristotelean Square of

---

A deductive argument is **valid**, if and only if, if all the premises are true, the conclusion must be true.

Opposition; the third is a new technique which has you construct a separate diagram for the conclusion. If you find yourself having trouble getting the correct answer using the Venn Diagram Method, these tips may help you see what you need to know.

TIP 1: VENN DIAGRAMS AND EXISTENCE COMMITMENTS

Sometimes recalling the equivalence relationships (i.e., the conversion, obversion, and contraposition forms) can help you interpret a Venn diagram. For example, consider the statement 'All P are M'. The Venn diagram looks like this:

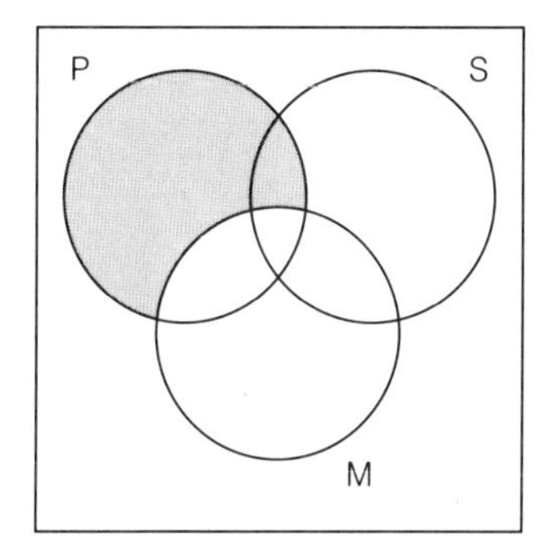

Students commonly ask why the P-circle is shaded *outside* the intersection of P and M. When you consider what 'All' means here, you may more readily understand.

The obverse of 'All P are M' is the statement 'No P are non-M'. In other words, 'Nothing is a P that is not an M'. Notice that the statement 'All P are M', does not commit us to the existence of any P's that are not M's. We are merely saying that if anything is a P (and it may be the case that nothing is), it is not an M. Consequently, we *shade* any P that is not an M (shading, you'll remember, designates a compartment as empty).

To make this clearer, let's use a common sense example. Recall our earlier comparison of the circles of the Venn diagram to corral fences. Suppose we are hovering over the Triple Bar Ranch in a helicopter and we're informed by radio that "All the calico-colored horses on the ranch are mares," and we decide to check this out. The calico horses are kept exclusively in the P-corral, and all the mares are kept in the M-corral. We fly overhead and this is what we see:

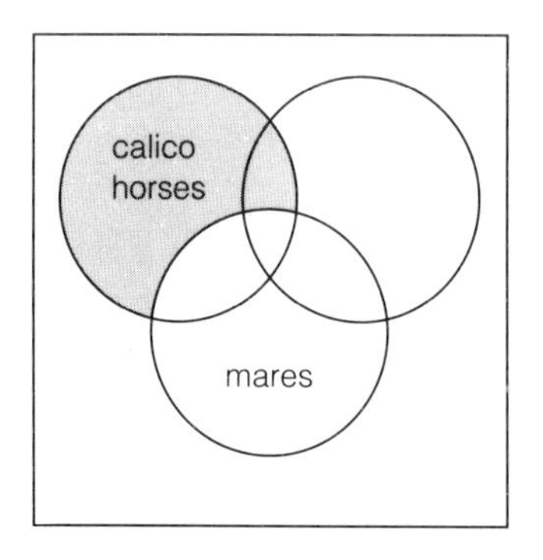

Because of freak weather, let's say, we can see clearly enough in part of the calico horse corral, but we can see nothing at all in the entire corral that holds mares. It doesn't matter, we see enough. We see that there are no horses at all outside the intersection in the calico horse corral, *so it doesn't matter that we can't see any mares.* Whether or not any mares are kept on the ranch, we know the statement must be true: "All calico-colored horses (if any exist at all!) *must* be mares." There's no other place they can be.

Suppose now we are required to do a bit of deduction. We're also informed that "All mares are yearlings." The yearlings are kept in the S-corral.

The radio tower now asks this question: "Are all of the calico-colored horses also yearlings?"

What's the answer?

We can now say definitively that the answer must be "Yes." Here's why.

Because of a break in the weather, we can now peer into the corral which holds mares and this is what we see:

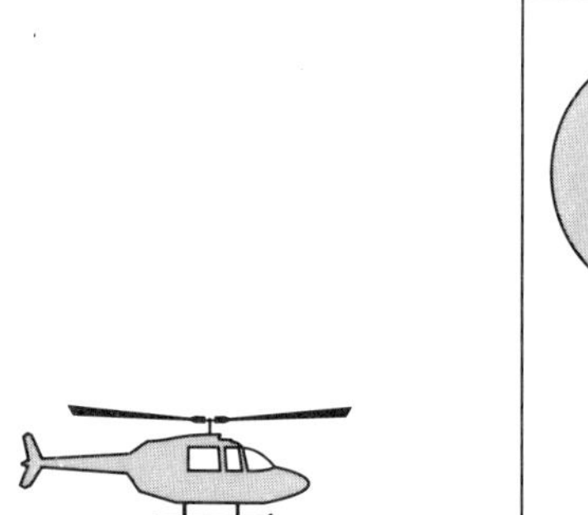

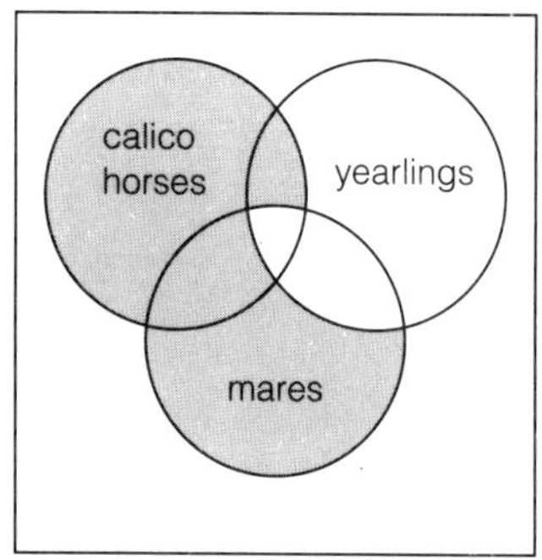

We see that most of the corral which normally holds mares is empty but we don't have a view into one compartment. That compartment is the intersection of the corrals which hold calico horses, mares, and yearlings in common. We don't know if any such horses exist, but *if they do, they must be yearlings.* Consequently, our deduction holds up: "All calico-colored horses (if any exist at all!) must be yearlings."

The same method holds good for universal negative statements. If, on another occasion at an entirely different ranch called "The Horseshoe Ranch," we're told that "No mares are yearlings," an overflight of the corral will show us this:

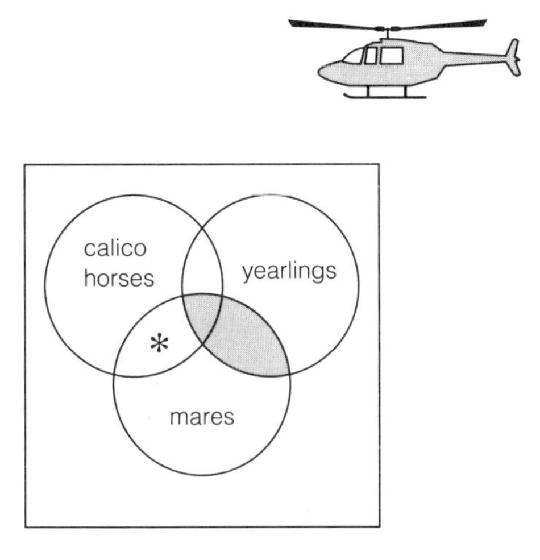

We don't have to look into the rest of the corral to know our statement must be true because we know that 'if anything is a mare, it is not a yearling'. Universal statements, whether they are universal affirmative or universal negative, do not commit us to the existence of anything. They merely commit us to the non-existence of a thing as of a certain type—in this case, the universal negative statement 'No mares are yearlings' commits us to the non-existence of mares that are also yearlings. No such animals are to be found on the ranch.

Notice, however, that particular affirmative and particular negative statements, which take asterisks, *do* imply existence and give a precise location in the corral.* Let's suppose we are now told that "Some calico-colored horses are mares." This is what we see:

*We would do well to recall here the discussion of 'weakly valid syllogisms' in Module 14.2 "A Date With Barbara." Some syllogisms *would* be valid if we were to add an additional assumption about the existence of certain terms. Of course, without such assumptions, such syllogisms are strictly speaking invalid—which is what the Venn diagrams here show.

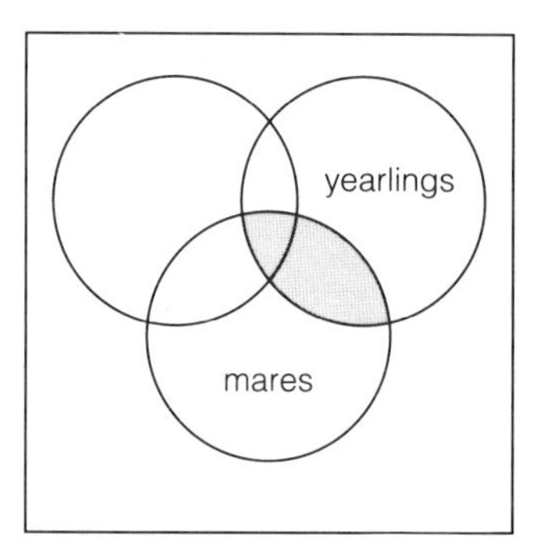

Looking down into the corral, we finally see a real horse (represented here by the asterisk).

The radio tower now poses a question to us in our helicopter: "Are some calico-colored horses also yearlings?"

The answer is, "We don't know." There's at least one calico horse compartment in which yearlings *could* be yearlings, but we're blanked out (weather conditions again). That compartment, for all we know, could even be empty. Therefore, on the basis of what we know, any deduction which holds the conclusion that some horses on the ranch are both yearlings and calico-colored is invalid.

TIP 2: THE "TURN-AROUND" QUESTION

Note: I have found that this method helps some students, but confuses others. It has helped enough students to justify explaining it here, but if you find it hopelessly confusing, try out some of the other tips and see if they help you more.

Some Venn diagrams more readily yield to interpretation when we know something about the *contradictories* of the conclusion statements, and the Aristotelean Square helps remind us of the contradictory relationships between the Aristotelean operators 'All', 'Some', and 'Not All'.

Remember, 'All' and 'No' are *not* contradictories (they're *contraries;* see Module 15.1). The contradictory of 'All' is 'Not All' and the contradictory of 'No' is 'Some'.

Recall our rule for interpreting validity: Test for invalidity by asking he question: "Given the diagram, is it possible for the conclusion to be *false*?"

Another way of asking this question is by turning it around (we'll call it the "turn-around" question):

"Given the diagram, is it possible for the *contradictory* of the conclusion to be true?"

The "Turn-Around" Question

Notice that we're asking the same question—this is an exact equivalent. Only now we're putting a new perspective on the matter.

Turning the question around like this can help us more clearly interpret some of the operators in a conclusion statement, especially for particular negative ('Not All') statements and universal negative ('No') statements. The other operators in conclusion statements ('Some' and 'All') yield to the original question in a straightforward manner, but the negations, 'Not All' and 'No', can be quite tricky sometimes.

See if you find this tip helpful with the following examples.

Take the following syllogism:

No M are P.
Some S are M.
———
Not all S are P.

Here is the completed Venn diagram:

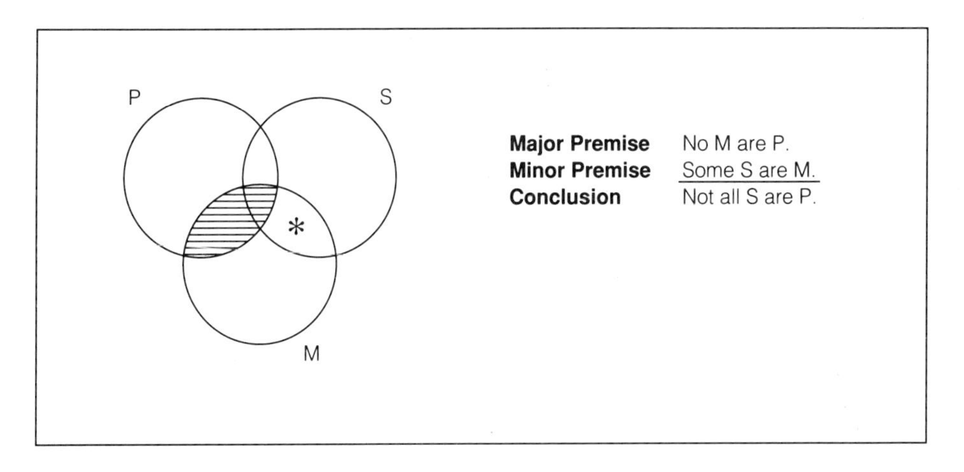

Is this syllogism valid or not? Posing the turn-around question may help us get a faster answer but to do this we first need to identify the contradictory of the conclusion. To do this consult the Aristotelean Square of Opposition (here's a "revised" version):

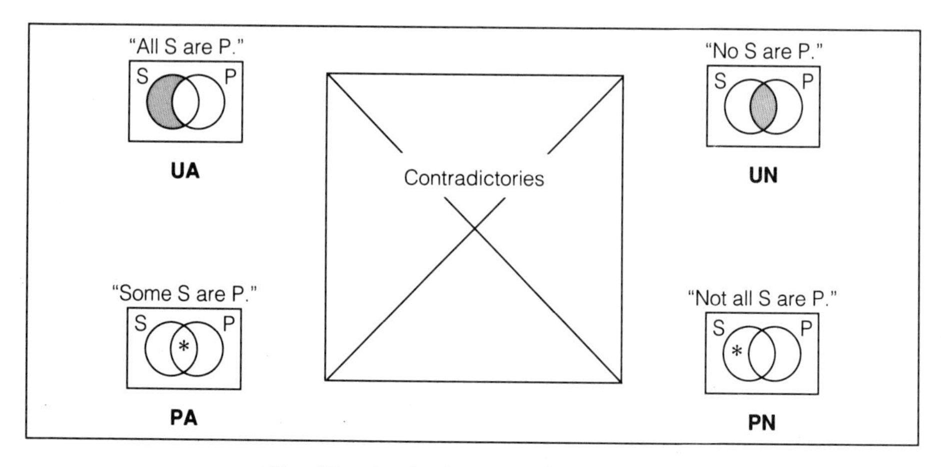

The "Revised" Square of Opposition

You'll notice that this Square of Opposition is revised to show the "MasterCards" (i.e., the diagrams for the categorical statements).

What is the contradictory for our conclusion operator 'Not All'? The answer, according to the square, is the operator 'All'. Therefore, for the purposes of employing the turn-around question, we must convert the conclusion from 'Not all S are P' to its contradictory '*All* S are P' and then ask, "Is it possible for the conclusion to be *true*?"

Given the diagram, is it possible for the contradictory of the conclusion to be true? In other words, given the diagram, is it possible for the statement 'All S are P' to be *true*? The answer is "No." Therefore, the reverse holds, that it is not possible for the (real) conclusion to be *false*. Therefore, the argument must be valid.

Let's try another example (an invalid one this time):

Not all M are P.

Some S are M.

No S are P.

Here's the completed diagram:

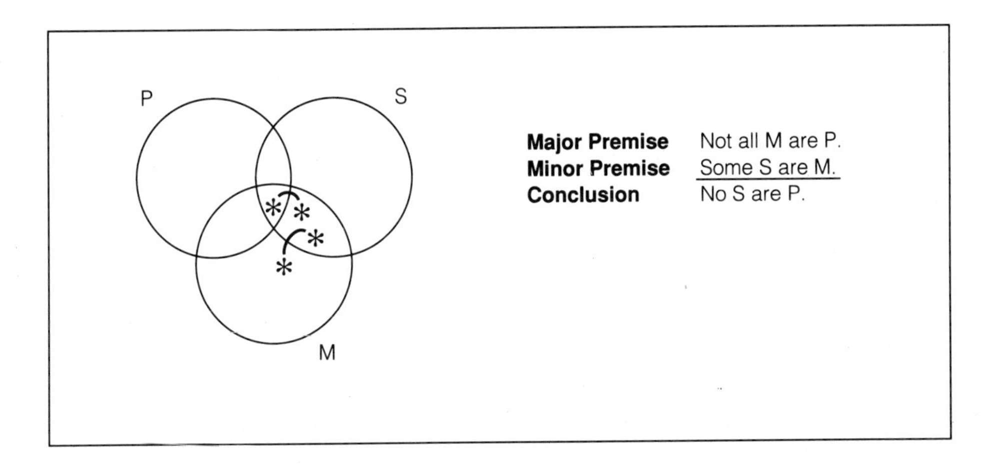

Bridged asterisks can make this sort of diagram very difficult to interpret, but asking the turn-around question can help cut through the difficulty.

We start by identifying the contradictory of the conclusion. The contradictory of 'No S are P' is (according to the Square of Opposition) 'Some S are P'.

Next, Let's pose the turn-around question: Given the diagram, is it possible for the *contradictory* of the conclusion to be *true*? In other words, is it possible for 'Some S are P' to be true?

The answer is decidedly, "Yes." But if it's possible for the contradictory of the conclusion to be true, then it follows that it's possible for the (real) conclusion to be *false*. Therefore, the argument is invalid.

To see why, it might be helpful to point an arrow at the compartment which shows it's possible for the contradictory of the conclusion to be true.

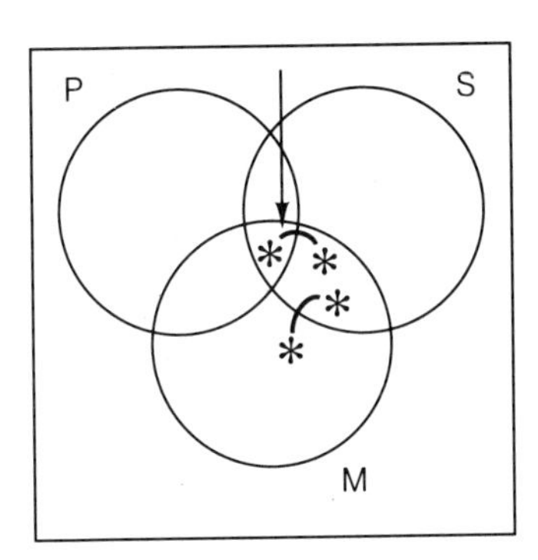

The arrow points to a bridged asterisk in the intersection of the S- and P-circles. Since this asterisk is bridged, we know that the compartment may be occupied or it may be empty. In other words, it *is* possible for 'Some S are P' to be true. Therefore, the original conclusion, 'No S are P', can be false, and the argument is invalid.

TIP 3: THE "CONCLUSION DIAGRAM" TEST

This is the third of three tips on how to interpret Venn diagrams, and it is probably the most powerful explanation of the three. Why save it for last? For a good reason: many students use it as a shortcut to working Venn diagrams. But sometimes students use it the way they do the Three Rules: as a solution manual, but without any real understanding of what validity is.

On the other hand, if your problem with Venn diagrams is in locating the compartments relevant to employing the Venn Diagram Method, then this may just be the tool you need.

The Conclusion diagram seemingly violates one of our earlier rules—*Never diagram the conclusion*! In fact, what we're doing is a separate diagram for the conclusion—a diagram distinct from the main Venn diagram for the premises. Keep that in mind. The Conclusion Diagram Method requires you to do *two* diagrams: one for the premises, and a second, separate one for the conclusion.

That's not so overwhelming a job as it might seem at first. Because of the Scholastic conventions introduced during the Middle Ages, every conclusion is legislated to be of the form: '[Operator] S are P'. That is, the S-term always precedes the P-term. (This was a convention adopted, and one we follow, to make rote memorization easier.) Since there are only four operators, we have exactly four possible Conclusion diagrams, one for each of:

'All S are P' **'No S are P'**

'Some S are P' and **'Not all S are P'**

The four possible Conclusion diagrams are as follows:

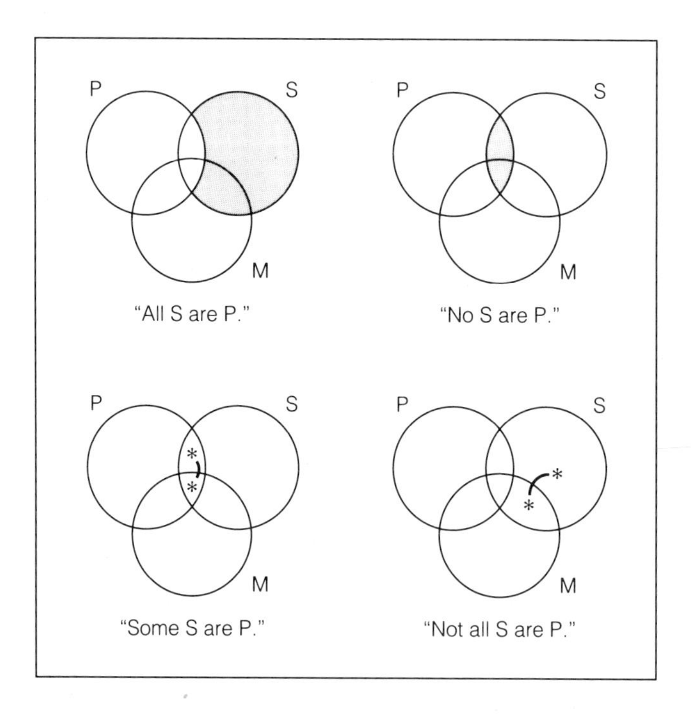

"All S are P." "No S are P." "Some S are P." "Not all S are P."

Here are some examples of how to use the Conclusion diagram:
Let's start with a simple example:

All M are P.

All S are M.

All S are P.

You'll recognize this as the familiar argument, 'Barbara'. The Venn diagram for the premises is as follows:

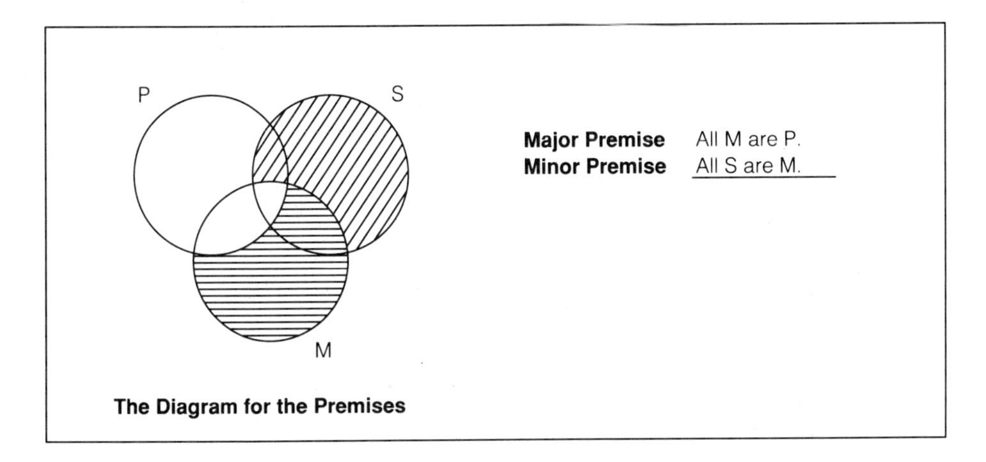

The Diagram for the Premises

On the basis of the Venn diagram, is the conclusion 'All S are P' *valid* or *invalid*? In order to determine this, we need to know which compartments are relevant to our interpretation. To simplify matters, let's number each of the compartments as we did in earlier examples:

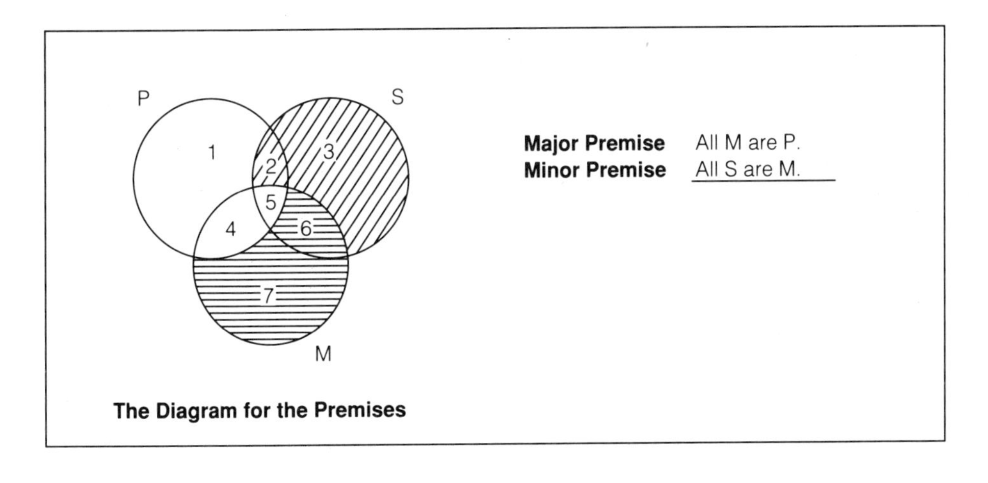

The Diagram for the Premises

Since the conclusion is 'All S are P', the compartments we need to examine are 3 and 6. Why is this so? To see why, let's construct a separate diagram for the conclusion:

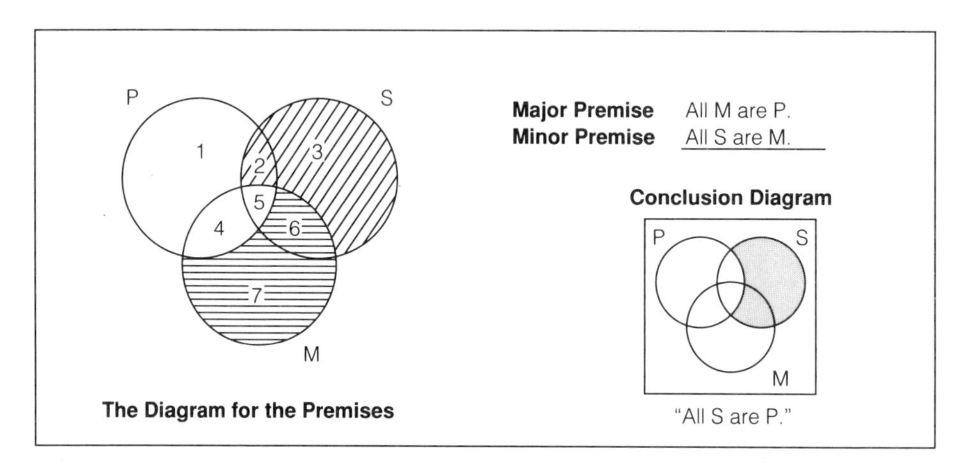

The Diagram for the Premises

The Conclusion diagram (which is included beside the regular Venn diagram for the premises) shows how compartments 3 and 6 are the ones we need to consider. The object, to show validity, is to get a match between the Conclusion diagram and the regular Venn diagram for the premises.* The rule for validity, using the Conclusion Diagram Method, is as follows:

Try to match the Conclusion diagram with the regular Venn diagram for the premises. If you get a match, the argument is *valid:* otherwise, it is *invalid*.

In the Conclusion diagram, compartments 3 and 6 are both *shaded.* When we look to the regular Venn diagram, we see that the same compartments are shaded as well. Therefore, we have obtained a match for the relevant compartments and the argument is *valid.*

Here's another example which employs a bridged asterisk; bridges are more difficult to interpret.

Suppose we're given this argument:

All M are P.

Some S are M.

Some S are P.

*This needs one important qualification, which shows that we should use Conclusion diagrams with caution: the case of certain particular conclusions where perhaps one bridged asterisk would have been shaded by the premises if we pay attention to them. See the explanation of bridged asterisks which follows.

The ordinary Venn diagram (i.e., the diagram for the premises *only*) will look like this:

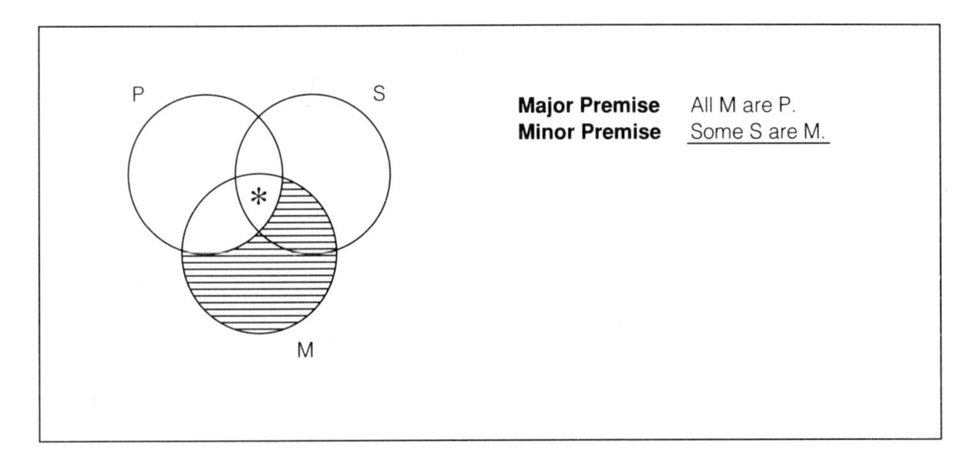

Is the conclusion 'Some S are P' *valid* or *invalid*? The way you find out is to locate the particular compartments that are relevant to the conclusion. Let's number the compartments once more to simplify our explanation:

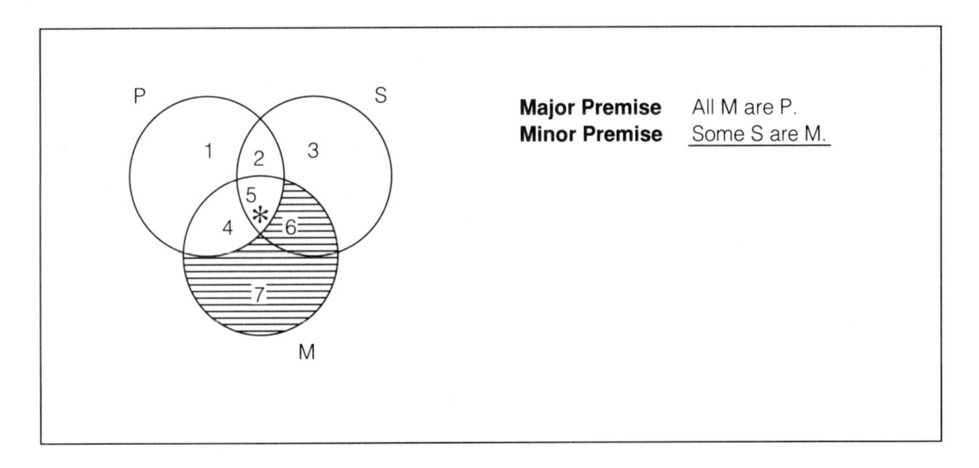

The separate Conclusion diagram shows us that compartments 2 and 5 are the ones we need to look at. We do the diagram for the conclusion separately, as in the following example:

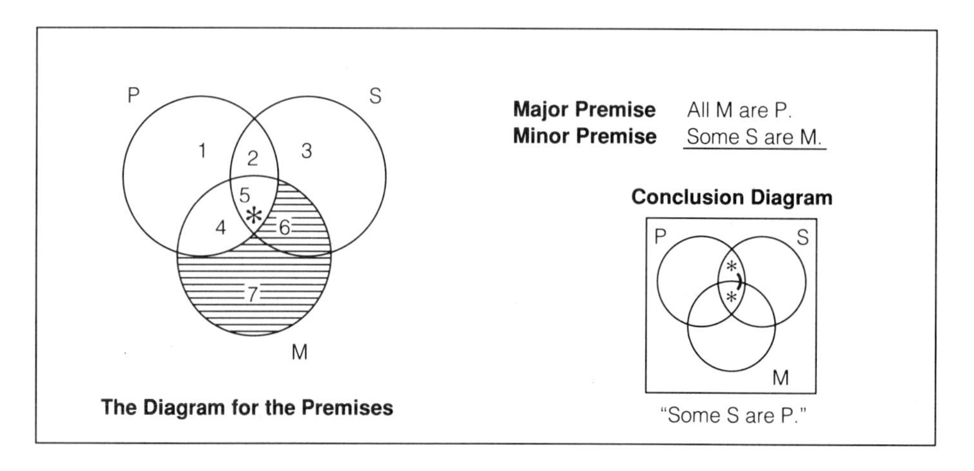

Notice that the Conclusion diagram shows a *bridged* asterisk for compartments 2 and 5. A bridged asterisk means that *one or the other or both* compartments must be occupied in order to make the argument valid. In fact, compartment 5 *is* occupied; it contains an asterisk. Therefore, the argument is valid.

The results we generated with the Conclusion Diagram Method can be extended to *any* regular Venn diagram with two premises. The method is quite simple: it requires you to work one of four diagrams for all possible conclusions.

The four Conclusion diagrams are worked out for you once more, but now they are labeled 'CD1'; 'CD2'; 'CD3'; and 'CD4' (for "Conclusion Diagram") respectively:

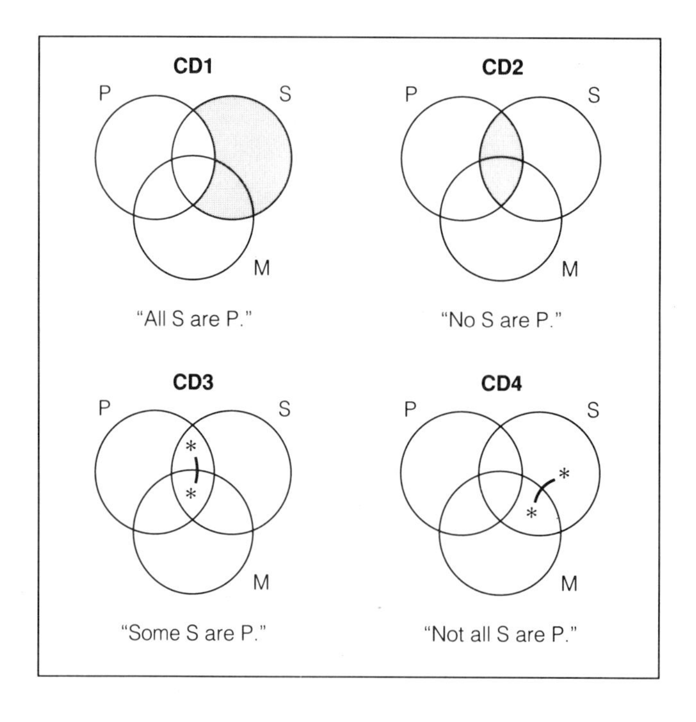

Now let's look at a final example of an argument with two premises, work the Venn diagram, and test *each* of the four Conclusion diagrams against it for validity.

Here's the argument:

No M are P.

Some S are M.

. . . And here's the worked out Venn Diagram:

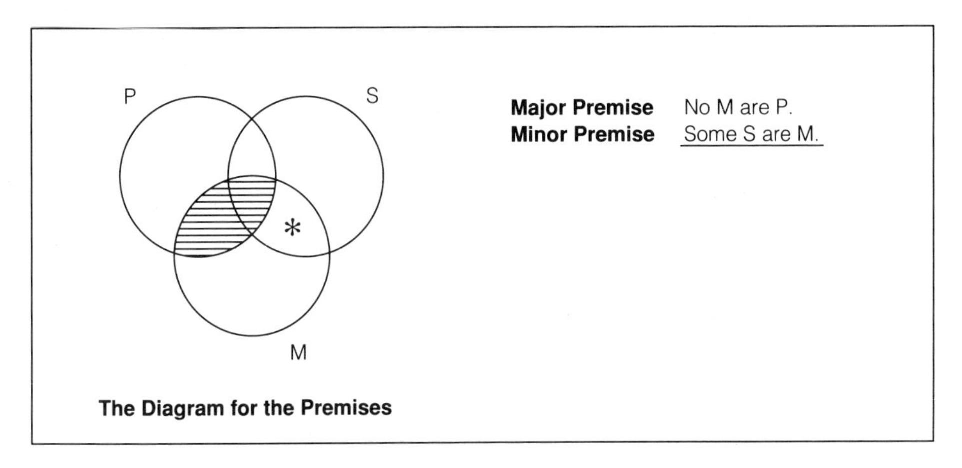

We have four possible conclusions. They're given below:

1. **'All S are P'.**
2. **'No S are P'.**
3. **'Some S are P'.**
4. **'Not all S are P'.**

Are any of these conclusions valid as tested against the Venn diagram? Let's see.

Conclusion 1 corresponds to Conclusion diagram "CD1."

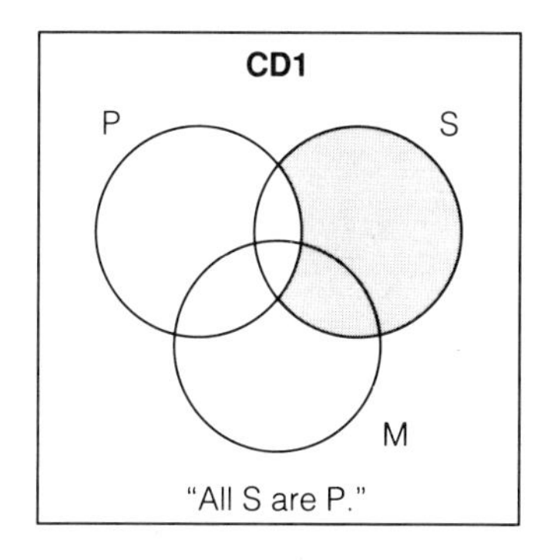

"All S are P."

Clearly, CD1 does not match the Venn diagram. In order to be valid, the Venn diagram should show that both compartments 3 and 6 are shaded. In fact, neither of them is. Therefore, the argument with 'All S are P' as its conclusion is *invalid.*

Let's consider the conclusion 'No S are P'. Does the regular Venn diagram for the premises match the Conclusion diagram for 'No S are P'? Here is the Conclusion diagram:

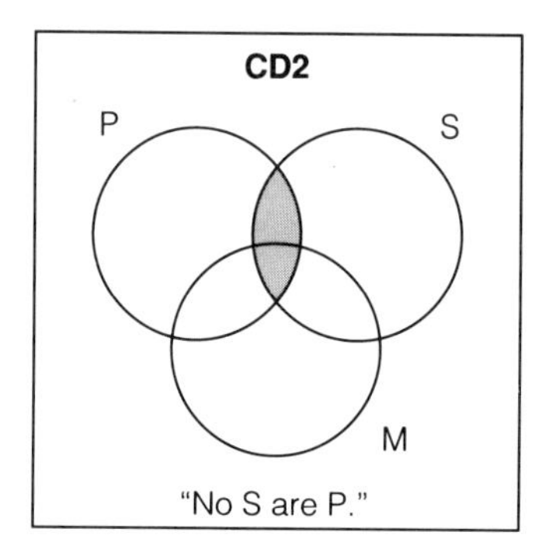

"No S are P."

If this argument were valid, then (as represented in CD2) both compartments 2 and 5 should be shaded in the Venn diagram. In fact, only compartment 5 is shaded in the regular Venn diagram for the premises. Consequently, we do *not* have a match, and the argument with 'No S are P' as the conclusion is *invalid.*

How does the argument fare with 'Some S are P' as the conclusion? Here is the Conclusion diagram:

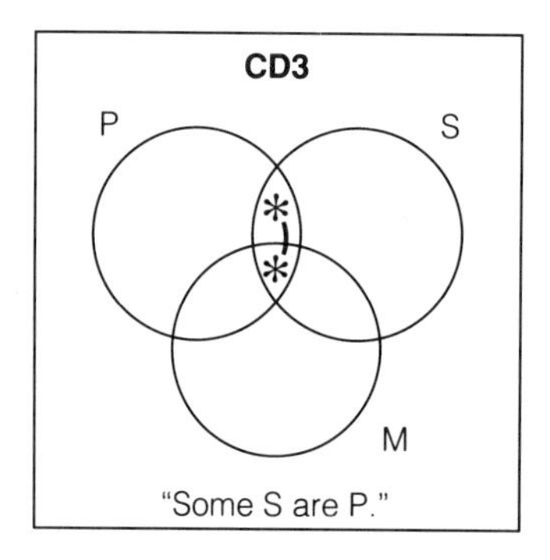

"Some S are P."

For the argument to be valid with 'Some S are P' as the conclusion, either compartment 2 or compartment 5 should contain an asterisk in the regular Venn diagram of the premises. In fact, *neither* compartment in the regular Venn diagram has an asterisk. Therefore, the argument with 'Some S are P' as the conclusion must be *invalid*.

Finally, let's look at the argument with 'Not all S are P' as the conclusion. Here is the Conclusion diagram:

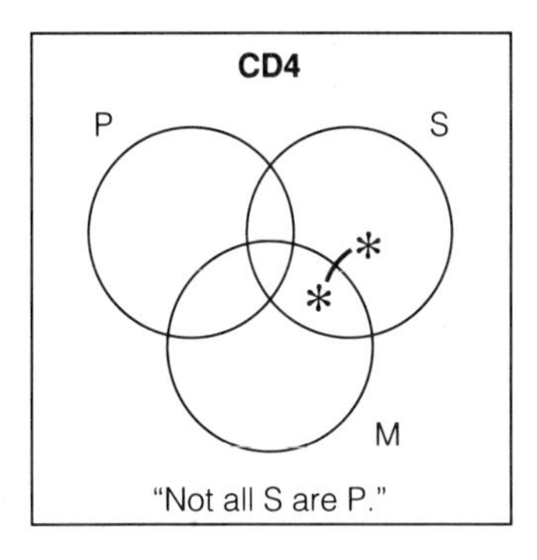

"Not all S are P."

If this argument with 'Not all S are P' as the conclusion were valid, then one or the other or both of compartments 3 and 6 should contain an asterisk in the regular Venn diagram for the premises. Here we do find a match. The Venn diagram shows that compartment 6 does indeed contain an asterisk. Consequently, the argument with 'Not all S are P' as its conclusion is indeed *valid*.

The Venn diagram and the Conclusion diagram are shown together below:

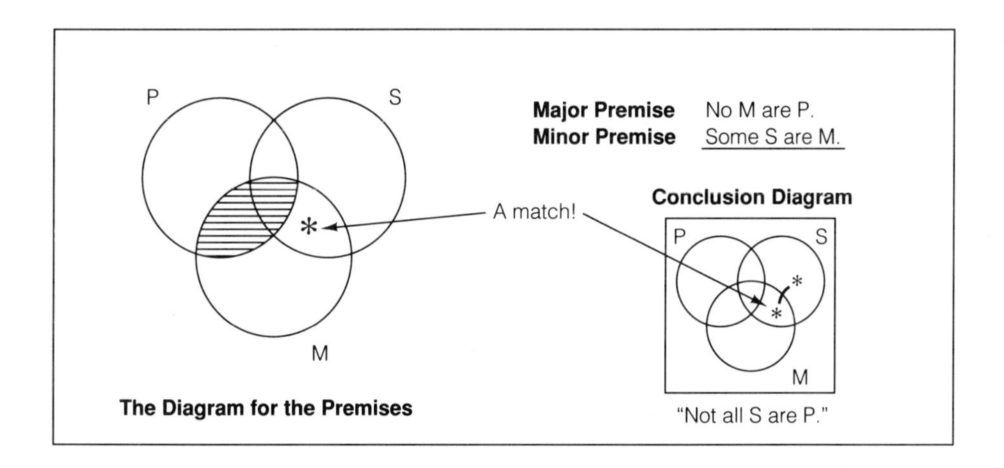

In sum: We have four possible conclusions for the argument with the premises:

No M are P.
Some S are M.

Three of the possible conclusions are *invalid;* one of them, the fourth, is *valid:*

Possible Conclusions

All S are P	*Invalid*
No S are P	*Invalid*
Some S are P	*Invalid*
Not all S are P	*Valid*

EXERCISES

DIRECTIONS: *Using the Conclusion Diagram Method, determine which (if any) of the four possible conclusions are valid when compared with the Venn diagram for the premises. The four Conclusion diagrams are reproduced for you below:*

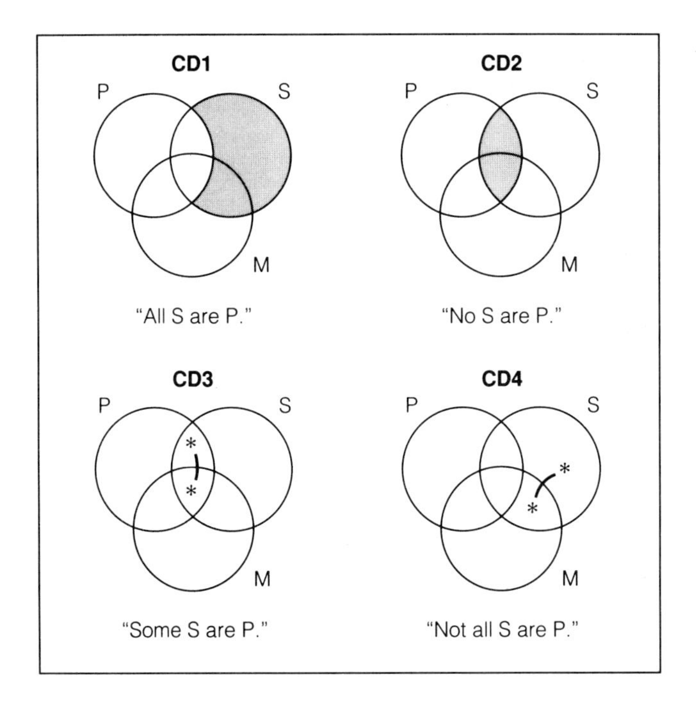

1.

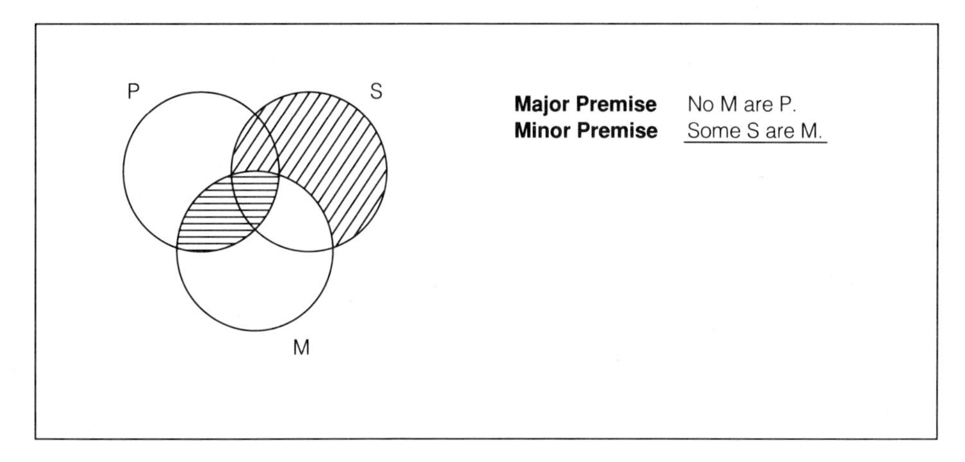

Possible Conclusions

All S are P ______
No S are P ______
Some S are P ______
Not all S are P ______

2.

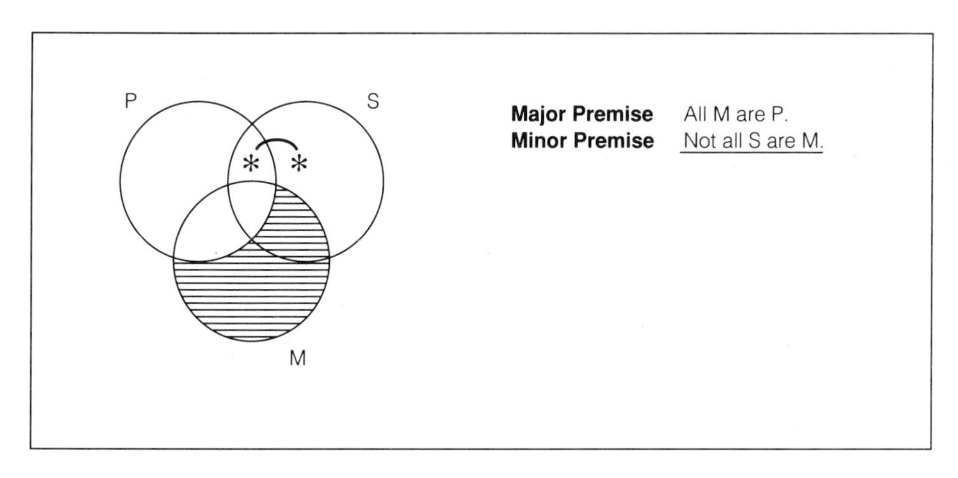

Possible Conclusions

All S are P ______
No S are P ______
Some S are P ______
Not all S are P ______

3.

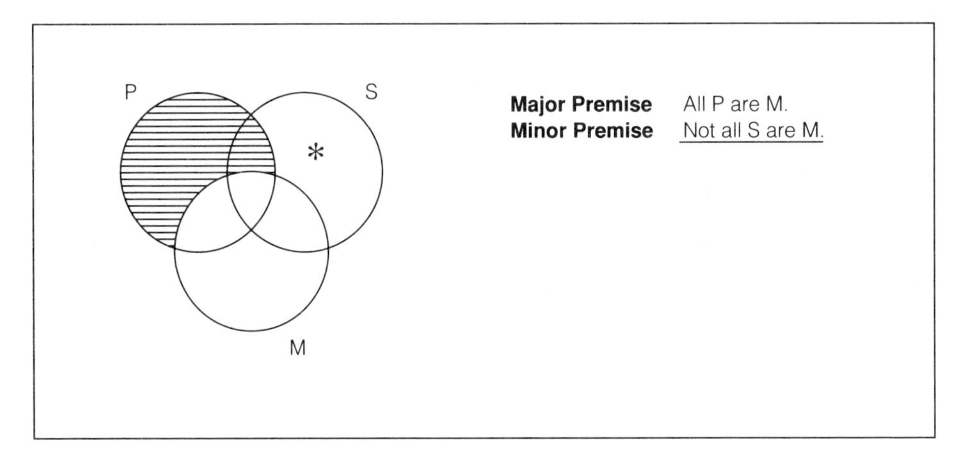

Possible Conclusions

All S are P ______
No S are P ______
Some S are P ______
Not all S are P ______

4.

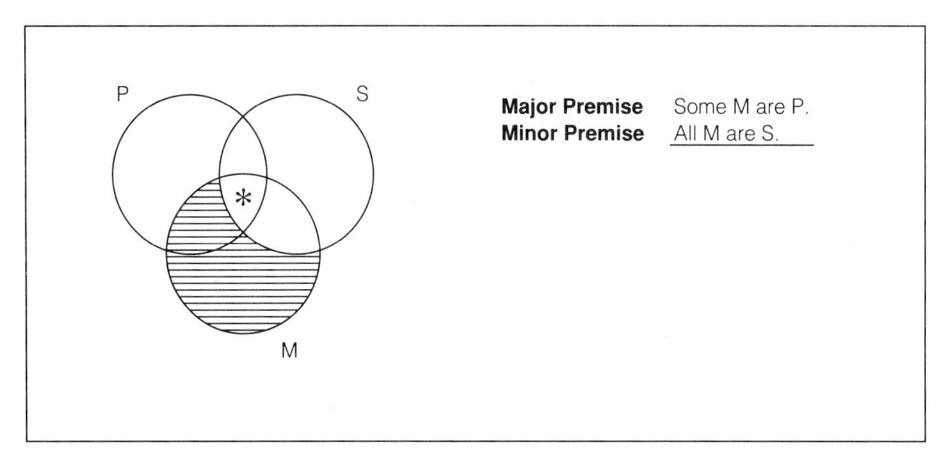

Possible Conclusions

All S are P ______
No S are P ______
Some S are P ______
Not all S are P ______

5.

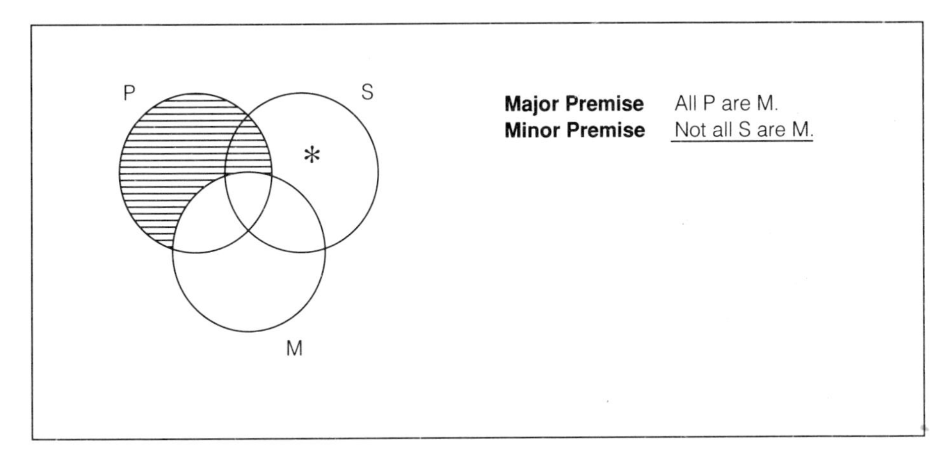

Possible Conclusions

All S are P ______
No S are P ______
Some S are P ______
Not all S are P ______

MODULE 22

APPENDIX TO UNIT TWO: A COMPLETE GUIDE TO VENN DIAGRAMS

Caution: Don't try to read this Module as you would the others. Instead, get familiar with it and use it to check your answers.

22.1 ALL THE VENN DIAGRAMS

Note: This section is an important resource. All possible Venn diagrams are reproduced here. Use this section to check all your work.

Recall that only fifteen strongly valid Scholastic schemata exist; correspondingly, we have exactly fifteen valid Venn diagram forms. (The "weakly" valid schemata are, strictly speaking, *invalid* on the Venn diagram test. This is because they need an existence assumption—a third premise—to make them valid and Venn diagrams work only with two premises.)

You might find it helpful to see *all* the possible Venn diagrams. The list is not very large; after all, you are limited to two premises. That means that there are only three terms and two moods possible for each Venn diagram since you diagram only the premises.

The exercises at the end of this section call on you to look at each of the completed Venn diagrams and see why the conclusions that are marked valid *must* be valid given the diagram, and why the conclusions marked invalid *must* be invalid. Work as many of these as you need to satisfy yourself that you see why they demonstrate validity or invalidity.

A deductive argument is **valid**, if and only if, if all the premises are true, then the conclusion must be true.

22.2 THE COMPLETE SET OF POSSIBLE VENN DIAGRAMS

Major Premise: A. (Universal Affirmative)
Minor Premise: A. (Universal Affirmative)

1.

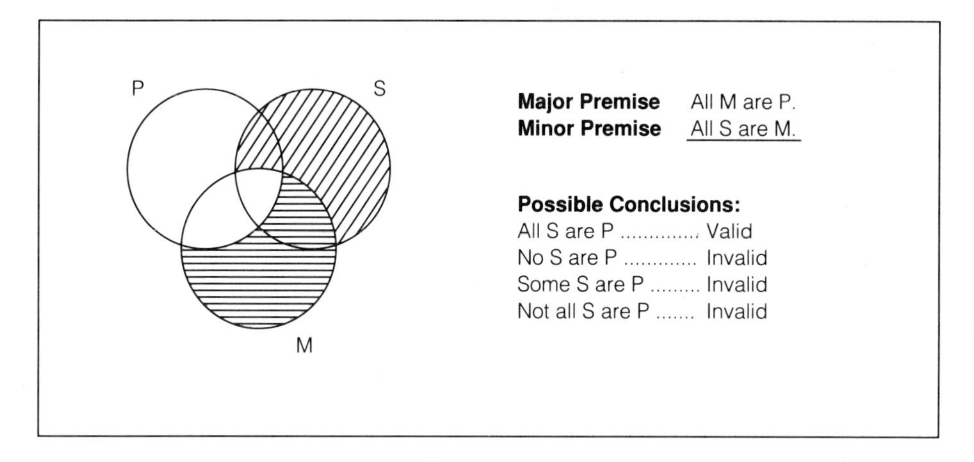

2.

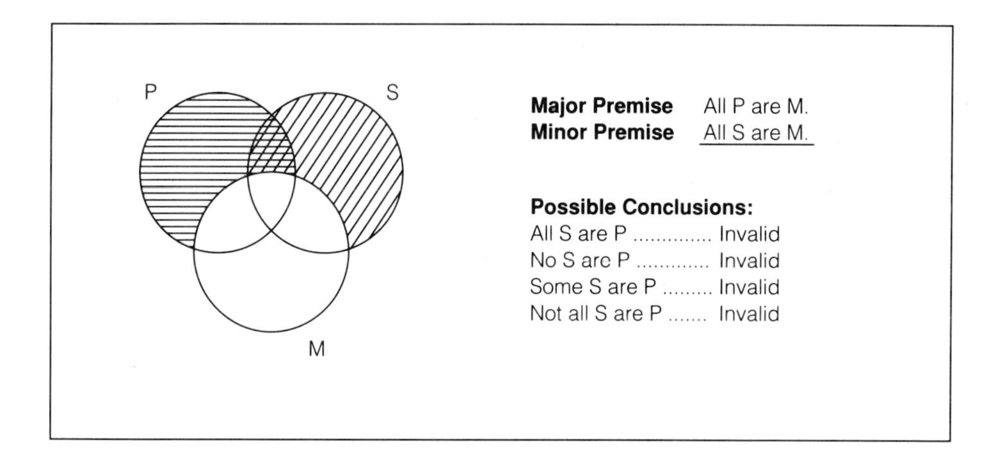

3.

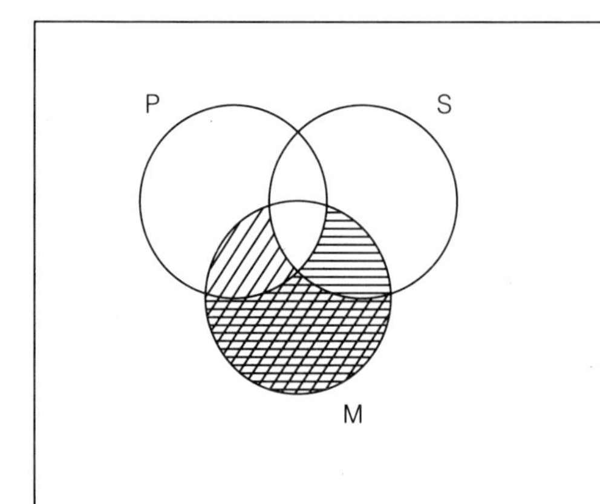

Major Premise All M are P.
Minor Premise All M are S.

Possible Conclusions:
All S are P .... Invalid
No S are P .... Invalid
Some S are P .... Invalid
Not all S are P .... Invalid

4.

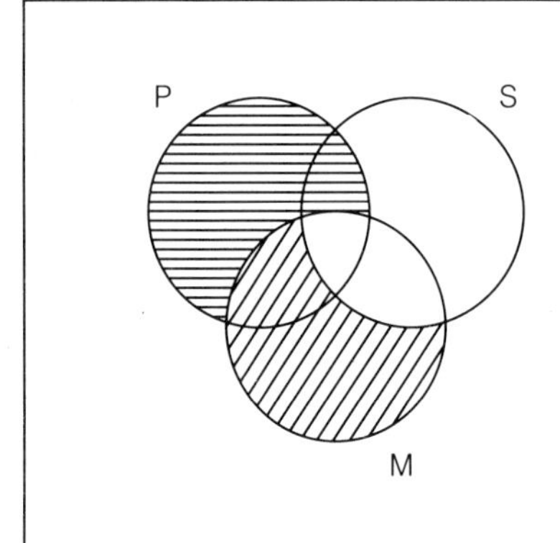

Major Premise All P are M.
Minor Premise All M are S.

Possible Conclusions:
All S are P .... Invalid
No S are P .... Invalid
Some S are P .... Invalid
Not all S are P .... Invalid

Major Premise: E. (Universal Negative)
Minor Premise: E. (Universal Negative)

5.

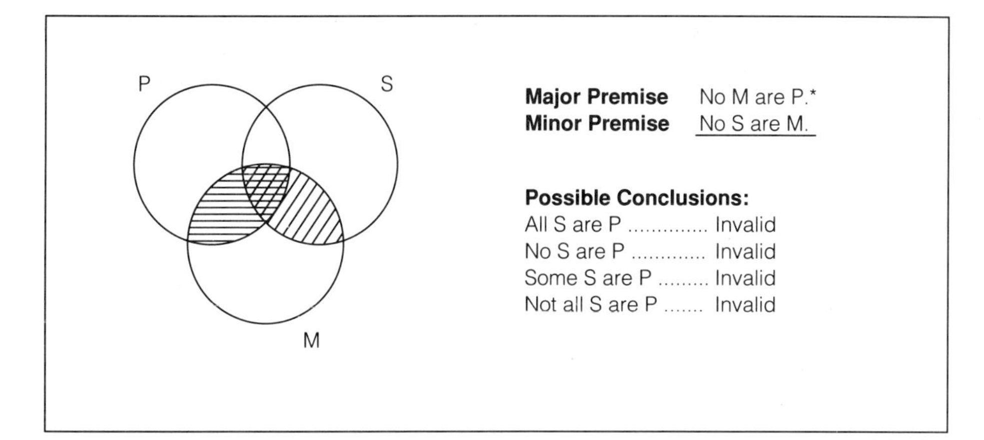

*Note: All syllogisms with two universal negative premises are *invalid*.

6. No P are M.
   No S are M.

Note: The diagram is identical to 5. All possible conclusions are *invalid*.

7. No M are P.
   No M are S.

Note: The diagram is identical to 5. All possible conclusions are *invalid*.

8. No P are M.
   No M are S.

Note: The diagram is identical to 5. All possible conclusions are *invalid*.

Major Premise: I. (Particular Affirmative)
Minor Premise: I. (Particular Affirmative)

9.

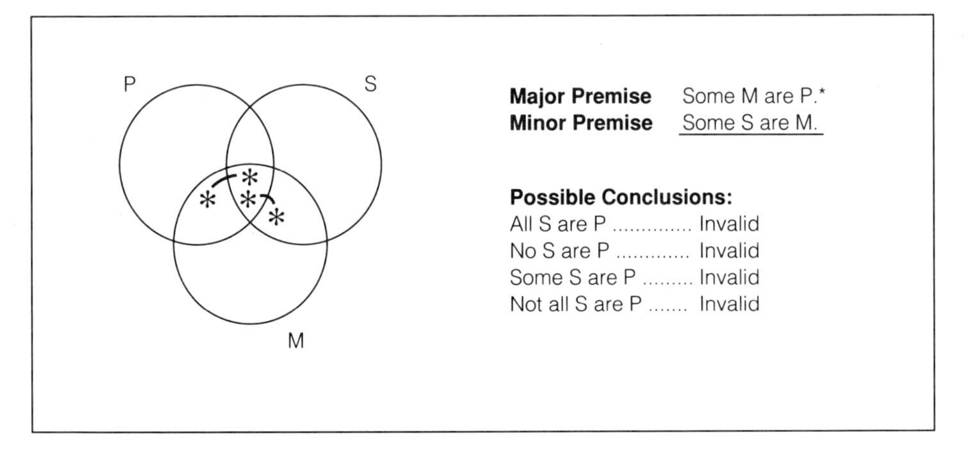

*Note: All syllogisms with two particular affirmative premises are *invalid*.

10. Some P are M.
    Some S are M.

Note: The diagram is identical to 9. All possible conclusions are *invalid*.

11. Some M are P.
    Some M are S.

Note: The diagram is identical to 9. All possible conclusions are *invalid*.

12. Some P are M.
    Some M are S.

Note: The diagram is identical to 9. All possible conclusions are *invalid*.

Major Premise: O. (Particular Negative)
Minor Premise: O. (Particular Negative)

13.

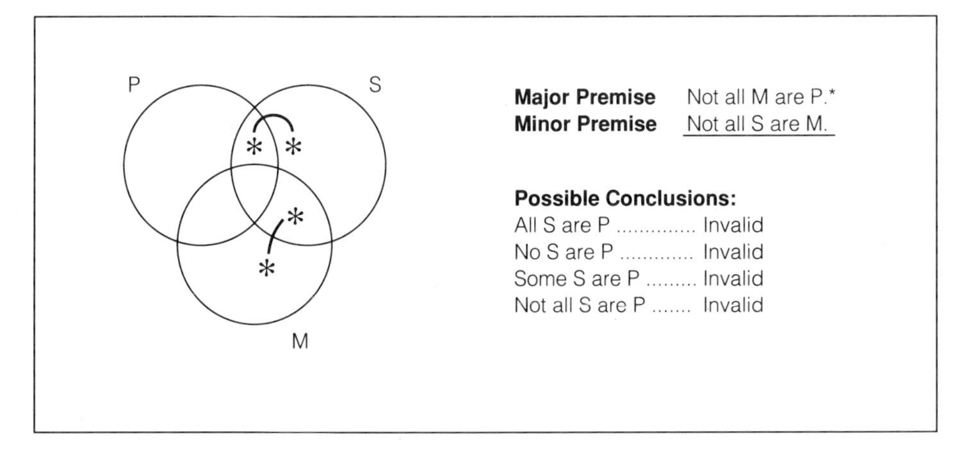

*Note: All syllogisms with two particular negative premises are *invalid*.

14.

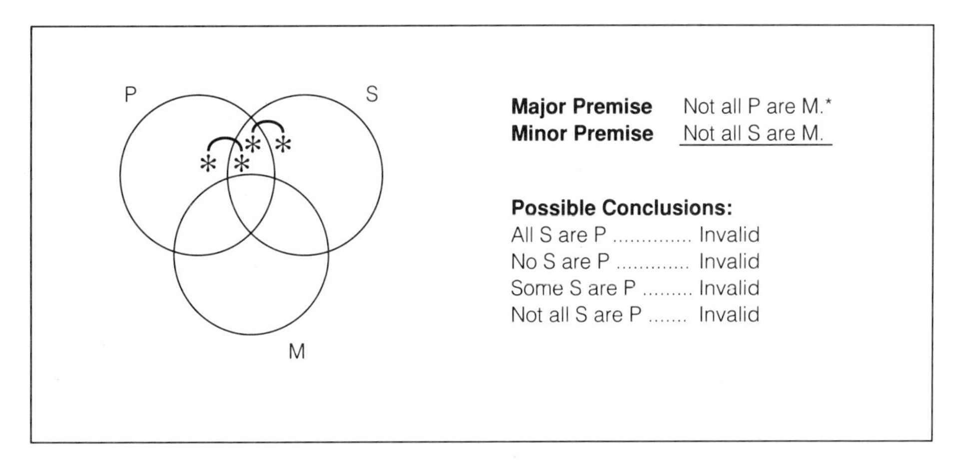

*Note: All syllogisms with two particular negative premises are *invalid*.

15.

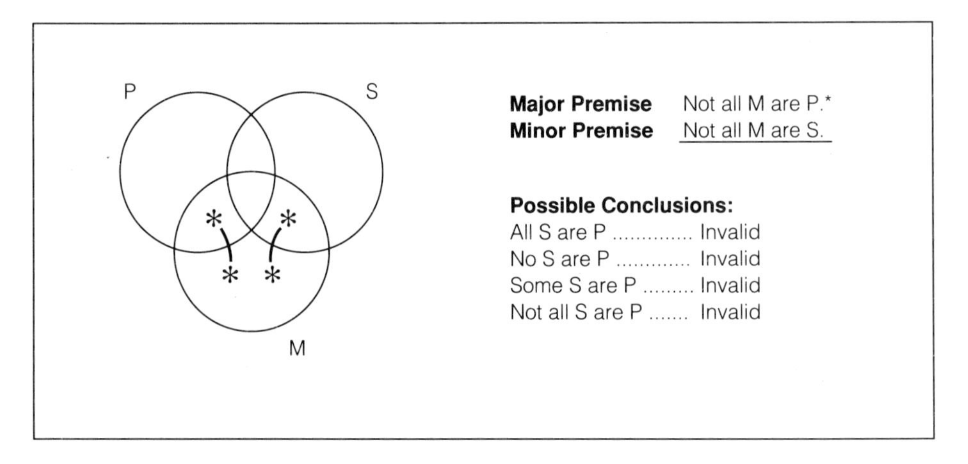

*Note: All syllogisms with two particular negative premises are *invalid*.

16.

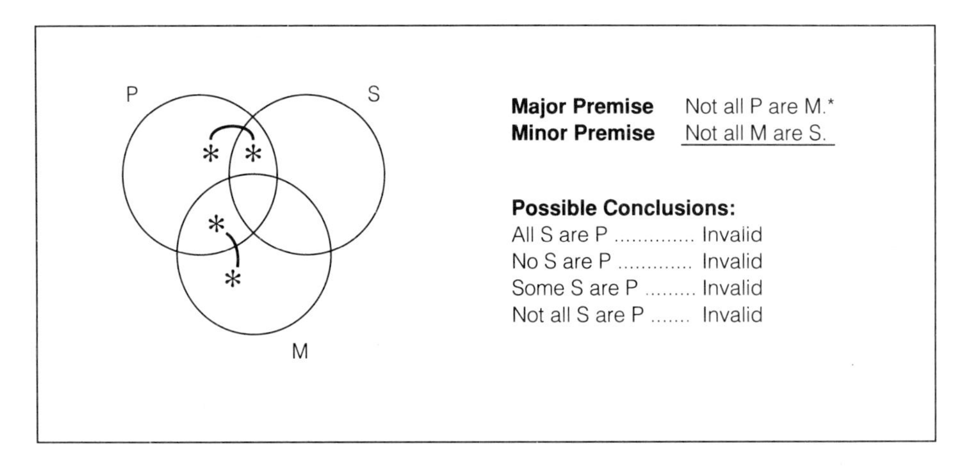

*Note: All syllogisms with two particular negative premises are *invalid*.

Major Premise: A. (Universal Affirmative)
Minor Premise: E. (Universal Negative)

17.

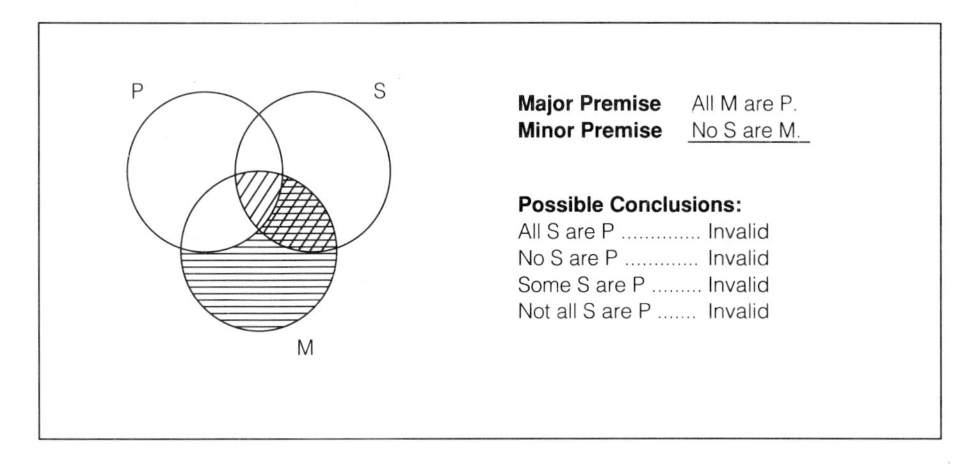

18.

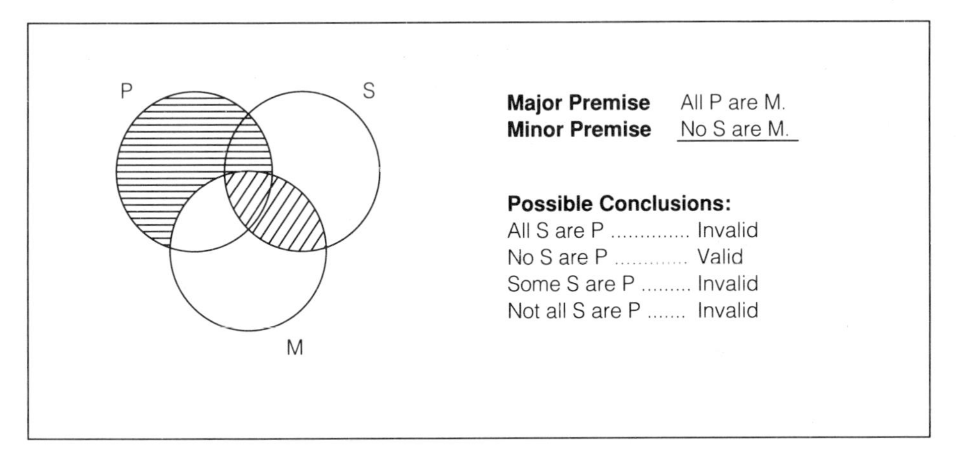

19. All M are P.
    No M are S.

Note: The diagram is identical to 17. All possible conclusions are *invalid*.

20. All P are M.
    No M are S.

Note: The diagram is identical to 18.

Possible Conclusions:

All S are P	Invalid
No S are P	Valid
Some S are P	Invalid
Not all S are P	Invalid

Major Premise: A. (Universal Affirmative) Minor Premise: I. (Particular Affirmative)

21.

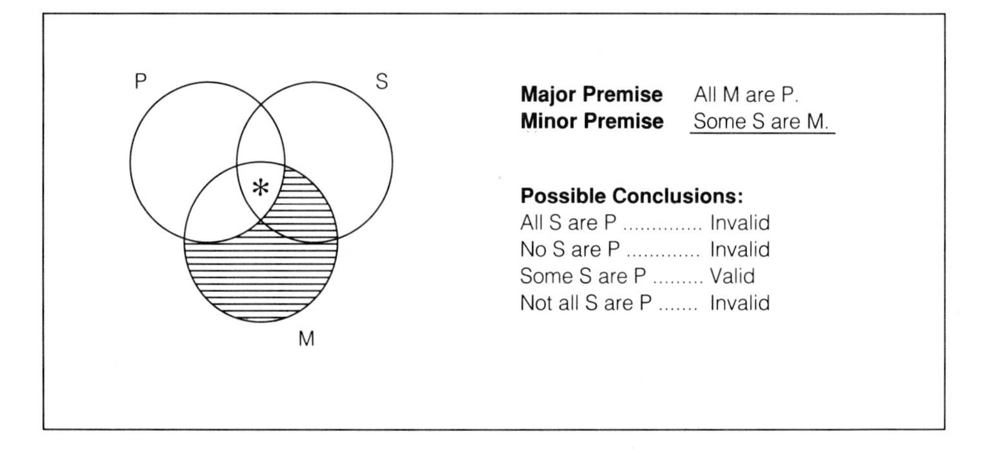

22.

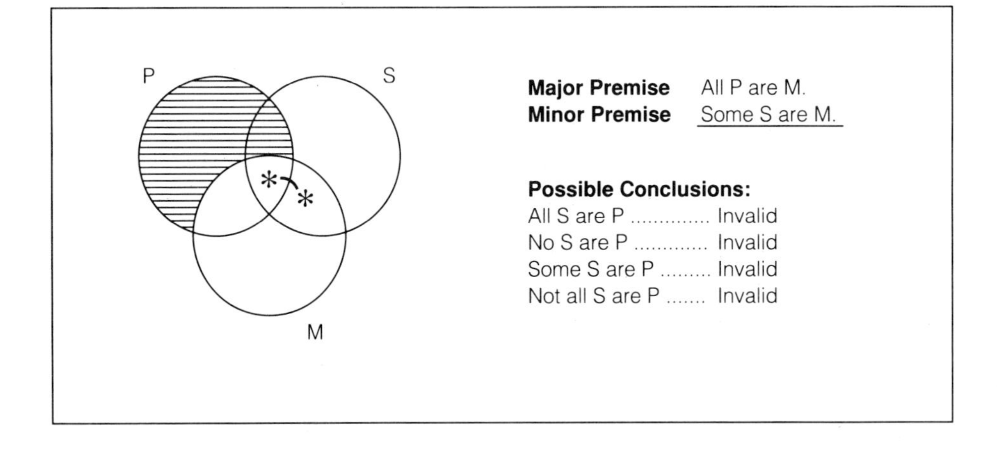

23. All M are P.
    Some M are S.

Note: The diagram is identical to 21.

Possible Conclusions:

All S are P	Invalid
No S are P	Invalid
Some S are P	Valid
Not all S are P	Invalid

24. All P are M.
    Some M are S.

Note: The diagram is identical to 22.

Possible Conclusions:

All S are P	Invalid
No S are P	Invalid
Some S are P	Invalid
Not all S are P	Invalid

Major Premise: A. (Universal Affirmative)
Minor Premise: O. (Particular Negative)

25.

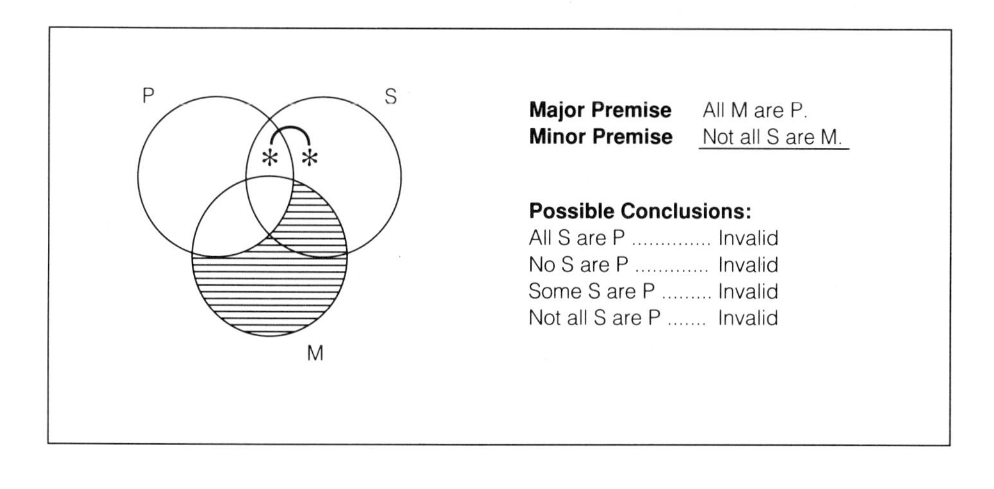

26.

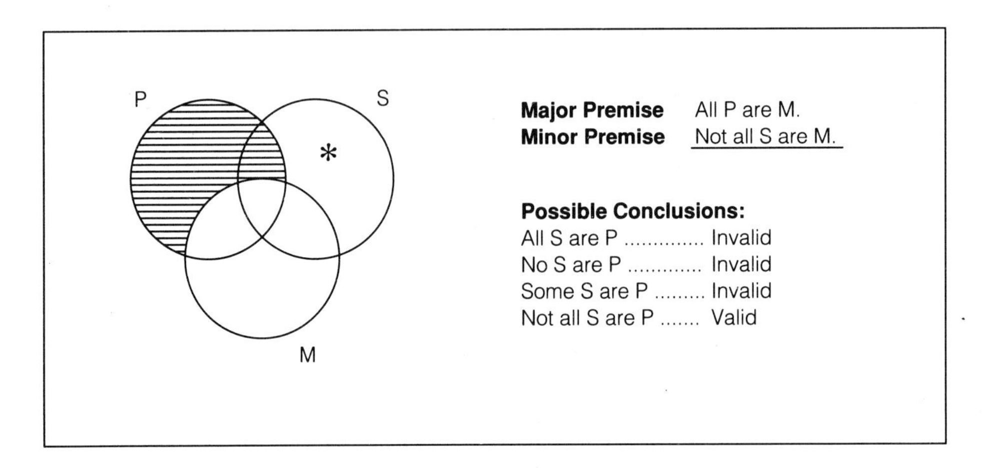
P
S
*
M
Major Premise All P are M.
Minor Premise Not all S are M.
Possible Conclusions:
All S are P .... Invalid
No S are P .... Invalid
Some S are P .... Invalid
Not all S are P .... Valid

27.

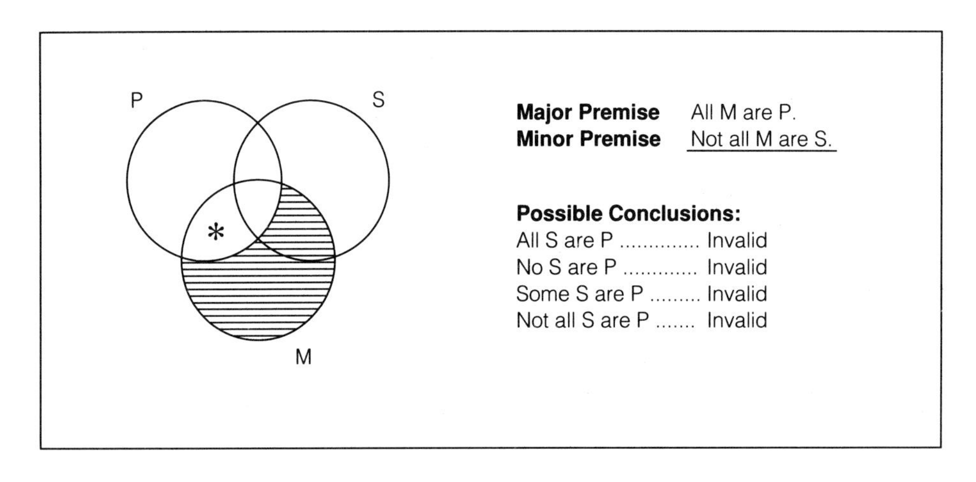
P
S
*
M
Major Premise All M are P.
Minor Premise Not all M are S.
Possible Conclusions:
All S are P .... Invalid
No S are P .... Invalid
Some S are P .... Invalid
Not all S are P .... Invalid

28.

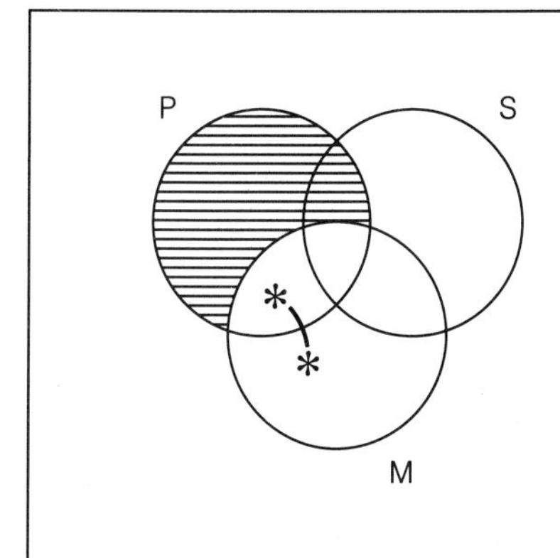

Major Premise All P are M.
Minor Premise Not all M are S.

Possible Conclusions:
All S are P .... Invalid
No S are P .... Invalid
Some S are P .... Invalid
Not all S are P .... Invalid

Major Premise: E. (Universal Negative)
Minor Premise: A. (Universal Affirmative)

29.

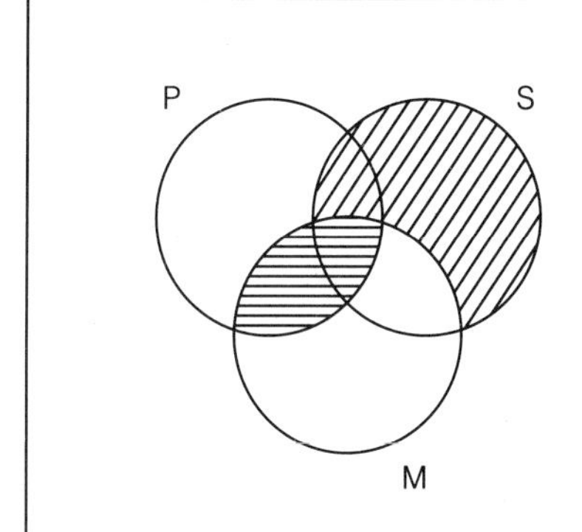

Major Premise No M are P.
Minor Premise All S are M.

Possible Conclusions:
All S are P .... Invalid
No S are P .... Valid
Some S are P .... Invalid
Not all S are P .... Invalid

30. No P are M.
All S are M.

Note: The diagram is identical to 29.
Possible Conclusions:

All S are P	Invalid
No S are P	Valid
Some S are P	Invalid
Not all S are P	Invalid

31.

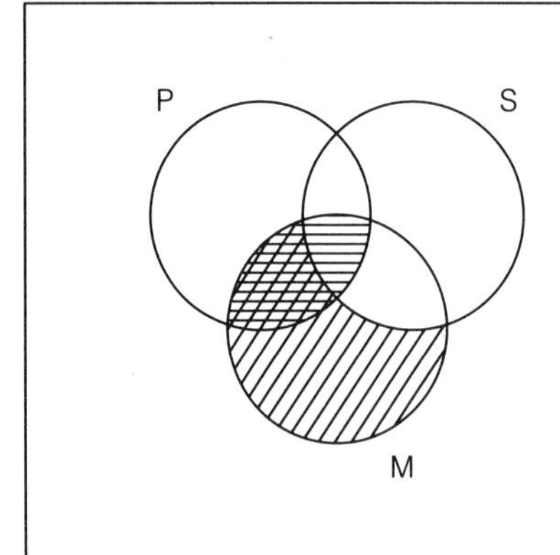

Major Premise No M are P.
Minor Premise All M are S.

Possible Conclusions:
All S are P .... Invalid
No S are P .... Invalid
Some S are P .... Invalid
Not all S are P .... Invalid

32. No P are M.
All M are S.

Note: The diagram is identical to 31.

Possible Conclusions:

All S are P	Invalid
No S are P	Invalid
Some S are P	Invalid
Not all S are P	Invalid

Major Premise: E. (Universal Negative)
Minor Premise: I. (Particular Affirmative)

33.

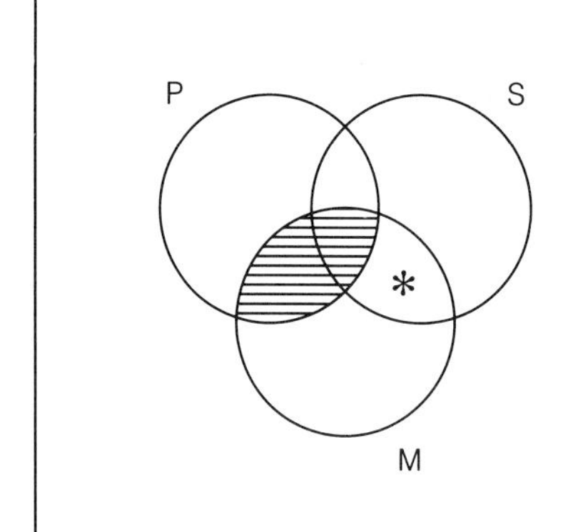

Major Premise No M are P.
Minor Premise Some S are M.

Possible Conclusions:
All S are P .... Invalid
No S are P .... Invalid
Some S are P .... Invalid
Not all S are P .... Valid

34. No P are M.
 Some S are M.

Note: The diagram is identical to 33.

Possible Conclusions:

All S are P	Invalid
No S are P	Invalid
Some S are P	Invalid
Not all S are P	Valid

35. No M are P.
 Some M are S.

Note: The diagram is identical to 33.

Possible Conclusions:

All S are P	Invalid
No S are P	Invalid
Some S are P	Invalid
Not all S are P	Valid

36. No P are M.
 Some M are S.

Note: The diagram is identical to 33.

Possible Conclusions:

All S are P	Invalid
No S are P	Invalid
Some S are P	Invalid
Not all S are P	Valid

Major Premise: E. (Universal Negative)
Minor Premise: O. (Particular Negative)

37.

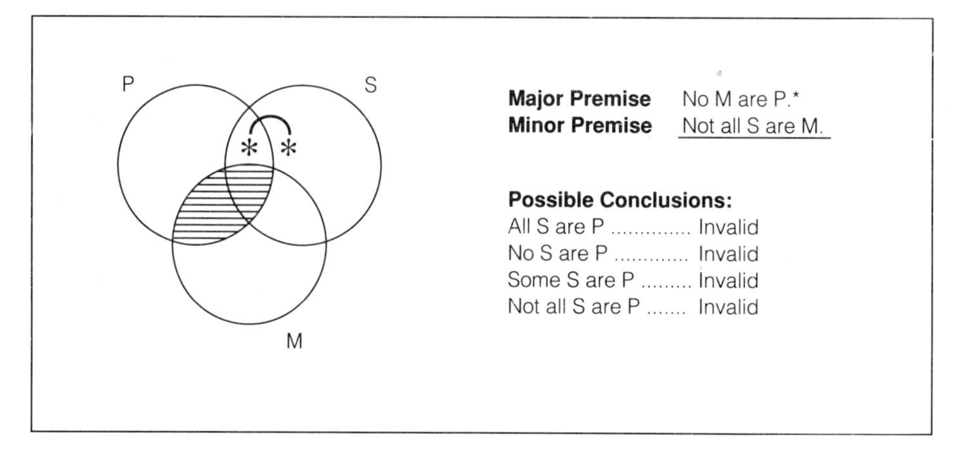

*Note: Any syllogism that has two negative premises is *invalid.*

38. No P are M.
    Not all S are M.

Note: The diagram is identical to 37.

Possible Conclusions:

All S are P	Invalid
No S are P	Invalid
Some S are P	Invalid
Not all S are P	Invalid

39.

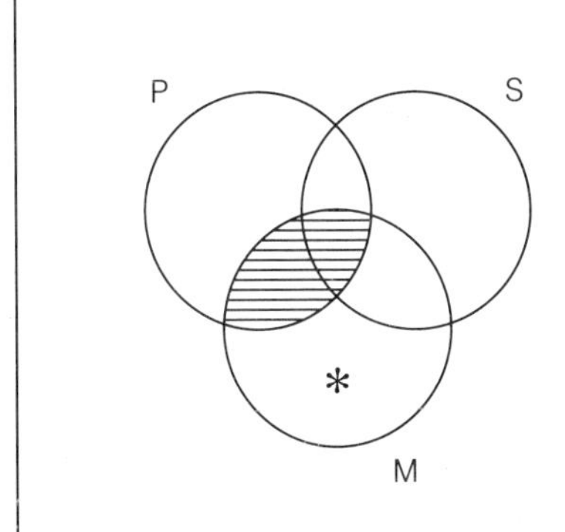

Major Premise No M are P.*
Minor Premise Not all M are S.

Possible Conclusions:
All S are P .... Invalid
No S are P .... Invalid
Some S are P .... Invalid
Not all S are P .... Invalid

*Note: Any syllogism that has two negative premises is *invalid.*

40. No P are M.
Not all M are S.

Note: The diagram is identical to 39.

Possible Conclusions:

All S are P	Invalid
No S are P	Invalid
Some S are P	Invalid
Not all S are P	Invalid

Major Premise: I. (Particular Affirmative)
Minor Premise: A. (Universal Affirmative)

41.

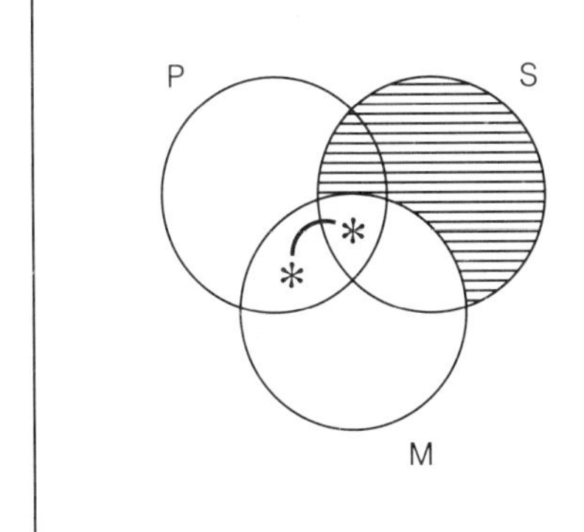

Major Premise Some M are P.
Minor Premise All S are M.

Possible Conclusions:
All S are P .... Invalid
No S are P .... Invalid
Some S are P .... Invalid
Not all S are P .... Invalid

42. Some P are M.
    All S are M.

Note: The diagram is identical to 41.

Possible Conclusions:

All S are P	Invalid
No S are P	Invalid
Some S are P	Invalid
Not all S are P	Invalid

43.

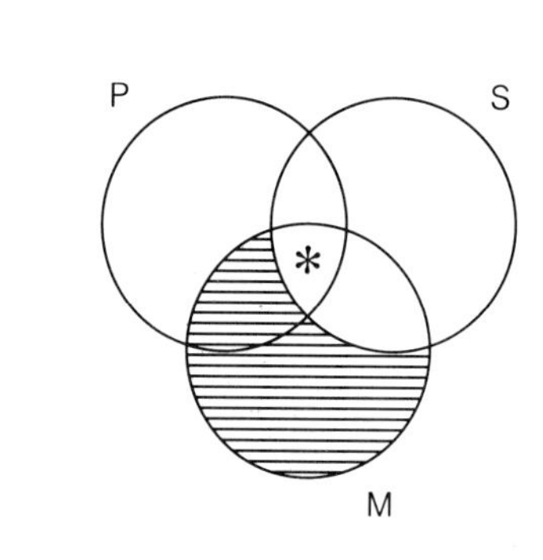

Major Premise Some M are P.
Minor Premise All M are S.

Possible Conclusions:

All S are P .... Invalid
No S are P .... Invalid
Some S are P .... Valid
Not all S are P .... Invalid

44. Some P are M.
    All M are S.

Note: The diagram is identical to 43.

Possible Conclusions:

All S are P	Invalid
No S are P	Invalid
Some S are P	Invalid
Not all S are P	Invalid

Major Premise: I. (Particular Affirmative)
Minor Premise: E. (Universal Negative)

45.

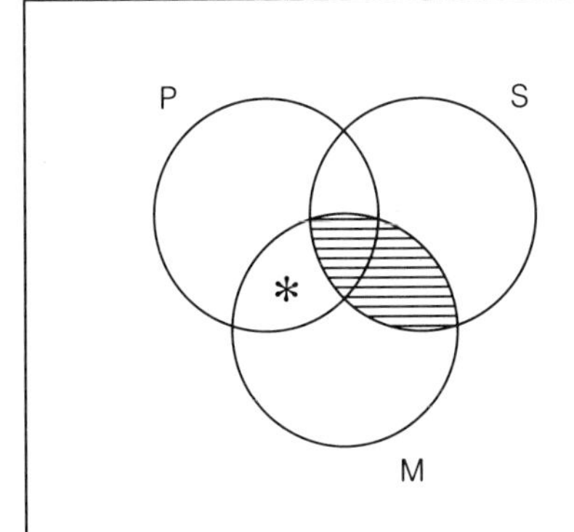

Major Premise Some M are P.
Minor Premise No S are M.

Possible Conclusions:

All S are P .... Invalid
No S are P .... Invalid
Some S are P .... Invalid
Not all S are P .... Invalid

46. Some P are M.
    No S are M.

Note: The diagram is identical to 45.

Possible Conclusions:

All S are P	Invalid
No S are P	Invalid
Some S are P	Invalid
Not all S are P	Invalid

47. Some M are P.
    No M are S.

Note: The diagram is identical to 45.

Possible Conclusions:

All S are P	Invalid
No S are P	Invalid
Some S are P	Invalid
Not all S are P	Invalid

48. Some P are M.
    No M are S.

Note: The diagram is identical to 45.

Possible Conclusions:

All S are P	Invalid
No S are P	Invalid
Some S are P	Invalid
Not all S are P	Invalid

Major Premise: I. (Particular Affirmative)
Minor Premise: O. (Particular Negative)

49.

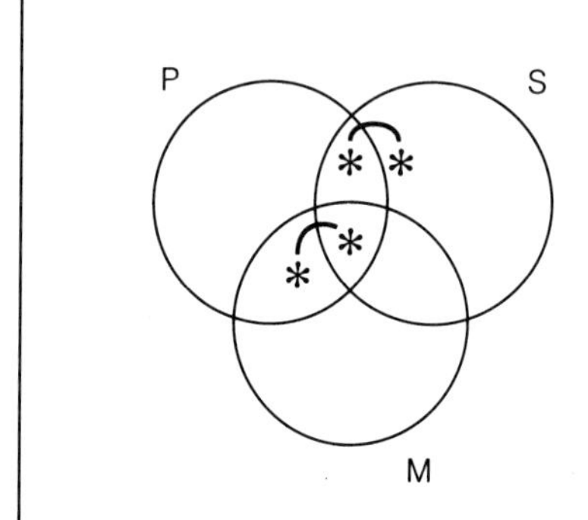

Major Premise Some M are P.*
Minor Premise Not all S are M.

Possible Conclusions:
All S are P .... Invalid
No S are P .... Invalid
Some S are P .... Invalid
Not all S are P .... Invalid

*Note: Diagrams with two bridges are always *invalid*.

50. Some P are M.
No S are M.

Note: The diagram is identical to 49.
Possible Conclusions:

All S are P	Invalid
No S are P	Invalid
Some S are P	Invalid
Not all S are P	Invalid

51.

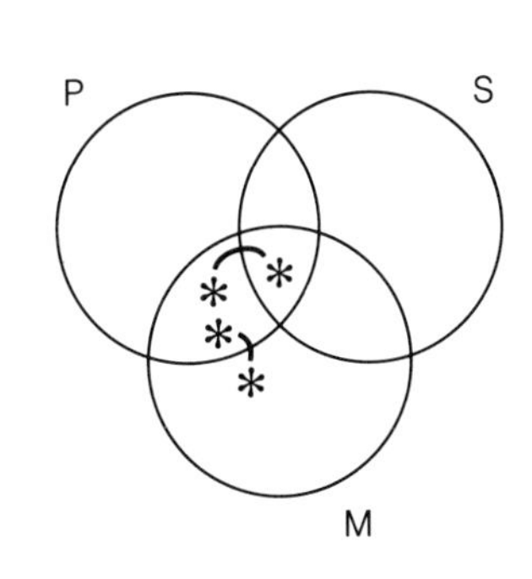

Major Premise Some M are P.*
Minor Premise Not all M are S.

Possible Conclusions:
All S are P .... Invalid
No S are P .... Invalid
Some S are P .... Invalid
Not all S are P .... Invalid

*Note: Diagrams with two bridges are always *invalid*.

52. Some P are M.
No M are S.

Note: The diagram is identical to 51.

Possible Conclusions:

All S are P	Invalid
No S are P	Invalid
Some S are P	Invalid
Not all S are P	Invalid

Major Premise: O. (Particular Negative)
Minor Premise: A. (Universal Affirmative)

53.

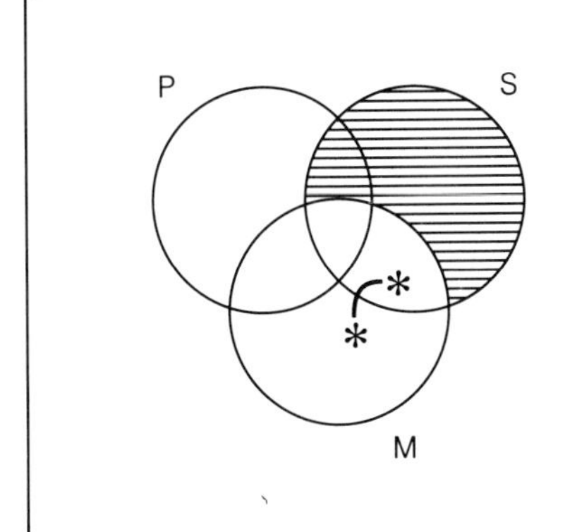

Major Premise Not all M are P.
Minor Premise All S are M.

Possible Conclusions:
All S are P .... Invalid
No S are P .... Invalid
Some S are P .... Invalid
Not all S are P .... Invalid

54.

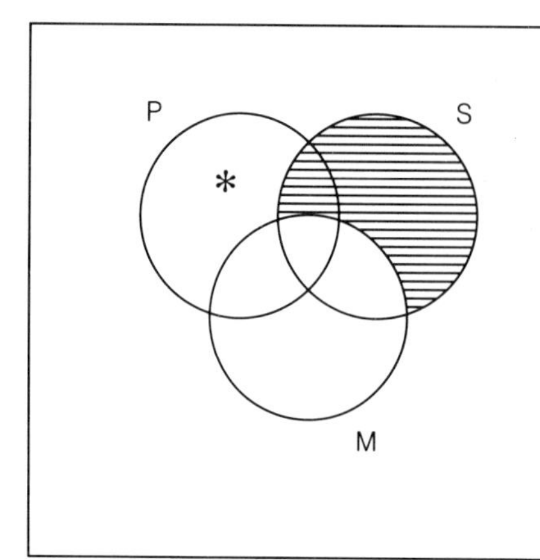

Major Premise Not all P are M.
Minor Premise All S are M.

Possible Conclusions:
All S are P .... Invalid
No S are P .... Invalid
Some S are P .... Invalid
Not all S are P .... Invalid

55.

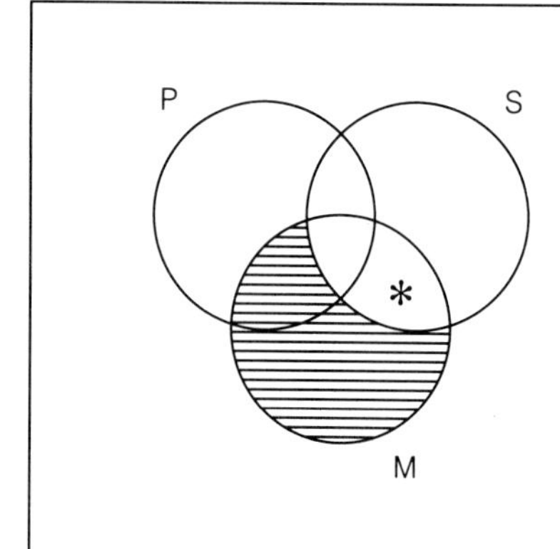

Major Premise Not all M are P.
Minor Premise All M are S.

Possible Conclusions:
All S are P .... Invalid
No S are P .... Invalid
Some S are P .... Invalid
Not all S are P .... Valid

56.

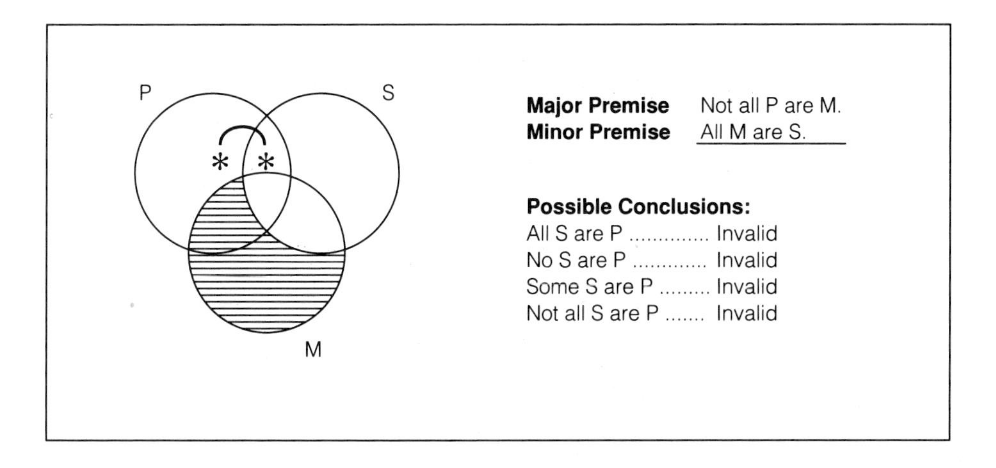

Major Premise: O. (Particular Negative)
Minor Premise: E. (Universal Negative)

57.

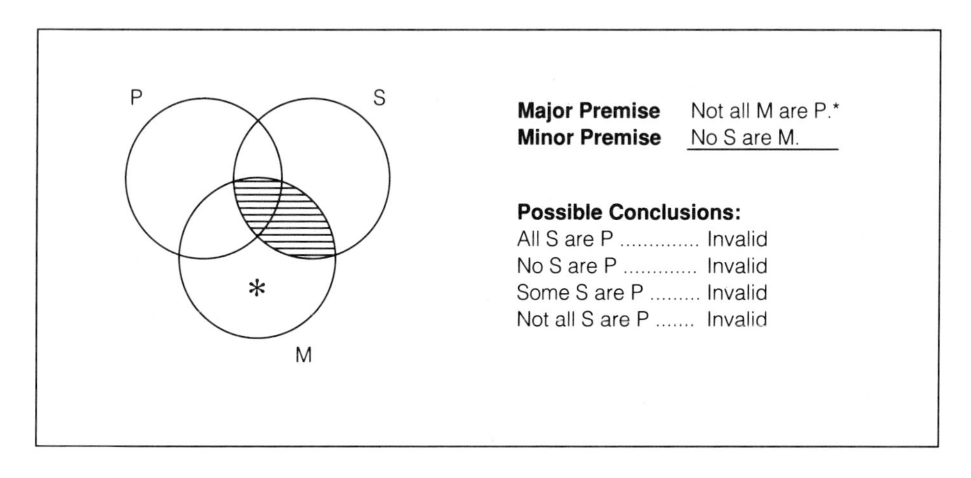

*Note: Syllogisms with two negative premises are always *invalid.*

58.

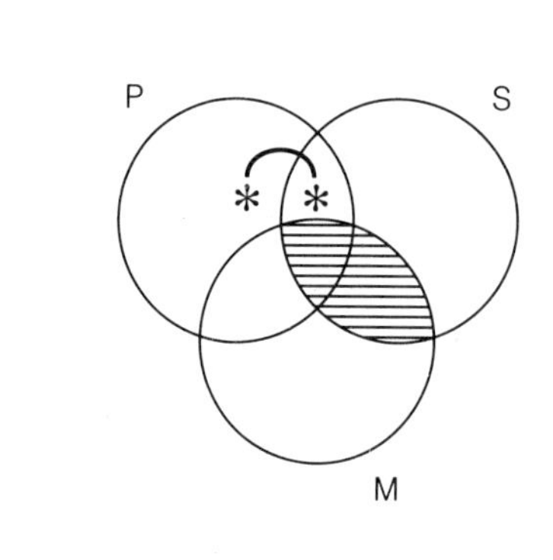

Major Premise Not all P are M.*
Minor Premise No S are M.

Possible Conclusions:
All S are P .... Invalid
No S are P .... Invalid
Some S are P .... Invalid
Not all S are P .... Invalid

*Note: Syllogisms with two negative premises are always *invalid.*

59. Not all M are P.
No M are S.

Note: The diagram is identical to 57.
Possible Conclusions:

All S are P	Invalid
No S are P	Invalid
Some S are P	Invalid
Not all S are P	Invalid

60. Not all P are M.
No M are S.

Note: The diagram is identical to 58.
Possible Conclusions:

All S are P	Invalid
No S are P	Invalid
Some S are P	Invalid
Not all S are P	Invalid

Major Premise: O. (Particular Negative)
Minor Premise: I. (Particular Affirmative)

61.

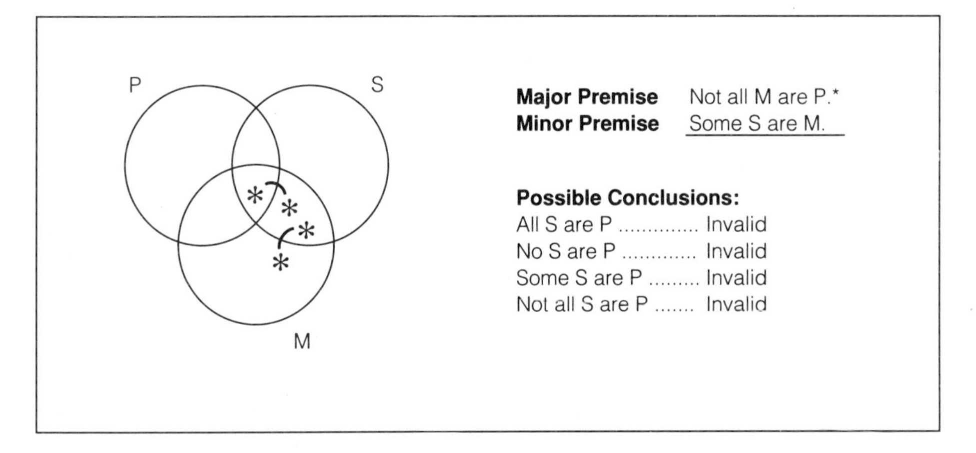

*Note: Diagrams with two bridges are always *invalid*.

62.

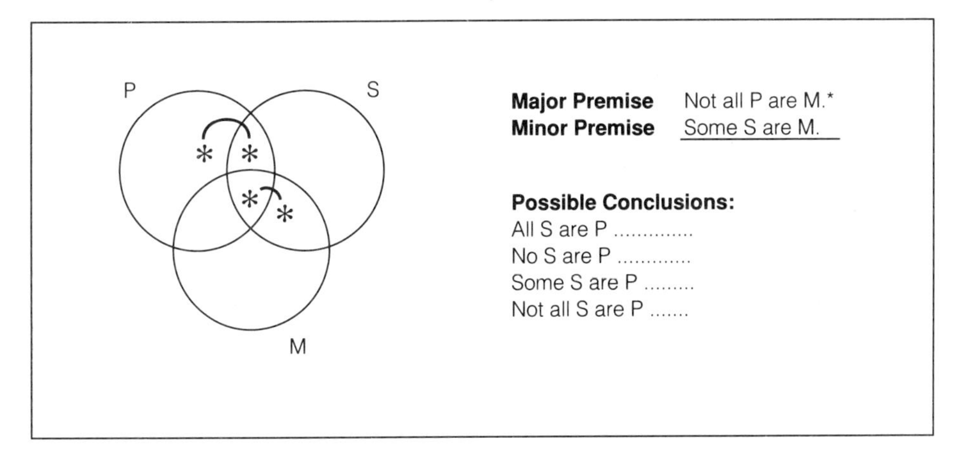

*Note: Diagrams with two bridges are always *invalid*.

63. Not all M are P.
Some M are S.

Note: The diagram is identical to 61.

Possible Conclusions:

All S are P	Invalid
No S are P	Invalid
Some S are P	Invalid
Not all S are P	Invalid

64. Not all P are M.
Some M are S.

Note: The diagram is identical to 62.

Possible Conclusions:

All S are P	Invalid
No S are P	Invalid
Some S are P	Invalid
Not all S are P	Invalid

EXERCISES FOR UNIT TWO

DIRECTIONS: *Determine whether the following syllogisms are* valid *or* invalid *by means of the Venn diagram tests and the Three Rules Method.*

1. All surfers are nonchalant.
Not all nonchalant people are agile.
∴ Not all surfers are agile.

2. Some cover lovers are bald.
Some cover lovers do not drive pickups.
∴ All bald people drive pickups.

3. No logic students wear high tops.
Some people who wear high tops are fleet-footed.
∴ Not all logic students are fleet-footed.

4. A dolphin is a finny friend.
Not all sharks are finny friends.
∴ Not all sharks are dolphins.

5. All Kit-Kat munchers have acne.
Some people who have acne eat Snickers.
∴ Some Snickers eaters are Kit-Kat munchers.

6. All earth bikes are speedy.
No speedy bikes are ten-speeds.
∴ No ten-speeds are earth bikes.

7. All light beers taste bad.
Some light beers are bottled in clear bottles.
∴ Some beers bottled in clear bottles are bad tasting beers.

8. A few hurricanes have wreaked havoc in North Carolina, but every hurricane is potentially dangerous to us so some potentially dangerous things are known to wreak havoc here.

9. Not all World Series have been dull, but they've all been televised. Therefore, some things on television have not been boring.

10. Some Bigfoot are snow creatures because all Bigfoot have thick coats and no creatures that wear thick coats are snow creatures.

SOLUTIONS TO EXERCISES FOR VENN DIAGRAMS

1.

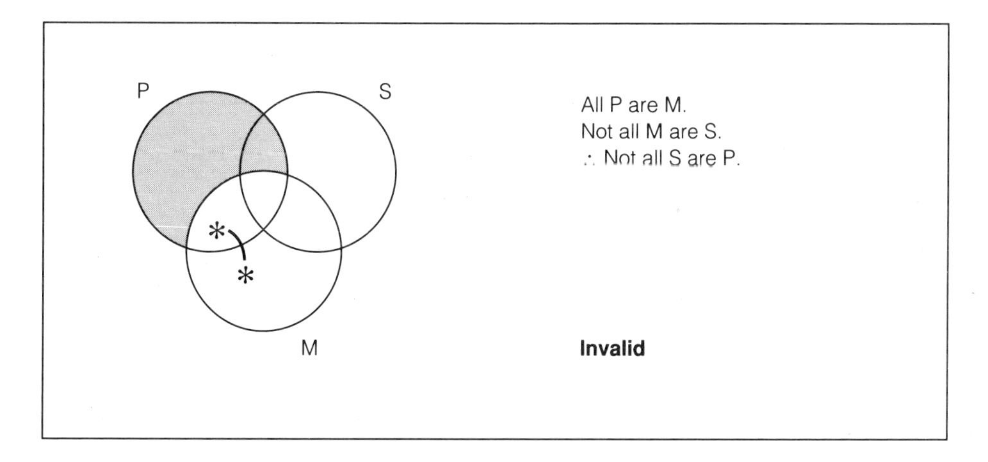

2.

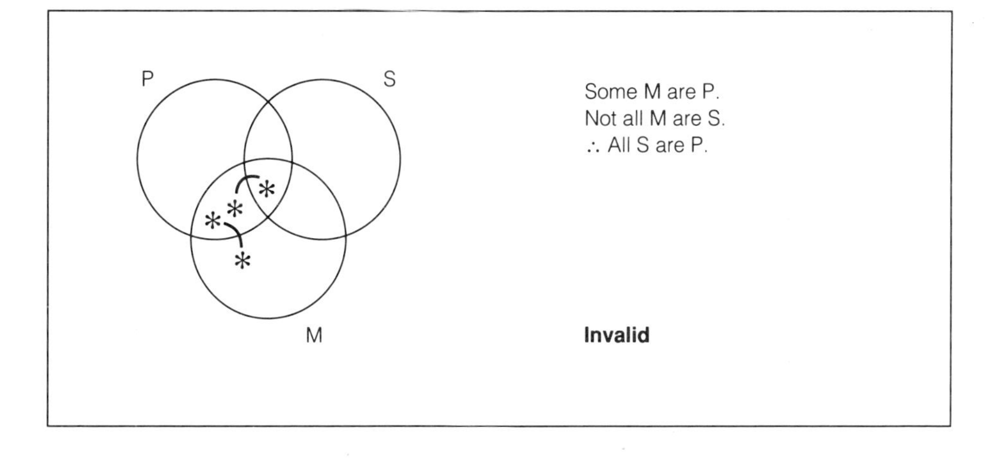

3.

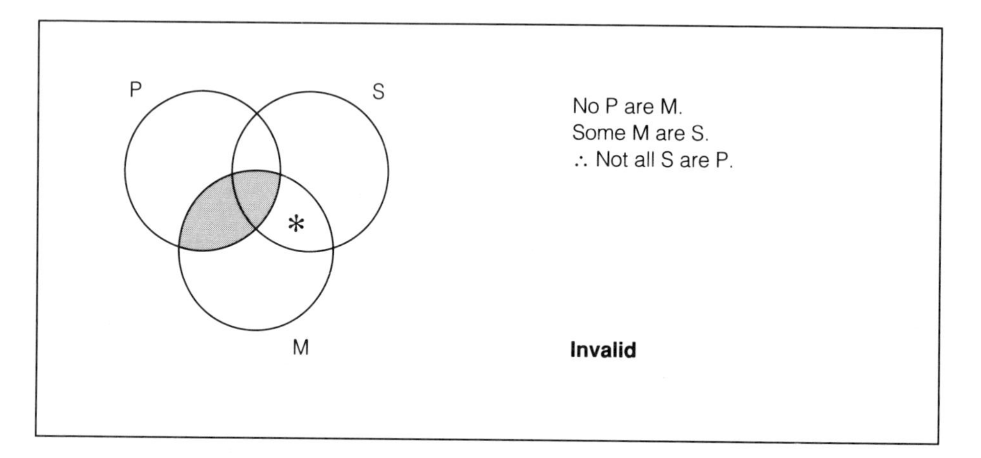
P
S
M
*
No P are M.
Some M are S.
∴ Not all S are P.
Invalid

4.

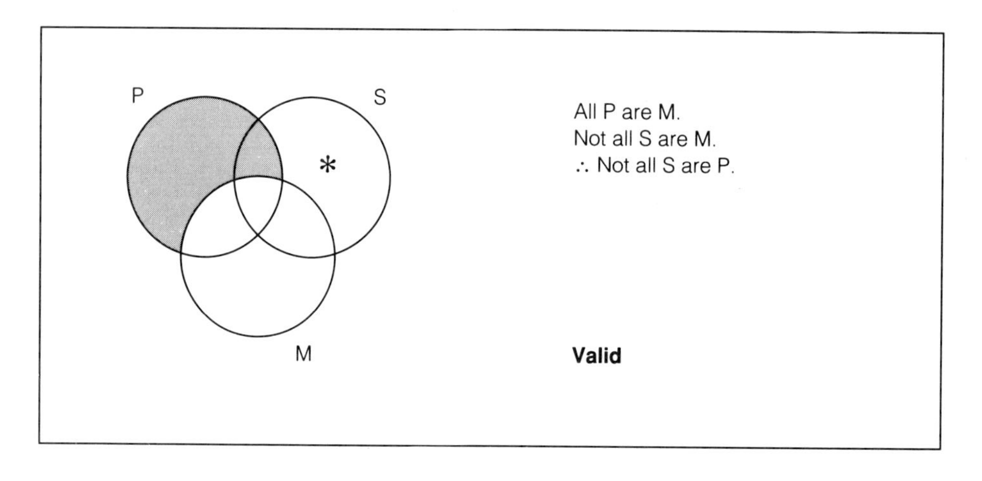
P
S
M
*
All P are M.
Not all S are M.
∴ Not all S are P.
Valid

5.

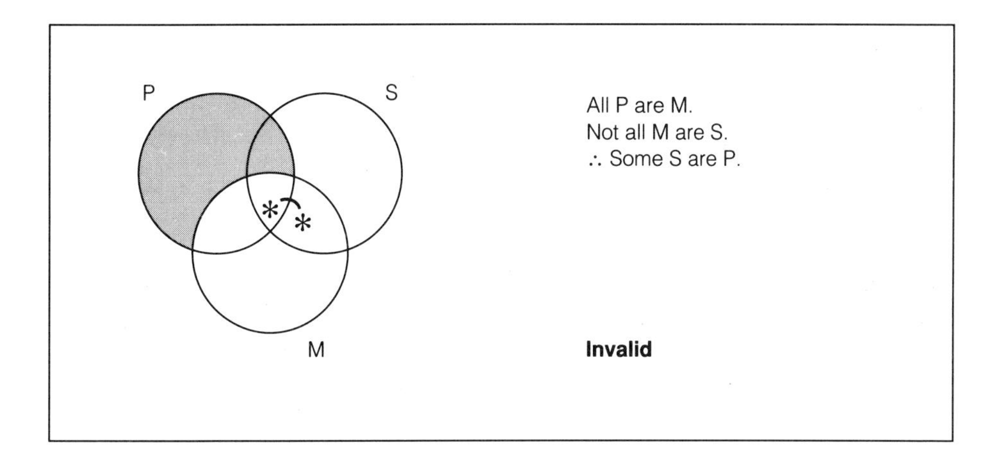
P
S
M
All P are M.
Not all M are S.
∴ Some S are P.
Invalid

6.

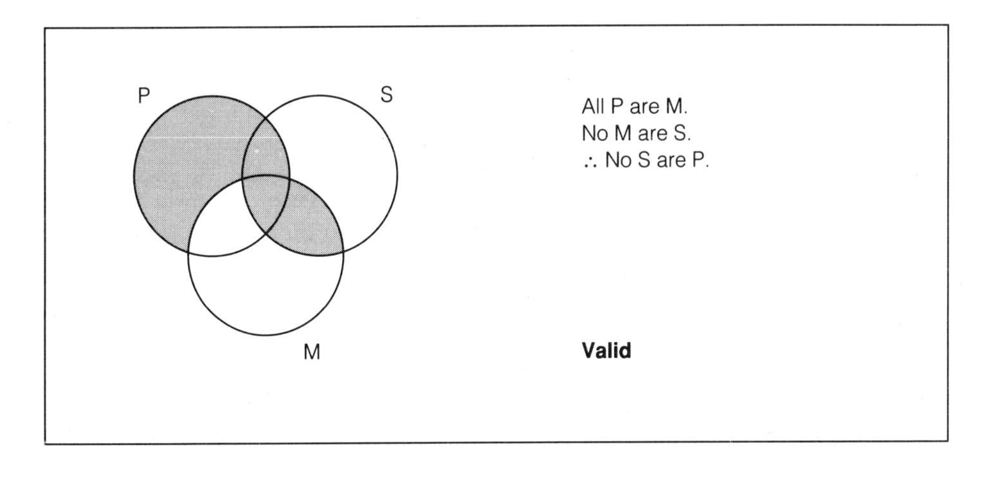
P
S
M
All P are M.
No M are S.
∴ No S are P.
Valid

7.

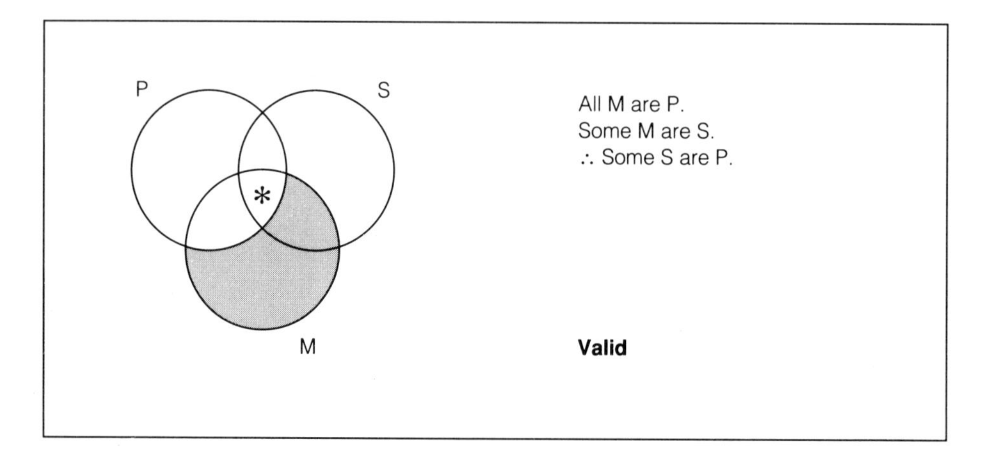
P
S
M
*
All M are P.
Some M are S.
∴ Some S are P.
Valid

8.

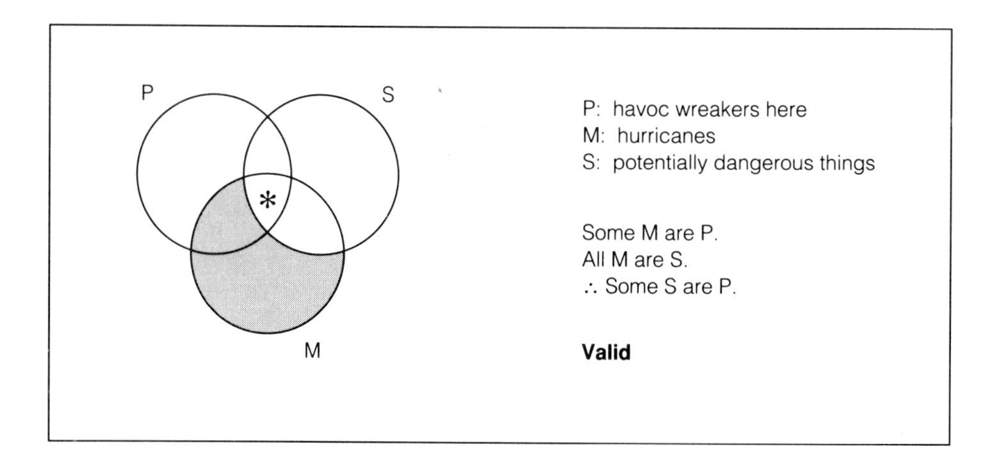
P
S
M
*
P: havoc wreakers here
M: hurricanes
S: potentially dangerous things
Some M are P.
All M are S.
∴ Some S are P.
Valid

9.

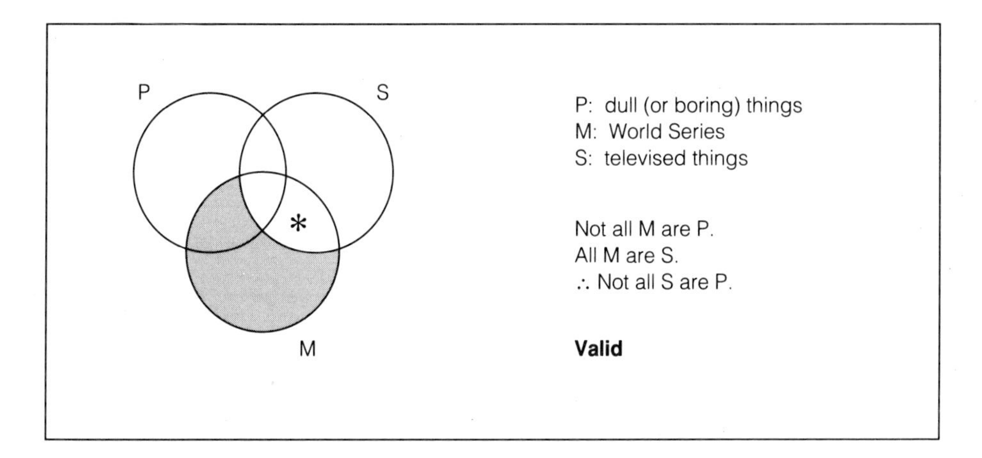

10.

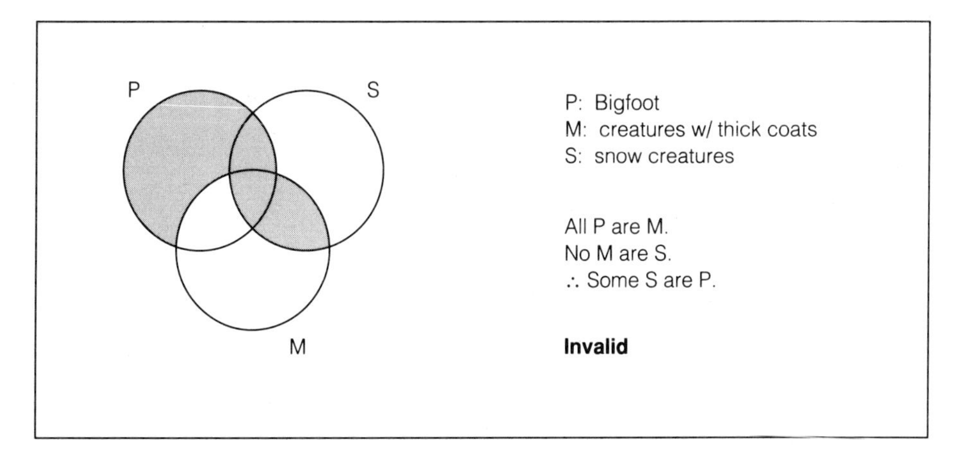

RECOMMENDED READING FOR UNIT TWO

Dumitriu, Anton. *History of Logic.* 4 vol. Abacus Press. Turbridge Wells, Kent, England, 1977.

John XXI. Tractatus Syncategorematum *and Selected Anonymous Treatises.* Translated by Joseph P. Mullally. Marquette University Press, Milwaukee, WI., 1964.

John XXI. *The* Summulae Logicales *of Peter of Spain.* Translated by Joseph P. Mullally. Publications in Mediaeval Studies, Notre Dame Press, Notre Dame, IN., 1945.

Kneale, William, and Kneale, Martha. *The Development of Logic.* The Clarendon Press, Oxford University Press, London, England, 1962.

Sherwood, William. *William of Sherwood's Introduction to Logic.* Translated by Norman Kretzmann. Minnesota Press, Minneapolis, 1966.

UNIT THREE

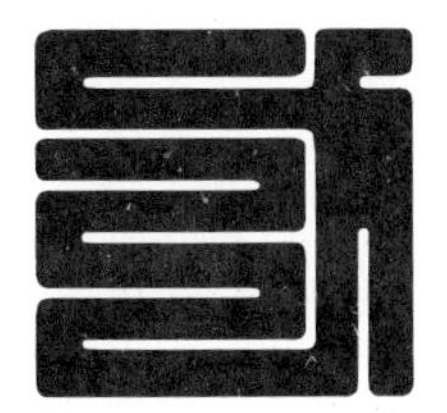

SYMBOLIC LOGIC

LOGICAL OPERATORS

The Logical Connectives in the Sentential Calculus

Operator	Operation	Russellian Notation	Stylistic Variants	Polish
it is not the case that . . .	negation	'~' [wave]	'–' also '¬'	N
both . . . and	conjunction	'&' [ampersand]	'•' also '∧'	K
either . . . or	disjunction	'∨' [wedge]	none	A
if . . . then	conditional [also called "hypothetical"]	'⊃' [horseshoe]	'→'	C
if and only if (iff)	biconditional [also called "equivalence"]	'≡' [triple bar]	'↔'	E

Quantifiers for Categorical (Aristotelean) Syllogisms

Operator	Operation	Scholastic Notation	Russellian Notation
All	universal affirmative	A	'(∀x)' or '(x)' "For all x . . ."
Some	particular affirmative	I	'(∃x)' "There is an x such that . . ."
No	universal negative	E	'~(∃x)' "It is not the case that there is an x such that . . ."
Not all	particular negative	O	'~(∀x)' or '~()' "It is not the case that, for all x . . ."

MODULE 23

SYMBOLS

23.1 IN GENERAL

The previous Unit was a study in the logic of general terms and the relations of quantities and qualities. Aristotelean logic is the logic of categorical syllogisms and, as we learned, a syllogism is a three-term argument with two premises, a conclusion, and logical operators that range over classes: 'all', 'no', 'some', and 'not all'.

However, the logic of the syllogism, strictly speaking, is very confining. When we encounter an argument in real life, it is rarely set up in such a way that we can readily adapt it to the syllogistic form. In fact, as we shall learn, many very good arguments do not faithfully follow the major premise-minor premise-conclusion regimen of the Aristotelean syllogism. Some good deductive arguments may consist of just one premise or a hundred premises. When these are arguments of the relation of quality and quantity, we can often put them into syllogistic form (sometimes by only the point of their elbows!) even if we need to construct whole concatenations of syllogistic chains to do it.

Some perfectly good arguments, however, do not adapt to the Aristotelean treatment at all. This was a discovery made by the classical Stoics. Consider, for example, the following argument:

> If the National Security Advisor directs special funding to the freedom fighters in El Chiquito, she will violate the division of powers specified

---

A **deductive argument** is valid, **if and only if, if** all the premises are true, **then** the conclusion must be true.

by the Constitution of the United States. And she has issued an order yesterday to direct such special funding to the freedom fighters. Therefore, she has violated a provision of the Constitution.

We can represent this argument another and more powerful way than we did in previous modules, for we are now studying the logic of whole statements rather than the relations between classes. Let the letter 'N' stand for "The National Security Advisor directs special funding to the freedom fighters in El Chiquito," and let the letter 'V' stand for "She will violate the division of powers specified by the Constitution of the United States." Then we can represent the argument as follows:

If N then V
And N
Therefore, V

We will now learn a whole new set of logical operators called "logical connectives" and a whole new system of symbolization for statements (instead of terms).

23.2 MOLECULAR ARGUMENTS

Let's say that any statement which can stand on its own and takes one of the values TRUE or FALSE is an **atomic statement.** From previous sections of this book, we know that a statement is any sentence, or part of a sentence, that is either true or false.

An example of a sentence which is a statement is W. C. Fields' famous epitaph: "On the whole I'd rather be in Philadelphia." We can see why this is a sentence which takes a truth value once we consider the fact that it is indeed his epitaph—it's surely true!

An example of a sentence which is *not* a statement is: "ARRGH! Cold coffee!" This sentence is neither true nor false because it is a *simple exclamation.*

Sometimes, however, some sentences which are statements get linked together to form **molecules.** When statements are linked together in such a way that they form an argument, we call the result a **molecular argument.**

Molecular arguments, in contrast to the categorical syllogisms of Aristotelean logic, are arguments made up of molecular statements. **Molecular statements** are two or more distinct statements connected by one or more logical operators which we call *logical connectives.* We distinguish the logical connectives from the complete set of logical operators (which also include the 'all', 'no', 'some', and 'not all' of Aristotelean logic) because these operators serve to fuse atomic statements into *molecules.* In doing so, we make an analogy to the atoms and molecules familiar to us from chemistry.

The list of logical connectives is brief. They are:

if . . . then
both . . . and
either . . . or
if and only if
it is not the case that . . .

Here are the names we give each of the logical connectives:

if . . . then	conditional
both . . . and	conjunction
either . . . or	disjunction
if and only if	biconditional
it is not the case that . . .	negation

The symbolic logic of molecular arguments is divided into two groups: the **Sentential Calculus** and the **Quantificational Calculus.** We use the name 'calculus' here to denote any method or system of analysis which employs a symbolic notation. The Sentential and Quantificational Calculi has little to do with strict mathematical differential or integral calculus (though we'll find some crossover when we explore the concept of the range of variables in Module 31, "Quantificational Calculus").

The Sentential Calculus (also called "the Propositional Calculus") represents the symbolic version of the logic of molecular arguments.

For over 2,000 years, we've treated categorical ("Aristotelean") logic and the logic of molecular arguments ("Stoic" logic) as though they were two distinct disciplines. Only at the end of the nineteenth and the dawn of the twentieth centuries, with the work of Charles Sanders Peirce, Gottlob Frege, Bertrand Russell, and others, were the Aristotelean and Stoic traditions finally merged together into one treatment, the Quantificational Calculus. As we shall see, the Aristotelean operators ('all', 'no', 'some', and 'not all') were added to the Stoic operators ('if . . . then', 'both . . . and', 'either . . . or', and 'it's not the case that . . .'). Together with a symbolic notation, they've fused into one coherent discipline.

However, we must start at the beginning, and the beginning is the Sentential Calculus.

In the Sentential Calculus, we will learn how to translate atomic statements into the symbolic notation of statement letters and how to translate the logical connectives of molecular arguments into symbols as well. Then we shall learn how to interpret and construct symbolic deductive arguments from their various combinations and how to decide their respective validity or invalidity.

We will learn four techniques for determining validity: the Truth Table test, the Semantic Trees test, Indirect Derivations, and Direct Derivations.

The most difficult of these methods of determining validity is direct derivations; the others are quite easy. However, we may find special reasons to learn the Direct Derivation Method. For one thing, it is the standard for constructing original deductions. It is usually introduced early in most logic books, though students usually find the other methods much more insightful and easier to use. All of the methods presuppose the Truth Table Method however, so we shall consider this first. The Semantic Tree Method presupposes the Indirect Derivation Method, but since it is a delight to use, we'll consider it next (later, we'll learn how it ties in with direct and indirect derivations).

As you'd expect, symbolic logic requires care and accuracy in translation. Before we consider any of the methods of testing for validity, we need to master symbolic technique. Note that any symbolic language is going to prove to be at most a shadow of the far-richer natural language (such as ordinary English). As such, we're exposed to the pitfalls of making our natural language fit a particular symbolic translation. Be aware that we are abstracting a symbolic language from our ordinary, everyday speech and writing so we can expect that some of the nuances will be lost. For example, tenses: in the Sentential Calculus, we generally render statements into the present tense. Getting a good translation is the hard part, but once we've done that the determinations of validity are generally easy to determine.

Signs and Symbols:
Negations

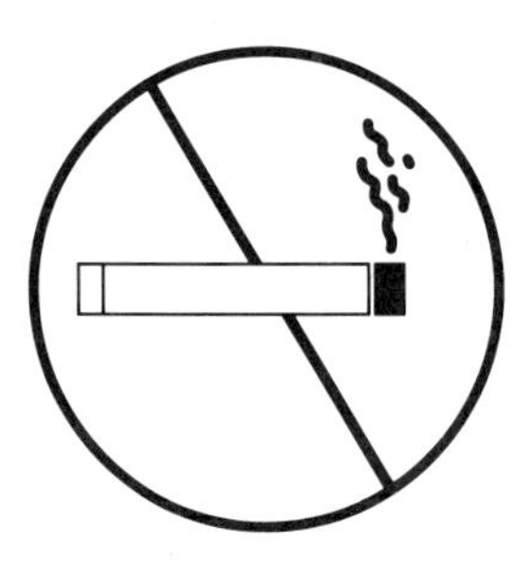

Symbols shouldn't intimidate or confound. Rather, their purpose is brevity, ease, and economy. At their best, symbols express complex or involved ideas in a striking and immediate way. Consider the internationally recognizable "No Smoking" sign. Think of the circle and slash configuration as a *negation* symbol. Looked at this way, the sign is shorthand for the statement: "It is not the case that smoking is permitted."

$$\sim S$$

In symbolic logic, we express negation with the symbol '~' (the wave or—as it is also called—the "tilde"). Let the statement letter 'S' stand for the statement "Smoking is permitted." We express the negation simply and economically as '~S'.

23.3 TRANSLATING STATEMENTS INTO SYMBOLS

Many different systems of symbolic notation are in current use. Perhaps the two main symbolic notations are the Polish, devised by Polish logician Jan Łukasiewicz, and the Russellian, named after Bertrand Russell. Although we will look at variations in notation in Module 26 (and the Polish notation will prove especially useful), the Russellian notation, which is derived in part from Boolean algebra, is usually considered standard. A slight variation on Russellian notation will be the standard notation for this book.

The characteristic feature of Russellian notation is its extensive use of parentheses. The way parentheses are used to set off and distinguish one symbolic expression from another will remind you of basic algebra. A key to mastering this notation is applying the parentheses attentively.

We symbolize two sets of expressions: the logical operators and the atomic statements. Logical connectives used in Sentential Calculus are distinguished from the total set of logical operators which includes both the logical connectives of molecular arguments and the logical operators of Aristotelean syllogisms.

Calling these operators 'logical connectives' has a special significance for molecular arguments. As mentioned, molecular arguments are the arguments derived from the old Stoic tradition which can be built up or broken down on an analogy to the molecules of chemistry. Atomic statements are the analogous component to chemical elements in molecules; the logical connectives are the bonds which hold chemical elements together. That's why they're called "connectives."

Using Russellian notation,* we symbolize the standard logical connectives in the following way:

Russellian Notation for the Logical Connectives

if . . . then	$\supset$	"the horseshoe"
both . . . and	&	"the ampersand"
either . . . or	$\vee$	"the wedge"
if and only if	$\equiv$	"the triple bar"
it is not the case that . . .	$\sim$	"the wave"

As you can see, the names for the symbols are descriptive—for example, the symbol '$\supset$' is called "the horseshoe" because it looks like a horseshoe turned on its side, and the symbol '$\sim$' is called "the wave" because it resembles a wave.

The second set of symbols which we need to learn are the symbols for the atomic statements in the molecular argument. Naturally enough, we call these **atomic statement letters.** They use capital letters of the English alphabet.

*In point of fact, Peano-Russell notation used the dot '•' instead of the ampersand '&'.

We will set aside the lowercase letters of the English alphabet for special purposes. For example, we will use some of these letters (p, q, r, s) to represent general statements about statements; we call these *statement-forms.* Any argument which is made up of statement-forms is thus an *argument-form*—i.e., a form of argument which represents many particular arguments. Statement-forms are neither true nor false; only particular statements which fall under the umbrella of these statement-forms are either true or false. When we assign a particular statement to a statement-form, we call the result a *substitution-instance.*

We also need one more symbol in our "vocabulary"—one that is probably already familiar to you from mathematics. We need the logical and mathematical symbol for "therefore" which is customarily rendered as a triangle of dots: '∴'.

You will recall from Module 2 that an asserted statement is any sentence or part of a sentence (or, in special circumstances, any combination of sentences) which takes a single truth-value (i.e., TRUE or FALSE). Obviously, in order to translate an argument into symbolic notation, you must first know what an asserted statement is. You may wish to review Module 2, "Statements," before proceeding any further.

Let's look at a simple example of a compound statement:

"If Los Angeles is smoggy, the mountains are hard to see."

Atomic statements contain no logical connectives. This example is a compound or *molecular* statement made up of two or more atomic statements, and such molecular statements may well contain logical connectives. 'If . . . then' is the logical connective here. Thus, we may assign the atomic statement letter 'L' to stand for "Los Angeles is smoggy." The atomic statement letter 'M' will stand for "The mountains are hard to see."

Thus, we have:

If L, then M.

Now we may symbolize the logical connective 'if . . . then' with the symbol '⊃'. We place the horseshoe (as it is called) '⊃' between the atomic statement letters to get this result:

L ⊃ M

To help set this off from other statements, premises, conclusions, etc., we wrap the whole expression in parentheses as follows:

(L ⊃ M)

And we are done symbolizing that asserted statement.

Here's a slightly more complicated example of a compound asserted statement. As we symbolize the asserted statement, we will lay out each of the steps in the process:

"If music be the food of love, play on."

This quotation from Shakespeare's *Twelfth Night* will help us review some special features of asserted statements. Notice first of all that this is a compound, made up of two simple statements: 'music be the food of love' and 'play on'. Although 'play on' is an imperative, we may treat it as a statement here (as in "you *ought* to play on"). Sometimes complex imperatives can be statements.*

However, only the whole conditional, '*If* music be the food of love, *then* play on', constitutes the *asserted* statement, and it is the asserted statement that is the unit of argument. How do we symbolize it? We need to do two things: we must pick out the logical connective (here it is 'if . . . then'); and we need to pick out each of the simple statements. We distinguish the two by underlining the logical connective(s) in the overall statement:

"<u>If</u> music be the food of love, [<u>then</u>] play on."

Notice that the 'then' of the 'if . . . then' has been added. The original Shakespearean quotation omits the 'then' for stylistic reasons (something which is also common in everyday speech).

STEP 1: We assign capital letters of the English alphabet to the simple statements:

Let 'M' stand for: "music is the food of love"
Let 'P' stand for: "play on" (or "you ought to play on")

. . .we have:

<u>If</u> M, <u>then</u> P.

STEP 2: Next, we insert the proper symbol for the logical connective. The symbol for the connective is the horseshoe, '⊃'. *To locate the horseshoe's proper place between simple statements, insert it wherever the 'then' ordinarily would appear.* Thus, we have:

M ⊃ P

STEP 3: Enclose the whole expression in parentheses (this helps us distinguish discrete atomic statements).

(M ⊃ P)

We are now finished symbolizing one asserted statement.
Let's look at the rest of the Shakespearean passage from *Twelfth Night:*

"If music be the food of love, play on;
Give me excess of it, that, surfeiting,
The appetite may sicken, and so die."

*See Module 2, "Statements."

This is not an argument (at least, not as it stands) but we can symbolize the statements nonetheless. We see that Shakespeare has his character assert the wish that he be given an excess of music on the condition that music is the food of love. Granting that, his hope is that his appetite (for love) "may sicken, and so die." (Obviously, this refers to an unhappy love affair.)

The second statement has no logical connectives. Thus, it may be symbolized using just one atomic statement letter. Since we have already reserved 'M' and 'P', let's designate this statement with the next letter in line, 'R', as follows:

Let 'R' stand for: "Give me excess of it, that, surfeiting, / The appetite may sicken, and so die." Notice that we don't need to employ parentheses around any atomic statement letter which stands alone on a line. Altogether then, we have:

(M ⊃ P)
R

EXERCISES

DIRECTIONS: *The following statements employ the logical connectives just learned. Translate the statements into atomic statement letters and show the proper placement of parentheses.*

______ **1.** The sun is at the center of the solar system and Earth is a planet that has life.

______ **2.** Either the sun is at the center of the solar system or the Earth is a planet that has life.

______ **3.** The sun is at the center of the solar system and the planet Mars orbits Jupiter.

______ **4.** Either the sun is at the center of the solar system or the planet Mars orbits Jupiter.

______ **5.** If Mars is the Red Planet, then Earth is a planet that has life.

______ **6.** If Mars is the Red Planet, then the moon has rings.

______ **7.** If the moon has rings, then Mars is the Red Planet.

______ **8.** Mercury is the closest planet to the sun if and only if Mars is the Red Planet.

______ **9.** Mercury is the closest planet to the sun if and only if the moon has rings.

______ **10.** The moon has rings if and only if Pluto is the closest planet to the sun.

23.4 WELL-FORMED FORMULAE ("WFFs")

In working symbolic logic, we are doing something very much like learning a language. We have a "vocabulary" of sorts (the atomic statement letters and the logical connectives), but we also need a "grammar" which will restrict us to formulations which make good sense.

You learned some procedures in the previous section for translating ordinary English language statements into symbolic notation, but this was only a rough guide. We need some rules which will assure us that we have constructed "good" and sensible symbolic formulae. When a formula is properly constructed so that it does make good sense, we call it a **well-formed formula,** or "wff" for short (pronounced "woof").

To put this another way, let's recall our analogy of molecular arguments to chemical molecules. In order for a molecule to "stick" together, it must be composed of certain elements and they must be bonded together. Some arrangements work, while others fall apart. Analogously, we are interested in only those symbolic molecular arguments which are arranged in a certain way so that the parts hold together. If the atomic statement letters and the logical connectives are in the right places, presto! we have a symbolic molecular argument.

Some arrangements just don't work. Obviously, the expression '⊃ ∨ &' doesn't work because it has no atomic statement letters. Similarly, the expression 'PQ R' likewise doesn't work because it has no logical connectives between the atomic statement letters. But these are extreme cases.

How about the expression '(P ⊃ ~ Q)'? Is this a well-formed formula?

We need only three rules to determine which are and which are not WFFs.

Rules for Well-Formed Formulae in Standard Russellian Notation

1. Any atomic statement letter is a WFF.*
2. Any WFF preceded by a wave ['~'] is a WFF.
3. The result of placing one of the symbols '⊃', '∨', '&', or '≡' between two WFFs and enclosing the whole in parentheses† is a WFF.

These rules are easy to apply. Let's take the example we just mentioned.

How about the expression '(P ⊃ ~Q)'? Is this a well-formed formula?

Let's start with 'P' and 'Q'. We see from rule 1 that "any atomic statement letter is a WFF." Thus, 'P' is a WFF; so is 'Q'. Here, however, they are part of a larger compound, so we must continue through the rules.

We see from rule 2 that "any [atomic statement letter] preceded by a wave ('~') is a WFF." We already determined that 'Q' is a WFF; thus, so is '~Q'.

*Sentence letters are the capital letters of the alphabet. By tradition, we usually assign sentence letters to statements by beginning with 'P', 'Q', 'R', 'S'. . . . (You may, however, assign letters to aid you in remembering what they stand for—e.g., "Let 'A' = 'The atmosphere actually assisted the asteroid's acceleration'.")

Should you exhaust the 26 letters of the alphabet in an exceptionally long argument, you may double up: e.g., 'AA', 'BB', 'CC' . . . etc.

†In a long string, you may substitute left and right brackets '[' and ']' or '{' and '}' provided that you do so consistently, e.g., '{[(P⊃ Q) ⊃ P] ⊃ P}'.

Rule 3 tells us that "the result of placing one of '⊃', '∨', '&', or '≡' between two WFFs and enclosing the whole in parentheses is a WFF." 'P' is a WFF; so is '~Q'; we find that a '⊃' has been inserted between the two WFFs and that the whole expression has been enclosed in parentheses. Therefore, the expression '(P ⊃ ~ Q)' meets the rules, and thus it is a well-formed formula.

The strategy we need to determine whether or not a given expression is a WFF is a negative one: try to determine which of the three rules is violated. If any one is, then the expression is not well-formed. If none is, it is a WFF.

Another example: the expression '(PQ)' is *not* well-formed. It violates rule 3 because it fails to contain a proper logical connective ('⊃', '&', '∨', or '≡').

Another example: the expression '(&RS)' is *not* a WFF. Again, it violates rule 3.

If '(P ⊃ ~Q)' is a WFF, then by rule 2 so is '~(P ⊃ ~Q)'.

All of the following are WFFs:

R	[rule 1]
~S	[rule 2]
~~P	[rule 2]
((P & Q) ∨ R)	[rule 3]
~((P & Q) ∨ R)	[rule 2]
(~((P & Q) ∨ R) ≡ (P ⊃ ~Q))	[rule 3]

EXERCISES

DIRECTIONS: *Determine which are and which are not WFFs. If you decide the formula is not a WFF, then explain why by referring to the rule that is violated.*

1. (P ⊃ Q)
2. AA
3. ~P
4. ~~B
5. ∴~(~Q ∨ R)
6. ~(~Q ∨ S
7. (P ⊃ (R ∨ (L & ~ S)))
8. (((P ⊃ Q) ⊃ P) ⊃ P)
9. (P ∨ R ≡ Q)
10. (PP & R) ≡ E
11. (AA & (R ∨ (~T ≡ SS))
12. &P

23.5 'IF . . . THEN'

Conditional ('if . . . then') statements can be expressed in a large variety of different ways in ordinary language. We call such variations of expression **stylistic variants.** We must be careful to symbolize them properly. In order to do that, we must distinguish between the *antecedent* and the *consequent* in the conditional statement.

As ordinarily written, the **antecedent** is the statement which follows the 'if' and precedes the 'then' of a conditional statement; the **consequent** is the statement which follows the 'then' of the conditional. Strictly speaking, the antecedent is the condition for the consequent.

We can illustrate this by substituting blanks for statement letters in a conditional statement. Thus, we have "if ____ then ____ ." The antecedent is the statement which falls in the first blank (the one after the 'if') and the consequent is the statement which falls in the second blank (the one after the 'then'). We may mark the blanks as follows:

if antecedent then consequent

In Ernest Hemingway's statement 'If you are lucky enough to have lived in Paris as a young man, then wherever you go for the rest of your life, it stays with you . . .', the antecedent is, 'you are lucky enough to have lived in Paris', and the consequent is, 'wherever you go for the rest of your life, it stays with you'. For this statement by Hemingway, the translation is: 'If A then B', where 'A' is the antecedent and 'B' is the consequent. We symbolize this symbolic statement with the "horseshoe" ('$\supset$') as follows: '$(A \supset B)$'.

For our purposes in this module, let 'α' stand for any antecedent statement here, and 'β' stand for any consequent. Thus, the standard form of the conditional is: '$(\alpha \supset \beta)$'.*

Very often, writers omit the 'then' in 'if . . . then' for stylistic reasons. They usually substitute a comma. For example, "If one has no heart, one cannot write for the masses." Here the antecedent is 'one has no heart,' and the consequent is, 'one cannot write for the masses'. Notice that these are negations; we may symbolize them as $(\sim H \supset \sim W)$.

Once again, the form is: '$(\alpha \supset \beta)$'.

This is obvious enough so far. Sometimes however, in ordinary discourse, the consequent may actually precede the antecedent. For example, consider this statement by D. H. Lawrence: "You were a lord if you had a horse." Remember: *The antecedent is the statement which follows the 'if'.* Applying this rule, we see that 'you had a horse' must be the antecedent (because it follows the 'if') and 'you were a lord' must be the consequent. We may symbolize the statement as $(H \supset L)$.

This symbolic statement follows the form '$(\alpha \supset \beta)$'.

There is a special case: **only if.** In ordinary English, the antecedent precedes the 'only if' and the consequent follows it. For example, "War is inevitable only if aggression continues." We may read this in a standard 'if . . . then' fashion as: "If aggression continues then war is inevitable." Once again, the antecedent is the statement which immediately follows the 'if'—in this case, the *if* in 'only if'. Thus, our rule continues to hold true.

*I use the Greek letters here as *special* statement-forms. It's useful to designate the antecedent with the Greek 'alpha' and the consequent with the Greek 'beta'.

By now, you will have noticed that, in a symbolic conditional, the antecedent ('α') always precedes the $\supset$ and the consequent ('β') always follows the '$\supset$'.
Here is a list of stylistic variants of the conditional form:

"(__α__ $\supset$ __β__)"

[CONDITIONAL]

SET 1*

if __α__ then __β__ [standard]

if __α__ , __β__ .

__α__ only if __β__ .

provided __α__ , then __β__ .

__α__ entails that __β__ .

__α__ implies that __β__ .

__α__ is a sufficient condition for __β__ .

SET 2*

__β__ if __α__ .

__β__ provided that __α__ .

__β__ in case that __α__ .

__β__ on the condition that __α__ .

__β__ given that __α__ .

__β__ is a necessary condition for __α__ .

The Stylistic Variants for the Conditional Form

23.6 STYLISTIC VARIANTS FOR 'EITHER . . . OR'

This one is easy. The logical connective '$\vee$' which stands for the standard English 'either . . . or' has only a few alternate renderings in English. The most common are:

*Notice that Set 2 of conditional stylistic variants places the consequent (β) before the antecedent (α), whereas Set 1 shows the normal order of antecedent before consequent. They are all, however, symbolized as ($\alpha \supset \beta$).

Stylistic Variants for Disjunction

"(______ $\vee$ ______)"
[DISJUNCTION]

either ______ or ______ [standard]

______ or ______

______ unless ______

______ otherwise ______

'Either . . . or' is called **disjunction** in logic. The blanks to either side of the '$\vee$' are to be filled in with atomic statement letters or other WFFs. In a disjunction, we call the WFFs that fill in the blanks **disjuncts.**

Here is an example of a stylistic variant (from a poem by e.e. cummings): "unless statistics lie he was / more brave than me. . . ." This statement can be rendered, less poetically, as "Either statistics lie, or he was more brave than me." We symbolize it as follows:

Let 'S' = "statistics lie"
Let 'B' = "he was more brave than me"
(S $\vee$ B)

23.7 STYLISTIC VARIANTS FOR 'BOTH . . . AND'

We call 'both . . . and' **conjunction** in logic; the atomic statement letters or WFFs which flank the '&' in conjunction are called its **conjuncts.** We take 'both . . . and' to be the standard English rendering of the logical connective '&', but many English variants exist.

The most common stylistic variant, and one we are likely to encounter even more often than the standard 'both . . . and' in ordinary discourse is the simple 'and'. The problem is that 'and' doesn't always signal a logical conjunction between statements, whereas 'both . . . and' always does. Consider the difference between these two statements:

"Roosevelt wants peace and Hitler wants war."

"Roosevelt and Churchill formed an alliance."

The first statement is a compound; it contains two independent statements. For that reason, this is a proper logical conjunction. It is symbolized as follows:

Let 'R' = "Roosevelt wants peace"
Let 'H' = "Hitler wants war"
(R & H)

The second statement is neither compound, nor a logical conjunction (even though it contains 'and'). 'Roosevelt and Churchill formed an alliance' cannot be broken into two statements because part of the meaning suggested by this sentence is that Roosevelt and Churchill formed an alliance (with each other). You see that if we tried to split up the two, we'd get "Roosevelt formed an alliance" and "Churchill formed an alliance" without any hint at the fact that they formed an alliance with each other at the same time. Thus, we render the symbolic transaction of this sentence, not as a conjunction, but as a single statement with a single atomic statement letter:

Let 'A' = "Roosevelt and Churchill formed an alliance"
A

When encountering an 'and' in a statement, you must always ask yourself the question, "Is this a compound sentence composed of two or more independent statements?" If the answer is "yes," then you symbolize it with the connective '&'; if the answer is "no," symbolize it as just one atomic statement letter.

We may find a long list of possible stylistic variants for 'both . . . and'. With a few exceptions, these include the whole list of what are also called "conjunctions" in grammar textbooks: 'and', 'but', 'although', 'so', etc. Grammar is less precise on this point than is logic, however, since grammarians also group 'or' and 'unless' as conjunctions, whereas we know in logic these are variants of logical disjunction.

Here is an abbreviated list of logical conjunctions:

Stylistic Variants for Conjunction

"(______ & ______)"
[CONJUNCTION]

both ______ and ______ [standard]
______ and ______
______ , but ______
______ , although ______
______ ; nevertheless ______
______ ; nonetheless ______
______ , even though ______
______ , though ______
______ and also ______
______ , whereas ______
not only ______ but also ______

23.8 STYLISTIC VARIANTS FOR 'IT IS NOT THE CASE THAT'

We call 'it is not the case that . . . ' by the name **negation** in logic (that's straightforward enough) and we symbolize it with the wave ('~'). The same sort of caution applies with negation as with logical conjunction. 'It is not the case that . . . ' is our standard English representation of negation because it clearly negates the whole sentence. However, we more commonly encounter a simple 'not' as the designation for negation, as in the sentence 'I am not a crook.' If we let 'C' = "I am a crook," then the sentence should be symbolized as the negation of 'C':

~C

. . . or "It is not the case that I am a crook."

Sometimes, however, 'not' (or prefixes that negate words, such as 'un-' or 'non-') don't negate the whole sentence. Here's an easy example:

"George is unruly" should be symbolized as the single atomic sentence letter, 'G'. (It makes no sense to say, "It is not the case that George is *ruly*"—no such word exists!)

Here's where it gets tricky. Suppose we encounter an argument that contains these two statements in the course of an argument:

"The Navy was prepared at Pearl Harbor."

"The Navy was unprepared at Pearl Harbor."

The statements are apparent contradictions. How do we represent that in symbols to show the contradiction? Obviously, we render the second statement as a logical negation, so we have the two statements symbolized as follows:

Let 'N' = "The Navy was prepared at Pearl Harbor."
N
~N

. . . and we thereby show the contradiction.

When eliminating 'not', 'non-', and 'un-' from sentences, let's follow this rough and ready rule for symbolization:

RULE: When you encounter *'not'* in a sentence, treat the sentence as a negated statement on the condition that the 'not' modifies an ordinary English auxiliary verb. Treat the sentence as a simple un-negated statement if the 'not', 'un-', or 'non-' modifies a predicate.

This is a bit involved, so let's get some practice. Take the sentence 'The guest of honor is unable to be here tonight.' Here, the 'un-' negates an auxiliary verb, so render the sentence as a negated statement: '~P' = "It is not the case that the guest is able to be here tonight."

Take the sentence 'The guest of honor is unruly tonight'. This should be rendered as an un-negated atomic statement letter: e.g., 'Q'. (Notice that it makes no sense to say, "it is not the case that the guest is *ruly*.")

Take the sentence 'Winston Smith became a non-person overnight'. Treat this sentence as an un-negated atomic statement letter (e.g., 'W') *unless* context reveals a sentence like this: "Winston Smith is a person still." That looks like a contradiction, and we'd want to symbolize the statements, one as '~W' and the other as 'W,' to show the contradiction.

Here is a partial list of stylistic variables for logical negation:

Stylistic Variants for Negation

"~ ______"
[NEGATION]

it is not the case that ______ [standard]
it is not true that ______
it is false that ______
not [usually]
un-[sometimes]
non-[sometimes]

We need to make one last remark about compound negations. As we know from the rules for well-formed formulae, sometimes negations may modify atomic statement letters, and sometimes negations may modify other compound WFFs. When we encounter a sentence like, "You can't have your cake and eat it too," what gets negated is the logical connective (in this case '&') and not the statement letters. To show this, we need to place the wave *outside* the parentheses, as shown below:

Let 'C' = "You can have your cake."
Let 'E' = "You can eat it."
~(C & E)

[Translation: "It is not the case that *both* you can have your cake *and* eat it."]

Notice that this is very different from a symbolic expression with the wave *inside* the parentheses, as in:

(~C & E)

[Translation: "Both it is not the case that you can have your cake and you can eat it."]

A more difficult expression to symbolize is 'neither . . . nor'. Consider the sentence 'Neither save your money nor spend it foolishly'. This is a negation of the logical connective 'either . . . or'. We symbolize it by placing the wave *outside* the parentheses:

Let 'M' = "Save your money"
Let 'F' = "Spend it foolishly"
~(M ∨ F)

23.9 STYLISTIC VARIANTS FOR 'IF AND ONLY IF'

The logical connective we call the **biconditional** (also called "equivalence") is an artificial logical construct. You won't often encounter 'if and only if' in ordinary language outside of logic and philosophy textbooks. However, this expression is very useful. It expresses, for example, an exact correspondence between both WFFs which is useful for definitional purposes (recall that we use an 'if and only if' in our formal definition of validity) and for the specification of exact criteria.

The symbol for 'if and only if' is the triple bar ('≡'). The biconditional may be thought of as a conditional that "goes both ways." The biconditional is true just on the condition that the truth values are the same for the expressions on each side of the triple bar ('≡').

Let's designate any well-formed formula that appears on the *left* side of the biconditional sign with the Greek letter 'φ' and any WFF which appears on the *right* side with the Greek letter 'δ'. The biconditional is true if 'φ' and 'δ' *both* have the truth values 'T' *or* if they both have the truth values 'F'; otherwise the biconditional is false.

What does this mean? The biconditional is an artificial construct, a conjunction of material conditionals, useful mainly for logicians and philosophers. Few everyday English sentences should be symbolized with the triple bar, for we restrict its use to those rare cases where we want to say that 'φ' will occur under exactly the same conditions under which 'δ' will occur. For example: "Water freezes if and only if its temperature drops to 0° C or lower."

The biconditional specifies both necessary and sufficient conditions. We say that a condition 'φ' is **necessary** for 'δ' when we mean that 'δ' requires 'φ'. We say that 'φ' is **sufficient** for 'δ' when the presence or occurrence of 'φ' is all we need for the presence or occurrence of 'δ'. When we have **necessary and sufficient conditions,** we have it both ways.

Recall that we already specified necessary conditions and sufficient conditions as stylistic variants of 'if . . . then'. The statement 'φ is a necessary condition for δ' is symbolized as '(φ ⊃ δ)'.

The statement 'φ is a sufficient condition for δ' is symbolized as '(δ ⊃ φ)' (where 'φ' and 'δ' stand for the same WFFs as before).

The statement 'φ is a necessary and sufficient condition for δ' is symbolized as '((φ ⊃ δ) & (δ ⊃ φ))'. The biconditional is shorthand for this long expression.

With the use of 'if and only if' we can shorten that expression to this compact symbolization: '(φ ≡ δ)'. That's what the triple bar means. To say that 'φ' is equivalent to 'δ' is to say that 'φ' is a condition for 'δ' *and* that 'δ' is a condition for 'φ'. In other words, if you have one, you have the other.

Because it is an artificial logical construct, only a few stylistic variants for 'if and only if' exist:

Stylistic Variants for Biconditionals

"(______ ≡ ______)"
[BICONDITIONAL]

______ if and only if ______ [standard]
______ if ______ [informal]
______ just in case that ______
______ is materially equivalent to ______
______ is a necessary and sufficient condition for ______

As a final example for this section, let's do a rather complicated symbolization. Take this famous statement by Edmund Burke and symbolize it: "The only thing necessary for the triumph of evil is for good men to do nothing." This is not obvious but the statement is indeed a biconditional. It is stating a necessary condition and it emphasizes that the condition is "the only thing" that is necessary for evil to triumph— in other words, it is also stating a sufficient condition. When a statement gives both necessary and sufficient conditions, it is a biconditional and is symbolized with the "triple bar" ('≡'). We may re-write the statement as "Evil triumphs if and only if good men do nothing." We may then symbolize it as follows:

Let 'E' = "evil triumphs"
Let 'G' = "good men do nothing"
(E ≡ G)

23.10 TRANSLATING WHOLE ARGUMENTS INTO SYMBOLS

So far, we've considered translating only statements and combinations of statements into symbols. We now need to symbolize whole arguments. Recall that an argument is a set of asserted statements offered in support of a conclusion. Symbolizing single statements is the difficult part; once you've mastered that, putting them together into symbolic arguments is easy. To do that, we need just one more symbol—"therefore" ('∴')

How Symbols Clarify Ordinary English

Symbolizing statements can often help us get clearer as to the meaning of an obscure or difficult passage—even passages that are not arguments (such as explanations). Where can symbolization of logical connectives be more helpful than with instructions for IRS tax returns?

Here is an example of a set of such instructions. The logical connectives (and other operators) have been highlighted for you.

"An individual may annually deduct up to 50% of his adjusted gross income (Line 15, Form 1040) for charitable contributions made to all qualified charitable, religious, educational, etc., organizations or to the U.S., state or local governments ***if*** the gift is made exclusively for public purposes. ***But*** the annual deduction ceiling is only 20% for contributions to nonoperating private charitable foundations that don't pass their contributions through to certain charities within specified time limits. In addition, ***if*** the donor makes contributions to a 50% donee in property which would result in a long-term capital gain ***if*** sold, that property is generally deductible ***only*** up to 30% of the donor's adjusted gross income, ***unless*** he specifically elects to give up part of his deductible contribution in exchange for bringing this property under the 50% ceiling." —*The Research Institute of America, Inc., 1976 Individual Tax Return Guide*, Grosset & Dunlap, p. 50

Here's a simple argument which we'll render into symbols:

"Either Professor Plum or Miss Scarlet committed the murder. If Professor Plum is the murderer, then he killed the victim with the lead pipe. However, further evidence reveals that the murder weapon was not the lead pipe, rather it was the knife. Therefore, we are led to the inexorable conclusion that Professor Plum is the murderer."

First, let's pick out the reason-indicators and the conclusion-indicators. We find one conclusion-indicator, which we'll underline:

"Either Professor Plum or Miss Scarlet committed the murder. If Professor Plum is the murderer, then he killed the victim with the lead pipe. However, further evidence reveals that the murder weapon was not the lead pipe, rather it was the knife. <u>Therefore, we are led to the inexorable conclusion that</u> Professor Plum is the murderer."

Next, let's pick out the logical connectives. (Notice that both the degree of conclusion force and the presence of the logical connectives assure us that this is a deductive, as opposed to inductive, argument.) The logical connectives are italicized:

"***Either*** Professor Plum ***or*** Miss Scarlet committed the murder. ***If*** Professor Plum is the murderer, ***then*** he killed the victim with the lead pipe. However, further evidence reveals that the murder weapon was ***not*** the lead pipe, ***rather*** it was the knife. Therefore, we are led to the inexorable conclusion that Professor Plum is the murderer."

The translation of each of the asserted statements into symbols is straightforward with the exception of 'rather' in the third sentence. A moment's reflection however will reveal that 'rather' is a stylistic variant of 'both . . . and' in this place. Thus, we translate the statements in the argument into atomic statement letters as follows:

Let 'P' = "Professor Plum is the murderer"
Let 'S' = "Miss Scarlet is the murderer"
Let 'L' = "The murder weapon was the lead pipe"
Let 'K' = "The murder weapon was the knife"

Notice two things: First, we took the 'not' out of 'The lead pipe was not the murder weapon'—that is the logical connective, 'it is not the case that . . .'. Second, we changed the sentences to the active voice in order to make their meaning clearer. Now we can symbolize the logical connectives and finish the translation of the whole argument:

PREMISE 1:	$(P \vee S)$
PREMISE 2:	$(P \supset L)$
PREMISE 3:	$(\sim L \ \& \ K)$
CONCLUSION:	$\therefore P$

The translation is complete. Notice how much clearer the argument is once it's translated into symbols. Once we learn something of the truth functions of the logical connectives, we will also be able to say whether or not it is a *valid* argument. If it is valid, we can declare the answer and win the game!

The discussion on the truth functional applications of the operators will be found throughout the rest of this Unit, but especially (and most clearly) in Module 24, "Truth Tables."

EXERCISES

Exercise 1

DIRECTIONS: *Symbolize the following statements. Some of the examples contain two or more statements—however, none of these examples are complete arguments.*

1. Silence gives consent, or a horrible feeling that nobody's listening." —Franklin P. Jones

2. "I don't even know what street Canada is on." —Al Capone

3. "We can lick gravity, but sometimes the paperwork is overwhelming." —Wernher von Braun

4. "This will never be a civilized country until we spend more money for books than we do for chewing gum." —Elbert Hubbard

5. "If advertising encourages people to live beyond their means, so does matrimony." —Bruce Barton

6. "War would end if the dead could return." —Stanley Baldwin

7. "If I cannot bend Heaven, I shall move Hell." —Virgil

8. "Guess if you can, choose if you dare." —Pierre Corneille

9. "Any of us can achieve virtue, if by virtue we merely mean the avoidance of vices that do not attract us." —Robert S. Lynd

10. "If we become two people—the suburban affluent and the urban poor, each filled with mistrust and fear of the other—then we shall effectively cripple each generation to come." —Lyndon B. Johnson

11. "I will not bathe my hands in the blood of the people of Mexico, nor will I participate in the guilt of those murders which have been and will hereafter be committed by our army there." —Joshua R. Giddings

12. "Heaven goes by favour. If it went by merit, you would stay out and your dog would go in." —Mark Twain

13. "If a politician murders his mother, the first response of the press or of his opponents will likely be not that it was a terrible thing to do but rather that in a statement made six years before he had gone on record as being opposed to matricide." —Meg Greenfield

14. "There was truth and there was untruth, and if you clung to the truth even against the whole world, you were not mad." —George Orwell

15. "Crime is contagious. If the government becomes a lawbreaker, it breeds contempt for the law." —Justice Louis D. Brandeis

Exercise 2

DIRECTIONS: *Symbolize the following arguments.*

1. "Our emotions cry for vengeance in the wake of a horrible crime, but we know that killing the criminal cannot undo the crime, will not prevent similar crimes by others, does not benefit the victim, destroys human life, and brutalizes society. If we are to still violence, we must cherish life." —Ramsey Clark

2. "Free enterprise ended in the United States a good many years ago. Big oil, big steel, big agriculture avoid the open marketplace. Big corporations fix prices among themselves and drive out the small entrepreneur. In their conglomerate forms, the huge corporations have begun to challenge the legitimacy of the state." —Gore Vidal

3. "The style is the man. Rather say the style is the way the man takes himself. If it is with outer seriousness, it must be with inner humor. If it is with outer humor, it must be with inner seriousness." —Robert Frost

4. "We are drowning our children in violence, cynicism and sadism piped into the living room and even the nursery. The grandchildren of the kids who used to weep because the Little Match Girl froze to death now feel cheated if she isn't slugged, raped and thrown into a Bessemer converter." —Jenkin Lloyd Jones

M O D U L E 24

TRUTH TABLES

24.1 ABOUT TRUTH TABLES

For molecular arguments, truth tables are the standard device for determination of the validity of arguments. Equally important, truth tables clarify our understanding of the *meaning* of deductive validity. As we shall see, they illustrate vividly the difference between *truth* and *validity.*

What is a truth table? A **truth table** is a table of all the possible assignments of truth values for a given argument.* This definition warrants some clarification.

Every argument must have two or more statements. Each statement must be either true or false. That means that the number of atomic statement letters in an argument determines how many possible assignments of truth values to statements and to combinations of statements and connectives there are in a given argument.

Let's begin with a single atomic statement letter, 'P'. How many possible truth value assignments will it take?

First of all, we need to determine how many possible arrangements of trues and falses we need to lay out in order to exhaust all the possible assignments

---

A deductive argument is **valid**, if and only if, if all the premises are true, then the conclusion must be true.

*Or a given statement. We shall consider truth tables for tautologies, contradictions, and contingencies.

(these are the truth-value assignments). Another way of asking this question is: How many rows do we need in the truth table?

The answer is obvious if we draw an analogy to flipping coins. If we have one coin, two results are possible: heads or tails. Likewise, if we have one and only one discrete atomic statement letter in a molecular argument, then two truth-value assignments are possible: true and false. In other words, the truth table has only two rows, one which gives the value of 'P' as TRUE, the other which gives the value of 'P' as FALSE, exhausting all the possibilities. Working the table for just the atomic statement letter 'P', we have:

P
T
F

Row 1 gives the possible value of 'P' as TRUE; row 2 gives the other possible assignment as FALSE—that exhausts all the possibilities.

This should be obvious enough, but what if we encounter two, three, or ten distinct atomic statement letters in a molecular argument? How do we make certain that we've exhausted all the possible truth-value assignments?

The analogy to flipping coins may be useful once more in making this problem intelligible. Suppose we flip two coins: on one flip, maybe the first coin will be heads, and the second tails; on another flip, maybe the first coin will be tails, and the other heads. On still other flips, maybe *both* coins will be heads, or both tails. So there are *four* possible arrangements, as shown in Figure 1.

So, we see that four *and only four* arrangements of the coins are possible on any given flip.

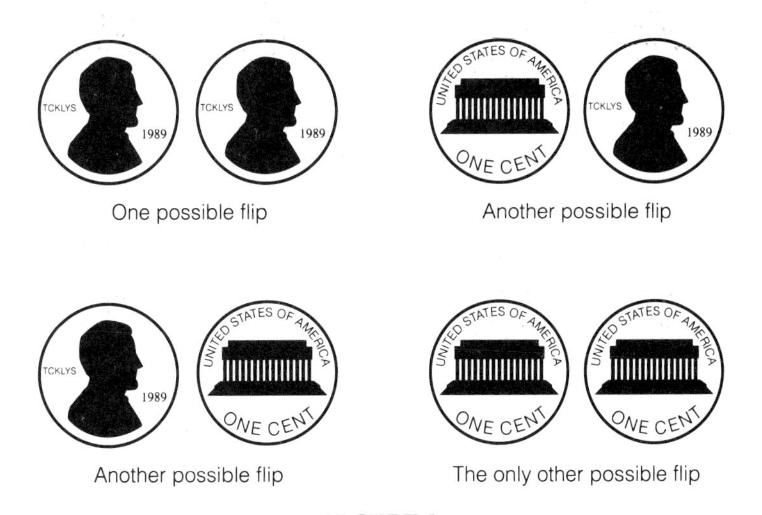

FIGURE 1

By analogy, if we have two discrete atomic statement letters in a molecular argument (let's call them 'P' and 'Q'), then *four* truth-value assignments are possible. In other words, we need four rows in the truth table. This is how the first step should look in setting up the truth table (though two atomic statement letters by themselves don't constitute an argument, so the table would need further finishing touches):

P	Q
T	T
T	F
F	T
F	F

Notice that if we substitute 'T' for "heads" and 'F' for "tails," we have a pattern identical to our chart for flipping coins in Figure 1.

What if we have three, four, or even ten discrete atomic statement letters in a molecular argument? Fortunately, we don't have to flip three, four, or ten sets of coins repeatedly to find the answer; we have a formula to do the job for us.

To find out how many rows of possible truth-value assignments there should be in a truth table, use the following formula:

The number of rows = 2^n (where 'n' = the number of discrete atomic statement letters in a symbolic sentence).

EXAMPLES:
The number of rows in the symbolic statement '~P' is <u>two</u> (or 2^1).
The number of rows in the symbolic statement '(P ⊃ Q)' is <u>four</u> (or 2^2, where 2 squared = 4).
The number of rows in the symbolic statement '((P ⊃ Q) & R)' is <u>eight</u> (there are three distinct atomic statement letters).
And the number of truth-value assignment rows for the truth table for '((P ⊃ P) & P)' is two (because there is only one discrete atomic statement letter, 'P').

24.2 TRUTH TABLE RULES

The table that follows gives the complete set of truth table rules. Notice that 'p' and 'q' are statement-forms so that it is really their substitution-instances which would be [TRUE] or [FALSE]. A proper substitution-instance of 'p' might be 'P' or it might be a larger WFF such as '(~ P ∨ (Q & S))'.

Read the table vertically and horizontally. Row 1, for example, says that if (a substitution-instance of) 'p' = 'T', then (the substitution-instance of) '~p' would be 'F'. Likewise, row 2 means when 'p' is 'T' and 'q' is 'F' then '(p ∨ q)' will be 'T'.

You will have occasion to consult this table frequently throughout this module, so learn the rules and keep this table handy.

Truth Table Rules

Row #	**p**	**q**	**~p**	**~q**	**(p & q)**	**(p ∨ q)**	**(p ⊃ q)**	**(p ≡ q)**
1.	T	T	F	F	T	T	T	T
2.	T	F	F	T	F	T	F	F
3.	F	T	T	F	F	T	T	F
4.	F	F	T	T	F	F	T	T

Row #	**~ (p & q)**	**~ (p ∨ q)**	**~ (p ⊃ q)**	**~ (p ≡ q)**
1.	F	F	F	F
2.	T	F	T	T
3.	T	F	F	T
4.	T	T	F	F

24.3 TRUTH TABLE ASSIGNMENT RULES

For '&': In a conjunction, the entire conjunction is TRUE only if both conjuncts are true; otherwise it is false.

For '∨':* In a disjunction, the entire disjunction is TRUE when one or the other (or both) disjuncts(s) is (are) true; it is false otherwise. (Note: one false case, where both disjuncts are false.)

For '⊃':* In a conditional, there is only one row where the entire conditional is FALSE, namely where the antecedent is true and the consequent is false; otherwise it is true.

For '≡': In equivalence (also called the "biconditional"), the equivalence is TRUE when both sides of the biconditional have equivalent truth values (where both are true or both are false); it is false otherwise.

In natural language, we encounter two distinct senses of the disjunction 'either . . . or'. Consider the following two ordinary language statements:

*A word about '⊃' and '∨': The Truth Tables for the material conditional ('⊃') and the inclusive disjunction ('∨') may look a bit peculiar to you at first glance, but it is important to note that both of these connectives have a formal, technical definition.

1. "Either Sam Spade was in Los Angeles or he was in New York at the time of the murder."

2. "You may read a book and/or watch television while you're waiting."

Statement 1 uses an **exclusive disjunction.** Sam Spade must be in one place or the other, but he can't be in both. In this sense, we mean the disjunction to exclude one possibility (called its "disjunct") or the other; we are saying: "At the time of the murder, either Sam Spade was in Los Angeles or he was in New York, *but not both.*" We often use disjunction in this way in natural language, but this is *not* the way we use the connective '$\vee$'.

Statement 2 exemplifies a **non-exclusive disjunction.** A disjunction in this sense allows for the possibility that *both* disjuncts may be true. In this case, we use the hopelessly bureaucratic-sounding construct 'and/or' to show the possibility that either one disjunct or the other or *both* may be true, but we often use the 'either . . . or' for the same purpose, as in the following example:

3. "Either of Dustin Hoffman or Glenn Close received Academy Awards this year."

We speak in this manner when we are not sure of the possibility that *both* were Oscar winners. This is non-exclusive disjunction, and we symbolize it with the logical connective '$\vee$'.

A simple rule: when you encounter an "either . . . or" statement, assume it's a non-exclusive disjunction unless the context indicates otherwise.

When the context *does* indicate otherwise (as in statement 1), you may symbolize the statement as follows:

'$((L \vee N) \,\&\, \sim (L \,\&\, N))$'

This literally translates as: "Either Sam Spade was in Los Angeles or he was in New York at the time of the murder, and it is not the case that he was in both at the time of the murder." That's clear enough: not even Sam Spade could be in two places at once.

Unless the context clearly indicates otherwise, we ordinarily symbolize an "either . . . or" statement as a non-exclusive disjunction. Thus, statement 2 should be symbolized as:

'$(B \vee T)$'

. . . and statement 3 should be symbolized as:

'$(H \vee C)$'

The **material conditional** ('$\supset$') has a peculiar feature: if the antecedent is true, then the whole asserted statement is true regardless of the value of the consequent. Perhaps that's not what we always mean by the ordinary hypothetical in natural language. However, despite the fact that the material conditional's truth-value assignments are slightly counterintuitive, there *are* times in natural language when we mean the whole conditional to be true even though the antecedent (i.e., the "condition") is false.

For example, we want to assert the truth of the whole statement:

> **"If the temperature drops below 32° F in the Mojave Desert today, then this water will freeze."**

—even though the temperature never dropped below 80° F. The condition may never be fulfilled, but we still want to assert the truth of the whole conditional.

Another reason for the seemingly peculiar truth conditions of the material conditional is that it can be defined reflexively in terms of the primitives '$\sim$' and '$\vee$' (negation and non-exclusive disjunction). What does '$(P \supset Q)$' mean? It can be translated as '$(\sim P \vee Q)$' because the truth tables for both expressions are equivalent.*

Thus, we call this sort of hypothetical the *material conditional,* and we symbolize its logical connective with the "horseshoe" ('$\supset$').

EXERCISES

DIRECTIONS: *The following statements employ the logical connectives just learned. Using the truth table rules, interpret the statements and their connectives and decide whether or not they are true.*

True **1.** The sun is at the center of the solar system and Earth is a planet that has life.
Solution: "The sun is at the center of the solar system" = TRUE; "Earth is a planet that has life" = TRUE. The entire assertion is a molecular statement and a conjunction. According to the truth table rules, in order for a conjunction to be true, both conjuncts must be true. Therefore, this assertion meets the rule.

______ **2.** Either the sun is at the center of the solar system or the earth is a planet that has life.

______ **3.** The sun is at the center of the solar system and the planet Mars orbits Jupiter.

______ **4.** Either the sun is at the center of the solar system or the planet Mars orbits Jupiter.

______ **5.** If Mars is the Red Planet then Earth is a planet that has life.

______ **6.** If Mars is the Red Planet then the moon has rings.

______ **7.** If the moon has rings, then Mars is the Red Planet.

______ **8.** Mercury is the closest planet to the sun if and only if Mars is the Red Planet.

*Construct truth tables for both expressions and see for yourself.

______ 9. Mercury is the closest planet to the sun if and only if the moon has rings.

______ 10. The moon has rings if and only if Pluto is the closest planet to the sun.

______ 11. If Mars has rings, then Pluto is the closest planet to the sun.

______ 12. Pluto is the closest planet to the sun if Mars has rings.

______ 13. If Mars is the Red Planet, then Pluto is the closest planet to the sun.

______ 14. Pluto is the closest planet to the sun if Mars is the Red Planet.

______ 15. Either Pluto is the closest planet to the sun or Mars has rings.

24.4 A TIP FOR CONSTRUCTING TRUTH TABLES

In order to make sure that your truth table is set up correctly, follow these rules:

1. Determine the number of rows in the truth table according to the formula on page 280.

2. Having determined the proper number of rows in the truth table, set out the atomic statement letters in the order in which you encounter them.

Thus, in the formula, '((P ⊃ Q) & (P ∨ R))', 'P' is the first atomic statement letter; 'Q' is the second; and 'R' is the third.

Thus:

P	Q	R	

3. Begin setting up your truth table by going to the column of the last discrete atomic statement letter and listing alternate T's and F's according to the number of rows in the formula.

Thus, in our example, there should be eight rows. So . . .

P	Q	R	
		T	
		F	
		T	
		F	
		T	
		F	
		T	
		F	

4. After you have done this, move to the next discrete atomic statement letter to the left. Then double the number of T's and F's in the alternation. In our example, the table should look like this:

P	Q	R	
	T	T	
	T	F	
	F	T	
	F	F	
	T	T	
	T	F	
	F	T	
	F	F	

Notice that we have doubled the alternation of T's and F's, so that the "Q" column reads (from top to bottom) TT FF TT FF (doubling the "R" column's T F T F). The same principle would apply if there were four, sixteen, or two hundred and fifty-six rows in the truth table.

5. Move to the next discrete atomic statement letter ('P') to the left, and double the alternation of T's and F's found in the previous column ('Q'). Continue this operation until all discrete sentence letters are accounted for.

Thus, in our example:

P	Q	R	
T	T	T	
T	T	F	
T	F	T	
T	F	F	
F	T	T	
F	T	F	
F	F	T	
F	F	F	

Notice that we have doubled the alternation again, so that the "P" column reads (from top to bottom) TTTT FFFF. The same principle would apply if there were four, sixteen, or two hundred and fifty-six rows in the truth table.

Testing for Validity Using Truth Tables

Let's recall from Module 12, "Truth and Validity" in Unit One, that a hypothetical argument with two premises and one conclusion can have seven possible valid arrangements among the truth values, TRUE and FALSE One other combination, and only one, is the hallmark of an invalid argument. Here is the schema again:

The Invalid Schema

Premise 1	true
Premise 2	true
Conclusion	false

Let's begin with a simple argument: one premise and one conclusion. Let's take an obviously invalid argument, where the premise is utterly irrelevant to the conclusion. Consider this argument then:

"We've just experienced the worst stock market crash in history. Therefore, the economy is healthier than ever."

This argument (absurd as it is) at least has the virtue of easy symbolization:

Let 'P' = "We've just experienced the worst stock market crash in history."

Let 'Q' = ". . . the economy is healthier than ever."

The argument is symbolized as:

P
∴ Q

Notice that even though we have two atomic statement letters standing alone on separate lines, they *do* constitute an argument here because the symbol '∴' means "therefore." This argument says: "[statement] P is true. Therefore, [statement] Q *must* be true." (As it stands, this is an invalid argument, and it's also a very *odd* argument, but that's not the point right now.) How do we set up a truth table which will demonstrate the validity of this argument? First of all, we need to determine how many possible arrangements of trues and falses we need to lay out in order to exhaust all the possible assignments; these are the truth-value assignments. As we saw above, another way of asking this question is: How many rows do we need in the truth table?

Applying the formula: the number of rows in a truth table = 2^n, we have two atomic statement letters, so n = 2. That gives us the answer *four.* We have four rows.

Row 1 gives the possible value of 'P' as TRUE and 'Q' as TRUE; row 2 gives another possible assignment of 'P' as TRUE and 'Q' as FALSE; row 3 yields 'P' as FALSE and 'Q' as TRUE; and the last row results in 'P' as FALSE and 'Q' as FALSE. That exhausts all the possibilities for the combination of 'P' and 'Q'. The first part of the truth table (i.e., the table for the atomic statement letters only) looks like this:

	P	Q		
1.	T	T		
2.	T	F		
3.	F	T		
4.	F	F		

Next, we need to work the truth table for the first premise of the argument. Since the first premise is just the atomic statement letter 'P', we need only repeat the table already completed:

	P	Q	P (Premise 1)	
1.	T	T	T	
2.	T	F	T	
3.	F	T	F	
4.	F	F	F	

The table for the conclusion is all that remains. Since the conclusion is no more than the atomic statement letter 'Q', we need only repeat the table for 'Q'. Thus, the completed truth table for the entire molecular argument is as follows:

	P	Q	P (Premise 1)	∴Q (Conclusion)
1.	T	T	T	T
2.	T	F	T	F
3.	F	T	F	T
4.	F	F	F	F

Now, how do we establish the invalidity of this argument using the Truth Table Method? To do that, we need to take a walk down invalidity row.

Down Invalidity Row

We recall that a deductive argument is valid, if and only if, if all the premises are true, then the conclusion *must* be true.

What does this mean? It means that, in order to find *in*validity, what we need to establish is that *at least one* row has all *true* premises and a *false* conclusion. That's all there is to it.

However, let's get clear about the strategy. When you are checking on the validity or invalidity of an argument, you don't try to prove validity. *The strategy is to try to prove* invalidity. If you succeed, naturally, the argument is *invalid;* if you fail to prove invalidity, it's obvious that the argument is *valid.*

As a result, we are interested in only the rows where *all* the premises have the value TRUE. Check them off as you go down the rows. Then, see if any of these rows with all true premises have a false conclusion. If you find one, the argument is *invalid.* If you find none, then the argument is *valid.*

Let's return to our obviously invalid argument from above, where the conclusion is utterly irrelevant to the premise.

			Premise 1	Conclusion
	P	**Q**	**P**	**∴Q**
1.	T	T	T	T
2.	T	F	T	F
3.	F	T	F	T
4.	F	F	F	F

Remember, our strategy is to find "invalidity row." Thus, our first step is to check off all the rows that have all true premises. Since we have just one premise, we need to find any "P" rows that have TRUE for values. We find two such rows, 1 and 2. Checking them off, we have:

				Premise 1	Conclusion
		P	**Q**	**P**	**∴Q**
✓	1.	T	T	T	T
✓	2.	T	F	T	F
	3.	F	T	F	T
	4.	F	F	F	F

We ignore any row that does not have all true premises. Next, we eliminate the rows that have all true premises but a true conclusion. Remember: we are looking for only those rows which have all true premises and a *false* conclusion. Our object is to prove *invalidity.*

Is there such a row? The answer is "yes;" row 2 has all true premises and a false conclusion. Just as our intuitions told us, this is an *invalid* argument.

			Premise 1	Conclusion	
	P	**Q**	**P**	**∴Q**	
✔ 1.	T	T	T	T	
✔ 2.	T	F	T	F	← **Invalid**
3.	F	T	F	T	
4.	F	F	F	F	

Now, let's look at the truth table for a *valid* argument. We'll construct a truth table for the following valid argument:

"If Manhattan is in New York state, then Manhattan residents pay federal taxes. Manhattan is in New York state. Therefore, Manhattan residents pay federal taxes."

We symbolize the argument as follows:

Let 'P' = "Manhattan is in New York state."
Let 'Q' = "Manhattan residents pay federal taxes."

(P ⊃ Q)
P
∴ Q

The truth table for this argument is shown completed below:

		Premise 1	Premise 2	Conclusion
P	**Q**	**(P ⊃ Q)**	**P**	**∴Q**
T	T	T	T	T
T	F	F	T	F
F	T	T	F	T
F	F	F	F	F

Our next step is to look exclusively at the premises and conclusion and determine if there are any rows in which all the premises are true.* In fact, there is one such row, the first row:

P	Q	(P ⊃ Q) Premise 1	P Premise 2	∴Q Conclusion	
T	T	T	T	T	✓
T	F	F	T	F	
F	T	T	F	T	
F	F	F	F	F	

In the next illustration, we see that the first row is shaded for the premises. Since there is but one row in which all the premises are true, we confine our attention to that row, and to that row alone:

P	Q	(P ⊃ Q) Premise 1	P Premise 2	∴Q Conclusion	
T	T	T	T	T	✓
T	F	F	T	F	
F	T	T	F	T	
F	F	F	F	F	

Next, we recall our definition of validity: *a deductive argument is valid, if and only if, if all the premises are true, then the conclusion must be true.* Remember that we're trying to prove *invalidity.* We're looking for a possible row in which all the premises are true and the conclusion is false. Since we have just one row in which all the premises are true—the first row—we look there to the truth value of the conclusion to see if it is false:

*In the event that we should construct a truth table without *any* rows in which the premises are all true, the argument is valid.

P	Q	(P ⊃ Q) Premise 1	P Premise 2	∴Q Conclusion
T	T	T	T	T ✓
T	F	F	T	F
F	T	T	F	T
F	F	F	F	F

In fact, it is true. Since this is the only possible assignment of truth values in which all the premises are true, and we see that the conclusion is true, we know that the argument must be *valid.*

Next, let's look at another example of an *invalid* argument. Consider the following (clearly invalid) argument:

"If Manhattan is in California, then Manhattan residents say 'Have a nice day!' It's not the case that Manhattan is in California. Therefore, Manhattan residents say 'Have a nice day.' "

Let 'P' = "Manhattan is in California."

Let 'Q' = "Manhattan residents say 'Have a nice day!' "

(P ⊃ Q)
~P
∴Q

We've completed the truth table for this argument below:

P	Q	(P ⊃ Q) Premise 1	P Premise 2	∴Q Conclusion
T	T	T	T	T
T	F	F	F	F
F	T	T	T	T
F	F	T	T	F

As before, we look for the rows in which *all* the premises are true. This time, we find two such rows:

P	Q	(P ⊃ Q) Premise 1	P Premise 2	∴Q Conclusion	
T	T	T	F	T	
T	F	F	F	F	
F	T	T	T	T	✓
F	F	T	T	F	✓

Remember that our strategy is to find one row, *any* row, in which all the premises are true and the conclusion is false. If we find such a row, then the argument is *invalid.* In fact, the last row does show all true premises and a false conclusion:

P	Q	(P ⊃ Q) Premise 1	P Premise 2	∴Q Conclusion	
T	T	T	F	T	
T	F	F	F	F	
F	T	T	T	T	✓
F	F	T	T	F	✓

Therefore, the argument is *invalid.* Notice that if we had stopped at row 3, we might have the made the mistake of thinking this was a valid argument. You must check *all* the rows to see if there's one in which all the premises are true and the conclusion is false. The very last row of this argument shows that.

We may construct truth tables with sixteen or thirty-two rows (or more!) and find several rows with all true premises. On some occasions, we may discover that the *last* row of a long truth table is the only one which shows invalidity. You must be thorough; check them all!

24.5 VALIDITY AND ARGUMENT FORMS

In the preceding section, we looked at truth tables for symbolic arguments which symbolized specific English arguments. We can likewise construct truth tables for an argument-form and then make certain cautious statements about the validity or invalidity of specific arguments *of that form*.

For example, we may construct a truth table for the argument-form:

$(p \supset q)$
p
$\therefore q$

The truth table will look like this:

p	q	(p ⊃ q) Premise 1	p Premise 2	∴q Conclusion
T	T	T	T	T
T	F	F	T	F
F	T	T	F	T
F	F	T	F	F

We know that a valid argument always derives from a valid argument-form. However, it is not the case that *invalid* arguments always derive from invalid argument-forms. What's important is that at least *some* substitution-instances of statements for statement-forms in invalid argument-forms result in invalid arguments.

24.6 FORMAL TAUTOLOGIES, CONTRADICTIONS, AND CONTINGENT STATEMENTS

Module 11 contains *informal* discussions of tautologies and contradictions. Informally speaking, by a 'tautology' we mean a statement which is true under all possible interpretations; a 'contradiction' is a statement which is false under all possible interpretations.

The terms 'tautology' and 'contradiction' are used in everyday discourse—especially the term 'contradiction'. However, natural language is so rich in meaning that genuine ordinary language tautologies and contradictions are hard to find. When the politician proclaims she is "pro-life" to an anti-abortion crowd in Wilkes-Barre and then turns around and says she is "pro-choice" to a pro-choice audience in Ann Arbor, we are certainly within our rights to suspect the politician's good faith because of an apparent contradiction in what she says. But is she really guilty of a contradiction? Formally speaking, that's hard to show. Natural language is ambiguous enough to permit some generous interpretations of what she said. (Perhaps she meant that she is basically opposed to abortion but believes that the moral decision whether or not to have an abortion is a matter of personal conscience. On the other hand, maybe she *really* is contradicting herself and saying only what she thinks her audience wants to hear."

Now, with the aid of symbolization and truth tables, we will be able to give a *formal* rendering of these notions. We now say that a tautology is a statement that is TRUE for all possible truth-value assignments on a truth table. A contradiction is a statement which is FALSE for all possible truth-value assignments on a truth table. A contingent statement is a statement which has a mix of TRUE and FALSE for its truth-value assignments on a truth table.

To illustrate this definition, let's look at the following example of a tautology—a symbolic statement-form which is called "the law of excluded middle":

'(P ∨ ~P)'

The truth table is easy to construct. We have just one atomic statement letter, 'P', so we know that the truth table has just two rows. The table looks like this:

P	**~P**	**(P ∨ ~P)**
T	F	T
F	T	T

We turn our attention to the final result—the truth-value assignments for the expression '(P ∨ ~P)' and its main connective, the '∨'. There, we see that the expression is true on all possible assignments of truth values (i.e., for all possible rows). The statement, therefore, is a formal tautology.

We can likewise make formal contradictions intelligible by virtue of the Truth Table Method. Consider this example:

(P & ~P)

Once more, the truth table is easy to construct:

P	**~P**	**(P & ~P)**
T	F	F
F	T	F

The contradiction is apparent in the truth-value assignment for the expression. We see that the truth-value assignment for the main connective (the '&')

is FALSE for all rows of the truth table. Therefore, the statement is a formal contradiction.

Statements other than tautologies and contradictions are called "contingent" (i.e., they are "contingently" TRUE or FALSE). The set of possible contingent statements is infinitely large, but we certainly encounter contingent statements more often in everyday discourse than tautologies or contradictions. These are the statements which "need to be checked out" in experience in order to determine whether or not they are true or false.

Take the following example of a contingent statement:

(P ∨ Q)

This table has four rows:

P	~Q	(P ∨ Q)
T	T	T
T	F	T
F	T	T
F	F	F

Since the value of '(P ∨ Q)' is FALSE in the fourth row, the statement is said to be formally contingent.

In the next module, we will learn how to construct truth tables for compound molecular statements (those with more than one logical connective). For example, to determine whether or not the following statement

(((P ⊃ Q) ⊃ P) ⊃ P)

is a tautology, a contradiction, or a contingency, we need to be able to identify the *main connective* and construct a truth table for it. We will learn how to do that in the next module.

EXERCISES

Exercise 1

DIRECTIONS: *Show that the material conditional is a construct of the primitives '~' and '∨' by constructing truth tables for both '(P ⊃ Q)' and '(~P ∨ Q)'. You will see that the truth-value assignments for both expressions are identical.*

Exercise 2

DIRECTIONS: *The following are symbolic arguments. Construct truth tables for each. Say whether they are valid or invalid. If a problem is invalid, identify the row or rows which indicate invalidity. (These are easy examples. More difficult arguments using larger com-*

pound statements can be found in the exercises for Module 25, "Atom Smashing Molecular Arguments."

1. C
   ∴(C ∨ C)
2. D
   ∴(D & D)
3. E
   ∴(E & G)
4. (P ∨ Q)
   ∴(Q ∨ P)
5. (M ⊃ R)
   R
   ∴M
6. (R ⊃ S)
   ~S
   ∴~R
7. (P & Q)
   ∴P
8. (A ≡ B)
   A
   ∴B
9. G
   ∴(G ∨ B)
10. (S ⊃ K)
    (K ⊃ B)
    ∴(S ⊃ B)
11. (A ≡ B)
    (S ⊃ R)
    ∴(A ∨ R)
12. ~(P & Q)
    ∴(~P ∨ ~Q)
13. (A ⊃ B)
    ∴(~B ⊃ ~A)
14. ~(P & Q)
    ∴(~P & ~Q)
15. ~(P ∨ Q)
    ∴(~P ∨ ~Q)

MODULE 25

ATOM SMASHING MOLECULAR ARGUMENTS

25.1 ATOM SMASHING

The arguments of the Sentential Calculus are called "molecular" arguments. This name is due to the fact that arguments made up of the operators, 'if . . . then', 'both . . . and', 'either . . . or', 'it is not the case that . . .', and 'if and only if', connect statements together (hence, we sometimes call these operators "logical *connectives*"). The asserted statements they connect are, figuratively speaking, the "atoms" of the molecule, whereas the logical connectives are the "bonding."

When constructing truth tables, we need to *deconstruct* the larger molecules and break them down into their component atoms. This is why I've called this process "atom smashing." When you've broken the argument into its simplest components, then and only then do you begin to work a truth table.

The next process is *reconstruction.* When you've split the molecule into its atoms, you then need to put the argument back together a step at a time. At each step, you need to work a truth table for the result. When you've completed reconstruction of the argument in its original form, then you have a completed truth table.

Let's take a relatively easy example:

PREMISE 1: $((P \supset \sim Q) \,\&\, (P \vee R))$

PREMISE 2: $(P \equiv R)$

CONCLUSION: $\therefore R$

A deductive argument is **valid**, if and only if, if all the premises are true, then the conclusion must be true.

Recall rule 2 in Module 24.4, "A Tip for Constructing Truth Tables." It reads: "Having determined the proper number of rows in the truth table, set out the atomic statement letters in the order in which you encounter them."

This is really the first step in atom smashing a molecular argument. The statement letters are the atoms. Just the process alone of identifying the discrete statement letters accomplishes a major part of your job. The trick comes in the re-assembly with the logical connectives.

Here then are the consecutive steps in "atom smashing" a molecular argument:

1. Set out the statement letters in the order in which you encounter them.

In our example, the discrete statement letters in the order in which we read them are 'P', 'Q', and 'R'. We know from our formula in Module 24.4 that the number of rows of the truth table = 2^n, where 'n' = the number of discrete statement letters. Since we have three discrete statement letters, the result is 2 × 2 × 2 = 8, or eight rows.

The first step in the truth table looks like this:

	P	Q	R
1.	T	T	T
2.	T	T	F
3.	T	F	T
4.	T	F	F
5.	F	T	T
6.	F	T	F
7.	F	F	T
8.	F	F	F

2. Begin reconstruction by identifying any *negated* statement letters.

In our example, we have one: '~Q'. We need construct only one table then: for '~Q'. Thus, we have:

	P	Q	R	~Q
1.	T	T	T	F
2.	T	T	F	F
3.	T	F	T	T
4.	T	F	F	T
5.	F	T	T	F
6.	F	T	F	F
7.	F	F	T	T
8.	F	F	F	T

3. After you've done all of the negated statement letters, next do the simplest molecule.

In order to find the simplest molecule, you must select the logical connective that's furthermost inside the premises. Take premise 1, for example: ((P ⊃ ~Q) & (P ∨ R)). Here, the **main connective**, or the connective which is furthermost *outside* the premises is the ampersand, '&'. Notice that this is so because the fewest parentheses surround the ampersand. Conversely, the horseshoe, '⊃', and the wedge, '∨', each have four parentheses surrounding them. They share the distinction of being the connective furthest inside. The simplest molecule, indifferent between these two cases, is the one that surrounds the '⊃ ' and also the one that surrounds the '∨'. Thus, we shall work a table for '(P ⊃ ~Q)' and another for '(P ∨ R)'.

	P	Q	R	~Q	(P ⊃ ~Q)	(P ∨ R)
1.	T	T	T	F	F	T
2.	T	T	F	F	F	T
3.	T	F	T	T	T	T
4.	T	F	F	T	T	T
5.	F	T	T	F	T	T
6.	F	T	F	F	T	F
7.	F	F	T	T	T	T
8.	F	F	F	T	T	F

Premise 2 is as simple as we need it, so we'll work the table for that next:

	P	Q	R	~Q	(P ⊃ ~Q)	(P ∨ R)	(P ≡ R)
1.	T	T	T	F	F	T	T
2.	T	T	F	F	F	T	F
3.	T	F	T	T	T	T	T
4.	T	F	F	T	T	T	F
5.	F	T	T	F	T	T	F
6.	F	T	F	F	T	F	T
7.	F	F	T	T	T	T	F
8.	F	F	F	T	T	F	T

4. Construct a truth table for the next largest molecule. In this case, that leaves only the completion of premise 1, ((P ⊃ ~Q) & (P ∨ R)). This is the table for the ampersand:

	P	Q	R	~Q	(P ⊃ ~Q)	(P ∨ R)	(P ≡ R)	((P ⊃ ~Q) & (P ∨ R))
1.	T	T	T	F	F	T	T	F
2.	T	T	F	F	F	T	F	F
3.	T	F	T	T	T	T	T	T
4.	T	F	F	T	T	T	F	T
5.	F	T	T	F	T	T	F	T
6.	F	T	F	F	T	F	T	F
7.	F	F	T	T	T	T	F	T
8.	F	F	F	T	T	F	T	F

Next, we work a table for the conclusion. However, we note that we've already worked a table for 'R'. This brings us to an important final step: a grouping rule.

5. The Grouping Rule: For clarity, group all your premises and your conclusion together at the end, even if this means you have to repeat a table, *and mark each clearly.*

Completing the table, we have:

Premise 1 $(P \equiv R)$	Premise 2 $((P \supset \sim Q) \& (P \vee R))$	Conclusion $\therefore R$
T	F	T
F	F	F
T	T	T
F	T	F
F	T	T
T	F	F
F	T	T
T	F	F

We see by row 3 the argument is *valid.*
Repeating the rules, we have:

1. Set out the statement letters in the order in which you encounter them.

2. Begin reconstruction by identifying any negated statement letters.

3. After you've done all of the negated statement letters, next do the simplest molecule.

4. Construct a truth table for the next largest molecule.

5. The Grouping Rule: For clarity, group all your premises and your conclusion together at the end, even if this means you have to repeat a table, *and mark each clearly.*

Follow This Recipe for Atom Smashing

Break Up Complex Expressions in This Order

1. First, break up expressions into statements.
2. Next, break up expressions into negated statement letters.
3. Next, break up parentheses. Take the innermost connective (the one farthest inside the parentheses).
4. Construct a truth table for each result.
5. Go to the next major connective (the one next farthest inside the parentheses).
6. Repeat 4.
7. Repeat 5.

EXERCISES

Exercise 1

DIRECTIONS: *The following are symbolic arguments. Construct truth tables for each. Say whether they are valid or invalid. If a problem is invalid, identify the row or rows which indicate invalidity.*

1. $(P \supset Q)$
$(P \,\&\, Q)$
$\therefore Q$
2. $(A \,\&\, (B \vee A))$
$\therefore (B \vee A)$
3. $\sim P$
$(\sim P \,\&\, \sim A)$
$(A \vee (P \equiv Q))$
$\therefore (\sim Q \vee P)$
4. $(A \vee (B \vee C))$
$(\sim A \,\&\, \sim B)$
$\therefore C$
5. R
S
$\therefore (R \,\&\, S)$
6. $(P \supset (Q \supset R))$
$((Q \supset R) \supset S)$
$\therefore (P \supset S)$
7. $\sim (P \,\&\, R)$
$\therefore (\sim P \,\&\, \sim R)$
8. $\sim (A \supset (H \equiv B))$
$\sim A \,\&\, ((H \supset B) \,\&\, (B \supset H)))$
$\therefore (B \supset H)$
9. $(R \supset S)$
$\therefore (R \supset (R \,\&\, S))$
10. $((P \supset Q) \,\&\, (R \supset S))$
$(P \vee R)$
$\therefore (Q \vee S)$
11. $(P \vee (Q \supset \sim A))$
$(Q \supset \sim A)$
$\therefore \sim P$
12. $((P \supset Q) \,\&\, (P \supset P))$
$\therefore (P \equiv Q)$
13. $(C \equiv (T \vee \sim R))$
$(\sim R \,\&\, \sim S)$
$(\sim S \supset T)$
$\therefore T$
14. W
$\therefore (W \vee G)$
15. $(A \supset B)$
$(B \supset C)$
$\therefore (A \supset C)$

Exercise 2

DIRECTIONS: *Construct a truth table for each of the following symbolic statements and determine whether they are tautologies, contradictions, or contingent.*

1. $(P \supset P)$
2. $(P \vee \sim P)$
3. $(P \,\&\, \sim P)$
4. $(((P \supset Q) \supset P) \supset P)$
5. $((P \supset Q) \supset P)$
6. $\sim (((P \supset Q) \supset P) \supset P)$
7. $(P \supset (Q \supset P))$
8. $(\sim P \supset (P \supset Q))$
9. $((P \equiv A) \equiv ((P \supset A) \,\&\, (A \supset P)))$
10. $((P \,\&\, S) \vee ((Q \supset P) \equiv (P \vee R)))$

MODULE 26

POLISH AND OTHER SYSTEMS OF SYMBOLIC NOTATION

26.1 IN GENERAL

Variations of symbolic notation abound, though most of them are variants of the standard Russellian notation which relies on the use of parentheses in the fashion of Boolean algebra. The most important exception to this is the Polish notation devised by the logician, Jan Łukasiewicz, which dispenses with the need for parentheses.

Two other systems of notation have importance from an historical point of view—that of the American logician, Charles Sanders Peirce, and that of the German logician, Gottlob Frege. Peirce and Frege were contemporaries, doing their major work in the latter part of the nineteenth century, and they may both have had a major influence on the work and notational system of Bertrand Russell. However, the systems of notation devised by Peirce and Frege are rarely used. Each may be worthy of study, but we will not look at them in this place.

Sometimes the variations between notations are very small and based on stylistic considerations. For example, some logicians prefer to use a dot ('•') in place of the ampersand ('&'), perhaps because many students find it difficult to

---

A deductive **argument** is valid, if and only if, if all the premises are true, then the conclusion must be true.

draw an ampersand (I routinely take a moment to let students practice drawing the figure). Other times, logicians may substitute an arrow ('→') for the horseshoe ('⊃') in conditional statements, perhaps because they think the arrow shows the direction of implication from antecedent to consequent more clearly. These changes are, however, not significant. They do not affect the substance of logic; they're merely a matter of stylistic preference.

One reason for looking at variations in symbolic notation has to do with the need for communication. At the moment, the varieties of symbolic notations are so numerous that they constitute something of a logician's "tower of Babel." Nearly every formal logic textbook introduces a stylistic change in notation (usually based on Russell's). If you want to do further work in symbolic logic, or even if you want to consult other textbooks to assist you in understanding the material explained in this book, you need to know what those differences are and what they mean. These are two reasons for showing the other notations here.

Yet, learning other notations can also help you better understand the operations of logic itself. The most beneficial of these variant notations is Łukasiewicz's Polish notation. Polish notation dispenses with the need for parentheses altogether. That has some important applications for computers: assembly languages are written in Polish notation, even though higher-level languages generally do not use Polish.

Dispensing with parentheses can reveal unexpected benefits. In the module on symbolization, you may have been somewhat confused about the placement of the negation symbol ('~') in symbolic formulae. For example, what is the difference between '~(P ∨ Q)' and '(~P ∨ ~Q)'? A very important difference does indeed exist! As we shall see, that sort of confusion doesn't even arise with Polish notation since it makes clear when a wave modifies an atomic statement letter and when it modifies a logical connective. It becomes obvious.

On the whole, I find Polish notation easiest of all, and the most illuminating as well. I've written this book in a close relative of Russellian notation, however, as a concession to the need for greater standardization in logical notation; but we will use Polish notation for some instructional purposes.

26.2 RULES FOR WELL-FORMED FORMULAE [WFFs] IN POLISH NOTATION

The small case letters of the alphabet, beginning (by convention) with 'p', 'q', 'r', 's' . . . and concluding with 'm', 'n', 'o', make up the atomic statement letters in Polish notation.

The capital letters, 'N', 'C', 'K', 'A', and 'E', are the symbols for the logical connectives for Polish notation. They stand for the following operations:

Operation's Name	Ordinary Language	Łukasiewicz's Name*	Polish Symbol
conjunction	"both . . . and"	**K**onjunktion	K
disjunction	"either . . . or"	**A**lternation	A
conditional	"if . . . then"	**C**onditional	C
biconditional	"if and only if"	**E**quivalence	E
negation	"it is not the case"	**N**egation	N

The rules for well-formed formulae ("WFFs") in Polish notation are:

1. Any atomic statement letter standing alone is a WFF.

2. Any WFF preceded by 'N' is a WFF.

3. The result of placing two WFFs after any one of 'C', 'K', 'A', or 'E' is a WFF.

Recognizing and constructing WFFs in Polish notation are truly easy operations. The one difference to remember is that it is the lowercase letters of the alphabet that stand for statements in Polish notation.

Let's consider a few examples. Decide which of the following are WFFs in Polish and which are not:

Polish Notation	Russellian Notation	WFF?
1. p	P	Yes
2. Np	~P	Yes
3. AN	v~	No
4. ApNp	(P ∨ ~P)	Yes
5. Cpq	(P ⊃ Q)	Yes
6. Emp	(M ≡ P)	Yes
7. ECpsKst	((P ⊃ S) ≡ (S & T))	Yes
8. ECpANqr	((P ⊃) ≡ (~Q ∨ R))	No
9. ∴NECpsANqr	∴~((P ⊃ S) ≡ (~Q ∨ R)	Yes
10. NApq	~(P ∨ Q)	Yes
11. ANpNq	(~P ∨ ~Q)	Yes

Problem 1, of course, is a WFF since it meets rule 1 for WFFs in Polish notation—it is an atomic statement letter standing alone on a line. Problem 2,

*Ironically, these names are in English and German, not Polish. That has to do with the greater universality of the English and German languages as compared with Polish. Łukasiewicz (whose name, by the way, is pronounced "Wook-ah-sev-itch") mainly wrote in these languages. Here, he obviously selected the German 'Konjunktion' for its initial 'K' to keep it separate from the 'C' in 'Conditional' (which is also translated with a 'K' in German, 'Konditional'). You can see that Łukasiewicz was indeed writing for a world audience. For disjunction, he obviously chose the English synonym 'Alternation' for its 'A' (the German word is "Trennung'). 'Negation' is the same in English and German.

'Np', is likewise easy. It meets rule 2. Problem 3 is *not* a WFF because it violates rule 3, and so forth. These are easy enough, but what about the compound WFFs we find later in the list?

To decide which are WFFs and which are not, underline the smallest molecules that make up WFFs (except for atomic statement letters) beginning from right to left. For example, consider problem 7: 'ECpsKst'. In English, this reads: "If p then s, if and only if, both s and t."

Beginning from right to left, we see that the 'Kst' is a WFF according to rule 3, so we underline it:

ECpsKst

Next, we see that the next expression (from right to left), 'Cps' is also a WFF, so we underline it as well:

ECps Kst

Finally, we see that the rest of placing two WFFs after an 'E' is also a WFF, so by rule 3 the whole expression is a WFF, and we underline it:

E Cps Kst

Notice that the main connective of the sentence is the sign for the biconditional, 'E' (because it comes first).

Naturally, we can apply the same principle to even more difficult WFFs. Consider 9, for example: '∴ NECpsANqr'.

Proceeding from right to left, we find that 'r' is a WFF [rule 1] and so is the compound expression 'Nq' [rule 2], which is also the first compound WFF we encounter so we underline it:

∴N E Cps ANqr

Once again, we start from right to left: 'r' is a WFF, and so is 'Nq' so the combination, 'ANqr', is also a WFF [rule 3]. Underline it:

∴N E Cps ANqr

Continuing from right to left, we encounter another WFF, 'Cps' [rule 3]. We underline it as well:

∴N E Cps ANqr

The operator 'E' followed by two WFFs is also a WFF [rule 3], so we underline it:

∴N E Cps ANqr

The connective **'N'** followed by a WFF is also a WFF [rule 2], so we underline it:

∴N E Cps ANqr

To see the evolution of our procedure in one glance, here is the whole process from top to bottom:

∴N E Cps ANqr
∴N E Cps ANqr
∴N E Cps ANqr
∴N E Cps ANqr
∴N E Cps ANqr
∴N E Cps ANqr

Let's see how we detect expressions which fail to be WFFs in Polish notation. Consider Problem 8: 'ECpANqr'. Checking this expression from right to left, we see it starts out very much like our previous example since the last part of the formula is likewise 'ANqr'. Thus, underlying the WFF, we have:

ECp ANqr
ECp ANqr

Next, we encounter a conditional. The **'C'** is followed by an atomic statement letter and by the WFF 'ANqr', so the conditional is likewise a WFF, and we underline it:

E Cp ANqr

Is the entire expression beginning with **'E'** a WFF? According to rule 3, **'E'** needs to precede two WFFs in order to be a WFF itself:

E Cp ANqr

But here is where we encounter trouble. The **'E'** (for equivalence) needs two WFFs to follow it in order to be a WFF. However, only one WFF appears—'CpANqr'. I've marked with a question mark the space where the second WFF might appear:

E ? Cp ANqr

Obviously, then, this is not a WFF.

We may apply the same procedures and principles and determine accurately each time whether any given expression is a WFF.

EXERCISES

DIRECTIONS: *Determine which of the following are, and which are not, WFFs in Polish notation. If a formula is not well-formed, say why it is not by pointing to the rule it violates.*

1. p
2. ∴Np
3. NNp
4. Nnp
5. Cpq
6. CPpq
7. CpApq
8. NCApqErs
9. AEpKrs
10. AENpqKrs

26.3 ATOM SMASHING AND TRUTH TABLES IN POLISH NOTATION

In the previous module, we considered how to dismantle WFFs in standard Russellian notation in order to prepare them for truth table operations. We called this "atom smashing" because the idea was to break up the molecular arguments into their simplest components, atomic statement letters, and then rebuild the expression step by step.

The process of atom smashing is much easier in Polish notation. You need only proceed from right to left to get the individual components out of the whole molecule.

Let's do a truth table for an argument in Polish notation. The argument is a simple one with only two atomic statement letters, thus it requires only four rows:

Cpq
Nq
∴Np

Obviously we construct a truth table by identifying the atomic statement letters, and then doing their negations. Here, the table is complete for the first atomic statement letter (reading right to left):

				Premise 1	Premise 2	Conclusion
p	**q**	**Np**	**Nq**	**Cpq**	**Nq**	**∴Np**
	T					
	F					
	T					
	F					

Next, we work a truth-value assignment for the first atomic statement letter, 'p':

p	q	Np	Nq	Cpq (Premise 1)	Nq (Premise 2)	∴Np (Conclusion)
T	T					
T	F					
F	T					
F	F					

Having done this, we work a table for 'Np' simply by negating each of the assignments (because 'N' is the negation symbol) in the column for 'p' working row by row—i.e., the table for 'Np' is exactly the opposite of the table for 'p':

p	q	Np	Nq	Cpq (Premise 1)	Nq (Premise 2)	∴Np (Conclusion)
T	T	F				
T	F	F				
F	T	T				
F	F	T				

. . . and the table for 'Nq' is exactly the opposite of the table for 'q':

p	q	Np	Nq	Cpq (Premise 1)	Nq (Premise 2)	∴Np (Conclusion)
T	T	F	F			
T	F	F	T			
F	T	T	F			
F	F	T	T			

Following the truth table rules for conditional statements, we get this result for 'Cpq':

				Premise 1	Premise 2	Conclusion
p	**q**	**Np**	**Nq**	**Cpq**	**Nq**	**∴Np**
T	T	F	F	T		
T	F	F	T	F		
F	T	T	F	T		
F	F	T	T	T		

We repeat the table for 'Nq', the second premise, and place it between the first premise and the conclusion, just to keep things organized. We also supply the table for the conclusion, 'Np', which is exactly the reverse of the assignment in the table for 'p':

				Premise 1	Premise 2	Conclusion
p	**q**	**Np**	**Nq**	**Cpq**	**Nq**	**∴Np**
T	T	F	F	T	F	F
T	F	F	T	F	T	F
F	T	T	F	T	F	T
F	F	T	T	T	T	T

Now the truth table is complete and we need only test for validity. Remember, the strategy is to test for *invalidity.* Thus, we look for the rows which have all true premises. We find two of them and check them off:

				Premise 1	Premise 2	Conclusion	
p	**q**	**Np**	**Nq**	**Cpq**	**Nq**	**∴Np**	
T	T	F	F	T	F	F	
T	F	F	T	F	T	F	
F	T	T	F	T	F	T	
F	F	T	T	T	T	T	✓

We discover that there is only one row with all true premises. That row has a true conclusion. Therefore, the argument must be valid:

p	q	Np	Nq	Cpq (Premise 1)	Nq (Premise 2)	∴Np (Conclusion)	
T	T	F	F	T	F	F	
T	F	F	T	F	T	F	VALID
F	T	T	F	T	F	T	
F	F	T	T	T	T	T	✓

Next, let's take a more difficult example to see how well atom smashing works in Polish notation.

Consider this argument:

ApNCpq
∴ECpNApqApNCpq

We read this argument in English as "Either p or it is not the case that if p, then q. Therefore, if p then it is not the case that either p or q, if and only if, either p or it is not the case that if p then q."

We atom smash this argument as follows:

ApNCpq
∴ECpNApq ApNCpq

Each of the underlines, of course, represents one compound WFF. We need to work a table for each of the compound WFFs, as well as for each of the atomic statement letters. We start with the statement letters and then we identify the compound WFFs, reading the original expression from right to left. Placing the result on a truth table, we find after evaluation that the argument is invalid:

INVALID

p	q	Cpq	NCpq	ApNCpq (Premise 1)	Apq	NApq	CpNApq	∴ECpNApqApNCpq (Conclusion)
T	T	T	F	T	T	F	F	F
T	F	F	T	T	T	F	F	F
F	T	T	F	F	T	F	T	F
F	F	T	F	F	F	T	T	F

Hand-Calculators Using Reverse Polish Notation ("RPN")

Some hand calculators, most notably Hewlett-Packard, use a variation on Polish notation known as "Reverse Polish Notation." RPN is to Boolean algebra as Polish notation is to Russellian symbolic logic.

RPN for hand calculators, like Polish notation for symbolic logic, is an acquired taste, but it makes many complex calculations easier to do. Perhaps you'll find, as have so many others before you, that once you've mastered Polish notation you'll regard the other notational systems as cumbersome. When does RPN make calculations easier to do? Here is a footnote from a Hewlett-Packard Owner's Manual:

"HP's operating logic is based on an unambiguous, parentheses-free mathematical logic known as 'Polish Notation,' developed by the Polish logician Jan Łukasiewicz (1878–1956). While conventional algebraic notation places the operators *between* the relevant numbers or variables, Łukasiewicz's notation places them *before* the numbers or variables. For optimal efficiency of [a memory] stack, we have modified that notation to specify the operators *after* the numbers. Hence the term *Reverse Polish Notation,* or *RPN.*" —Hewlett-Packard, *RPN Scientific Calculator Owner's Manual,* Model PH-32 S, p. 35

RPN uses '+' and '−' as signs instead of as operators. This way, you can sign the number as you enter it into a calculation (the '=' sign would then become the operator and on a hand calculator the function is entered after the signed numbers are entered).

Here is a slightly oversimplified example. Suppose you want to tally some statistical data. Often, the numbers have negative values. RPN makes such calculations more intuitive, as in this example:

To perform this calculation

$$\begin{array}{r} -3 \\ +2 \\ \hline =? \end{array}$$

. . . press these buttons in this order:

[−] [3]   [+] [2]   [=]

The shaded boxes show the two rows where *all* (in this case, we have just one) the premises are 'T'. In both of those rows, we see a false conclusion whereas only one would be enough to prove invalidity.

EXERCISES

Exercise 1

DIRECTIONS: *The following are symbolic arguments. Construct truth tables for each and determine whether they are valid or invalid. If they are invalid, indicate which row shows invalidity.*

1. Cpq
   Kpq
   ∴q
2. KaAba
   ∴Aba
3. Np
   KNpNa
   AaEpq
   ∴ANqp
4. AaAbc
   KNaNb
   ∴c
5. r
   s
   ∴Krs
6. CpCqr
   CCqrs
   ∴Cps
7. NKpr
   ∴KNpNr
8. NCaEhb
   KNaKChbCbh
   ∴Cbh
9. Crs
   ∴CrKrs
10. KCpqCrs
    Apr
    ∴Aqs
11. ApCqNa
    CqNa
    ∴Np
12. KCpqCqp
    ∴Epq
13. EcAtNr
    KNrNs
    CNst
    ∴t
14. w
    ∴Awg
15. Cab
    Cbc
    ∴Cac

Exercise 2

DIRECTIONS: *Construct a truth table for each of the following symbolic statements and determine whether they are tautologies, contradictions, or contingencies.*

1. Cpp
2. ApNp
3. KpNp
4. CCCpqpp
5. CCpqp
6. NCCCpqpp
7. CpCqp
8. CNpCpq
9. EEpaKCpaCap
10. AKpsECqpApr

MODULE 27

SEMANTIC TREES

27.1 IN GENERAL

The Truth Table Method for testing for validity has some great strengths. For one thing, it is an easy method to learn and to apply, and no other method can show the meaning of validity so well. However, truth tables also have a significant drawback. If more than three or four statement letters are involved, truth tables can become prohibitively long. Imagine a truth table with 256 rows! And yet, a table need have just eight atomic statement letters to generate that many rows.

Truth tables, then, have a decided disadvantage. For this reason, even the most introductory logic course should introduce Semantic Trees.

Semantic Trees are often called "Smullyan Trees" because the logician Raymond Smullyan is credited with devising them as an adaptation of the "semantic tableaux" of logician Evert W. Beth. Others credit logicians Jaakko Hintikka and Gerhard Gentzen with inspiring their creation. The trees of the method have also been called "Consistency Trees," "Truth Trees," and even "Sequential Trees." Whatever their true origin, at least logicians agree on calling them "trees."

You'll find that once you have mastered the rules, Semantic Trees are fast, easy to use, and powerful. A little bit of understanding where they come from is also a wonderful way to understand the basic semantics of logic (perhaps that's why I prefer calling them "Semantic" Trees).

A deductive argument is **valid**, if and only if, if all the premises are true, then the conclusion must be true.

Moreover, Semantic Trees have few of the limitations of the other methods of deduction we'll learn later in this unit. Not only can they quickly decide the validity or invalidity of arguments, but they can also determine when a statement is tautological, contradictory, or contingent, and they can even be used in creative deductions. We'll have some fun along those lines in Section 27.7.

Semantic Trees have one drawback of their own. While they are fast, they can get sloppy very easily. You may find that you'll need very large sheets of paper to solve some of the trickier deductions because Semantic Trees swell and grow in all directions, like amoebae.* Usually, I find an 8½ × 14 yellow legal pad sufficient for most of my work with Semantic Trees but the rapid way they grow may make you sometimes wish you could use the table cloth instead.

Be careful when you work Semantic Trees to keep them neat and legible; allow yourself plenty of space (in all directions!). We will learn some strategies that may help to keep them down in size.

27.2 THE SENTENTIAL CALCULUS

It may be useful to review the symbols of the Sentential Calculus for a moment:

Symbols Used

'(' ')'	left/right parentheses
P, Q, R . . . M, N, O	atomic statement letters (beginning with 'P' in the alphabet)

The Logical Connectives (standard Russellian notation)

'∼'	"not" (Negation)
'&'	"both . . . and" (Conjunction)
'∨'	"either . . . or" (Disjunction)
'⊃'	"if . . . then" (Conditional)
'≡'	"if and only if" (Equivalence)

You may also find a brief review of the truth tables on page 315 helpful.

27.3 SEMANTIC TREES

Where 'p' and 'q' represent any discrete atomic statement letters or discrete well-formed formulae:

*We will refrain from calling them "Amoebae Trees" as strongly as I am tempted!

The number of rows of a truth table is calculated according to the following formula: Where 'n' is the number of discrete atomic statement letters in an argument, then the number of rows in the truth table equals 2^n.

p	q	~p	~q	(p & q)	(p ∨ q)	(p ⊃ q)	(p ≡ q)
T	T	F	T	T	T	T	T
T	F	F	T	F	T	F	F
F	T	T	F	F	T	T	F
F	F	T	T	F	F	T	T

Where 'p' and 'q' represent any discrete atomic statement letters or discrete well-formed formulae.

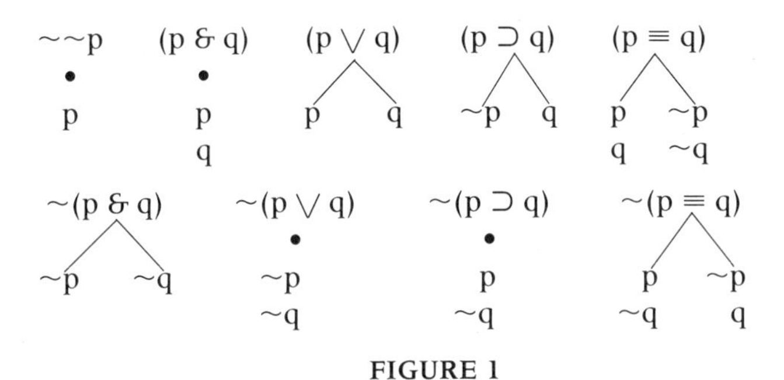

FIGURE 1

Rules for Semantic Trees

27.4 APPLYING THE RULES FOR SEMANTIC TREES

The rules for Semantic Trees correspond exactly to the rules for truth tables. In fact, they are a shortcut derived from what is usually called "the method of assigning truth values," a strategy designed to search for one row of a truth table that will prove invalidity. Unfortunately, this strategy is not mechanical and is easily confusing. Semantic Trees, however, *are* a mechanical routine: you plug in the rules and you get the proper result every time.

In fact, Semantic Trees are so mechanical a routine that we can devise a computer-like "flowchart" to map our procedure in solving any problem in the Sentential Calculus.

Before explaining how or why this flowchart works, let's just take some time to try it out and apply it to some easy problems. Consider this argument—is it valid or not?

(P ∨ Q)
(Q ⊃ R)
∴ (P ∨ R)

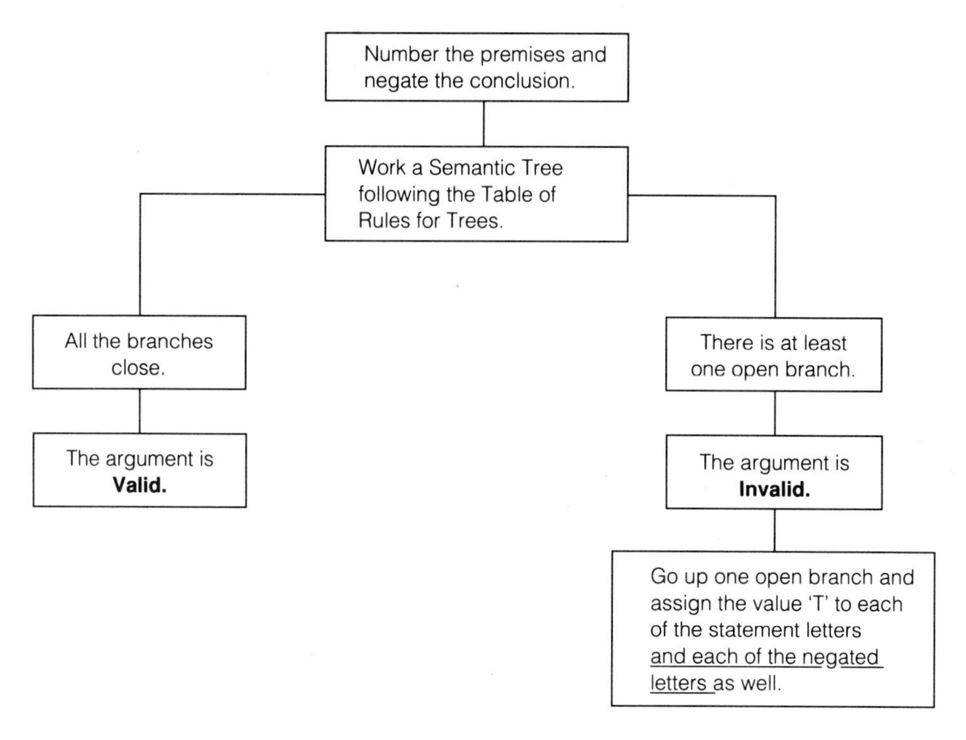

FIGURE 2

Flowchart to Test Arguments for Validity in the Sentential Calculus

To find out the answer, we'll follow each of the steps outlined in Figure 2.

STEP 1. **"Number the premises and negate the conclusion." (The negation of the conclusion is an assumption for the Indirect Derivation Method called, *reductio ad absurdum*, explained in Module 30. We obtain this result:**

1. $(P \vee Q)$
2. $(Q \supset R)$
3. $\sim(P \vee R)$

Numbering each of the steps allows us to see from where we have derived each of the succeeding steps. This procedure is called the "justification." We will now show the justifications for each of these lines:

Argument	**Justification**
1. $(P \vee Q)$	Premise
2. $(Q \supset R)$	Premise
3. $\sim(P \vee R)$	Assumption for Trees

STEP 2. **The next step is to begin "decomposing" each of the compound statement molecules into its component atomic statement let-**

ters. This is a procedure directly analogous to what we called "atom smashing" for truth tables (described in Module 25, "Atom Smashing Molecular Arguments"). To do this, we follow the "Rules for Trees" given in Figure 1.

Our next step is to decompose those expressions that result in columns instead of branches. One molecular statement does decompose into a column, the conclusion '~(P ∨ R)', which was negated according to our assumption for Semantic Trees in step 1. After decomposing the molecular statement, place a checkmark after the molecular statement to show that it was decomposed.

Numbering the new line as '4', the result is:

Argument	**Justification**
1. (P ∨ Q)	Premise
2. (Q ⊃ R)	Premise
3. ~(P ∨ R)✓	Assumption for Trees
4. ~P	
~R	From: 3

We have no other molecular statements that decompose into columns, and so we'll begin working branches.

STEP 3. Decompose the premise in line 2 into a branched statement (see Figure 1):

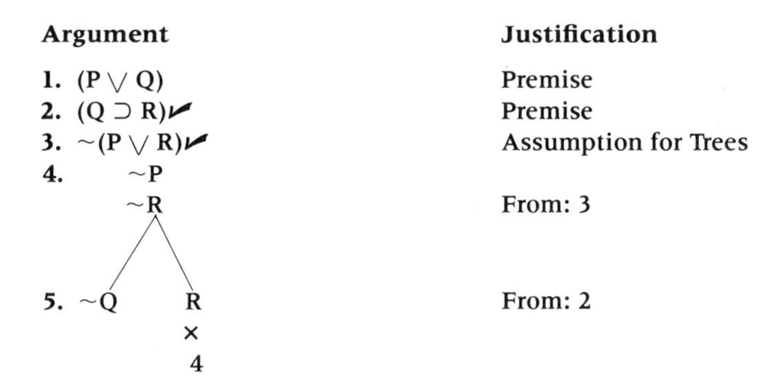

We numbered the line as '5' and showed 'From: 2' as justification. Notice that the branch goes directly underneath the column that we already worked.

Notice also that we placed an '×' and a numeral '4' underneath the 'R' in line 5 of the Tree. This means that we have found a contradiction between 'R' in line 5 and '~R' in line 4. That closes the one side of the branch. '×' = "closed" and '4' here means that the branch closed on line 4. We need not perform any further operations to the closed branch.

STEP 4. Now we decompose the final expression—the molecular statement in line 1. This also results in a branch, but it needs to be placed under the only remaining open column:

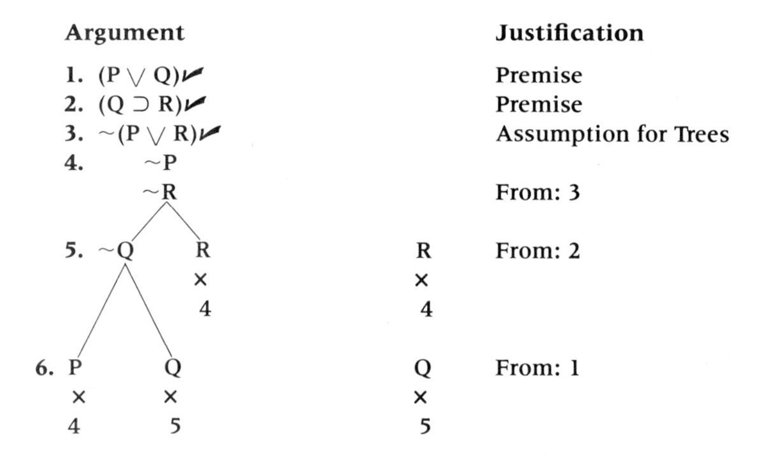

In line 6, the 'P' closes on line 4 and the 'Q' closes on line 5, so all branches have closed. According to the flowchart in Figure 2, if all the branches close, then the argument is *valid*.

Here is another example of a Semantic Tree. The argument is as follows:

(P ∨ Q)
(Q ⊃ R)
~(P ∨ R)
~P
∴R

Strategies for "Constructing" Semantic Trees

Tip: When constructing Semantic Trees, decompose those expressions first that result in *columns* instead of *branches*. This saves a lot of work and much needed space. These are the expressions that result in *columns* when decomposed (there are only four):

~~p	(p & q)	~(p ∨ q)	~(p ⊃ q)
•	•	•	•
p	p	~p	p
	q	~q	~q

All the other expressions decompose into branches.

This is also a valid argument, as the Semantic Tree shows (fill in the justifications yourself as an exercise):

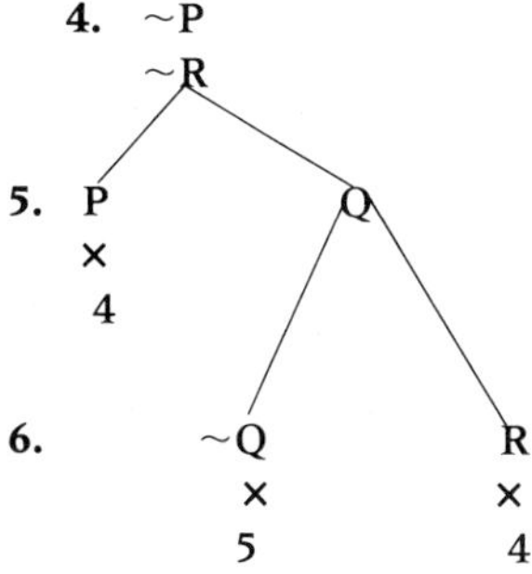

Here is an example of the application of a Semantic Tree for an *invalid* argument. This is useful, since it will demonstrate how to reclaim truth values from the open columns and branches and that will permit us to calculate which row of a truth table would prove it invalid.

Here's the argument:

(A ≡ B)
(B & (C ∨ A))
∴(R ⊃ C)

Again, repeating step 1 in the flowchart in Figure 2, we number each line and negate the conclusion, as follows:

Argument	**Justification**
1. (A ≡ B)	Premise
2. (B & (C ∨ A))	Premise
3. ~(R ⊃ C)	Assumption for Trees

Now we can proceed with decomposing each statement and apply the rules for Semantic Trees. Thus, we have:

Argument	**Justification**
1. (A ≡ B)✔	Premise
2. (B & (C ∨ A))✔	Premise
3. ~(R ⊃ C)✔	Assumption for Trees
4. **R**	
~C	From: 3

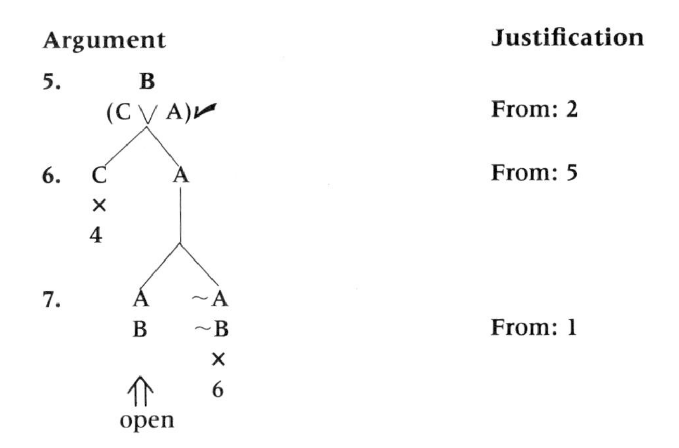

In line 7, the '~B' closed in the 'B' in line 6, but we can find no contradictions (i.e., no case of an atomic statement letter and the presence of its negation—'A', '~A', 'B', '~B') anywhere up the side of the branch designated by a double arrow '⇑'. That means that this branch remains open.

To see why this is so, follow the arrow all the way up the open branch and upwards through the column to which it is connected. To illustrate this "path," I've put into **boldface** all the atomic statement letters (and *negated* atomic statement letters) you'll encounter in tracing this path:

Argument	Justification
1. (A ≡ B)✓	Premise
2. (B & (C ∨ A))✓	Premise
3. ~(R ⊃ C)✓	Assumption for Trees
4. **R**	
~C	From: 3
5. **B**	
(C ∨ A)✓	From: 2
6. C ×4 **A**	From: 5
7. **A** ~A	
B ~B ×6	From: 1
⇑	

In no place along this path will you find a statement letter and its contradiction. Consequently, the branch is open, and according to the flowchart in Figure 2, the argument is *invalid*.

We might well find it valuable to reclaim the truth values for each of the atomic statement letters in the open path so we can say which row of a truth table would show it to be invalid. We do this according to the box in the flowchart in Figure 2 that reads:

Go up one open branch and assign the value 'T' to each of the statement letters and each of the negated letters as well.

We encounter 'A', 'B', '~C', and 'R'. Thus, we assign TRUE to each of these as their value, as follows:

'A' = TRUE
'B' = TRUE
'~C' = TRUE
'R' = TRUE

Notice that if '~C' is TRUE then its affirmative 'C' must be FALSE. Thus, the row of the truth table which shows this argument to be invalid must be the row which gives each of the un-negated atomic statement letters the following truth-value assignment:

'A' = T
'B' = T
'C' = F
'R' = T

A moment's reflection will tell you that row 3 must be the row of the truth table which proves invalidity. Why? We lay out the atomic statement letters in the order in which we encounter them in the argument. That order happens to be 'A' then 'B' then 'C' then 'R'. Recall the formula: "The number of rows in the truth table = 2^n." Since there are four atomic statement letters, then 'n' = 4 and we calculate that we have eight rows in the truth table. Here are the first three rows in that table:

A	B	C	R	(A ≡ B)	(B & (C ∨ A))	∴ (R ⊃ C)
T	T	T	T	T	T	T
T	T	T	F	T	T	T
T	T	F	T	T	T	F

As you can see, row 3 gives the desired assignment of truth values and it is indeed the row that proves the argument to be invalid.

27.5 CREATIVE DEDUCTIONS USING SEMANTIC TREES

We can extend the results we just learned to the exploration of valid conclusions given any argument and even the creation of new deductions altogether. This feature makes the Semantic Tree Method unparalleled in its combination of scope and ease of use.

Let's use the Semantic Tree Method in the solution of a murder mystery. Suppose we have uncovered some evidence of the murder at Castle Clue. We know that either Professor Plum committed the murder only if he used the revolver or else the lead pipe was the murder weapon. Moreover, later evidence has determined that the lead pipe was not the murder weapon. What can we deduce?

First, let's symbolize the argument as follows:

Let 'P' = "Professor Plum committed the murder."
Let 'R' = "The revolver was the murder weapon."
Let 'L' = "The lead pipe was the murder weapon."

We symbolize the premises in the argument as follows:

$((P \supset R) \vee L)$
$\sim L$

To work a Tree in which no conclusion is supplied, we do not negate the conclusion, of course (we have none) so we merely number the premises and then apply the Semantic Tree rules in Figure 1 in order to see which (if any) branches close.

The result is as follows:

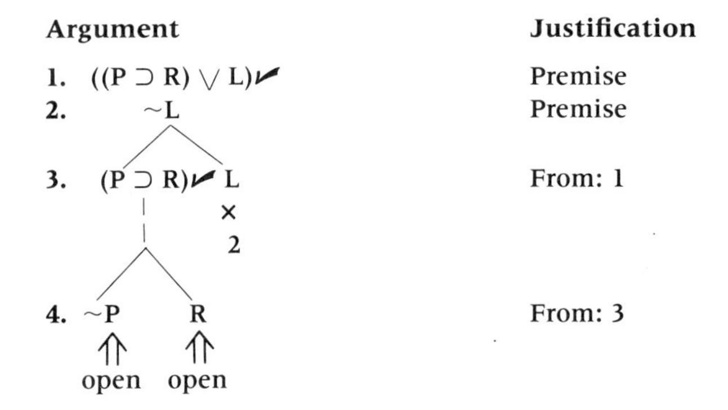

We see that only one branch closed; two others are open. What does that mean? It means that two possibilities are left us regarding the interpretation of the evidence. Let's look at each of these possibilities in turn.

First possibility: Assigning truth values for the first open path, we see that we have the following result:

~P = TRUE
~L = TRUE

. . . or alternately:

P is false
L is false

That means that one possible interpretation of the evidence is that Professor Plum didn't commit the murder and that the lead pipe was not the murder weapon.

That, however, is not the only interpretation of the evidence. We have a second possibility, based on the other open branch:

The second possibility: Assigning truth values for the second open path, we have the following result:

R = TRUE
~L = TRUE

. . . or alternately:

R is true
L is false

That means that the other possible interpretation of the evidence is that the revolver was indeed the murder weapon and not the lead pipe. However, neither interpretation allows us to validly deduce that Professor Plum is the murderer (in the second interpretation, someone else could have used the revolver on the murder victim). We just don't have enough evidence to convict Professor Plum.

The Diagnosis

Now for a more difficult example, let's extend the principles we just learned.

Suppose that you're an attending physician and that a patient has consulted you about a severe spider bite; however, she doesn't know what kind of spider bit her. You're concerned because you believe that such a reaction may be due to the bite of a black widow or a violin (or "brown") spider.

You consult *The Merck Manual of Diagnosis* and this is what you learn:

> "Necessary and sufficient conditions for the bite of *Loxosceles* (the 'violin' or 'brown' spider) include both the eventual development of a high fever and the development of a bleb at the site of the bite, which comes to take on the appearance of a bull's eye. If the patient was bitten by *Lactrodectus*, a 'black widow' spider, then the bite immediately gives rise to a sharp, pinprick-like pain."

You have questioned the patient and learned that a bleb (or small blister) has in fact developed into a welt resembling a bull's eye and that the patient did not feel any pain at the site of the bite. However, you don't think you have enough data to make an informed diagnosis (you haven't ruled out the possibility yet that something else entirely is causing the conditions the patient describes), so you send her for tests.

Later you learn that the patient is developing a high fever. Can you make an accurate diagnosis?

Let's see. We need to symbolize the argument:

Let 'V' = "The patient was bitten by a violin spider."
Let 'H' = "The patient develops a high fever."
Let 'B' = "A small bleb forms which comes to take on the appearance of a bull's eye."
Let 'W' = "The patient was bitten by a black widow spider."
Let 'P' = "The bite gives rise to an immediate, sharp, pinprick-like pain."

We then symbolize the argument as:

(V ≡ (H & B))
(W ⊃ P)
(B & ~P)
H

What can we deduce? Let's find out, using the Semantic Tree Method:

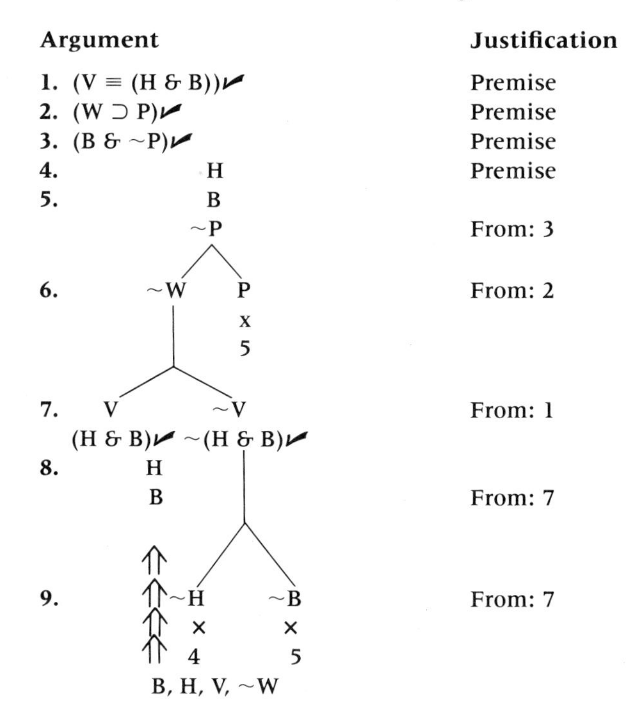

Only one branch is open (as indicated by the double arrows). Retracing that path, we discover four of the atomic statement letters or negated atomic statement letters: 'B', 'H', 'V', and '~W'.

Assigning truth values, we have:

'B' = TRUE
'H' = TRUE
'V' = TRUE
and
'~W' = TRUE
so . . .
'W' = FALSE

Thus, we have learned that it is false that the patient was bitten by a black widow spider and that it is indeed true she was bitten by a violin spider.

Our diagnosis made, we may now begin treatment.

27.6 TAUTOLOGIES, CONTRADICTIONS, AND CONTINGENT STATEMENTS

Semantic Trees can also be used to detect whether a given statement is a tautology, contradiction, or contingent (i.e., "neither" tautology nor contradiction).

In order to do so, consult the following flowchart:

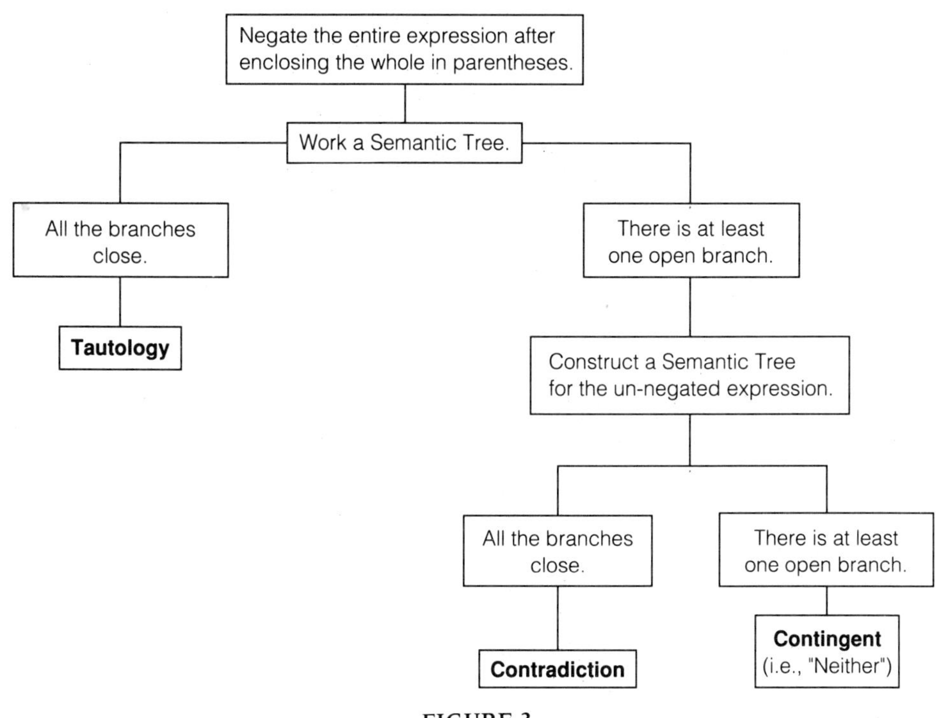

FIGURE 3
Flowchart for Tautologies

NOTE: This procedure is a test for *tautologies*. If you suspect that the statement is a contradiction, you may save yourself some work by beginning with that test instead. (The test for contradictions begins by leaving the statement un-negated in the first box.)

A simple example: Test '$(P \lor \sim P)$' [the Law of Excluded Middle] to see if it is a tautology.

Argument	**Justification**
1. $\sim(P \lor \sim P)$	Assumption for Trees
2. $\sim P$	
$\sim\sim P$	From: 1
×	

All the branches close (one branch). Therefore, '$(P \lor \sim P)$' is indeed a tautology.

Let's test for a contradiction. Test '$(P \equiv \sim P)$'. As we noted, we can save some time by testing this right away to see if it is a contradiction because we already suspect it is. The procedure for testing for contradictions omits the usual assumption for Semantic Trees. We leave the expression un-negated, as follows:

Argument	**Justification**
1. $(P \equiv \sim P)$	Given
P $\sim P$	From: 1
$\sim P$ P	
× ×	

Both statements close, so we know it is indeed a contradiction.

In practice, you may well notice that most ordinary statements prove to be contingent (with some open branches and some closed branches).

Here's a more complicated example:

Prove that '$(((P \supset Q) \supset P) \supset P)$' is a tautology.

This is a theorem of logic that we call, Peirce's Law.*

The solution is as follows: first, we number and negate the statement, then we proceed with the rules:

Argument	**Justification**
1. $\sim(((P \supset Q) \supset P) \supset P)$✓	Assumption for Trees
2. $((P \supset Q) \supset P)$✓	
$\sim P$	From: 1
3. $\sim(P \supset Q)$✓ P	From: 2

*Named after the incomparable logician, mathematician, philosopher, scientist, and founder of the school of philosophy known as "American Pragmatism." Peirce did his greatest work from 1870 to 1910. He resided in Milford, Pennsylvania from the 1890s until his death in 1914.

Argument		Justification
4. P	×	
~Q	2	From: 3
×		
2		

Both branches close so this is a tautology.

27.7 WHY DO SEMANTIC TREES WORK?

Every expression in logic can be reduced to combinations of '~', '&', and '∨'. In effect, Semantic Tree rules reduce every expression to these constituent parts: the branch represents the '∨'; the column (designated by the dot) represents the '&'; and '~' is represented by itself.

Now, note the following equivalences (which reduce all logical expressions to those three operators while preserving truth values):

(p ⊃ q)	::	(~p ∨ q)*	(p ≡ q)	::	((p & q) ∨ (~p & ~q))
~(p & q)	::	(~p ∨ ~q)	~(p ∨ q)	::	(~p & ~q)
~(p ⊃ q)	::	(p & ~q)	~(p ≡ q)	::	((p & q) ∨ (~p & q)

The thesis that every expression can be so reduced to the constituent operators '~', '&', and '∨' (and, thus, to the wave, branch, and column) is subject to a formal logical demonstration.† Furthermore, on the basis of another formal demonstration, Semantic Trees have been shown to be logically complete and consistent (that is to say that they capture the whole of logic).

EXERCISES

Exercise 1

DIRECTIONS: *Determine whether these arguments are valid or invalid by means of the Semantic Tree test.*

1. (P ⊃ Q)
   (P & Q)
   ∴Q

2. (A & (B ∨ A))
   ∴(B ∨ A)

*(::) is the symbol for replacement. See Module 28.3 for explanation.

†You might like to check this out for yourself in a rough and ready way. Take an argument you have determined to be valid (or invalid) using the Semantic Tree Method, and construct a truth table for it. You will come to see the ways in which Semantic Tree columns work like a truth table conjunction between atomic statements and the way in which Semantic Tree branches work like truth table disjunctions between statements.

3. $\sim P$
   $(\sim P \mathbin{\&} \sim A)$
   $(A \lor (P \equiv Q))$
   $\therefore(\sim Q \lor P)$

4. $(A \lor (B \lor C))$
   $(\sim A \mathbin{\&} \sim B)$
   $\therefore C$

5. R
   S
   $\therefore(R \mathbin{\&} S)$

6. $(P \supset (Q \supset R))$
   $((Q \supset R) \supset S)$
   $\therefore(P \supset S)$

7. $\sim(P \mathbin{\&} R)$
   $\therefore(\sim P \mathbin{\&} \sim R)$

8. $\sim(A \supset (H \equiv B))$
   $(\sim A \mathbin{\&} ((H \supset B) \mathbin{\&} (B \supset H)))$
   $\therefore(B \supset H)$

9. $(R \supset S)$
   $\therefore(R \supset (R \mathbin{\&} S))$

10. $((P \supset Q) \mathbin{\&} (R \supset S))$
    $(P \lor R)$
    $\therefore(Q \lor S)$

11. $(P \lor (Q \supset \sim A))$
    $(Q \supset \sim A)$
    $\therefore\sim P$

12. $((P \supset Q) \mathbin{\&} (Q \supset P))$
    $\therefore(P \equiv Q)$

13. $(C \equiv (T \lor \sim R))$
    $(\sim R \mathbin{\&} \sim S)$
    $(\sim S \supset T)$
    $\therefore T$

14. W
    $\therefore(W \lor G)$

15. $(A \supset B)$
    $(B \supset C)$
    $\therefore(A \supset C)$

Exercise 2

DIRECTIONS: *Say whether the following statements are tautologies, contradictions, or neither on the basis of the Semantic Tree test.*

1. $(P \supset P)$
2. $(P \lor \sim P)$
3. $(P \mathbin{\&} \sim P)$
4. $(((P \supset Q) \supset P) \supset P)$
5. $((P \supset Q) \supset P)$
6. $\sim[((P \supset Q) \supset P) \supset P]$
7. $(P \supset (Q \supset P))$
8. $(\sim P \supset (P \supset Q))$
9. $[(P \equiv A) \equiv ((P \supset A) \mathbin{\&} (A \supset P))]$
10. $((P \mathbin{\&} S) \lor ((Q \supset P) \equiv (P \lor R)))$

Exercise 3

Creative Deductions
Using the Semantic Tree Method*
FIND THE ATOM BOMB!

DIRECTIONS: *USING THE SEMANTIC TREE METHOD, USE THE AVAILABLE INFORMATION TO DEDUCE THE LOCATION OF THE THERMONUCLEAR BOMB. The argument is already translated into symbols for you. Give your answer to the mystery in the appropriate space.*

BACKGROUND: You are James Bond, a secret agent for Her Majesty's Secret Service. Your liaison officer has just wired you information, based on reliable leads, that

*I am indebted to Jack Nelson of Temple University for inspiring the form of these exercises.

the villain Goldfinger has placed a 20 megaton thermonuclear device in one of four strategic locations. You have twenty minutes to find it and disarm it.

GOOD LUCK, 007. AND HURRY!

TO: JAMES BOND
FOR YOUR EYES ONLY
MESSAGE FOLLOWS

IF S.P.E.C.T.R.E. IS NOT SHREWDLY CONCEALING THE BOMB IN YOUR LONDON FLAT, THEN EITHER IT IS IN THE BRIEFCASE YOU ARE HOLDING OR THE MUNITIONS OFFICER "Q" MUST BE HANDING IT TO YOU RIGHT NOW DISGUISED AS AN INKPEN. IF THE BOMB IS NEITHER IN YOUR ASTON-MARTIN DB-V TOURING CAR NOR IN YOUR BRIEFCASE, THEN CHECK THE INKPEN "Q" IS HANDING TO YOU BEFORE IT'S TOO LATE. BUT IF IT ISN'T THE CASE THAT IF IT ISN'T IN YOUR ASTON-MARTIN THEN IT SURELY IS IN YOUR BRIEFCASE. S.P.E.C.T.R.E. IS CONCEALING NOTHING.

REGARDS, "M"

END MESSAGE.

Assign the statement letters to the appropriate statements in the text:

S:

B:

Q:

A:

The symbolic translation:

PREMISE 1. $(\sim S \supset (B \vee Q))$

PREMISE 2. $(\sim(A \vee B) \supset Q)$

PREMISE 3. $\sim(\sim A \supset \sim\sim B)$

PREMISE 4. $\sim S$

THEREFORE, THE TWENTY MEGATON THERMONUCLEAR BOMB IS ____________.

THINK FAST!

Exercise 4

FIND THE HIDDEN TREASURE

DIRECTIONS: *USING THE SEMANTIC TREE METHOD, USE THE AVAILABLE INFORMATION TO DEDUCE THE LOCATION OF THE BURIED TREASURE (IT WAS BURIED BY JEAN LAFFITE, THE PIRATE.) The argument is translated into symbols for you. Give your answer to the mystery in the appropriate space.*

Here is your secret treasure map:

"If either the treasure is on the island of Barbados or it's on Martinique, then Jean Laffite was drunk. If it's not on Martinique, then either it's buried on Barbados or on Guadalupe or on Oak Island. However, if Jean Laffite was drunk,

then Bluebeard was in the vicinity. If the treasure is buried on Oak Island, then Bluebeard was *not* in the vicinity. The treasure is buried on Martinique if and only if Captain Hook can swim, and everybody knows that Captain Hook can't swim to save his life. If the treasure was buried on Guadalupe, then it's not the case that Bluebeard was in the vicinity, but Bluebeard was indeed in the region at the time."

Therefore, the treasure is buried on __________.

Let B: the treasure is buried on the island of Barbados.

Let M: the treasure is buried on the island of Martinique.

Let O: the treasure is buried on Oak Island.

Let G: the treasure is buried on Guadalupe.

Let J: Jean Laffite was drunk.

Let V: Bluebeard was in the vicinity.

Let C: Captain Hook can swim.

The symbolic translation:

PREMISE 1: $((B \vee M) \supset J)$

PREMISE 2: $({\sim}M \supset (B \vee (G \vee O)))$

PREMISE 3: $(J \supset V)$

PREMISE 4: $(O \supset {\sim}V)$

PREMISE 5: $((M \equiv C)\ \&\ {\sim}C)$

PREMISE 6: $((G \supset {\sim}V)\ \&\ V)$

MODULE 28

RULES OF INFERENCE

28.1 IN GENERAL

In other modules, we have assumed the existence and application of a set of rules used to derive valid arguments known as *rules of inference.* We have defined valid arguments, for instance, as arguments which faithfully follow rules of inference without exception. We've also defined *formal* fallacies as errors in reasoning which involve the violation of a rule of inference.

Until this point, we needed only a rough idea of what rules of inference were in order to understand topics discussed in previous sections. Now, however, we are ready to embark on derivational methods of determining validity for symbolic arguments. In order to construct derivations, we now need to say precisely what those rules of inference are.

28.2 ARGUMENT FORMS AND "WFF" VARIABLES

By now, we have become accustomed to working with atomic statement letters which we know to be the capital letters of the English alphabet beginning with 'P', 'Q', 'R', and so forth, which stand for asserted statements, the units of argument. We also know by now that a *well-formed formula ("WFF")* is any

---

A deductive argument is **valid,** if and only if, if all the premises are true, then the conclusion must be true.

symbolic formula made up of a set of combinations of logical connectives and atomic statement letters, creating a meaningful expression in symbolic language analogous to a grammatically correct sentence in English. WFFs can be single atomic statement letters, or they can be long and complex expressions made up of many atomic statement letters in combinations with properly placed logical connectives. In other words, WFFs are the *molecules* of molecular logic. (See Module 23.3)

We have also said that we will use the lowercase letters of the English alphabet beginning with 'p', 'q', 'r', etc., to stand for general statements about symbolic statements. We now need to make this concept more precise. Let's now call these letters **WFF variables** and say that they stand for WFFs.

WFF variables will become useful when we list the rules of inference in the next section because they allow us to generalize these rules for argument-forms, where an **argument-form** may be said to be any argument consisting of an arrangement of WFF variables that allows us to substitute any WFF for a WFF variable, provided that we do so uniformly and consistently.

Why are argument-forms important? The reason is that valid arguments are valid by virtue of their form. (Recall our discussion of truth tables in Module 24.) Should we determine that a particular argument is valid, we are able to generalize its form and determine that any other argument which matches that form is likewise valid. For example, consider the following particular argument:

> "If you leave milk on the table, then it will spoil. You left the milk on the table all right, and—sure enough—it spoiled."

We may symbolize this argument as follows.

Let 'P' = "you leave milk on the table"
Let 'Q' = "it will spoil"

(P ⊃ Q)
P
∴Q

We may easily determine that this argument is valid (by using truth tables, for example). We know it is valid by virtue of its form, so we've learned a valuable thing about arguments in general. Any argument which uniformly substitutes WFFs (even complex WFFs) for these atomic statement letters would likewise be valid. By coming to this realization, we then know something about very many arguments. We know, for example, that the following argument is valid as well:

Let 'S' = "a full moon is out"
Let 'T' = "people act crazy"

(S ⊃ T)
S
∴T

Thus, the corresponding English argument is also valid: "When a full moon is out, people act crazy. There's a full moon tonight, so we can expect that people will act crazy."

When a WFF variable occurs in an argument-form, it becomes a placeholder for any WFF whatsoever. For example, the WFF variable 'p' can stand for either 'P' in the first argument or for 'S' in the second; the WFF variable 'q' can stand for the 'Q' in the first argument or for the 'T' in the second. We can thus generalize to an argument-form:*

Statement 1

(p ⊃ q)
p
∴q

. . . and we know that any uniform and consistent substitution of atomic statement letters for these WFF variables will yield a valid argument. Even more importantly, the WFF variables 'p' and 'q' can stand for *complex WFFs* and we know that any argument made up of complex WFFs uniformly and consistently substituted for the WFF variables in the argument-form in statement 1 will be valid.

Let's consider an example where WFF variables stand for complex WFFs (instead of for WFFs made up of single atomic statement letters as above). We know that the expression '~(A ∨ (B ⊃ ~C))' is a WFF, so we will let 'p' stand for that WFF. We know that the expression '(~G ≡ H)' is a WFF, so we will let 'q' stand for it. Following the valid argument form in statement 1, and making uniform substitutions of these WFFs for the WFF variables 'p' and 'q', we know that the following result is likewise an instance of the valid argument-form, and so we know it also to be valid:

Statement 2

{~(A ∨ (B ⊃ ~C)) ⊃ (~G ≡ H)}
~(A ∨ (B ⊃ ~C))
∴(~G ≡ H)

We used the brackets {','} to stand for the left and right parentheses in the WFF variable expression '(p ⊃ q)'. We will now introduce two styles of brackets {,} and [,] and say that these count as parentheses whenever they help to make an expression clearer.

To see how statement 2 counts as a substitution-instance of the argument-form in statement 1, consider the diagram on the next page.

*We call this argument-form *Modus Ponens*. This is a traditional Latin name which literally translates as "the mode of 'putting forth' or 'proposing.'" Since the name translates badly into English, we have retained the traditional Latin.

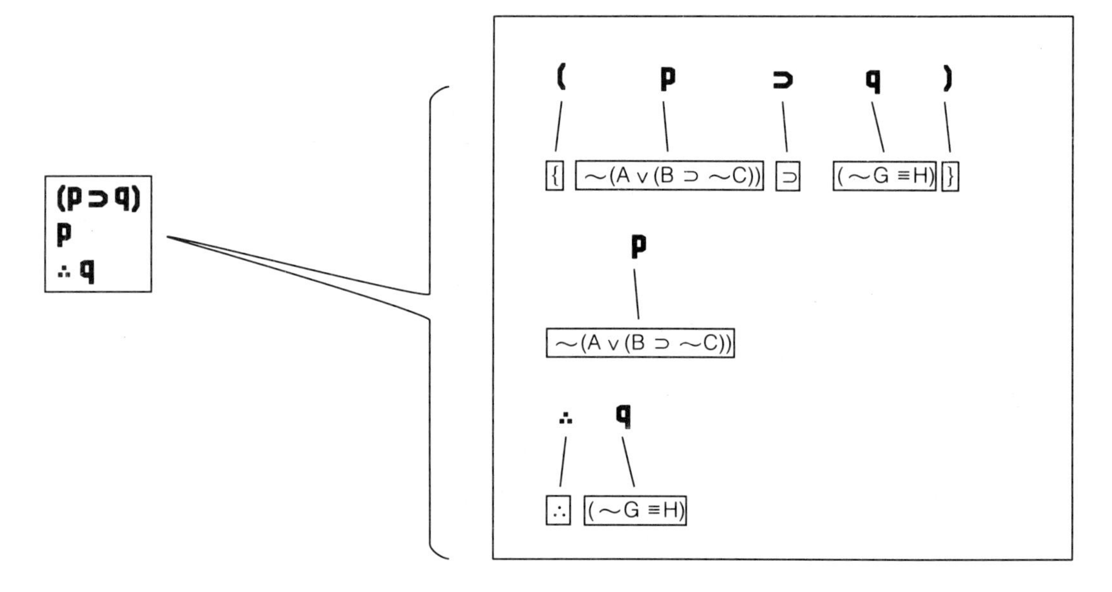

Let's consider some new examples. Here is a different valid argument-form* using the WFF variables 's' and 't':

Statement 3

(s ⊃ t)
~t
∴~s

We may now substitute the following WFFs for their respective WFF variables:

Let 's' = '(G & B)'
Let 't' = '~D'

Now, if we consistently and uniformly replace 's' and 't' with the WFFs given above, we get the result at the top of page 335.

Finishing the substitution, we have the following symbolic argument as the result:

{(G & B) ⊃ ~D}
~~D
∴~(G & B)

Notice that '~t' becomes the double negation '~~D' when '~D' is substituted for the 't' in our argument-form.

*We call this argument form *Modus Tollens,* which literally means, "the mode of 'taking away' or 'removing.' "

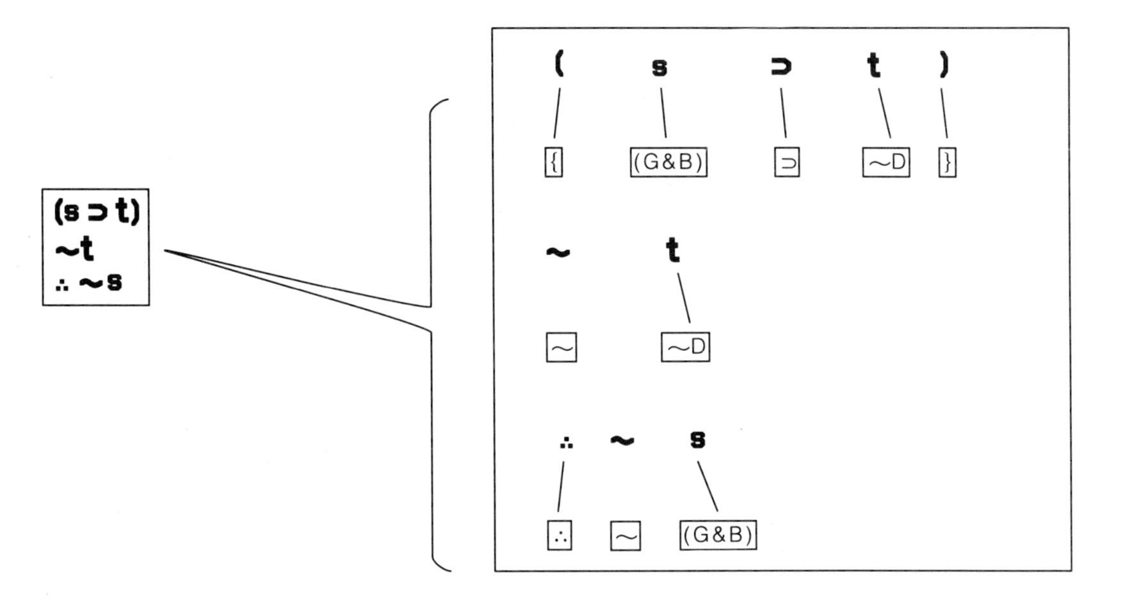

WFF'N PROOF
The Game of Modern Logic

WFF'N PROOF is a game of symbolic logic developed at Yale by a group headed by Layman E. Allen, a legal scholar and sociologist, in 1962. The game, which is sold by WFF'N PROOF ASSOCIATES, PO Box 71, New Haven, CT 06501, is designed to develop both skills in formulating well-formed formulae (WFFs) and skills in symbolic logical deduction.

The game consists of two sets of die: 18 capital letter cubes (9 red, 9 blue); 18 lowercase letter cubes (9 red, 9 blue); 3 playing mats; an instruction booklet; and a timer. A total of twenty-one games can be mastered; when learned sequentially, they extend the player's abilities through most phases of the Sentential Calculus.

The game has one odd feature: it's played in Polish notation. The reason? To dispense with the need to form parentheses when throwing the die. One set of dice is marked with the symbols for the logical connectives (in Polish) and the other contains some lowercase atomic statement letters in the Polish fashion.

About the name of the game, Layman Allen writes:

"Giving the name 'WFF'N PROOF' to a game that was developed at Yale has misled a few imaginative souls into thinking that a new activity for teaching music has been devised by a certain famous undergraduate singing group known to congregate occasionally at the tables down at Mory's.

Actually, the name 'WFF'N PROOF' sprang from a somewhat less rhythmical source. Until 1961 there were two sets of games: one called the 'WFF Games' and the other called the 'Proof Games'—both designed to teach something about mathematical logic. 'WFF' and 'Proof' are technical terms in mathematical logic."

You should have the idea now, so try your hand at a few simple exercises before proceeding to the next section.

EXERCISES

DIRECTIONS:

Let 'p' = '(~P & ~(Q ∨ R))'
Let 'q' = '(S ≡ T)'
Let 'r' = '~P'
Let 's' = '(P ⊃ (P ⊃ (P ⊃ Q)))'

Make substitutions for the following argument-forms:

1. (p ⊃ q)
   p
   ∴q
2. ~~p
   ∴p
3. p
   ∴~~p
4. p
   q
   ∴(p & q)
5. p
   ∴(p ∨ q)
6. (p ⊃ q)
   (q ⊃ r)
   ∴(p ⊃ r)
7. ((p ⊃ q) & (r ⊃ s))
   (p ∨ r)
   ∴(q ∨ s)
8. ~(p & q)
   ∴(~p ∨ ~q)
9. (p ⊃ q)
   ∴(~q ⊃ ~p)
10. p
    ∴(p ∨ p)

28.3 THE RULES OF VALID DEDUCTION

We may now define a rule of inference: a **rule of valid deduction** is an argument-form which is valid. Any "instantiation" (i.e., a consistent and uniform substitution of WFFs for WFF variables) of a valid argument-form always yields a valid symbolic argument. Any uniform substitution of asserted statements for the atomic statement letters in any valid symbolic argument always yields a valid argument in natural language.

We may distinguish between two kinds of valid argument forms: **rules of inference** and **rules of replacement.** Strictly speaking, the rules of replacement are also rules of valid deduction, but because they express equivalences between two sets of symbolic expressions, replacement rules have a special feature—we can use them to work *inside* of parentheses and still derive valid deductions. In other words, we can use rules of replacement to *replace* any expression by an equivalent form during the course of a deduction. Any deduction that proceeds according to rules of inference or rules of replacement explicitly named is called a **derivation.**

Any valid argument-form could be a rule of valid deduction, but by convention and practice we restrict their number to a handful of the most useful examples. It is to one subset of valid argument-forms that we give the title, "rules of inference;" to a second subset of the rules of valid deduction, we give the name of "rules of replacement." Rules of replacement show equivalent expressions.

Here are the rules of inference and the rules of replacement which together make up the rules of valid deduction:

The Rules of Inference

1. *Modus Ponens* (MP)
   (p ⊃ q)
   p
   ∴q
2. *Modus Tollens* (MT)
   (p ⊃ q)
   ~q
   ∴~p
3. Disjunctive Syllogism (DS)
   (p ∨ q)
   ~p
   ∴~q
4. Simplification (Simp)
   (p & q)
   ∴p
5. Conjunction (Conj)
   p
   q
   ∴(p & q)
6. Addition (Add)
   p
   ∴(p ∨ q)
7. Hypothetical Syllogism (HS)
   (p ⊃ q)
   (q ⊃ r)
   ∴(p ⊃ r)
8. Dilemma (Dil)
   (p ⊃ q)
   (r ⊃ s)
   (p ∨ r)
   ∴(q ∨ s)
9. Absorption (Abs)
   (p ⊃ q)
   ∴(p ⊃ (p & q))
10. Repetition (Rep)
    p
    ∴p

The rules of replacement are themselves part of the set of rules of valid deduction. For the sake of economy, we introduce a new symbol, the double colon ('::') which separates one WFF from the other. The double colon tells us that we may replace the WFF on the one side with the WFF on the other.

Here are ten of the most common rules of replacement:

The Rules of Replacement

11. Double Negation (DN)	$p :: \sim\sim p$
12. Commutation (Com)	$(p \vee q) :: (q \vee p)$ $(p \,\&\, q) :: (q \,\&\, p)$
13. Association (Assoc)	$(p \vee (q \vee r)) :: ((p \vee q) \vee r)$ $(p \,\&\, (q \,\&\, r)) :: ((p \,\&\, q) \,\&\, r)$
14. De Morgan's Laws (DeM)	$\sim(p \,\&\, q) :: (\sim p \vee \sim q)$ $\sim(p \vee q) :: (\sim p \,\&\, \sim q)$
15. Distribution (Dist)	$(p \,\&\, (q \vee r)) :: ((p \,\&\, q) \vee (p \,\&\, r))$ $(p \vee (q \,\&\, r)) :: ((p \vee q) \,\&\, (p \vee r))$
16. Equivalence (Equiv)	$(p \equiv q) :: ((p \supset q) \,\&\, (q \supset p))$
17. Transposition (Trans)	$(p \supset q) :: (\sim q \supset \sim p)$
18. Material Implication (Imp)	$(p \supset q) :: (\sim p \vee q)$
19. Exportation (Exp)	$((p \,\&\, q) \supset r) :: (p \supset (q \supset r))$
20. Duplication (Dup)	$p :: (p \vee p)$ $p :: (p \,\&\, p)$

Thus, you have a total of twenty rules to learn: ten rules of inference and another ten rules of replacement.

Each of the rules of replacement can be expressed as a rule of inference merely by inserting a '∴' between an expression on either side of the double colon ('::'), preserving validity whether the formula is read as "left expression + '∴' + right expression" or "right expression + '∴' + left expression."

For example, consider the rule of replacement for 'Duplication (Dup)'. The following is a valid deductive argument form:

p
$\therefore (p \vee p)$

So is its reverse:

$(p \vee p)$
$\therefore p$

So why make a whole new set of rules called "rules of replacement?" One reason is that rules of replacement can be applied *inside* of parentheses without loss of truth content; the same cannot be said for rules of inference. To put this another way, a rule of replacement can be applied in *part of a line* without loss of truth content, whereas in order to avoid problems about preserving truth content a rule of inference must be applied to the *whole line.*

Consider, for example, the rule of replacement called "Double Negation":

Double Negation (DN) p :: ∼∼p

This says that whenever we have 'p' we may substitute its double negation '∼∼p' (or *vice versa*) without loss of meaning. Should we encounter the following symbolic statement (which contains an instance of double negation):

(R ∨ (S ⊃ ∼∼P))

. . . we may use the rule 'DN' to eliminate the two waves without loss of truth content:

(R ∨ (S ⊃ P)) **DN**

. . . even though the expression '∼∼P' occurs well inside the parentheses surrounding the formula.

We can also express both rules of replacement and rules of inference as theorems, in the sense of "mathematical theorems." A theorem in mathematics is a proven proposition, derived from a set of assumed axioms. A **theorem in symbolic logic** is a single well-formed formula (WFF) which is necessarily true (i.e., tautological).

Turning rules of replacement into symbolic logic theorems is easy and obvious once we realize the reason for the introduction of the double colon ('::') convention. The double colon is a "meta-logical" convention introduced here to save us from having to add more parentheses. Informally speaking, it stands for the biconditional and so we may substitute the triple bar ('≡') for the double colon ('::') wherever it occurs without loss of truth content provided that we then enclose the whole expression with parentheses or brackets.

For example, recall the rule of replacement called "Equivalence":

Equivalence (Equiv) (p ≡ q) :: ((p ⊃ q) & (q ⊃ p))

We may re-write it as a theorem provided that we replace the ('::') with a triple bar ('≡') and then enclose the whole formula in parentheses or brackets. The result is as follows:

Equivalence Theorem {(p ≡ q) ≡ ((p ⊃ q) & (q ⊃ p))}

This symbolic statement, as we may easily demonstrate, is a tautology.

We can also render rules of inference as tautological symbolic statements. To do so, we have to realize what the 'therefore' which separates premises from conclusion really means: it tells us that the conjunction of the premises *logically implies* the conclusion! That logical implication is the implication expressed by the horseshoe ('⊃'). Thus, follow this formula to render a rule of inference as a single symbolic statement:

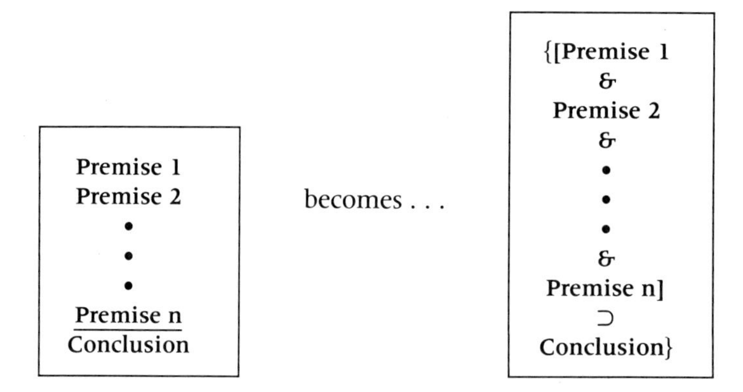

FIGURE 1

Written out as a single symbolic statement, the rule of inference becomes one long symbolic formula: {[*Premise* 1 & *Premise 2* & *Premise 'n'*] ⊃ *Conclusion*}. . . . where 'n' represents an indefinite number of premises. Be careful to follow conventions for parentheses when conjoining premises here.

Let's take an example using the rule of inference called *Modus Ponens.*

Modus Ponens (MP)
(p ⊃ q)
p
∴q

We can turn this argument-form into a tautological symbolic statement by applying the formula in Figure 1. The result is the symbolic statement: '{((p ⊃ q) & p) ⊃ q)', which is a tautology and so necessarily true.

MODULE 29

DIRECT DERIVATIONS

29.1 IN GENERAL

Direct derivations are a mainstay of symbolic logic. Unfortunately, they tend to prove very lengthy and difficult. In a previous module, we learned to use Semantic Trees, which are a powerful and easy way to determine validity or invalidity. Yet, direct derivations serve a special purpose.

Because other and easier methods have been discovered for determining whether or not a given argument in the Sentential Calculus is valid or invalid, you may wonder why we continue to study the Direct Derivation Method. One reason is that direct derivations probably model the way we think when we think logically. Another is that the Direct Derivation Method demonstrates precisely how the rules of inference and rules of replacement (jointly called "the rules of valid deduction") work, and we have defined both validity and fallacies elsewhere in terms of rules of inference. Remember, we said that a valid deductive argument is one which follows rules of inference faithfully, and that a formal fallacy is what violates a rule of inference.)

Yet an even more important reason exists for the study of direct derivations: they are a powerful way to generate *new* conclusions. They are the key to logical discovery. And they probably mimic the way we do original thinking when we think carefully and systematically.

A deductive argument is **valid**, if and only if, if all the premises are true, then the conclusion must be true.

Thus, Semantic Trees will probably be the method you will use most often when attempting to determine validity. But if you want to generate new arguments, the Direct Derivations Method is the way to go.

I recommend that when you are engaged in a process of systematic and original thinking, you use the Direct Derivation Method to carry you along. But then go back and check your deduction by means of Semantic Trees. You'll find that they serve as a good double check on your reasoning.

29.2 DIRECT DERIVATIONS (USING RULES OF INFERENCE)

We may now work some simple deductions (called "derivations") using the rules of inference.

Consider this example:

FROM: '(P ⊃ Q)' and 'P'
DEDUCE: '(Q ∨ R)'

This argument involves the application of two rules of inference. First, we use *Modus Ponens*. Once again, the rule for *Modus Ponens* is as follows:

Statement 1

Modus Ponens (MP)
(p ⊃ q)
p
∴q

Remember: this is an argument-form; we need to substitute the WFFs in our problem for the WFF variables in the argument-form given in statement 1. This is obvious enough:

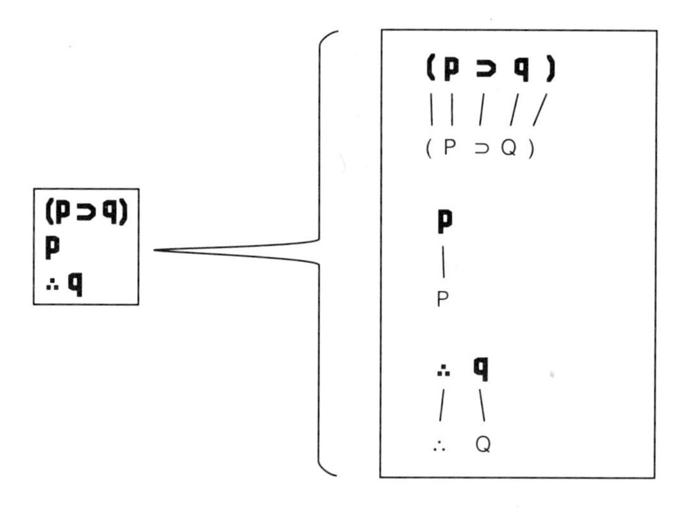

The result is placed in standard derivation form as follows:

DEDUCE: (Q ∨ R)

1. (P ⊃ Q)	Premise
2. P	Premise
3. Q	1,2 MP

Notice that each line in the derivation is numbered. Also, notice the marginal notes. After each line in the derivation, we need to show a justification in the margin. The marginal notes for lines 1 and 2 tell us that the symbolic statements are given to us as premises. The marginal note in line 3 gives us a location as well as a justification: '1,2' tells us that we used the symbolic statements in lines 1 and 2 to get the result in 3; 'MP' is the justification—*Modus Ponens.*

But we're not done yet. We have to show '(Q ∨ R)' validly follows. To do this, we apply the rule of inference known as 'Addition' (abbreviated as 'Add'):

---

6. Addition (Add):
p
∴(p ∨ q)

Once more, we need to substitute WFFs for WFF variables in the argument-form for 'Addition', as follows:

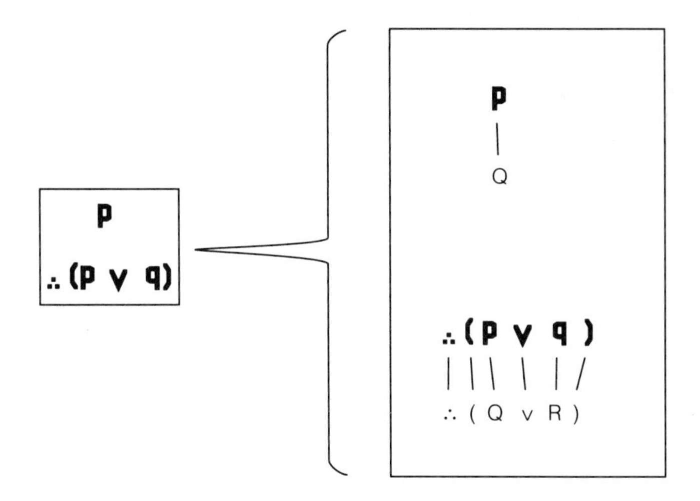

Notice that the argument-form for 'Addition' tells us that we may substitute *any* WFF whatever for the 'q' in '(p ∨ q)'. A moment's reflection will tell you why this is so: if the statement, 'Tuition is too expensive here', is true, you may imagine a student saying in exasperation, "Either tuition is too high here . . . or I'm a monkey's uncle!" The truth table for disjunction ('∨') tells us that only one disjunct needs to be true in order to make the entire disjunction true, so we can add whatever disjunct we like to a true statement (perhaps, as here, for emphasis) and preserve truth.

We may now return to our derivation with the following result:

DEDUCE: (Q ∨ R)

1. (P ⊃ Q)	Premise
2. P	Premise
3. Q	1,2 MP
4. (Q ∨ R)	3, Add

The derivation, which proves the argument to be valid, is now complete. By the way, we call this a **direct derivation** because we proceeded through the derivation exclusively by means of rules of valid deduction without any reliance on special assumptions added to the premises already given. Direct derivations can often get complicated and lengthy, however.

Here is a slightly more difficult direct derivation using only rules of inference:

FROM: ((A & B) ⊃ (C ⊃ D))
A
B
(D ⊃ F)
DEDUCE: (C ⊃ F)

To begin, we need a word on strategy: sometimes you may find it helpful to work *backwards** from the conclusion to see what rule of valid deduction you can use to get the desired result. Here, we see that if only we could get the expression '(C ⊃ D)' free, we could use the rule for 'Hypothetical Syllogism' to get the desired result:

(C ⊃ D)
(D ⊃ F)
(C ⊃ F)

Here's our concluding strategy. To obtain the line '(C ⊃ F)', we'll apply the rules of 'Conjunction' and *Modus Ponens* as follows:

DEDUCE: (C ⊃ F)

1. ((A & B) (C ⊃ D))	Premise
2. A	Premise
3. B	Premise
4. (D ⊃ F)	Premise
5. (A & B)	2,3 Conj

*Working backwards in this way is something you do on "scratch paper," of course. You shouldn't show it in your formal presentation of the derivation.

Obtaining line 5 is straightforward, using the rule called "Conjunction" to conjoin 'A' [from line 2] and 'B' [from line 3]. Let's continue the derivation, using the result in line 5 and the premise in line 1 acccording to the rule of *Modus Ponens*. The chart below shows the uniform and consistent substitution of the WFFs in our arguments for the WFF variables in the *Modus Ponens* argument-form, as follows:

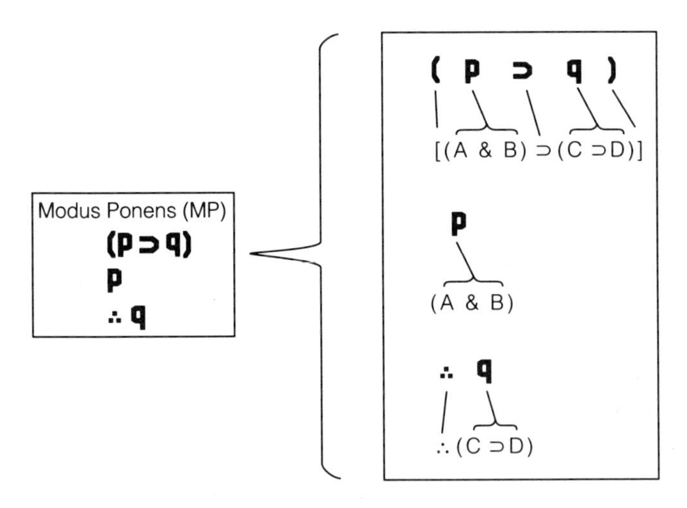

The result becomes line 6 of the derivation:

DEDUCE: (C ⊃ F)

1. ((A & B) ⊃ (C ⊃ D))	**Premise**
2. A	**Premise**
3. B	**Premise**
4. (D ⊃ F)	**Premise**
5. (A & B)	**2,3 Conj**
6. (C ⊃ D)	**5,1 MP**

We may now complete the derivation by using the rule, 'Hypothetical Syllogism':

DEDUCE: (C ⊃ F)

1. ((A & B) ⊃ (C ⊃ D))	**Premise**
2. A	**Premise**
3. B	**Premise**
4. (D ⊃ F)	**Premise**
5. (A & B)	**2,3 Conj**
6. (C ⊃ D)	**5,1 MP**
7. (C ⊃ F)	**6,4 HS**

The argument is valid.

29.3 DIRECT DERIVATIONS (USING RULES OF REPLACEMENT)

In a fashion analogous to the previous section, we may use rules of replacement to derive valid conclusions.

Here is a simple example:

FROM: (~P ∨ ~Q)
DEDUCE: ~(Q & P)

The first step is to use the rule of replacement called "De Morgan's Laws (DeM)." Notice that we are given *two* such laws—one for the connective '∨' and another for the connective '&'. We need the second expression in the first law: '~(P & Q) :: (~P ∨ ~Q)'.

Thus, for '(~P ∨ ~Q)' we may validly infer that '~(P & Q)', substituting the WFFs 'P' and 'Q' for the WFF variables 'p' and 'q' respectively. We then obtain the following partial derivation:

DEDUCE: ~(Q & P)

1. (~P ∨ ~Q)	Premise
2. ~(P & Q)	1, DeM

We complete the derivation using the replacement rule for 'Commutation (Com)':

12. Commutation (Com)	(p ∨ q) :: (q ∨ p) (p & q) :: (q & p)

Notice that 'Com' legitimately works *inside* the parentheses, leaving the wave ('~') intact in its function of negating the whole expression, as follows:

DEDUCE: ~(Q & P)

1. (~P ∨ ~Q)	Premise
2. ~(P & Q)	1, DeM
3. ~(Q & P)	2, Com

The argument is valid.

Now, for a slightly more difficult example:

FROM: (~~(A & B) ∨ (D ∨ C))
DEDUCE: (((A & B) ∨ C) ∨ D)

We use three rules of replacement: Double Negation (DN), Commutation (Com), and Association (Assoc). First, we work inside the parentheses to obtain the following lines of the derivation:

DEDUCE: (((A & B) ∨ C) ∨ D)

1.	**(~~(A & B) ∨ (D ∨ C))**	**Premise**
2.	**((A & B) ∨ (D ∨ C))**	**1, DN**

. . . according to the rule, 'DN':

---

11. Double Negation (DN)	**p :: ~~p**

. . . where 'p' = '(A & B)' and '~~p' = '~~(A & B)'.

Next, we need to commute places for 'D' and 'C' in the expression '(D ∨ C)':

DEDUCE: (((A & B) ∨ C) ∨ D)

1.	**(~~(A & B) ∨ (D ∨ C))**	**Premise**
2.	**((A & B) ∨ (D ∨ C))**	**1, DN**
3.	**((A & B) ∨ (C ∨ D)**	**2, Com**

The final step is to apply the rule for 'Association (Assoc)':

---

13. Association (Assoc)	**(p ∨ (q ∨ r)) :: ((p ∨ q) ∨ r)**
	(p & (q & r)) :: ((p & q) & r)

. . . where 'p' = '(A & B)'
'q' = 'C'
and 'r' = 'D'
. . . with the result:

DEDUCE: (((A & B) ∨ C) ∨ D)

1.	**(~~(A & B) ∨ (D ∨ C))**	**Premise**
2.	**((A & B) ∨ (D ∨ C))**	**1, DN**
3.	**((A & B) ∨ (C ∨ D)**	**2, Com**
4.	**((A & B) ∨ C) ∨ D)**	**3, Assoc**

The derivation shows that the inference is valid.

29.4 DIRECT DERIVATIONS (USING MIXED RULES)

We can now use combinations of rules of inference and rules of replacement to work derivations.

We take it as given that every valid deductive argument in the Sentential Calculus can be shown to be valid using the Direct Derivation Method from rules of inference together with rules of replacement. We also take it as given that every proper application of rules of inference and rules of replacement will yield only valid arguments in the Sentential Calculus. We also assume that every valid inference proven by direct derivation in the Sentential Calculus can be shown to be valid by the method of truth tables (though the truth tables may be prohibitively long) or by means of the Semantic Trees Method. These assumptions can be formally demonstrated for the Sentential Calculus.

Here is a simple example of a derivation using a mix of inference rules and rules of replacement:

((A ⊃ B) & A)
∴B

We can see right away that this resembles the rule of inference known as *Modus Ponens.* However, to use MP we need first to extract the '(A ⊃ B)' from ((A ⊃ B) & A) to extract the second 'A' from ((A ⊃ B) & A) and show it as a separate premise. We may do this by applying the rule of inference called "Simplification":

DEDUCE: B

1. ((A ⊃ B) & A)	Premise
2. (A ⊃ B)	1, Simp

To free up the atomic statement letter 'A' in order to use it on a *Modus Ponens* operation, we need to turn the expression in line 1 around. Thus, we apply the replacement rule 'Commutation (Com)':

12. Commutation (Com)	(p ∨ q) :: (q ∨ p) (p & q) :: (q & p)

. . . and then apply the rule of inference, 'Simplification (Simp)'. The result of these operations is as follows:

DEDUCE: B

1. ((A ⊃ B) & A)	Premise
2. (A ⊃ B)	1, Simp
3. (A & (A ⊃ B))	1, Com
4. A	3, Simp

The last step is the obvious application of *Modus Ponens:*

DEDUCE: B

1. ((A ⊃ B) & A)	Premise
2. (A ⊃ B)	1, Simp

3. (A & (A ⊃ B))	1, Com
4. A	3, Simp
5. B	2,4 MP

The derivation shows that the inference from (((A ⊃ B) & A) to 'B' is valid.

Here is another "mixed" derivation which uses three inference rules (*Modus Ponens, Modus Tollens* and 'Conjunction') plus two replacement rules ('Association' and 'De Morgan's Laws'). The derivation is fairly straightforward, but notice that the direct derivations are beginning to get more involved and less obvious as more rules are introduced.

FROM: ((A ∨ B) ∨ Q)
(~A & ~B)
(Q ⊃ M)
DEDUCE: (~A & (~B & M))

The solution follows:

DEDUCE: (~A & (~B & M))

1. ((A ∨ B) ∨ Q)	Premise
2. (~A & ~B)	Premise
3. (Q ⊃ M)	Premise
4. ~(A ∨ B)	2, DeM
5. Q	1,4 DS
6. M	3,5 MP
7. ((~A & ~B) & M)	2,6 Conj
8. (~A & (~B & M))	7, Assoc

Because the derivations do get increasingly more involved when using the Direct Derivations Method, logicians have hit upon other methods which are its equivalent in power and completeness, but which are usually easier to use. Some of these other methods involve making strategic assumptions in the course of a derivation (such as the Indirect Derivation and Conditional Proof Methods) whereas others (such as the Semantic Trees Method) offer dramatically new shortcuts. We will look at these methods in the remaining sections of this unit.

EXERCISES

Exercise 1

DIRECTIONS: *Practice direct derivations using both rules of inference and rules of replacement. Fill in the missing steps of the derivation according to the justifications given in the margins. THE STRATEGY: Work backward from the conclusion and forward from the premises to see what symbolic statements are needed to get the conclusion.*

The First Penitent

The following anecdote may illustrate the contention that direct derivations using rules of inference do indeed mimic the way we think when we think systematically and carefully:

Fr. Murphy was about to retire after many years of faithful service to his parish and his community, so the Holy Name Society decided to hold a banquet in his honor. Members of the parish contributed to the purchase of a brand new, black Cadillac, and Mayor Smith—the most prestigious person in the town—was asked to present the keys to Fr. Murphy at the conclusion of the banquet.

The dinner was nearly over, the speeches were about to begin and the Master of Ceremonies noticed that Mayor Smith hadn't yet arrived. The humorists were nearly finished telling their jokes when the M.C. was brought to the telephone. It was Mayor Smith who explained he had car problems but to keep the banquet going, he would be there as soon as he could.

The speeches were nearly finished and the mayor had not yet shown. The M.C. looked around the banquet hall to find additional public speakers to prolong the banquet when he realized that Fr. Murphy himself was an accomplished speaker.

"Father Murphy," he asked. "Would you care to say a few words?" The priest was received with a standing ovation, and he began to speak. Fr. Murphy was getting old and he tended to ramble in his later years. He spoke about how drafty the rectory was and how kind his housekeeper had been through the years. He explained how hard it was to be a priest nowadays, and then he added:

"But it's always been hard to be a priest. I remember my first penitent. I was hearing confessions in this very town, when my first penitent confessed to a brutal murder."

The audience murmured in shock, but Fr. Murphy went on to talk about more mundane problems, including the draft in the rectory again. Just then, the mayor arrived. The M.C. thanked Fr. Murphy for speaking and then revealed that the parish had collected money for a gift for the old priest. Mayor Smith tearfully presented the keys to the Cadillac, and the audience cheered.

Never one to miss an opportunity to speak, Mayor Smith addressed the crowd.

"I'm so pleased to be here and see you all again," he said. "I'm especially pleased to see Fr. Murphy once more after all these years. He probably doesn't remember," the Mayor explained, "but I was his first penitent."

In order to get the punch line, you need to work a simple deduction based on the rule of inference known as *Modus Ponens*. You do it automatically, without reflection, but the process is similar to the Direct Derivations Method. A version of this joke was cited by the philosopher A.C. Ewing in *The Fundamental Questions of Philosophy* to make the point that logic does indeed lead to discoveries.

1. DEDUCE: $\sim(R \equiv S)$

1.	$(P \supset Q)$	Premise
2.	$(((R \equiv S) \supset ((P \,\&\, \sim Q) \,\&\, R))$	Premise
3.		1, Imp
4.		3, Add
5.		4, Assoc
6.		5, DeM
7.		6, DN
8.		7, DeM
9.		8, Assoc
10.	$\sim(R \equiv S)$	9, 2 MT

2. *In these examples, you will need to insert justifications and marginal notes as well as lines of the derivation.*

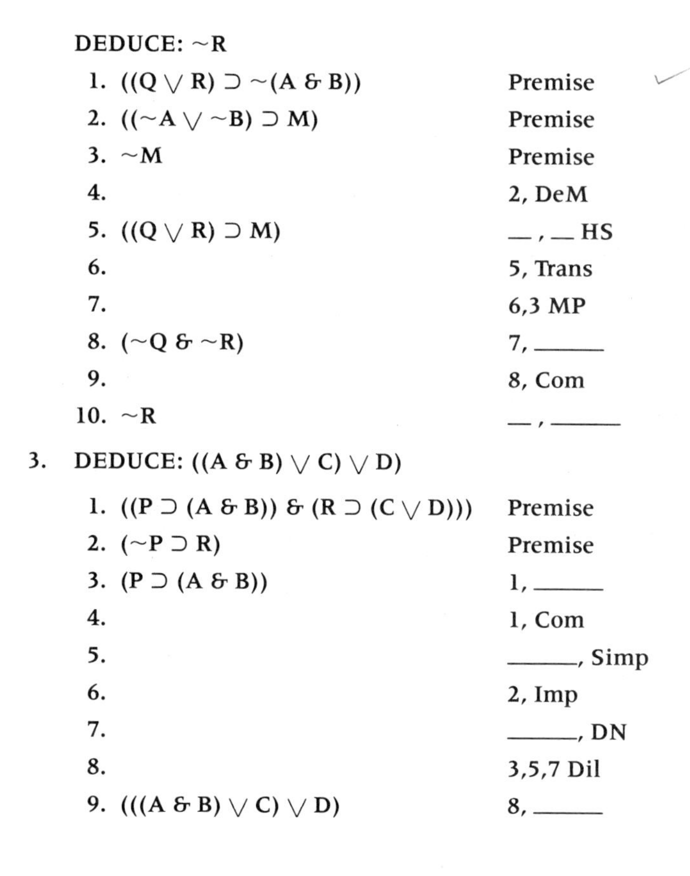

DEDUCE: $\sim R$

1.	$((Q \vee R) \supset \sim(A \,\&\, B))$	Premise
2.	$((\sim A \vee \sim B) \supset M)$	Premise
3.	$\sim M$	Premise
4.		2, DeM
5.	$((Q \vee R) \supset M)$	__, __ HS
6.		5, Trans
7.		6,3 MP
8.	$(\sim Q \,\&\, \sim R)$	7, _____
9.		8, Com
10.	$\sim R$	__, _____

3. DEDUCE: $((A \,\&\, B) \vee C) \vee D)$

1.	$((P \supset (A \,\&\, B)) \,\&\, (R \supset (C \vee D)))$	Premise
2.	$(\sim P \supset R)$	Premise
3.	$(P \supset (A \,\&\, B))$	1, _____
4.		1, Com
5.		_____, Simp
6.		2, Imp
7.		_____, DN
8.		3,5,7 Dil
9.	$(((A \,\&\, B) \vee C) \vee D)$	8, _____

Exercise 2

DIRECTIONS: *Using the Direct Derivation Method, construct a proof for each of the following problems.*

1. DEDUCE: (~Q ∨ ~(A ≡ P))

   1. ((A ≡ P) ⊃ R) — Premise
   2. (~R ⊃ ~Q) — Premise

2. DEDUCE: (A & (B & C))

   1. (A & B) — Premise
   2. (B ⊃ C) — Premise

3. DEDUCE: (~R ⊃ (Q ∨ S))

   1. R — Premise

4. DEDUCE: (T ∨ W)

   1. P — Premise
   2. (Q & S) — Premise
   3. ((Q & P) ⊃ T) — Premise

5. DEDUCE: ((Q & S) & R)

   1. (Q & D) — Premise
   2. (S & M) — Premise
   3. (F & (H & R)) — Premise

MODULE 30

INDIRECT DERIVATIONS AND CONDITIONAL PROOFS

30.1 IN GENERAL

In this section, we will study two more methods of derivation: *indirect derivations* and *conditional proofs*.

Both of these methods introduce assumptions in the course of a derivation according to carefully prescribed guidelines. The Method of Indirect Derivation assumes the negation of the conclusion (or a line of the deduction—a subconclusion) for the sake of argument. The Method of Conditional Proof works on lines of the derivation (especially the conclusion) when that symbolic statement has a horseshoe ('⊃') as its main connective.

We have already presupposed the Method of Indirect Derivation when we learned how to use Semantic Trees in Module 27. This is the rationale behind the 'Assumption for Trees' where we negate the conclusion.

Indirect derivations and conditional proofs are usually far easier to use than the Direct Derivation Method; however, they have drawbacks. These methods are generally not as easy to use for exploring or for discovering new conclusions that follow from the premises.

A deductive argument is **valid**, if and only if, if all the premises are true, the conclusion must be true.

30.2 INDIRECT DERIVATIONS (ALSO CALLED "INDIRECT PROOFS")

Indirect derivations have a special history. The Latin name for them (which still has current usage) is revealing: *reductio ad absurdum,* which literally means "reduce to absurdity." The idea is that we will get absurd results (contradictions) if the *negation* of the conclusion of a valid argument is assumed to be true. Informally, we may describe ordinary arguments as *reductios* even when they do not lead to explicitly contradictory results—as when a course of argument leads us to a conclusion that our common sense or deepest convictions cannot readily accept.

More formally, we say that a *reductio* occurs when an explicit contradiction occurs within the lines of a formal deduction. When one is confronted with a conclusion that her common sense or deepest convictions cannot readily accept, no formal contradiction is present (although she may have cause for worry about the truth of her convictions). This does not constitute a formal *reductio,* although it still raises problems with her belief structure. A formal *reductio* occurs, however, when we make an assumption that causes us to derive contradictory results.

We now introduce a new technique: the technique of *boxing* derivations which proceed from the assumption for *indirect proof.* The **Assumption for Indirect Proof** is the negation of the conclusion. If we encounter a contradiction within the confines of the box, we may close the box and declare a *reductio.*

Boxes come into play most importantly when we work indirect proofs and conditional proofs because the boxes for these methods will house some assumptions which need to be made with care.

To see the difference between a direct derivation on the one hand and an indirect derivation on the other, compare the following schemata: the first is the schematic form of a direct derivation, the second, the schematic form of an indirect derivation:

DEDUCE: Φ

.
.
.
.
Φ

Direct Derivation

For the purposes of indirect derivation, we will say that a contradiction is present should we encounter any WFF and its negation in the course of an unboxed derivation. Thus, the presence of 'P' and its negation, '~P', within the confines of the same box constitutes a contradiction, and then we are entitled to

DEDUCE: Φ

~Φ
.
.
R
.
.
~R

Indirect Derivation

close that box and declare our conclusion. Of course, we may encounter contradictions between larger WFFs than atomic statement letters. For example, should we encounter the molecular WFF '(P ∨ Q)' and its negation '~(P ∨ Q)' within the confines of the same box, we may declare a *reductio* and close the box.*

Up to this point, we had few complications regarding boxed derivations. Boxes for elementary direct derivations are straightforward: when you get the desired result, you close the box to show that the derivation is complete and you have successfully demonstrated that the conclusion does follow from its premises or that a statement is tautologically true. However, once we work with indirect derivations and conditional proofs, we have to make certain guarded assumptions. The box then takes on an indispensable role: think of the box as the "guardian" of those "guarded assumptions."

Boxes and Assumptions

Think of a box as a way of fencing off the results you obtain when making assumptions for indirect derivations or conditional proofs. In other words, the box contains the details of the indirect proof or the conditional proof and, once proved, you are *not* entitled to take any lines from inside the box and repeat them or apply them in the derivation outside of the box. The boxed-off result is *sealed off* from the rest of the derivation.

To say that "a box is closed" is to declare that the assumption has been used in a way to successfully demonstrate the desired result of an indirect proof or a conditional proof. If a box remains open, then the desired result of the indirect or conditional proof has not yet been obtained (and perhaps it's unattainable—the argument may not be valid; the statement may not be tautological; or your strategy may be an improper one for the particular proof).

---

*A stylistic preference: some logicians prefer to show a contradiction as a conjunction of a WFF and its negation—e.g., they conjoin 'P' and '~P' on one line as '(P & ~P)' and declare the rule of conjunction as its justification. You may do this if you prefer, but it is not necessary.

Here is an example of an argument which uses the Method of Indirect Derivation (which we annotate as '**IP**' ['Indirect Proof'] as our justification). The argument is as follows:

((P ∨ Q) ⊃ R)
(Q & ~R)
∴ ~P

We set up the derivation as follows:

DEDUCE: ~P

1.	((P ∨ Q) ⊃ R)	Premise 1
2.	(Q & ~R)	Premise 2

Next, we introduce the Assumption for indirect proof ('IP') which requires us to negate the conclusion:

DEDUCE: ~P

1.	((P ∨ Q) ⊃ R)	Premise 1
2.	(Q & ~R)	Premise 2
3.	~~P	Assumption for IP

Since the conclusion (~P) is a negation, the Assumption for IP results in a double negation.

The next move will be to use the rule of replacement, 'Double Negation (DN)', to obtain line 4.

DEDUCE: ~P

1.	((P ∨ Q) ⊃ R)	Premise 1
2.	(Q & ~R)	Premise 2
3.	~~P	Assumption for IP
4.	P	3, DN

The strategy is to derive the antecedent for the conditional in line 1. We can do that by using the rule of inference, 'Addition (Add)', to obtain '(P ∨ Q)' in line 5:

DEDUCE: ~P

1.	((P ∨ Q) ⊃ R)	Premise 1
2.	(Q & ~R)	Premise 2
3.	~~P	Assumption for IP
4.	P	3, DN
5.	(P ∨ Q)	4, Add

Now we have derived the antecedent '(P ∨ Q)' of line 1 '((P ∨ Q) ⊃ R)' and we can apply the rule of inference, *Modus Ponens* (MP):

DEDUCE: ~P

1.	((P ∨ Q) ⊃ R)	Premise 1
2.	(Q & ~R)	Premise 2
3.	~~P	Assumption for IP
4.	P	3, DN
5.	(P ∨ Q)	4, Add
6.	R	1,5 MP

That allows us to show 'R' on line 6. Next, we turn the expression in line 2—'(Q & ~R)'—around by the rule of replacement called "Commutation (Com)":

DEDUCE: ~P

1.	((P ∨ Q) ⊃ R)	Premise 1
2.	(Q & ~R)	Premise 2
3.	~~P	Assumption for IP
4.	P	3, DN
5.	(P ∨ Q)	4, Add
6.	R	1,5 MP
7.	(~R & Q)	2, Com

The remaining strategy is obvious: we can obtain '~R' from line 7 by application of the rule of inference, 'Simplification (Simp)':

DEDUCE: ~P

1.	((P ∨ Q) ⊃ R)	Premise 1
2.	(Q & ~R)	Premise 2
3.	~~P	Assumption for IP
4.	P	3, DN
5.	(P ∨ Q)	4, Add
6.	R	1,5 MP
7.	(~R & Q)	2, Com
8.	~R	7, Simp

Therefore, we have the desired contradiction for our *reductio:* '~R' on line 8 is the contradiction of 'R' on line 6. The next step is, strictly speaking, unnecessary, but we will repeat 'R' from line 6 to place it directly below the '~R' result obtained in line 8 by means of the rule of inference called "Repetition (Rep)." We do this just to make our *reductio* clear:

DEDUCE: ~P

1.	((P ∨ Q) ⊃ R)	Premise 1
2.	(Q & ~R)	Premise 2
3.	~~P	Assumption for IP
4.	P	3, DN
5.	(P ∨ Q)	4, Add
6.	R	1,5 MP
7.	(~R & Q)	2, Com
8.	~R	7, Simp
9.	R	6, Rep

Now we box off the result in lines 3 through 9 to show that the *reductio* is complete:

DEDUCE: ~P

1.	((P ∨ Q) ⊃ R)	Premise 1
2.	(Q & ~R)	Premise 2
3.	~~P	Assumption for IP
4.	P	3, DN
5.	(P ∨ Q)	4, Add
6.	R	1,5 MP
7.	(~R & Q)	2, Com
8.	~R	7, Simp
9.	R	6, Rep

Line 10 shows that we have obtained the desired result: we have derived '~P' from the premises. The justification, a *reductio,* is found in lines 3 through 9, indirect proof. We box off the result:

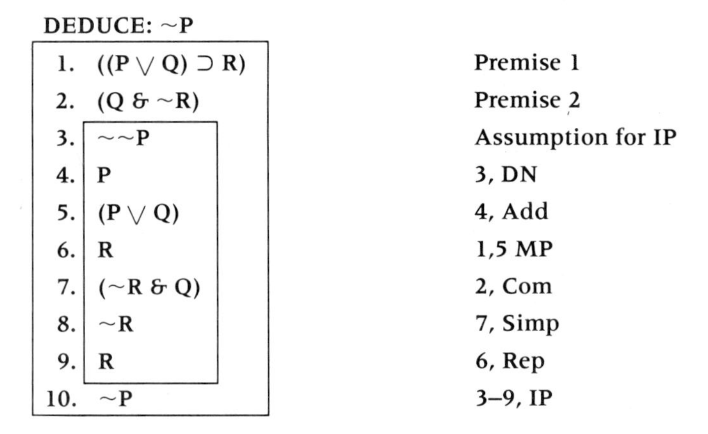

DEDUCE: ~P

1.	((P ∨ Q) ⊃ R)	Premise 1
2.	(Q & ~R)	Premise 2
3.	~~P	Assumption for IP
4.	P	3, DN
5.	(P ∨ Q)	4, Add
6.	R	1,5 MP
7.	(~R & Q)	2, Com
8.	~R	7, Simp
9.	R	6, Rep
10.	~P	3–9, IP

This problem illustrates the following rule:

Rule: You may re-introduce or work operations on any previous line of an unboxed derivation, but you may *not* re-introduce or work operations on any line which occurs in a part of the derivation which is already boxed.

This rule is illustrated by line 6, where 'R' is derived from line 1 and line 5 by *Modus Ponens.* It is also illustrated by line 7, where '(~R & Q)' is obtained from line 2 by the rule of commutation.

Here is another problem we can solve using IP (Indirect proof):

(A ≡ (B ∨ C))
B
(C & D)
∴ A

To begin the Method of Indirect Derivations, we make the Assumption for Indirect Proof (we negate the conclusion). The conclusion is 'A' so the assumption requires us to negate the expression, with the result: '~A'.

The box which surrounds the derivation after we make the Assumption for Indirect Proof is closed once we discover a contradiction. That signals that the derivation is complete and that the affirmation of the conclusion validly follows from the premise. Such an argument is valid:

The derivation is worked for you below:

DEDUCE: A

1.	~A	**Assumption for IP**
2.	(A ≡ (B & C))	**Premise 1**
3.	B	**Premise 2**
4.	(C & D)	**Premise 3**
5.	C	**4, Simp**
6.	(B & C)	**3,5 Conj**
7.	((A ⊃ (B & C)) & ((B & C) ⊃ A))	**2, Equiv**
8.	((B & C) ⊃ A)	**7, Com**
9.	A	**6,8 MP**
10.	~A	**1, Rep**

You will notice that the indirect derivation proceeds otherwise just like any direct derivation, except that a special assumption has been added together with a boxed-off result. The justification for line 1 is our Assumption for Indirect Proof. As soon as the assumption is made, we need to box off whatever we infer from that assumption.

Steps 1–10 show an indirect derivation. When a contradiction is derived between 'A' (line 9) and our assumption '~A' (line 1), we have found the desired *reductio ad absurdum*. As a matter of clarity, we use the rule of "repetition" to repeat the entry of '~A' and carry it to line 10 to show more clearly that the contradiction between 'A' and '~A' had been derived.

30.3 CONDITIONAL PROOFS

Obviously enough, the Method of Conditional Proofs is used where the conclusion of an argument is a material conditional (i.e., where the '⊃' is the main connective). The conditional proof (annotated as '**CP**') can help us work derivations more easily than the Method of Indirect Proof on some conclusions that are conditionals.

The Method of Conditional Proof is also especially valuable in constructing derivations for many formal tautologies that are (or may be interpreted as) conditionals.

The schema for the Method of Conditional Proof is as follows:

DEDUCE: (Φ ⊃ Ψ)

Φ
.
.
.
Ψ

Conditional Proof

We interpret this schema to mean that where we are instructed to demonstrate a conditional statement, we assume the antecedent and see if the consequent follows from the application of rules of inference and rules of replacement. If the consequent does follow from the antecedent by means of the derivation, we box off the result and the demonstration is successful.

In an argument, where the conclusion is a conditional (or where it can be rendered as a conditional by means of rules of replacement); if the consequent of the conditional follows from the antecedent by justified and repeated applications of the rules of inference and replacement to the premises and the Assumption for Conditional Proof, the argument is said to be valid.

When a statement standing by itself may be demonstrated in a derivation by means of a conditional proof, then we say that the statement is a tautology.

Here is the proof that the symbolic statement '(~P ⊃ ~(P & Q))' is a tautology by means of **CP:**

DEDUCE: (~P ⊃ ~(P & Q))

1. ~P		Assumption for CP

Line 1 is the Assumption for Conditional Proof: we assume the antecedent, '$\sim P$', is true. Next, we apply the rule of inference, 'Addition (Add)', to '$\sim P$'. Recall that according to the rule of 'Addition', we are entitled to declare any WFF we want to flank the '$\vee$' and the '$\sim P$'. In this case, we choose the WFF '$\sim Q$':

DEDUCE: $(\sim P \supset \sim(P \& Q))$

1. $\sim P$ — Assumption for CP
2. $(\sim P \vee \sim Q)$ — 1, Add

Line 3 shows the application of 'DeMorgan's Laws' to the result obtained in the previous line:

DEDUCE: $(\sim P \supset \sim(P \& Q))$

1. $\sim P$ — Assumption for CP
2. $(\sim P \vee \sim Q)$ — 1, Add
3. $\sim(P \& Q)$ — 2, DeM

That gives us the result we wanted: '$\sim(P \& Q)$', the consequent, follows from the antecedent, '$\sim P$'. Therefore, we made good on the Assumption for Conditional Proof, and we are entitled to close off the box:

DEDUCE: $(\sim P \supset \sim(P \& Q))$

1. $\sim P$ — Assumption for CP
2. $(\sim P \vee \sim Q)$ — 1, Add
3. $\sim(P \& Q)$ — 2, DeM

Finally, we show the whole formula, '$(\sim P \supset \sim(P \& Q))$', and we declare that it was derived from lines 1 through 3 by means of 'CP'. We box off the whole result, and the derivation is complete: '$(\sim P \supset \sim(P \& Q))$' is indeed a tautology:

DEDUCE: $(\sim P \supset \sim(P \& Q))$

1. $\sim P$ — Assumption for CP
2. $(\sim P \vee \sim Q)$ — 1, Add
3. $\sim(P \& Q)$ — 2, DeM
4. $(\sim P \supset \sim(P \& Q))$ — 1–3CP

Let's look at another argument which uses CP (Conditional Proof). The argument is as follows:

$(P \supset (Q \supset R))$
$\therefore (P \supset R)$

We set up the problem in just the way we learned to set up problems for direct derivations:

DEDUCE: (P ⊃ R)

1.	(P ⊃ (Q ⊃ R)	Premise 1

Next, we make the Assumption for Conditional Proof. Here's the difference: we assume the antecedent of the conclusion:

DEDUCE: (P ⊃ R)

1.	(P ⊃ (Q ⊃ R)	Premise 1
2.	P	Assumption for CP

Next, we apply the rule of inference *Modus Ponens* (MP):

DEDUCE: (P ⊃ R)

1.	(P ⊃ (Q ⊃ R)	Premise 1
2.	P	Assumption for CP
3.	(Q ⊃ R)	1,2 MP

Applying the rule of replacement, 'Commutation (Com)', we get:

DEDUCE: (P ⊃ R)

1.	(P ⊃ (Q ⊃ R)	Premise 1
2.	P	Assumption for CP
3.	(Q ⊃ R)	1,2 MP
4.	(R & Q)	3, Com

Next, we apply the rule of inference, 'Simplification (Simp)', to the result:

DEDUCE: (P ⊃ R)

1.	(P ⊃ (Q ⊃ R)	Premise 1
2.	P	Assumption for CP
3.	(Q ⊃ R)	1,2 MP
4.	(R & Q)	3, Com
5.	R	4, Simp

That gives us the result we wanted: From 'P' and the premise '(P ⊃ (Q ⊃ R))' we have shown we can derive the consequent of the conclusion, 'R'. We can now close off the box:

DEDUCE: (P ⊃ R)

1.	(P ⊃ (Q ⊃ R)	Premise 1
2.	P	Assumption for CP
3.	(Q ⊃ R)	1,2 MP
4.	(R & Q)	3, Com
5.	R	4, Simp

Next, we show the justification: lines 2 through 5, conditional proof:

DEDUCE: (P ⊃ R)

1.	(P ⊃ (Q ⊃ R)	Premise 1
2.	P	Assumption for CP
3.	(Q ⊃ R)	1,2 MP
4.	(R & Q)	3, Com
5.	R	4, Simp
6.	(P ⊃ R)	2–5 CP

The derivation is complete.

30.4 MIXED IP AND CP DERIVATIONS

Sometimes a mixture of **IP** and **CP** can team up to solve especially difficult problems. Now we have to be especially on our guard to see that we don't illegitimately repeat a line from another boxed-off result. We must keep the assumptions for IP and CP separate.

When we use a combination of direct derivations, indirect derivations, and conditional proofs in the course of a derivation, we call the result a "mixed" derivation. Now the boxes we used before as the "guardians" of our guarded assumptions become even more important in order to help us distinguish the direct derivation from the indirect derivation from the conditional proof. We may have "boxes in boxes" (analogous to your childhood toy "Chinese boxes"). When one box occurs inside of another box, we say that box is "nested."

Once more, we repeat the "Rule of Boxes:"

Rule: You may re-introduce or work operations on any previous line of an unboxed derivation, but you may *not* re-introduce or work operations on any line which occurs in a part of the derivation which is already boxed.

Peirce's Law

One hundred years ago, the great American logician, Charles Sanders Peirce, demonstrated that '(((P ⊃ Q) ⊃ P) ⊃ P)' is a tautology. The direct derivation that proves this symbolic statement to be a tautology is quite difficult,* but, as we shall see, the application of mixed assumptions for IP and CP, properly used, can make the proof quite straightforward.

Here is the complete solution:

*You may try it if you like. It *can* be done.

DEDUCE: (((P ⊃ Q) ⊃ P) ⊃ P)

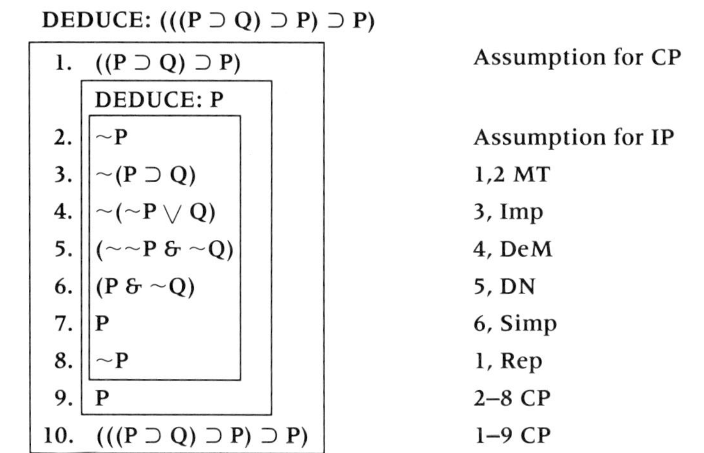

Let's look at each of the steps in turn. Step 1 is the application of the assumption for **CP:** we assume the entire antecedent:

DEDUCE: (((P ⊃ Q) ⊃ P) ⊃ P)

1.	((P ⊃ Q) ⊃ P)	Assumption for CP

Now, here's the *pièce de résistance:* we assume the negation of the conclusion, using the Indirect Proof Method. Now, this may look like an illegitimate move to you at first glance but, remember, we are going to box off this result from the assumption we made for conditional proof.

DEDUCE: (((P ⊃ Q) ⊃ P) ⊃ P)

1.	((P ⊃ Q) ⊃ P)	Assumption for CP
	DEDUCE: P	
2.	~P	Assumption for IP

The deduction proceeds through line 6, applying the rules for *Modus Tollens* (MT), 'Implication (IMP)', 'DeMorgan (DeM)', and 'Double Negation (DN)'. The result is as follows:

DEDUCE: (((P ⊃ Q) ⊃ P) ⊃) P)

1.	((P ⊃ Q) ⊃ P)	Assumption for CP
	DEDUCE: P	
2.	~P	Assumption for IP
3.	~(P ⊃ Q)	1,2 MT
4.	~(~P ∨ Q)	3, Imp
5.	(~~P & ~Q)	4, DeM
6.	(P & ~Q)	5, DN

With '(P & ~Q)' obtained in line 6, the rest of the strategy should be obvious. We apply the rule of inference, 'Simplification (Simp)', as follows:

DEDUCE: $(((P \supset Q) \supset P) \supset P)$

1.	$((P \supset Q) \supset P)$	Assumption for CP
	DEDUCE: P	
2.	$\sim P$	Assumption for IP
3.	$\sim(P \supset Q)$	1,2 MT
4.	$\sim(\sim P \vee Q)$	3, Imp
5.	$(\sim\sim P \mathbin{\&} \sim Q)$	4, DeM
6.	$(P \mathbin{\&} \sim Q)$	5, DN
7.	P	6, Simp

We've succeeded in deducing 'P'. We apply the rule, 'Repetition (Rep)', to show the contradiction between 'P' and '~P' more clearly. Now (and this is very important,) we must box off the result for the Assumption for Indirect Proof, from lines 2 through 8. Thus, we have:

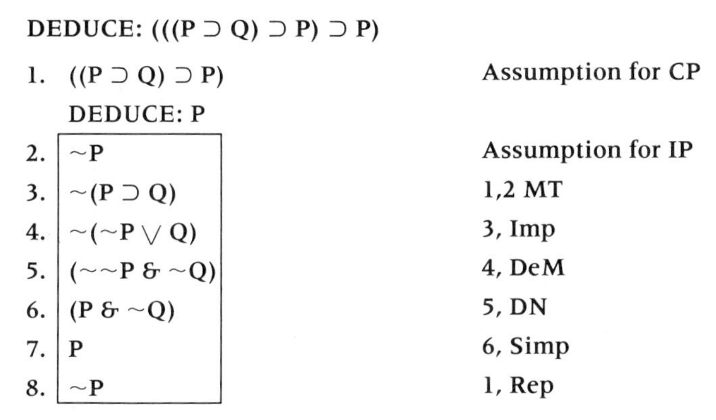

DEDUCE: $(((P \supset Q) \supset P) \supset P)$

1.	$((P \supset Q) \supset P)$	Assumption for CP
	DEDUCE: P	
2.	$\sim P$	Assumption for IP
3.	$\sim(P \supset Q)$	1,2 MT
4.	$\sim(\sim P \vee Q)$	3, Imp
5.	$(\sim\sim P \mathbin{\&} \sim Q)$	4, DeM
6.	$(P \mathbin{\&} \sim Q)$	5, DN
7.	P	6, Simp
8.	$\sim P$	1, Rep

Since we have legitimately deduced 'P' according to the conditions for indirect proof, we show it on the next line with the justification, IP from lines 2 through 8:

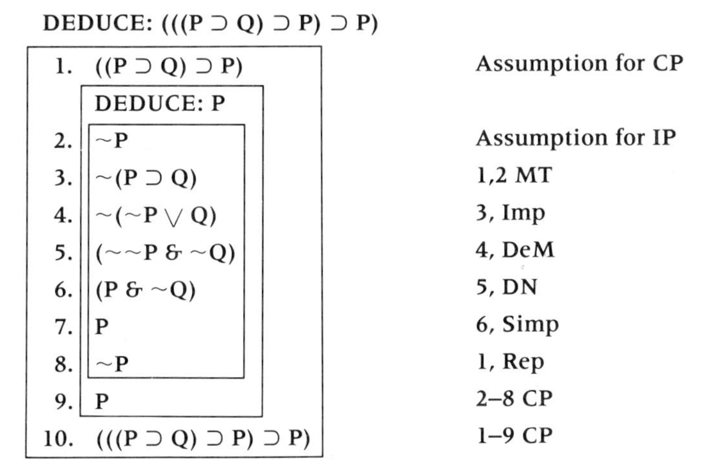

DEDUCE: $(((P \supset Q) \supset P) \supset P)$

1.	$((P \supset Q) \supset P)$	Assumption for CP
	DEDUCE: P	
2.	$\sim P$	Assumption for IP
3.	$\sim(P \supset Q)$	1,2 MT
4.	$\sim(\sim P \vee Q)$	3, Imp
5.	$(\sim\sim P \mathbin{\&} \sim Q)$	4, DeM
6.	$(P \mathbin{\&} \sim Q)$	5, DN
7.	P	6, Simp
8.	$\sim P$	1, Rep
9.	P	2–8 CP
10.	$(((P \supset Q) \supset P) \supset P)$	1–9 CP

We add line 10, close off the box for the conditional proof, and we're done.

EXERCISES

Exercise 1

DIRECTIONS: *Use the Method of Indirect Proof (IP) to show that the following arguments are valid:*

1. $(A \& C)$
   $(C \supset P)$
   $\therefore P$
2. $(\sim(M \supset S) \& (B \supset L))$
   $(L \supset S)$
   $\therefore \sim B$
3. $((\sim W \vee S) \& N)$
   $(\sim S \vee \sim N)$
   $\therefore \sim W$
4. $(P \vee (P \& Q))$
   $(P \supset R)$
   $\therefore R$
5. $(P \supset (D \supset \sim P))$
   $(P \equiv D)$
   $\therefore (\sim P \vee \sim D)$

Exercise 2

DIRECTIONS: *Use the Method of Conditional Proof (CP) to show that the following arguments are valid:*

1. $(A \supset (A \supset B))$
   $\therefore (A \supset B)$
2. $((A \& B) \supset C)$
   $\therefore (A \supset (B \supset C))$
3. $((P \vee S) \supset T)$
   $((T \supset G) \& (P \& S))$
   $\therefore (S \equiv T)$
4. $((\sim G \vee \sim B) \supset \sim P)$
   $\therefore (P \supset G)$
5. $((M \vee Z) \supset (Z \& M))$
   $\therefore (Z \equiv M)$

Exercise 3

DIRECTIONS: *Use the Method of Conditional Proof (CP) to show that the following symbolic statements are tautologies:*

1. $((\sim P \supset P) \supset P)$
2. $((\sim Q \supset \sim P) \supset (P \supset Q))$
3. $\sim((P \& Q) \& \sim P)$
4. $((P \supset (Q \supset R)) \supset ((P \supset Q) \supset (P \supset R)))$
5. $(P \supset (\sim P \supset R))$

Exercise 4

DIRECTIONS: *Put the following arguments into symbolic form, and prove their validity by means of the Method of Indirect Proof (IP).*

1. Either Marvin is a felon or Harvey has framed him. If Harvey has not framed him, then Marvin is a felon. Thus, Marvin is a felon. *M, H.*

2. If Neil Armstrong was jittery, then the crew of Apollo 11 was jittery too. If his crew had the jitters, then Apollo 11 had a bumpy ride to the moon, only if Neil Armstrong was jittery. Hence, the crew of Apollo 11 was jittery. *A, C, B.*

3. Either Slimer slimed Egon or Peter Venkman lost his power pack, only if both Ray Stanz and Egon escaped the containment vessel. Ray Stanz never did escape the containment vessel. Therefore, Slimer did not slime Egon. *S, P, R, E.*

4. Pennsylvania is not beautiful if Philadelphia is the most overrated city in America. Pennsylvania is beautiful or California is tropical. California is not tropical. Thus, Philadelphia is not overrated. *I, H, E, C.*

5. If you are Russian, you claim that Igor Sikorsky invented the airplane, only if Americans claim the Wright brothers did. Therefore, either Sikorsky did not invent the airplane or Americans claim the Wright brothers did. *R, S, W.*

6. Leonardo is a Teenage Ninja Mutant Turtle only if Tim is, or Michelangelo. If Michelangelo is not a Teenage Ninja Mutant Turtle, then Leonardo is. Therefore, if Michelangelo is a Teenage Ninja Turtle, then Tim is too. *L, T, M.*

7. Bob bobbed in the boat and Stan ran like a man, or Steve weaves the sleeve and Stan ran like a man. If Stan ran like a man, then Bob hadn't bobbed in the boat. Thus, Steve weaves the sleeve. *B, S, T.*

8. If Marcy speaks Farsi, then Sally dallies only if Trudy is moody. Either Marcy speaks Farsi, or if Sally dallies then Trudy is moody. Sally dallies. Therefore, Trudy is moody. *M, S, T.*

9. Logic makes one lethargic but of the making of books, there is no end, only if you are trying to think up examples for exercises. If you are trying to think up examples for exercises, you get dizzy after a while. Of the making of books, there is no end, but you don't get dizzy after a while. Hence, logic does not make one lethargic. *L, B, E, D.*

10. Banana spiders are not poisonous but black widows are, or rattlesnakes are. If black widows are not poisonous, then banana spiders are not poisonous either and rattlesnakes are. Hence, black widow spiders are poisonous. *B, W, R.*

11. If Maria loves Tony, then the Jets will rumble tonight only if Officer Krupski is down on his knees. It is not the case that either Officer Krupski is down on his knees, or Maria loves Tony. Thus, the Jets will not rumble tonight. *M, J, K.*

12. If Malenkov became Premier in 1953 then Bulganin replaced him. Khrushchev would be Premier provided that Stalin did not die in 1953. Hence, if Malenkov became Premier in 1953 and Khrushchev would not be Premier, then Bulganin replaced Malenkov and Stalin died in 1953. *M, B, K, S.*

13. Either that's an aardvark or it's an armadillo and armadillos have silly names like aardvarks. If that's an aardvark, then aardvarks bark. Armadillos have silly names like aardvarks provided that aardvarks bark. Therefore, armadillos have silly names like aardvarks. *A, R, S, B.*

14. If the Presidium is part of the Supreme Soviet, then it must be controlled by the Communist party. If the Presidium is not part of the Supreme Soviet, then Gorbachev must have dissolved it. If Gorbachev dissolved it, then it no longer includes a small group of top Communist leaders. If the Presidium includes more than thirty members, then it must not be controlled by the Communist party. Therefore, either the Presidium contains more than thirty members, or it no longer includes a small group of top Communist leaders. *P, C, G, T, M.*

15. The shrine of Our Lady at Czestochowa is called "The Black Madonna" or it is the shrine at Szczecinek only if there is no shrine at Wloclawek or at Bialystok. If there is no shrine at Bialystok or at Bydgoszcz, then there is no shrine at Wloclawek. Therefore, if the shrine of Our Lady at Czestochowa is called "The Black Madonna," then there is no shrine at Wloclawek. *C, S, W, B, Y.*

16. Snakes shake if and only if rakes break. If lakes slake thirst, then snakes shake. Either snakes do not shake or cakes wake, and rakes do not break. Hence, lakes slake thirst. *S, R, L, C.*

MODULE 31

THE QUANTIFICATIONAL CALCULUS

31.1 IN GENERAL

Perhaps the greatest advance in the history of modern logic was the unification of the molecular logic of the Stoics with the syllogistic logic of Aristotle into one symbolic notation. We call this the **Quantificational Calculus.** It is a form of calculus because we need to make inferences based on the range of variables. It is called 'quantificational' because it adds the operators or quantifiers of Aristotelean logic ('all', 'no', 'some', 'not all') to the set of operators in Stoic logic ('it is not the case that', 'if . . . then', 'either . . . or', 'both . . . and', and 'if and only if'). The result is a powerful and comprehensive symbolic logic, perhaps as important an achievement in the development of knowledge as the invention of the airplane was in the advance of technology.

To whom do we give credit for this achievement? Most histories of logic credit the German logician and mathematician Gottlob Frege (1848–1925) with the perfection of a logic of quantifiers (with an occasional nod to Bertrand Russell in Great Britain), but that is at the very least oversimplified and possibly false. In fact, the American logician Charles Sanders Peirce (who was a disciple of George Boole, the founder of "Boolean" algebra) may well have been the first. Some evidence exists that the American philosopher Alfred North Whitehead brought knowledge of the Peircean quantifier to Bertrand Russell which became a mainstay in the development of their masterwork, *Principia Mathema-*

A deductive argument is **valid**, if and only if, if all the premises are true, the conclusion must be true.

*tica.** However, Peirce frequently objected to taking algebraic logic very seriously and so he may have done his work with quantifiers as an exercise in pure research. Perhaps his protests about algebraic logic were sufficient to cause us to neglect his inestimable role in its later development.

In any case, quantificational logic seems to be "an idea whose time had come." Frege worked independently with an obscure notation on the same project at much the same time. Russell and Whitehead had need of a theory of quantifiers at the same time that Peirce and Frege were doing their work. Great ideas are often like that: they emerge suddenly but simultaneously (or almost so) in a community of thoughtful people. Anyone who has studied the history of aviation knows how oversimplified the histories are which credit the Wright brothers with the development of the first airplane; that too was an "idea whose time had come" (which was about the same time, by the way, that Peirce, Frege and Russell were developing the roots of modern symbolic logic!).

This module will follow approximately the same plan as Module 13, "Aristotelean Logic," to show how Aristotelean syllogisms can be rendered into symbolic logic. First, however, we will need to introduce some new symbols and conventions.

31.2 THE SYMBOLIC NOTATION

In addition to the symbols in the Sentential Calculus, we add the following Russellian† symbolic operators in place of the Aristotelean operators:

Statement letters: All of the capital letters of the English alphabet

Predicate letters: All of the capital letters of the English alphabet

Variables: 'x', 'y', 'z', 'w', 'v', 'u'

Term-letters: 'a', 'a^1', 'a^2', 'a^3', b, c, d, etc.

Quantifiers: '(∀ __)'—the universal quantifier, "All;" also '(__)'‡

'(∃__)'—the existential quantifier, "Some"

We need to say a word about the introduction of some of these conventions. Recall that in the Sentential Calculus of molecular arguments we allowed the use of *all* of the capital letters of the alphabet as placeholders for statements. In the Quantificational Calculus we also use all of the capital letters beginning with

*Hilary Putnam, "Peirce the Logician," *Historia Mathematica* 9 290–301(1982). Special thanks to Robert Burch for bringing this to my attention.

†Note: We use the same symbolic quantifiers in Polish notation.

‡Strictly speaking, the left and right parentheses '( )' are used in Russell-Whitehead notation. The "upside down 'A' " (∀) is a more recent innovation.

'F', 'G', 'H', etc. to stand for *predicates*. These **predicate letters** as they are called take variables ('x', 'y', etc.) or term-letters ('a', 'a^1', 'a^2', or 'a', 'b', 'c', etc.) in order to compose a well-formed formula. To keep them distinct from the atomic statement letters, remember that the predicate letters must take either a variable (e.g., 'Fx'), or a term-letter (e.g., 'Fa').

For now, we will consider only those cases in which predicate letters take a single variable. (In Module 32, we will consider cases of *relational* predicates that take two or more variables.) Notice also that we retain statement letters from the Sentential Calculus, but that they do not take variables or term-letters.

The connection between logic at this level and algebra is very clear. A variable, we know from algebra, stands for an unknown quantity. We read the formula "Fx" as: "some unknown quantity 'x' is an 'F' " where 'F' can stand for any predicate whatever.

Consider a simple example:

"Everything changes."

In this case, we will let 'C' stand for "things that change." The operator is 'all' and so we assign the symbol '(∀__)' to stand for the universal affirmative operation. Note that the blank space in the parentheses after the '∀' is reserved for a variable. Thus we symbolize the statement above as:

(∀x) Cx

This reads as, "For all x, x is a C" or:

"For all x, x is a thing that changes."

Below we have a table for converting each of the Aristotelean operators into symbolic notation in the Quantificational Calculus.

Converting Aristotelean Operators to Symbolic Notation

	ARISTOTELEAN		RUSSELLIAN
Operation	**Operator**	**Term**	**Quantifier**
Universal Affirmative	ALL	A	(∀X)
Universal Negative	NO	E	~(∃X)
Particular Affirmative	SOME	I	(∃X)
Particular Negative	NOT ALL	O	~(∀X)

. . . where 'X' stands for any variable.

We said before that the variables 'x', 'y', etc. stand for "unknown quantities" (or, strictly speaking, *ranges* of quantities). When we know that a particular thing falls under the range of a quantifier, we assign a term-letter such as 'a'. Then we will have given a name to the something that falls under the range of the variable.

Following is a guide on converting the symbolized quantifier into English.

Quantifier	Read As . . .
(∀X)	"For all X . . ."
(∃X)	"There is an X, such that . . ."
~(∀X)	"It is not the case that, for all X . . ."
~(∃X)	"It is not the case that there is an X, such that . . ."

. . . where, once more, 'X' stands for any variable.

Now we need to say something about well-formed formulae with regard to the new symbols and operators we've introduced in quantificational logic. To do this, we extend and revise the rules for well-formed formulae laid down in Module 23, "Symbols":

Rules for Well-Formed Formulae for the Quantificational Calculus

1. **Any atomic statement letter is a WFF.**
2. **Any predicate letter followed by a variable is a WFF.**
3. **Any WFF preceded by a wave ('~') is a WFF.**
4. **The result of placing one of the symbols '⊃', '∨', '&', or '≡' between two WFFs and wrapping the whole in parentheses* is a WFF.**
5. **If 'Φ' is a WFF and 'X' is a variable, then**
   (∀X) Φ,
   (∃X) Φ
   are also WFFs,

. . . where 'X' stands for any variable and 'Φ' stands for any WFF. Notice that condition 5 permits us to write both '(∀x) Fx' as well as '(∀x) (Fx ⊃ Gx)'. Analogously, we are permitted to write both '(∃y) Oy' and '(∃y) (Oy & Ly)'.

Term-letters are a special case of the "instantiation" of variables, where we identify one thing which falls under the range of the variable. Each of the following are WFFs under the conditions laid down for the Quantificational Calculus:

1. Fa
2. Fa^1
3. (∀x) Fa
4. (∃y) (Fa^2 & Gy)

Note that in 3, the term-letter 'a' is free of the influence of the quantifier which universalizes the variable 'x'. This brings us to an important discussion about the distinction between *free* and *bound* variables.

When a variable falls under the range of the quantifier, we call it a **bound variable.** Thus, the 'x' in '(∀x) (Fx ⊃ Gx)' is said to be *bound.* When a variable fails to fall under the range of a quantifier, it is said to be a **free variable.** Thus, the 'y' in '(∃x) (Gx ⊃ Hy)' is said to be *free.*

*In a long string, you may substitute left and right brackets '[' and ']' or '{' and '}' provided that you do so consistently, e.g., '{[(P ⊃ Q) ⊃ P] ⊃ P}'.

In a useful analogy, the philosopher and logician Hans Reichenbach compares the act of binding variables to the running of a watch. He writes:

> If we call such variables *bound variables* this name may be conceived as meaning 'bound to run'. We may compare such a statement to a watch, which works only as long as it is running; once the spring is wound it is bound to run through all its states, and in doing so it tells the time. Similarly, a bound variable is a 'nonstop variable' which tells us something by its running. A free variable, on the other hand, can be 'stopped' by the insertion of a special value of 'x'; it is only then that the expression tells us something, i.e. becomes a statement.*

The means we have for "stopping the watch" so to speak is the insertion of a term-letter. The term-letter such as 'a' is a way of representing the free variable so that the variable is given an interpretation for an individual statement. Thus, in the example '(∃y) (Fa^2 & Gy)' the variable 'y' in 'Gy' is bound by the quantifier but the term letter 'a^2' is free. Strictly speaking, 'a^2' is not a free *variable* but in our presentation it can be seen as taking the place of a free variable. Ordinarily, we will not find variables such as 'Fx' without any quantifier binding the 'x' occurring in a formula, though we will of course have need for the term-letters such as 'Fa^2' which necessarily are freed from the boundaries of the quantifier.

31.3 TRANSLATION IN THE QUANTIFICATIONAL CALCULUS

Now we are in a position to begin doing some elementary translations into the Quantificational Calculus. To do so, we will go back over the examples given in Unit Two, "Traditional Logic," and show how categorical statements can be translated into symbols.

Let's begin by updating the traditional Aristotelean Square of Opposition to show the relation of the symbolic operators to one another:

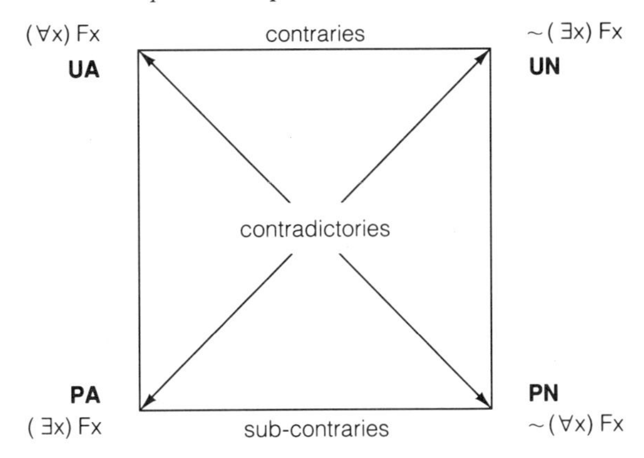

The Symbolic Square of Opposition

*Hans Reichenbach, *Elements of Symbolic Logic,* New York: The Free Press, 1947, pp. 88–89.

Notice that the symbolic translations show the contradictories even more clearly than the ordinary language. In ordinary English, for most students, 'not all' is not obviously the contradictory of 'all'. Yet, in symbolic form, the relation is clear: the contradictory of '$(\forall x)$' is the negation '$\sim(\forall x)$'.

We also need to look at some important equivalences. You may recall that in Unit Two, "Traditional Logic," we studied some basic equivalences between categorical statements. For example, 'All S are not P' is equivalent to 'No S are P'. To say that two expressions are **equivalent** is to say that one may be substituted for the other in an argument without loss of truth value.

Here are some basic equivalences in the Quantificational Calculus.

$(\forall X)\varphi X$	::	$\sim(\exists X)\sim\varphi X$
$(\forall X)\varphi X$	::	$\sim(\forall X)\sim\varphi X$
$\sim(\exists X)\varphi X$	::	$(\forall X)\sim\varphi X$
$\sim(\forall X)\varphi X$	::	$(\exists X)\sim\varphi X$

. . . where 'X' stands for any variable and 'φ' stands for any atomic predicate letter. The correspondence between the symbolic notation in the Quantificational Calculus and the ordinary language phrasing of traditional logic should be very clear. I've added the English equivalents as they would be expressed in traditional logic below:

$(\forall X)\varphi X$ (All S are P)	::	$\sim(\exists X)\sim\varphi X$ (No S are not P)
$(\exists X)\varphi X$ (Some S are P)	::	$\sim(\forall X)\sim\varphi X$ (Not all S are not P)
$\sim(\exists X)\varphi X$ (No S are P)	::	$(\forall X)\sim\varphi X$ (All S are not P)
$\sim(\forall X)\varphi X$ (Not all S are P)	::	$(\exists X)\sim\varphi X$ (Some S are not P)

As you can see, these relationships are much easier to express and comprehend in symbolic notation than they are in ordinary English. This ease of expression is one more reason to adopt the symbolic notation over the ordinary English of the traditional logic approach.

Now we are in a position to lay down some basic conventions in translating ordinary English statements into the Quantificational Calculus.

Symbolization is the first step in working deductions in the Quantificational Calculus; it is also the most important (and usually) the most difficult step. In this module, we will work our deductions using the Semantic Tree module, and most of that is quite easy, once we've laid down a few rules. Yet no deduction can proceed unless we have properly translated the ordinary English into a proper translation.

Among the chief aims of symbolic logic are precision and clarity. Natural languages, like English, are very rich, but they often contain statements which are slightly ambiguous. Symbolic translation requires us to remove the ambiguity. This has some dangers. For one thing, we may give an inaccurate translation and in that event whatever work we do on the deduction is useless. More problematic is the case where a statement is ambiguous and we may choose the interpretation that does not yield a valid deduction. We must be sensitive to these cases. The overall goal of logic is to reveal the truth. To do this, we must always strive to give the translation that most likely will result in a valid argument. That, of course, takes practice and a deep understanding of the nuances of ordinary language. However, here are some guidelines which will help you to give the most appropriate translation of a given ordinary language statement.

Here, then, are standard symbolic translations of statements containing the four Aristotelean operators:

Let 'F' = "is a palomino"

Let 'G' = "is a horse"

Let 'H' = "is a zebra"

Statement	**Symbolic Translation**
"All palominos are horses."	(∀x) (Fx ⊃ Gx)
"Some horses are palominos."	(∃x) (Gx & Fx)
"No horses are zebras."	~(∃x) (Gx & Hx)
"Not all horses are palominos."	~(∀x) (Gx ⊃ Fx)

You may find that whenever the '∀' quantifier is present, the standard translation takes the horseshoe '⊃' and whenever the '∃' quantifier is present the standard translation takes the ampersand '&'. There is good reason for this.

Conditionals are a bit weaker than conjunctions from a truth-functional point of view, and the existence commitments of '∀' are also weaker than those which employ '∃'. When we say, "All palominos are horses," we mean, "if anything at all is a palomino, then it is also a horse." Nothing commits us to the existence of palominos in our corral. All we need to know is that if anything *were* a palomino it would also have to be a horse.

Likewise, when we say, "Not all horses are palominos" in symbolic translation we are saying, "It is not the case that if anything is a horse then it's a palomino." Nothing is said about whether there are or are not any horses in our corral, but we do deny that if any horses are there, all of them are palominos.

Sometimes however the ampersand is more appropriate to use for the '∀' quantifier than is the conditional, as in the statement, 'Everything is both bright and beautiful'. This translates as, "For every x, x is bright and x is beautiful." In symbolic notation, where 'B' is the predicate letter for 'is bright' and 'E' is the predicate letter for 'is beautiful,' we have:

(∀x) (Bx & Ex)

Likewise, sometimes the horseshoe is more appropriate to use for the '∃' quantifier than is the ampersand, as in 'If something is rough then it's ready'. This translates as, "There is an x such that if x is rough then x is ready." In symbolic notation, where 'R' is the predicate letter for 'is rough' and 'E' is the predicate letter for 'is ready', we have:

(∃x) (Rx ⊃ Ex)

Spotting these exceptions to standard form is a matter of judgment and fluency of language use.

Here is a sampling of easy translations:

"All unicorns have horns."	$(\forall x)\ (Ux \supset Hx)$
"No unicorns have stripes."	$\sim(\exists x)\ (Ux \,\&\, Sx)$
"Some unicorns are horses."	$(\exists x)\ (Ux \,\&\, Hx)$
"Not all unicorns are horses."	$\sim(\forall x)\ (Ux \supset Hx)$
"Nothing is forever."	$\sim(\exists x)\ (Fx)$
"Everything is beautiful in its own way."	$(\forall x)\ (Bx)$
"Everything is beautiful, each in its own way."	$(\forall x)\ (\forall y)\ (Bx \,\&\, Wy)$
"Socrates is a man. [read: "Everything identical with Socrates is a man."]	$(\forall x)\ (Sx \supset Mx)$
"Everything is either matter or energy."	$(\forall x)\ (Mx \vee Ex)$
"A dolphin is a mammal."	$(\forall x)\ (Dx \supset Mx)$
"A dolphin saved my life."	$(\exists x)\ (Dx \,\&\, Lx)$
"Not all birds fly."	$\sim(\forall x)\ (Bx \supset Fx)$
"A bachelor is a never-married male."	$(\forall x)\ (Bx \supset Nx)$
"Every action has an equal and opposite reaction."	$(\forall x)\ (Ax \equiv Rx)$
"All birds are not mammals."	$(\forall x) \sim(Bx \supset Mx)$ or: $\sim(\exists x)\ (Bx \,\&\, Mx)$

The last example needs some clarification. 'All birds are not mammals' is really a universal negation. The two symbolizations '$(\forall x) \sim(Bx \supset Mx)$' and '$\sim(\exists x)\ (Bx \,\&\, Mx)$' are truth-functionally equivalent. One may be substituted for the other. However, the second example states the universal negation more clearly.

EXERCISES

DIRECTIONS: *Here are some fairly easy statements to translate into the Quantificational Calculus. To the right of each example a suggested set of predicate letters appears in parentheses. More difficult or tricky examples are starred (*).*

1. Everything is wrong! (Wx)

2. No man is an island. (Mx, Ix) —John Donne
3. Some men are peninsulas. (Mx, Px)
4. Not all men are peninsulas. (Mx, Px)
5. A man is a god in ruins. (Mx, Gx) —Ralph Waldo Emerson
6. All hell broke loose. (Hx, Lx) —John Milton
*7. All quiet on the Western front. (Tx, Qx) —Erich Maria Remarque
8. All men are created equal. (Mx, Ex) —Thomas Jefferson
*9. All hope abandon, ye who enter here. (Hx, Ex) —Dante
10. Every man is the architect of his own fortune. (Mx, Fx) —Gaius Sallustius Crispus
*11. All aglow is the work. (Ax, Wx) —Virgil
*12. All that glitters is not gold. (Gx, Hx) —ancient proverb, perhaps originating with Aristotle
13. Some are born great. (Bx, Gx) —Shakespeare
14. Some come to take their ease / And sleep an act or two. (Fx: x is a person; Ex, Sx) —Shakespeare
15. Each of us bears his own hell. (Fx: x is a person; Hx) —Virgil
*16. I will fear no evil. (Ix, Ex) —Psalms 23:4
17. Some quick to arm. (Qx) —Ezra Pound
18. Not every man has gentians in his house / in soft September, at slow, sad Michaelmas. (Mx, Gx) —D.H. Lawrence
19. No lot is altogether happy. (Lx, Hx) —Horace
*20. There is no new thing under the sun. (Nx, Sx) —Ecclesiastes 1:9
21. There is no remembrance of former things. (Rx, Fx) —Ecclesiastes 1:11
*22. Something there is that doesn't love a wall. (Lx) —Robert Frost
*23. Every man for himself. (Mx, Fx) —saying
*24. No one conquers who doesn't fight. (Px: x is a person, Cx, Fx) —Gabriel Biel
*25. No picture is made to endure nor to live with. (Px, Ex, Lx) —Ezra Pound

---

31.4 ARGUMENTS IN THE QUANTIFICATIONAL CALCULUS

After we've had experience translating statements into the Quantificational Calculus, translations of whole arguments are an obvious next step—and a relatively easy one, since we need to learn no new principles to do it. We just need to be able to recognize an argument with quantifiers when we see one.

The remainder of this module will be devoted to a straightforward and easy task: we will translate simple Aristotelean syllogisms into symbolic language with quantifiers. Then we will work some simple deductions. The purpose of this module is to introduce students to quantificational theory in as simple a manner as possible, so we will not work any direct derivations on quantificational logic for now (we'll save those details for the following module, on

"Advanced Quantificational Logic"). Instead, we'll introduce some simple modifications on the rules for Semantic Trees and emphasize this method.

Here is an easy first example of a syllogism which we'll translate into the Quantificational Calculus:

No office ducks are campus roosters.
Some campus roosters are classroom chickens.
∴ Not all classroom chickens are office ducks.

The syllogism doesn't make much sense unless you're a student at Cal State/ Bakersfield, where ducks, chickens, and roosters wander in and out of office and classroom buildings, but that isn't important.

This is a valid argument (as you can easily enough see if you worked Unit Two on "Traditional Logic"). However, if the argument were translated into the Sentential Calculus (without using the new, more powerful apparatus introduced in this module for the Quantificational Calculus), it would falsely appear to be *invalid.* In the Sentential Calculus, the argument might be translated as:

P
Q
∴ R

This would be invalid because the conclusion appears to be entirely irrelevant to the premises. However, by introducing the apparatus of the Quantificational Calculus, we can now introduce the sort of details into our symbolic notation which will show the argument to be valid after all, thereby preserving our original intuitions on the matter.

The translation is easy enough:

Let 'F' = "is an office duck"
Let 'G' = "is a campus rooster"
Let 'H' = "is a classroom chicken"

$\sim(\exists x)\ (Fx\ \&\ Gx)$
$(\exists x)\ (Gx\ \&\ Hx)$
$\therefore\ \sim(\forall x)\ (Hx \supset Fx)$

31.5 SEMANTIC TREES IN THE QUANTIFICATIONAL CALCULUS

In order to work Semantic Trees in the Quantificational Calculus, we must first remove the quantifiers from the expressions and substitute terms for variables. When a symbolic statement in the Quantificational Calculus appears without quantifiers and with all terms instead of variables for the predicate letters, we say it is in **prenex normal form.***

*'Prenex' comes from the Latin *praenexus* and means literally, "tied up" or "bound in front." 'Prenex normal form' then suggests that the "binding" of the quantifiers is removed and the expression appears in "normal" form.

Why do we put expressions into prenex normal form while working derivations? Remember what the quantifiers mean. For example, take the expression: "Everything has its place." We symbolize this as:

(∀x) Px (where 'P' stands for "has a place")

Now, let's consider the scope of this quantifier. Here is a list of household things: hairbrush, ruler, frying pan, scissors. It follows from our formula that each of these *has its place*—for example, the hairbrush on the bureau, the ruler and scissors in the desk drawer, the frying pan on the pot hook. This process whereby we represent a universal by a concrete instance is called **instantiation.**

We may show a symbolic list of instantiations of the formula '(∀x) Px' by using term-letters, as follows:

'a' has its place (the hairbrush)

'a^1' has its place (the ruler)

'a^2' has its place (the frying pan)

'a^3' has its place (the scissors)

and so on, for every *thing*.

Prenex normal form, then, is the instantiation of variables for concrete things (represented by term-letters).

We can now test for validity. Notice that we are not using truth tables for the Quantificational Calculus. There is an excellent reason for this. Truth tables are useful only if the number of rows needed to determine validity is eight or sixteen—after that, truth tables become unwieldy. Consider then that (most) arguments in the Quantificational Calculus have *infinite* numbers of subformulas. If we consider all the instantiations of our previous example, the formula '(∀x) Px', we have an infinite number of things:

'a' has its place (the hairbrush)

'a^1' has its place (the ruler)

'a^2' has its place (the frying pan)

'a^3' has its place (the scissors)

•

•

•

•

•

. . . ad infinitum!

Thus, most truth tables for the Quantificational Calculus would be infinitely long and infinitely wide.

Fortunately, humans can often detect patterns in infinite regresses which may show a quantificational argument to be valid or invalid, even if they cannot always lay it out in a truth table. Now we'll see how direct derivations permit us to make decisions on the validity or invalidity of arguments in the Quantificational Calculus without grinding out every possible instance of a universal.

In order to use Semantic Trees to determine whether or not arguments in the Quantificational Calculus are valid, we need to introduce four new rules which we will add to the other rules for Semantic Trees in the Sentential Calculus. Two of these new rules have important restrictions—be sure to be careful to take note of them. Included are the rules for all the trees in the Sentential Calculus (where 'X' stands for any variable, 'φ' for any predicate letter, and 'T' for any term):

Semantic Trees for the Quantificational Calculus

(∀X) φ	~(∃X) φ	*(∃X) φ	* ~(∀X) φ
•	•	•	•
φ (T/X)	~φ(T/X)	φ(T/X)	~φ(T/X)

__Restrictions:__ For '(∃X) φ' and '~(∀X) φ', 'T' must be a term new to everything on the pathway through the branches above it.

These rules need some explanation. The expressions in parentheses with the quantifiers, '(∀X)' for example, are called the *operators*. The expressions represented by 'φ' are called the *operands*. The operand of a formula may be simple or complex. If it is simple, as in 'Fx', then we don't need to surround the operand with parentheses; we do, however, when the operand is complex, as in '(Fx ⊃ Gx)'. The rules instruct us to substitute a variable for a term-letter or a term-letter for a variable in the operand of the formula. Thus, if we have a formula '(∀y) (Hy ⊃ Ty)' the first rule instructs us to drop the operator '(∀y)' and substitute a term-letter for 'y' in the operand: e.g., '(Ha ⊃ Ta)'. This substitution should be uniform and complete, so that each instance of 'y' is uniformly substituted by the term letter 'a'.

Let's work a simple deduction according to the Semantic Tree Method. We already know that the formulae '~(∀x) (Fx ⊃ Gx)' and '(∃x) (Fx & ~Gx)' are equivalent. That means that the one should be deducible from the other. In the following example, we put that assumption to the test, using the Semantic Tree Method.

DEDUCE: (∃x) (Fx & ~Gx)	**From:**
1. ~(∀x) (Fx ⊃ Gx)✓	**Given**
2. ~(∃x) (Fx & ~Gx)✓	**Assumption**

*Once more, for the sake of mellifluous simplication, I do not here distinguish between general terms ('a^1', 'a^2', 'a^3') and individual constants (a, b, c, d). I leave that discussion to Module 32, "Advanced Quantificational Logic." As stated, these rules are adapted from Hugues Leblanc and William Wisdom, *Deductive Logic,* Boston: Allyn & Bacon, 1972.

DEDUCE: $(\exists x)(Fx \;\&\; {\sim}Gx)$	**From:**
3. $\sim(Fa \supset Ga)$ ✓	1
4. $\sim(Fa \;\&\; {\sim}Ga)$ ✓	2
5. Fa $\sim Ga$	3
6. $\sim Fa$ $\quad$ $\sim\sim Ga$	4
7. x (5) $\quad$ Ga x (5)	6

Since all the branches close the argument is valid.

Now we can see how to work an ordinary syllogism symbolized in the Quantificational Calculus. Let's return to our earlier example:

No office ducks are campus roosters.
Some campus roosters are classroom chickens.
∴ Not all classroom chickens are office ducks.

We know that this argument is symbolized as:

Let 'F' = "is an office duck."
Let 'G' = "is a campus rooster."
Let 'H' = "is a classroom chicken."

$\sim(\exists x)(Fx \;\&\; Gx)$
$(\exists x)(Gx \;\&\; Hx)$
$\supset \sim(\forall x)(Hx \supset Fx)$

We prove the validity of the argument using Semantic Trees as follows:

DEDUCE: $\sim(\forall x)(Hx \supset Fx)$	**From:**
1. $\sim(\exists x)(Fx \;\&\; Gx)$ ✓	Premise
2. $(\exists x)(Gx \;\&\; Hx)$ ✓	Premise
3. $\sim\sim(\forall x)(Hx \supset Fx)$ ✓	Assumption (for IP)
4. $(\forall x)(Hx \supset Fx)$ ✓	3
5. $\sim(Fa \;\&\; Ga)$ ✓	1
6. $(Ga \;\&\; Ha)$ ✓	2
7. Ga Ha	6
8. $(Ha \supset Fa)$ ✓	4
9. $\sim Ha$ x (7) $\quad$ Fa	8
10. $\sim Fa$ x (9) $\quad$ $\sim Ga5$ x (7)	5

Since all the branches close the argument is valid.

Now, let's look at an invalid argument. Suppose we are asked to deduce '(∀x) (~Fx ⊃ Gx)' from '(∃x) (Fx & Gx)'. We work a Semantic Tree and the result is:

DEDUCE: (∀x) (~Fx ⊃ Gx)	**From:**
1. (∃x) (Fx & Gx)✓	Given
2. ~(∀x) (~Fx ⊃ Gx)✓	Assumption
3. Fa & Ga✓	1
4. Fa	
Ga	3
5. ~(~Fa^1 ⊃ Ga^1)	2 [Restriction on '~(∀X)']
6. ~Fa^1	
~Ga^1	5
OPEN	

Because of the restriction on '~(∀X)', we are unable to use the term-letter 'a' (which was already introduced on line 4).* Thus, we have to use the next term-letter 'a^1'. The result is an open branch. Still, we have other strategies open to us—in fact, we have an *infinite* number of such strategies left to us. We can continue the Semantic Tree as follows:

DEDUCE: (∀x) (Fx ⊃ Gx)	**From:**
1. (∃x) (Fx & Gx)✓	Given
2. ~(∀x) (Fx ⊃ Gx)✓	Assumption
3. Fa & Ga✓	1
4. Fa	3
Ga	
5. ~(~Fa^1 ⊃ Ga^1)✓	2 [Restriction on '~(∀X)']
6. ~Fa^1	
~Ga^1	5
7. ~(~Fa^2 ⊃ Ga^2)✓	2 [Restriction on '~(∀X)']
8. ~Fa^2	
~Ga^2	7
OPEN	

But this branch is also open. In fact, it should be obvious that whatever term-letter we introduce, the restriction on '~(∀X)' will always prevent us from obtaining the term-letters we need to find the contradiction which the Semantic Tree Method requires to close the branches and prove the argument valid. Therefore, it's obvious that the argument must be *invalid*.

This is obvious to us, but it is not at all obvious to a computer. A computer will be programmed by a method of trial and error. Every time it encounters

*For a discussion of this use restriction, see Module 32, "Advanced Quantificational Logic."

such an open branch, it will turn to the next possible term-letter and "plug it in" into the Semantic Tree. In fact, unless it is (somewhat arbitrarily) programmed to stop after so many attempts at finding a contradiction in the Semantic Tree, it will continue forever in a vain attempt to prove the argument valid.

Formal Logic: Its Scope and Limits

We could program a computing machine to test inferences for validity by the tree method. Presented with a valid inference, the machine would eventually inform us of its validity, and presented with an invalid inference, the machine would never erroneously classify it as valid. In this sense, the tree method adequately formalizes logic.

But in another sense, the method is inadequate. The process of building a tree to test an invalid inference may go on forever, so that there need be no finite number of steps after which the machine classifies a given invalid inference as valid. Thus, presented with the invalid inference*

$$\frac{(x)(\exists y)xLy}{aLa}$$

the machine will never finish the tree and will therefore never give a "yes" or "no" answer to the question "Is this inference valid?" If at some point the machine could predict that it will never finish the tree, it could tell us at that point that the correct answer is "no;" but as matters stand, we have an adequate "yes" machine which is inadequate as a "no" machine.

In this sense, the tree method is inadequate, but in this sense, *every mechanical routine is inadequate*; there can be no adequate "no" machine for quantificational validity. It follows that there can be no adequate mechanical routine for determining whether inferences have finite or infinite trees. Confronted with any particular inference whose tree is infinite, we may be able to recognize, after some finite number of steps, that the tree will never stop growing. But there is no uniform mechanical procedure for doing this; a procedure that works for some inferences must always fail for others, so that there is always a place for human ingenuity in the matter of recognizing invalidity.

*Richard C. Jeffrey, *Formal Logic: Its Scope and Limits,* New York: McGraw-Hill, pp. 195–196, 1967. To put this in our own notation, the formula would read:

$$(\forall x)\ (\exists y)\ Lxy$$
$$\therefore La\ a^1$$

Notice that this formula uses relational predicates and dyadic quantifiers. These will be discussed in the next module.

Even if the computer is programmed to stop after so many failed attempts at validity, it will never "see" what you see—a recurring pattern that makes *any* attempt at validity futile. In this respect, you are superior to any computer. You can see the result. You can understand *why* this argument can never be valid.

EXERCISES

DIRECTIONS: *Determine whether the following syllogisms are valid or invalid by means of the Semantic Tree Method for the Quantificational Calculus. First, you need to put these syllogisms into symbolic notation for the Quantificational Calculus. Next, work a tree.*

Notice that these are some of the same exercises you worked in Unit Two, "Traditional Logic." If you did that section, you will have already determined which are valid and which are not. Make your results in the Quantificational Calculus tally with those results.

1. All surfers are sharkbait.
   Some sharkbait are seals.
   ∴ All surfers are seals.
2. Some backpackers are bearbait.
   Not all bearbait are berries.
   ∴ Not all backpackers are berries.
3. Some software is copy-protected.
   No science is copy-protected.
   ∴ Software is no science.
4. Some meanders are frets.
   All frets are ornamental borders.
   ∴ All meanders are ornamental borders.
5. All accents are mellifluous.
   Some mellifluous sounds are heard in Brooklyn.
   ∴ All accents are heard in Brooklyn.
6. All droves are herds.
   Not all herds are packs.
   ∴ Not all droves are packs.
7. Some old movies are colorized.
   No colorized movies are any good.
   ∴ Some old movies are not any good.
8. All scorpions are arachnids.
   Some arachnids have venomous stings.
   ∴ Not all scorpions have venomous stings.
9. No shrewdness is a skulk.
   Some skulks are sounders.
   ∴ Not all shrewdness are sounders.
10. All wisps are snipes.
    Some bevies are snipes.
    ∴ Some wisps are bevies.

11. A kindle is a brood.
Some broods are dogs.
∴ No dogs are kindles.

12. All gams are schools.
Some gams are shoals.
∴ Some schools are shoals.

13. No swarms are coveys.
Some conveys are bevies.
∴ Not all bevies are swarms.

14. All drifts are droves.
Some drifts are hogs.
∴ Some droves are hogs.

15. All falls are coveys.
No conveys are nides.
∴ No nides are falls.

MODULE 32

ADVANCED QUANTIFICATIONAL LOGIC

32.1 DIRECT DERIVATIONS FOR QUANTIFICATIONAL LOGIC

In the previous module, we considered a powerful and effective method of deduction for quantificational logic: the Semantic Tree test. Introducing this method first has many advantages. For one thing, with a little bit of applied ingenuity, it is all you need to determine validity or invalidity in the Quantificational Calculus. Even though strictly speaking no mechanical test for such determinations exists, this is the closest thing to it. Secondly, it is the *easiest* method of determining validity or invalidity in the Quantificational Calculus. We need far fewer restrictions on the rules for the Semantic Trees than we do for derivations in quantificational logic.

Nevertheless, many logicians believe that direct derivations more closely model the way we think than any of the other methods of deduction, including Semantic Trees. Moreover, it may be the case that learning and applying the many and sometimes complicated restrictions on the rules for quantificational logic can help us better understand why a given deduction is valid or not. Direct derivations may also lead to new discoveries of what follows from the preceding premises. Such explorations can be done with Semantic Trees but they are more cumbersome than direct derivations.

A deductive argument is **valid,** if and only if, if all the premises are true, then the conclusion must be true.

For the sake of simplicity, we will restrict our discussion to Direct Derivations in the Quantificational Calculus. Certainly, Indirect Derivations are possible (the Semantic Tree Method is a spin-off of that method) and so are Conditional Proofs. However, Indirect Proof and Conditional Proof require a few more restrictions than Direct Derivations do because of the nature of the assumptions they make. Thus, we will not consider them here.

In order to work direct derivations, we must be careful to apply the following set of rules for converting quantified expressions into prenex normal form. Some of these rules have special restrictions. *Be careful to regard those restrictions scrupulously.* Otherwise, you may wind up incorrectly "proving" some invalid arguments valid.

Here are the rules. These rules and the accompanying terminology will be explained shortly.

INTELIM RULES* for Direct Derivation

Universal Instantiation **UI:**

$(\forall X)(\varphi X)$
$\therefore \varphi T/X$

. . . where 'X' is any variable and 'T' is any term-letter substituted for 'X'. (Here 'T/X' means a term 'T' is substituted for a variable 'X'.)

Existential Generalization **EG**:

φT
$\therefore (\exists X)(\varphi X/T)$

(Here 'X/T' means X is substituted for T.)

Restriction:

(1) 'X' must be a variable that does not already appear in 'φT'.

Universal Generalization **UG**:

φT
$\therefore (\forall X)(\varphi X/T)$

*'Intelim' is a "portmanteau word" designating rules of introduction and elimination. A portmanteau is a type of suitcase that breaks into two parts. A 'portmanteau word' is a word that likewise breaks into two parts—for example, 'smog' is a word made up of *smoke* and *fog*. Lewis Carroll, the author of *Alice in Wonderland* and an esteemed logician in his own time, coined the term 'portmanteau word'.

Restrictions:
(1) 'T' cannot be generalized if it is an *individual constant*, or if 'T' is a general term that already appears in '$(\forall X)(\varphi X/T)$' or in any premise or in any preceding line gotten by EI.
(2) Each occurrence of a term-letter 'T' in 'φT' must be uniformly and consistently replaced with the same variable 'X'.

Existential Instantiation **EI:**

$(\exists X)(\varphi X)$
$\therefore \varphi X/T$

Restrictions:
(1) 'T' is an *individual constant* that does not already appear in '$(\exists X)(\varphi X)$' nor anywhere else in the preceding direct derivation.
(2) 'T' does not appear in the conclusion to be derived.
(3) Each occurrence of 'X' in 'φX' must be uniformly and consistently replaced with the same term-letter 'T'.

In addition to these rules, we have one special Rule of Replacement for Quantifiers:

Quantifier Negation **QN:**

$\sim(\forall X)\ \varphi :: (\exists X) \sim(\varphi)$
$\sim(\exists X)\ \varphi :: (\forall X) \sim(\varphi)$

The Intelim Rules and the additional rule of replacement, Quantifier Negation, together assume the whole set of rules of inference and rules of replacement of the Sentential Calculus discussed earlier in Module 28 in this Unit.

Notice that two of the rules, Universal Instantiation (**UI**) and Existential Instantiation (**EI**), eliminate the quantifiers. We call these "rules of Quantifier Elimination." The other two rules, Universal Generalization (**UG**) and Existential Generalization (**EG**), introduce quantifiers for expressions in prenex normal form. Correspondingly, we call these "rules of Quantifier Introduction." Hence, the title: "Intelim Rules" = Introduction/Elimination Rules.

Here's a simple example of Universal Instantiation (to help you understand the chart above):

Suppose you are given this formula: $(\forall x)Bx$. Here, the small case variable 'x' replaces the capital letter 'X' in the chart, and the predicate letter 'B' replaces the placeholder 'φ' in the chart. **UI** is without restrictions and the chart tells us we are entitled to show the following:

Ba

. . . where 'a' is a term-letter ('T') which instantiates the variable 'x'. Notice that 'T/X' means "replace the variable with a term-letter."

General Terms and Individual Constants

We must now distinguish between two handy instruments that we carry in our toolbox for advanced quantificational logic: the general term and the individual constant. (The expression 'term-letter' comprises both *general terms* and *individual constants*). **General terms** designate everything and not any one thing in particular. Thus, when we are confronted with the formula, '(∀X)Bx'—"For every *x, x* is beautiful"—we can use Universal Instantiation (**UI**) to apply to any general term at random.

However, an **individual constant** functions as a proper name for one designated thing or it functions as a "definite description." It is understood to refer to a single individual only—perhaps a person or some particular thing we have in mind, such as the wife of Ulysses. Thus, from '(∀x)Bx'—"Everything is beautiful"—we may apply **UI** to yield the following result:

Bp

. . . where 'p' designates Penelope, the wife of Ulysses.

What is the upshot of this distinction? It serves a very important purpose in formulating our restrictions on rules of universal instantiation and generalization. Notice that, from "Everything is beautiful," we can legitimately deduce that *anything at random is beautiful* (designated by a general term) or that *one thing in particular is beautiful if anything is* (e.g., "Penelope")—designated by an individual constant.

Going the other way, if anything we can think of fits the predicate, '. . . is beautiful', so that

$\mathbf{Bc^1}$

. . . then we can validly deduce, "Everything is beautiful." But we cannot, from an individual constant (e.g., from 'Bc'), validly deduce a universal statement, (e.g., '(∀x)Bx'). To attempt to do so runs the risk of a fallacy of Hasty Generalization.

Why? The answer is obvious. From the fact that Penelope is beautiful, we are not entitled to deduce that *everyone* is beautiful, for it doesn't follow that Andre the Giant is beautiful, nor the Wicked Witch of the West, nor Boris Karloff. Where an individual is named or exclusively designated, it doesn't normally make sense to assign general superscripts ('a^1', 'a^2', 'a^3' . . .) just as it normally makes no sense to designate differing versions of Penelope, the wife of Ulysses ('Penelope^1', 'Penelope^2', 'Penelope^3' . . .), clones and Doppelgangers notwithstanding.

To make this distinction clear in practice, we will always use the superscript with a term-letter to indicate that the term in use is a general term (e.g., 'a^{1}'); when a term-letter is without a superscript (e.g., 'a'), that means we are using it as an individual constant.

Should we be presented with an expression containing an individual constant, the application of **UG** is restricted to the constant itself (as in 'S' = "All things identical with Socrates," a class containing just one member). Note that our restrictions on **UG** do reflect the fine distinction between general terms and individual constants, but the restrictions are guidelines only; this discussion is meant to give you a fuller understanding.

Direct Derivations Using UI

The rule of Universal Instantiation (**UI**) is as follows:

$(\forall X)(\varphi X)$
$\therefore \varphi X/T$

. . . where 'X' is any variable and 'T' is any term (i.e., individual constant or general term).

Let's consider the following argument: "Everything is fine. Everything is beautiful. Therefore, anything we may take at random is fine and beautiful too."

Symbolizing, we have

FROM:	$(\forall x)Fx$	"Everything is fine."
	$(\forall x)Bx$	"Everything is beautiful."
DEDUCE:	$(Fa^{1} \& Ba^{1})$	"Anything we may take at random is fine and beautiful too."

DEDUCE: $(Fa^{1} \& Ba^{1})$

1. $(\forall x)Fx$	Premise
2. $(\forall x)Bx$	Premise
3. Fa^{1}	1, **UI**
4. Ba^{2}	2, **UI**
5. $(Fa^{1} \& Ba^{1})$	3,4 Conj

Since we have no restrictions on the instantiation of term-letters under the rule **UI** outside of the very general proviso that each occurrence of 'x' in 'Fx' must be uniformly and consistently replaced with 'a^{1}', we can re-introduce the general term-letter 'a^{1}' for the instantiation of '$(\forall x)Bx$' at step 4 as well. This yields a valid argument.

Notice that **UI** allows us to instantiate to either general terms or individual constants, but not to variables. If *Everything is going well* then it follows that anything we can name is going well, including progress in logic class, *glasnost* and *perestroika,* and trade negotiations to improve the Philadelphia Phillies.

Returning to our previous example, what do we do if we want to generalize our results to *all* things and say that "Everything is fine and beautiful *too*?" Will this also yield a valid argument? To make that determination, we need to appeal to a new rule which does have some restrictions: the rule of Universal Generalization (**UG**).

Direct Derivations Using UG

The rule of Universal Generalization (**UG**) is basically as follows (ignoring any restrictions for now):

φT
∴ (∀X)(φX/T)

. . . which says that a variable 'X' may be substituted for a term 'T'.

Now, let's return to our earlier example and apply **UG** to obtain the new, more general conclusion: "Everything is fine and beautiful too."

One again, the premises are:

FROM:	(∀x)Fx	"Everything is fine."
	(∀x)Bx	"Everything is beautiful".

Now we can continue our deduction as shown:

DEDUCE: (∀x)(Fx & Bx)

1. (∀x)Fx	Premise
2. (∀x)Bx	Premise
3. Fa^1	1, **UI**
4. Ba^1	2, **UI**
5. $(Fa^1$ & $Ba^1)$	3,4 Conj
6. (∀x)(Fx & Bx)	5, **UG**

Why does this work? It works because we have selected at random anything we can conjure up to instantiate the universals, "Everything is fine," and "Everything is beautiful." Those instantiations are designated with general terms. Anything we can think of will fit: Persian cats or pedigree dogs, winter constellations, or crescent moons. Thus, the instantiation of '(∀x)Fx' can be expanded indefinitely to cover anything and everything. This also holds true for the statement, "Everything is beautiful." Because this is so, we are entitled to conclude with the conjunction, "Everything is fine and beautiful too," by an application of the rule **UG**.

The rule of Universal Generalization, however, cautions us to observe two restrictions: **(1)** 'T' cannot be generalized if it is an individual constant or a term that already appears in '(∀X)(φX/T)' or in any premise or preceding line gotten by **EI**, and **(2)** each occurrence of a term letter 'T' in 'φT' must be uniformly and consistently replaced with the same variable 'X'.

Let's consider restriction **(1)**. Remember that 'φT' is a placeholder for a formula which contains predicate letters and terms that can be expanded to a very large and very complex well-formed formula, so 'φT' can stand for a simple atomic formula such as "Fa^1" or it can be as complex an expression as "(∀x)(∃y)(Fa^1 & (Bx ∨ Cy))"—where 'x' and 'y' are bound variables in this instance.

Restriction **(1)** says that we cannot use **UG** to generalize from (∀x)(∃y)(Fa^1 & (Bx ∨ Cy)) to (∀x)(∃y)(Fx & (Bx ∨ Cy)) because the variable 'x' already appears in the formula, modifying 'B' in "Bx." The rule is that you must select a *new* variable to continue the deduction properly.

Here's an example of a violation of restriction **(2)**:

DEDUCE: (∀x)(Fx & Bx)

1. (∀x)Fx	Premise	
2. Ba̲	Premise	
3. Fa^1	1, **UI**	
4. (Fa^1 & Ba)	3,4 Conj	
6. (∀x)(Fx & Bx)	5, **UG**	violates restriction (2)

The violation occurs because the 'a' in "Ba" is an *individual constant,* not a general term. One may not substitute a single variable for both a constant and a general term under the provisions of the Intelim Rules.

Notice that restriction **(1)** on **UG** also says that 'X' cannot be a variable from any preceding line gotten by **EI.** If we did not have that restriction on **UG,** then we might be tempted to make the outrageously invalid deduction that:

DEDUCE: (∀x)(Px & Wx)

1. (∃x)(Px & Wx)	Premise	
2. (Pa & Wa)	1, **EI**	
3. (∀x)(Px & Wx)	2, **UG**	violates restriction (1)

If we translate the premise '(∃x)(Px & Wx)' as "There is at least one person who was the wife of Ulysses," then the conclusion fallaciously asserts that *everyone* is the wife of Ulysses, an outcome which would make the hero of the Battle of Troy the most outlandish bigamist of all time! Just think what the faithful Penelope would have to say about that application of **UG**!

If that example strikes you as silly (it should!) then don't be intimidated by the restrictions on these rules. For the most part, they are painfully obvious if you think about them at all.

Direct Derivations Using EI

The rule of Existential Instantiation (**EI**) makes the severest restriction of them all. This is the rule which states that:

(∃X)(φX)
∴ φX/T)
Restrictions:
(1): 'T' is an *individual constant* that does not already appear in '(∃X)(φX)' nor anywhere else in the preceding direct derivation.
(2): 'T' does not appear in the conclusion to be derived.
(3): Each occurrence of 'X' in 'φX' must be uniformly and consistently replaced with the same term-letter 'T'.

Let's consider for a moment what this rules means in ordinary language. When we say that '(∃x)Dx', or "There is an x such that x is a D," we are saying that there is (at least) one thing that is a D. For example, let us say that there is at least one person that is the wife of Ulysses. We symbolize that expression as:

(∃x)(Px & Wx) **"There is an *x* such that *x* is a person and *x* is the wife of Ulysses."**

According to **EI**, we are entitled to deduce that "(Pp & Wp)," where 'p' is an individual constant that designates one individual (namely, Penelope).

Here's a sample of a simple derivation which showpieces **EI:**

DEDUCE: (Ea)	"The number 2 is even."	
1. (∀x)(Ex ⊃ Ea)	Premise	"If any number is even, the number 2 is."
2. (∃x)Ex	Premise	"At least one number is even."
3. Eb	2, **EI**	"One number is even."
4. (Eb ⊃ Ea)	1, **UI**	"If some number or other is even, then the number 2 is."
5. Ea	3, 4 MP	"The number 2 is even."

Step 3 is legitimate because 'Eb' was obtained by **EI** before '(Eb ⊃ Ea)' was obtained by **UI**, and no restrictions apply to Universal Instantiation. We could not have worked the derivation in the opposite direction, deriving '(Eb ⊃ Ea)' by **UI** and then showing 'Eb' as a subsequent line derived by **EI** because that would have been prohibited by our restrictions.

Those restrictions on **EI** are ones with which we are already familiar from our discussion of the distinction between general terms and individual constants. The restriction says: "The term-letter 'T' must be an *individual constant* that has not previously appeared anywhere in the derivation." That means that the following derivation is disallowed:

DEDUCE: (Pa & Wa)

1. (∃x)(Px & Wx)	Premise	
2. Ga	Premise	
3. (Pa & Wa)	1, **EI**	violates restriction **(1)**

Here is another example of why this restriction is so important to the proper application of **EI**.

1. (∃x)Mx	Premise	"Some men exist."
2. (∃x)Wx	Premise	"Some women exist."
3. M$\underline{a}$	1, **EI**	"Some particular person is a man."
4. W$\underline{a}$	2, **EI** [violation!]	"The same person is a woman too!"

If this fallacious deduction were allowed to proceed, we would end up with the untenable conclusion that something 'a' can be both a man and a woman at the same time, which is an impossibility even for modern surgical techniques.

Let's consider why restriction **(2)** is so important. That restriction prohibits us from using a term-letter that appears in the conclusion we are attempting to derive. For example, consider this very bad argument:

(∃x)Nx	Premise	"At least one thing is a skyscraper."
∴ Ni	1, **EI**	violates restriction **(2)**

While it is true that a whole number of things qualify as skyscrapers—the Empire State Building, the Chrysler Building, the World Trade Towers and so on—the conclusion 'Ni' may be interpreted to say that *Independence Hall* is a skyscraper, when it is in fact a small building in a city that has an ordinance against skyscrapers.

Without restriction **(2)** we are in danger of committing the fallacy of Begging the Question*—the fallacy of assuming the very thing we are attempting to prove. Imagine someone who mistakenly but stubbornly believes that Independence Hall is a skyscraper and argues: "I know it is! You've got to admit that there are a whole lot of skyscrapers in the big cities of this country. Independence Hall is a building in a big city. It's only reasonable that Independence Hall should be a skyscraper too!" With restriction **(2)** in place, you are entitled to respond that the speaker's contention is *not* reasonable.

This restriction on Existential Instantiation allows us to recognize that something actually does exist without pinning us down to the existence of something that we have assumed meets the condition. In other words, using individual constants in this way is equivalent to making an *assumption* the way we did in conditional proofs and indirect derivations. We cannot rest with that assumption as the terminating line of our derivation without the risk of assuming the very

*For further discussion of this fallacy, see Unit Four, Module 36, "Fallacies of Presumption."

thing we intend to prove. Doing so would sometimes lead to false conclusions from true premises.

Restriction **(3)** will prove especially important once we introduce dyadic quantifiers and relational terms later in this module, but for now we can make the point with a simple application. Take the following faulty argument as an example:

DEDUCE: (Ga & Hx)

1. (∃x)(Gx & Hx)	Premise	
2. (Ga & Hx)	1, **EI**	violates restriction (3).

This example not only illustrates restriction **(3)** on **EI** but it also shows one reason why we prohibit free variables from appearing in our system. Where the premise '(∃x)(Gx & HX)' is concerned, the derivation of '(Ga & Hx)' is not permitted. As the Maine woodsman says: "You can't get there from here."

Direct Derivations Using EG

The rule of Existential Generalization (**EG**) states that:

φT
∴ (∃X)(φX/T)

Restriction:
(1) 'X' must be a variable that does not already appear in 'φT'.

This rule harbors the assumption that any subject worthy of a name or a designation can be generalized to existence. No restrictions apply to **EG,** but this rule of quantifier introduction may well be the most curious and counterintuitive of all the quantifier rules we've explained thus far.

First, let's consider a simple example of an application of **EG** by working a direct derivation for the following argument:

All rock concerts are loud.
Some rock concerts are bad.
Therefore, some loud things are bad.

In symbols, this becomes:

(∀x)(Rx ⊃ Lx)
(∃x)(Rx & Bx)
∴ (∃x)(Lx & Bx)

We work the derivation as follows:

DEDUCE: (∃x)(Lx & Bx)

1.	(∀x)(Rx ⊃ Lx)	Premise
2.	(∃x)(Rx & Bx)	Premise
3.	(Ra & Ba)	2, **EI**
4.	(Ra ⊃ La)	1, **UI**
5.	Ra	3, Simp
6.	La	4, 5 MP
7.	(Ba & Ra)	3, Com
8.	Ba	7, Simp
9.	(La & Ba)	6, 8 Conj
10.	(∃x)(Lx & Bx)	9, **EG**

Judging by this application alone, you may protest that the rule takes us from the obvious to the superfluous, for the step which takes us from '(La & Ba)' to '(∃x)(Lx & Bx)' is painfully obvious. Yet, as we shall see later in this unit, when we have introduced polyadic quantifiers and relational predicates, **EG** plays an important role in working derivations. Perhaps the most important application of **EG** has to do with what we call "definite descriptions." A **definite description** is a phrase which uniquely designates one thing without explicitly naming it.

For example, **EG** permits us to make the following deduction:

The highest mountain in North Carolina is 6,684 feet high.
Therefore, the highest mountain in North Carolina exists.

We will wait until the end of this module to symbolize this argument because, in order to do so, we need to know something about polyadic quantifiers, definite descriptions, and relations of identity; but intuitively you should be able to see the need for **EG** just on the basis of the argument in English.

Direct Derivations Using QN

The one rule of replacement we've provided here for quantificational logic is the rule of Quantifier Negation (**QN**). Once more, the rule is:

$\sim(\forall X)\ \varphi :: (\exists X) \sim(\varphi)$
$\sim(\exists X)\ \varphi :: (\forall X) \sim(\varphi)$

A simple derivation using **QN** might look like this:

1.	(∀x) ~Px	Premise
2.	~(∃x)Px	1, **QN**

Quantifier Negation is often very useful in reducing the length of what might otherwise be very complicated derivations. For example, consider the following Direct Derivation:

DEDUCE: $\sim$**Qa**

1. $\sim(\exists x)(Px \supset Qx)$	Premise
2. $(\forall x) \sim(Px \supset Qx)$	1, **QN**
3. $\sim(Pa \supset Qa)$	2, **UI**
4. $\sim(\sim Pa \vee Qa)$	3, Impl
5. $(\sim\sim Pa \ \&\ \sim Qa)$	4, DeM
6. $(\sim Qa \ \&\ \sim\sim Pa)$	5, Com
7. $\sim Qa$	6, Simp

Throughout this module, we have used the Intelim Rules and the rule of Quantifier Negation only to work direct derivations. If we intend to use the Intelim Rules on indirect derivations or conditional proofs, then we must observe a few more restrictions. We will not spell out those additional restrictions here, but the complete set of restrictions for **IP** and **CP** is provided for you in the Appendix.

32.2 RELATIONAL PREDICATES

We now distinguish between two types of predicates. Up to this point, we have looked only at those predicates which assign a property to a single entity. For example:

Statement	**Predicate**	**Symbolization**
Conan is a barbarian	("... is a barbarian.")	Bc
Schwartzenegger doesn't talk much.	("... is someone who talks much.")	~Ts
Dirty Harry packs a magnum.	("... is someone who packs a magnum.")	Md
Rocky can take a punch.	("... is someone who can take a punch.")	Pr
Who you gonna call? Ghostbusters!	("... is who you're gonna call.")	Cg

Now we will look at a second kind of predicate, the **relational predicate,** which is a predicate that assigns a relation between two or more things.

"Chuck Norris is tougher than Sylvester Stallone."

"The Joker is far more interesting a character than Batman."

"Freddy Kruger is more violent than Jason."

How do we get such relational predicates into standard symbolic form? A two-place predicate may be rendered according to the following schema:

Φ__ __

. . . where 'Φ' is any relational predicate (any capital letter of the alphabet not already assigned to serve as statement letters, ordinary predicates, or other relational predicates) and the blanks are placeholders for variables, general terms, or individual constants. Thus, the statement "Tristan loves Isolde" may be rendered in symbols as:

Lti

. . . or "t loves i," where 't' = Tristan and 'i' = Isolde.

The statement "Tristan loves himself" is thus rendered as:

Ltt

. . . or "t loves t."

"Everybody loves somebody" becomes:

(∀x)(∃y)Lxy

. . . or "For all *x*, a *y* exists such that *x* loves *y*" (which doesn't preclude loving oneself as well).*

A three-place predicate is rendered in schematic form as:

Φ__ __ __

. . . where Φ once more is any relational predicate (any capital letter of the alphabet not already assigned to serve as statement letters, ordinary predicates, or other relational predicates) and the blanks are placeholders for variables, general terms, or individual constants. Thus, the statement "Tristan loves Juliet more than Isolde" may be rendered as:

Ltji

. . . or "t loves j more than i," where 't' = Tristan, 'j' = Juliet, and 'i' = Isolde.

"*Tristan loves himself more than Juliet*" may be symbolized as:

Lttj

. . . or "t loves t more than j."

How do we translate those statements which we introduced at the beginning of this section? Here are the suggested symbolic translations together with a few other examples that contain relational predicates:

"Chuck Norris is tougher than Sylvester Stallone."

Tcs

"The Joker is far more interesting a character than Batman."

[(Cj & Cb) & Ijb]

*Alternatively, this may be symbolized as '(∀x)[(∃y)Lxy]', which is equivalent.

. . . "The Joker is a character and Batman is a character and the Joker is a far more interesting character than Batman."

"Freddy Krueger is more violent than Jason."

Vfj

The love of money is the root of all evil. —*The Book of Proverbs*

(∀x)(Ex ⊃ Rlx)

. . . where 'l' = "the love of money" and 'Rxy' = "x is the root of y."

God helps them that help themselves. —*Poor Richard's Almanac*

(∀x)[(Px ⊃ (Hxx ⊃ Hgx)]

What does not kill me makes me stronger. —*Nietzsche*

(∀x)(~Kxm ⊃ Sxm)

Stolen waters are sweet. —*The Book of Proverbs*

(∀x)[(Sx & Wx) ⊃ Ex] "For all x, if x is stolen (S) and x is water (W), then x is sweet (E)."

Hope deferred makes the heart sick. —*The Book of Proverbs*

(Ax)[(Hx & Dx) ⊃ Sxh]

Having introduced relational predicates, we must now consider the possibility that we may want to introduce more than one quantifier into a symbolic expression. We turn to that task in the next section.

32.3 DYADIC AND POLYADIC QUANTIFIERS

The material we've studied thus far is sometimes called the **Monadic Predicate Calculus.** The word 'monadic' comes from Latin and literally means "single" or "pertaining to one."* Thus, the "Monadic Predicate Calculus" refers to symbolic formulae involving just one quantifier, as in the following examples:

(∀x)(Gx ⊃ Mx)	**"All Greeks are mortals."**
(∃y)(Cy & Fy)	**"Some fat cats exist."**

However, we can greatly extend our range of symbolization if we introduce two *or more* quantifiers into our symbolic expressions. When two quantifiers appear in a symbolic formula, we call the expression 'dyadic' meaning, literally, "pertaining to two." If we introduce two *or more* quantifiers into a formula, we call the expression 'polyadic' meaning, "pertaining to many."

Adding polyadic quantification to our symbolic baggage allows us to symbolize many more natural language expressions than would be possible other-

*'Monadic' also refers to one-place predicates.

wise. Such an addition can also greatly complicate our derivations. However, polyadic quantification has two main benefits: (1) deductions that are very complicated and confusing in natural language very often can be worked more easily in polyadic symbolic form; and (2) symbolization, now as before, tends to resolve ambiguous expressions of natural language and give them precise formulation.

How do we translate the following statement into the Quantificational Calculus?

"Everything interacts with everything."

Obviously, we can't do it using only one quantifier. We need two:

$(\forall x)(\forall y)\ Ixy$

Now let's introduce a personal element. Suppose some thirty-something Yuppies are networking with each other at a party, and one of them says, "Everybody is interacting with everybody!" How do we symbolize that? We introduce 'P' for "person":

$(\forall x)(\forall y)[(Px \,\&\, Py) \supset Ixy]$

The placement of the quantifiers is exceedingly important. For example, we may translate "Everything is either colorful or big" as:

$(\forall x)(Cx \vee Bx)$

However, the English statement "Either something is hot or something is cold" must be translated with two quantifiers, as below:

$[(\exists x)\ Hx \vee (\exists y)\ Cx]$

We introduce different variables 'x' and 'y' here for the obvious reason that the same something can't be both hot *and* cold at the same time.

Notice however that these symbolic expressions are equivalent:

$(\forall x)(\forall y)[(Px \,\&\, Py) \supset (Lxy \supset Ux)]$

. . . and

$(\forall x)[(Px \supset (\forall y)(Py \supset (Lxy \supset Ux))]$

We may translate either of these to read "People who love people* are the luckiest people." In this example, the difference is only a matter of style, not of content.

*Read: *all* people. By convention, where no qualification is offered, a group term like 'people' should be interpreted as meaning *all* people.

Since we're feeling lyrical at the moment, let's translate "Everybody loves somebody sometime;" this will give us an opportunity to look at three quantifiers in action. I suggest the following translation:

(∀x)[Px ⊃ (∃y)(∃z)[((Py & Tz) & Lxy)]

For all *x*, if *x* is a person then a *y* exists such that *y* is a person and a *z* exists such that *z* is a particular time and *x* loves *y* at z.

The most difficult—and the most controversial—part of quantificational logic, especially where polyadic quantifiers and relational predicates are combined, may well be the proper translation of natural language statements into symbolic formulae. 'Proper' is possibly too strong a word here since many qualified logicians often heatedly debate the way a given statement should be symbolized. These disputes are often a matter of style, since many logically equivalent translations are possible, but sometimes these debates transcend stylistic disputes and involve authentic matters of content, as when a statement is ambiguous in English. We will consider some of these problems in more detail momentarily. Some conventions have been widely adopted however, so we will now consult a list of more-or-less standard translations.

"Everything interacts with everything."

(∀x)(∀y) Ixy

"Everything interacts with something (or other)."

(∀x)(∃y) Ixy

"Everything interacts with some (one) particular thing."

(∃x)(∀y) Ixy

"Something interacts with everything."

(∃x)(∀y) Ixy

"Something interacts with something."

(∃x)(∃y) Ixy

"Everybody is known by everybody."

(∀x)(∀y)[(Px & Py) ⊃ Kyx]—'Kyx' because x is *known by* y and 'Kxy' means 'x *knows* y'.

"Somebody is known by somebody."

(∃x)(∃y)[(Px & Py) ⊃ Kyx]

"There is somebody who knows everybody."

(∃x)[(Px & (∀y)(Py ⊃ Kxy)]

"Somebody is known by everybody."

(∃x)[Px & (∀y)(Py ⊃ Kyx)]

32.4 DEFINITE DESCRIPTIONS AND IDENTITY

In the first section of this module, we considered that the rule of Existential Generalization (EG) allows us to make the following deduction:

The highest mountain in North Carolina is 6,684 feet high.
Therefore, the highest mountain in North Carolina exists.

'The highest mountain in North Carolina' is a definite description, that is, an expression which uniquely refers to one individual entity without explicitly naming the individual entity. To see the difference between names and definite descriptions, consider that 'Mt. Mitchell' is the name of the entity which the expression 'the highest mountain in North Carolina' describes.

Many different definite descriptions can designate the same unique entity, but notice that a proper definite description refers to just one thing. For example, 'the highest mountain in North Carolina' is a definite description which refers to Mt. Mitchell, but so do each of the following descriptions:

1. **the highest mountain in the Southern states.**
2. **the highest mountain in the Eastern United States.**
3. **the highest mountain in the Appalachian Range.**

Notice, however, that the following description may not uniquely refer to Mt. Mitchell and strictly speaking is not, therefore, a definite description:

4. **the mountain that is 6,684 feet high.**

Description (4) does not uniquely identify Mt. Mitchell because some other mountains in the world may be exactly that high. For example, many mountains in the Sierra Nevada Range of California, some so comparably insignificant that they're nameless, come very close to the exact elevation of Mt. Mitchell. Nevertheless, we shall treat (4) as a definite description because we may suppose that the speaker intends to refer to Mt. Mitchell alone but is unaware that other mountains that high may exist. We may say that the speaker has constructed a faulty definite description.

Intended definite descriptions like (4) may be characterized as cases of mistaken description. The same holds true for descriptions which fail to designate a unique entity because no such entity in fact exists. Such is the case with the definite description 'the wife of Ulysses' in "The wife of Ulysses is both faithful and brave," if we are right in assuming that both Ulysses and his wife, Penelope, are fictional characters who did not actually live.*

How do we treat definite descriptions of mistaken reference? The matter is controversial, but we shall follow the favored treatment proposed by philosopher Bertrand Russell and say that statements which contain definite descriptions of mistaken reference are *false*.† Each of the following sentences contains a

*Of course, we can't be sure. We have archaeological evidence that the sack of Troy actually took place so a general named 'Ulysses' or 'Odysseus' may well have presided over the battle. Such uncertainties show the advantage of treating definite descriptions with mistaken references as possibly false, rather than as non-statements or meaningless.

†See Bertrand Russell, "On Denoting," *Mind*, vol. 14, 1905, pp. 479–493. For a dissenting argument on how to treat definite descriptions of mistaken reference, see P. F. Strawson, "On Referring," in Anthony Flew (ed.) *Essays in Conceptual Analysis*, London: Macmillan, 1956.

definite description of mistaken reference, and we are to regard each of these as meaningful, but false statements:

1. **The only mountain in the world that is 6,684 feet high is in North Carolina.**

2. **The wife of Ulysses is faithful and brave.**

3. **The hand that counts more than a Royal Flush in Poker belongs to me.**

4. **The only logic textbook on the market is** *Logic for an Overcast Tuesday.*

Each of these statements is to be treated as false for our purposes, but they may be legitimately employed in argument even so. Such arguments may turn out to be valid but since they contain a false statement, they will always be unsound.

This is part of the rationale behind the rule of **EG.** Existential Generalization is a valid operation even for statements which contain a definite description of mistaken reference. However, you must keep in mind that the result contains a false statement which may still be used to obtain a valid conclusion in a properly executed derivation; such an argument is, of course, necessarily unsound.*

Now we may turn to the matter of the proper symbolization of such definite descriptions. In order to do such symbolizations, we need to introduce another operator, the identity symbol ('='). We will use the identity symbol for such dyadic relations as "x is identical with y" ('x = y', not 'Ixy') or "x is the same as y" (also 'x = y', not 'Sxy').

Suppose we want to reveal Batman's alter ego. We may do so by means of the symbolic expression:

'b = w'

. . . where 'b' stands for "Batman" and 'w' stands for "Bruce Wayne." That's easy enough. Now let's introduce a more complicated example. In the movie "Dracula," two hansom cab drivers are approaching Dracula's castle amid some very weird goings on, and one of the cab drivers says to the other:

"Everybody in this world is crazy but you and me."

*As already acknowledged, the whole matter of definite descriptions with mistaken references is a hotly debated issue. A full treatment of that debate is outside the scope of this textbook, but a word or two about possible complications is certainly in order in a footnote. Consider the following sentence: "Penelope is the wife of Ulysses." We may detect an exception to our rule here. That sentence concerns a fictional character but, according to Homer's epic, the statement is true. The issue is one of aesthetics as well as logic. One might want to protest that the sentence isn't *true,* but it has the literary quality of *verisimilitude* (i.e., it is *true* of Homer's account). In practice, however, I regard the sentence as contextually true, and the following sentence as contextually false: "Captain Kirk is a Vulcan with pointed ears." That sentence, I maintain, is a statement, and it is false in the context of the "Star Trek" series.

For an earlier discussion of some of these points in this book, see Module 2, "Statements," especially p. 16.

The identity operator allows us to symbolize this statement in the following way:

$$(\forall x)[(Px \,\&\, (\sim(x = a) \,\&\, \sim(x = b))) \supset Cx]$$

. . . where 'P' stands for "a person in this world," 'a' stands for "you," 'b' stands for "me," and 'C' stands for "is crazy." The symbolic statement is read as follows: "For all x, if x is a person in this world, and it is not the case that x is identical with you and it is not the case that x is identical with me, then x is crazy." In other words, everybody else is crazy.

Now we have the tools we need to symbolize the argument with which we opened this section:

The highest mountain in North Carolina is 6,684 feet high.
Therefore, the highest mountain in North Carolina exists.

We symbolize the argument as follows:

Let 'Mx' = "x is a highest mountain in North Carolina."
Let 'Hx' = "x is 6,684 feet high."

$$(\exists x)[Mx \,\&\, (\forall y)(My \supset (x = y)) \,\&\, Hx]$$
$$\therefore (\exists x)[Mx \,\&\, (\forall y)(My \supset (x = y))]$$

In other words, the first premise reads: "There is an x such that x is a mountain in North Carolina and, for all y, if y is the highest mountain in North Carolina then x is identical with y, and x is 6,684 feet high."

The conclusion reads, "There is an x such that x is a mountain in North Carolina and, for all y, if y is a mountain in North Carolina then x is identical with y." The conclusion, simply translated, means that the highest mountain in North Carolina exists, and the justification for the conclusion is the rule **EG.**

EXERCISES

Exercise 1

1. What advantages do direct derivations have over Semantic Trees? Semantic Trees over direct derivations?

2. What are the two main benefits of adding relational predicates and polyadic quantifiers to our symbolic hardware in quantificational logic?

Exercise 2

DIRECTIONS: *Translate the following natural language statements into the symbols of quantificational logic. Some will include relational predicates; some polyadic quantifiers; some combinations; and still others will include none of these (i.e., they're monadic). Supply the appropriate symbolizations.*

1. What is done cannot be undone.
2. A drowning man will catch at a straw.
3. The wicked flee when no one pursues.
4. One is known by her company.
5. None are so blind as those who will not see.
6. One believes what one wishes to believe.
7. A rotten apple spoils its companion.
8. If wishes were horses then beggars would ride.
9. Odd numbers are lucky.
10. Nothing is difficult to willing minds.
11. He is happy that learns by other men's harms.
12. A bow long bent loses its spring.
13. Absence makes the heart grow fonder.
14. Little strokes fell great oaks.
15. A penny saved is a penny earned.
16. Hope deferred makes the heart sick.
17. A soft answer turneth away wrath.
18. If the cap fits, wear it.
19. A bird in the hand is worth two in the bush.
20. Fling him in the Nile and he'll come up with a fish in his mouth.
21. Forewarned is forearmed.
22. Idleness is the beginning of all psychology.
23. All truth is simple.
24. Man does not strive after happiness, only the Englishman does that.
25. Even a worm will turn.

Exercise 3

DIRECTIONS: *Translate the following natural language arguments into symbols, then construct direct derivations to show that they're valid.* [*These arguments are monadic and do not have relational predicates.*]

1. All gams are schools.
   Some gams are shoals.
   ∴ Some schools are shoals.
2. No swarms are coveys.
   Some coveys are bevies.
   ∴ Not all bevies are swarms.
3. All drifts are droves.
   Some drifts are hogs.
   ∴ Some droves are hogs.
4. All falls are coveys.
   No coveys are nides.
   ∴ No nides are falls.

5. A dolphin is a finny friend.
   Not all sharks are finny friends.
   ∴ Not all sharks are dolphins.
6. All earth bikes are speedy.
   No speedy bikes are ten-speeds.
   ∴ No ten speeds are earth bikes.
7. All light beers taste bad.
   Some light beers are bottled in clear bottles.
   ∴ Some beers bottled in clear bottles are bad-tasting beers.

8. A few hurricanes have wreaked havoc in North Carolina, but every hurricane is potentially dangerous to us so some potentially dangerous things are known to wreak havoc there.

9. Not all World Series have been dull, but they've all been televised. Therefore, some things on television have not been boring.

Exercise 4

DIRECTIONS: *Using the Semantic Tree Method, show that the following expressions are equivalent.*

$(\forall x)(\forall y)[(Px \& Py) \supset (Lxy \supset Ux)]$

. . . and

$(\forall x)[Px \supset (\forall y)(Py \supset (Lxy \supset Ux))]$

32.5 A FINAL WORD ON TRANSLATION

At just about this level of logic, whether because natural language is so rich that it can never adequately be captured by symbolic formulae or because we haven't succeeded in doing so yet, subtle nuances and ambiguities of natural language make symbolization an uncertain science, if it is a science at all.

People who are given to certitudes will be dissatisfied with a rendering of symbolic translations that allows for a grey area of debate. However, despite the fact that many symbolic formulae are clearly faithful translations of natural language and whereas many other attempts at translation are clearly inappropriate, that grey area of debate is real and it is vast. The reader should note that we are only approaching the level of sophistication in Quantificational Logic at which that area of greyness exists; the symbolizations we have studied thus far are quite uncontroversial. Nevertheless, you should be aware that you are approaching the frontiers where such brushfires begin, and that the disputes are legitimate ones. I am told that this is a "gloomy" outlook. I disagree. I find such disputes exciting. They truly reflect the boundaries of established knowledge. The unexplored terrain is just ahead of you.

There can be no hard and fast rules for translation into symbols. To translate a statement from ordinary English is to produce a symbolic rendition which has the same or very nearly the same force—i.e., the same truth conditions—as the original. So to translate a statement from the vernacular is to produce one in logical notation which would be true and would be false under the same or very nearly the same conditions as would the original. The truth conditions of statements in English are familiar to fluent speakers of the language (since being familiar with them is a large part of what fluency is).

—HUGUES LEBLANC AND WILLIAM A. WISDOM, *Deductive Logic,* Boston: Allyn and Bacon, Inc., 1974, p. 131.

The danger of pure abstraction is sterility; that is an important theme of this logic book, for symbolic logic can distort natural language or skeletalize the richness of natural language. But logic can also help us refine our thinking and make our everyday speech more precise and less liable to misinterpretation. Let's consider an example.

In Module 17, "All Banks are Not the Same," we found an example of a common natural language expression that is highly ambiguous from a logical point of view. "All banks are not the same" can be translated into symbols in a variety of ways. For example, we can translate it as:

$(\forall x)(\forall y)[Bx \supset (By \supset \sim Sxy)]$

"For all x, for all y, if x is a bank then if y is a bank then x is not the same as y."

Notice that such a translation is compatible with the possibility that every bank is different.

However, in common speech we may use such constructions as "All are not . . ." in two distinct ways. For example, we may say:

"All capital gains are now not tax deductible."

$(\forall x)(Cx \supset \sim Dx)$

. . . or, by the rule of **QN**:

$\sim(\exists x)(Cx \,\&\, Dx)$

. . . which clearly has the sense of a Universal Negation. If you interpreted this as a Particular Negation (**O.**), supposing that maybe *some* capital gains are still deductible, you might well find yourself in trouble with the IRS. But some uses of the construction "All are not . . ." clearly do have the sense of a Particular Negation, as in the case of this quotation from Shakespeare:

"All that glitters is not gold."

$\sim(\forall x)(Gx \supset Ox)$*

The statement *"All banks are not the same"* is surely meant to be interpreted in that way, as a Particular Negation:

$\sim(\forall x)(\forall y)[Bx \supset (By \supset (x = y))]$†

"It is not the case that for all x and for all y, if x is a bank then if y is a bank, x is identical with (i.e., 'the same as') y."

This translation allows for the possible interpretation that some banks are different, and that is surely what the advertisers want to assert. However, as you can see, this is a complicated and convoluted rendering of what should be a very plain and straightforward statement. We have to use some very complex machinery to get out the intended meaning.

We said before that the study of symbolic logic should help us in becoming clearer and more precise in our everyday speaking and writing. Knowing how tangled the intended meaning of "All banks are not the same" turns out in symbolic notation and knowing how ambiguous the statement is as it stands, we would do well to reserve the construction "All are not . . ." for Universal Negative statements so that we can always render them as '$\sim(\exists x)\varphi$' and thus avoid the complications and the ambiguity.

One more comment: the example "All banks are not the same" shows how natural language can be ambiguous. Even though considerations of elegance in translation warrant the adoption of a convention to reserve translations of "all are not . . ." to Universal Negations (i.e., to expressions of the form '$\sim(\exists x)\ \varphi$') nevertheless, such constructs in everyday discourse are often meant to be Particular Negative (i.e., expressions of the form '$\sim(\forall x)\ \varphi$'). We must adopt the **Principle of Charity** by making every effort to cast the arguments of others in the way they intend. Strive to interpret or translate arguments in such a way that they have the best chance of turning out valid.

The Principle of Charity is very important in logic, where the object is to attain truth and clear thinking as opposed to rhetoric, where sometimes the object is just to win the argument. That means we should always attempt to put the other person's argument in the best light, even when that requires generosity in translation.

*This is equivalent to $(\exists x)(Gx \,\&\, \sim Ox)$.

†This is equivalent to $(\exists x)(\exists y)[(Bx \,\&\, By) \,\&\, \sim Sxy)]$.

EXERCISES FOR UNIT THREE

Exercise 1

DIRECTIONS: *On each page, you will find an argument. Some of them are already in symbolic form. If the problem is not given in symbols, then you must symbolize it. Next, construct a truth table. Then determine whether it is* valid *or* invalid. *If* invalid, *tell which row (or rows) show(s) invalidity. Put your completed truth table in the space below each problem.*

1. [In the classic 1930 film "Der Blaue Engel"]:* Either Professor Immanuel Rath (played by Emil Jannings) will marry Lola (played by Marlene Dietrich) or he won't, but if he does then he will be degraded and led to his destruction through his infatuation with the heartless cafe entertainer. Professor Rath (eventually) is degraded and led to his destruction by Lola. Therefore, he must have married the heartless cafe entertainer.

Let P = ______________________________

Let D = ______________________________

The argument in symbolic form is: **((P ∨ ~P) & (P ⊃ D))**
D
∴ P

VALID OR INVALID?____________ If invalid, which row(s)?________________

(Put your truth table here):

2. [In the 1983 film "Liquid Sky"]: New Wave Manhattan model Margaret (played by Anne Carlisle) serves as the primary attraction for a U.F.O. which lands atop her penthouse if and only if both she has heroin in her apartment and her sexual encounters provide the chemical nourishment that the aliens need. If (indeed) Margaret both has heroin in her apartment and her sexual encounters provide the chemical nourishment that the aliens need, then the film was made by ex-patriot Russian immigrants who were denied such pleasures in their arid native land. The film was in fact made by ex-patriot Russian filmmakers denied such pleasures. Therefore, Margaret serves as the primary attraction for a U.F.O. which lands atop her penthouse.

M = ______________________________

H = ______________________________

S = ______________________________

R = ______________________________

In symbolic form?

*"The Blue Angel."

VALID OR INVALID?__________ If invalid, which row(s)?__________________
(Put your truth table here):

3. [In the 1984 film "The Terminator"]: It is not the case that the Terminator (played by Arnold Schwarzenegger) can succeed in his mission to destroy the mother of the hero of the year 2029 by going back in time to 1985 Los Angeles if Arnold Schwarzenegger can act. In fact, Arnold Schwarzenegger can act provided he plays one of two roles: either a robot or a barbarian. In this movie, Arnold Schwarzenegger plays a cyborg—a robot that is part man, part machine. Therefore, the Terminator can succeed.

Let T = ______________________________

Let A = ______________________________

Let R = ______________________________

Let B = ______________________________

In symbolic form: **$(A \supset \sim T)$**

$((R \vee B) \supset A)$

R

$\therefore T$

VALID OR INVALID?__________ If invalid, which row(s)? ______________
(Put your truth table here):

Exercise 2

DIRECTIONS: *Work a Semantic Tree for each of the problems in Exercise 1.*

UNIT FOUR

ARGUMENTS AND FALLACIES

A LIST OF FALLACIES TO BE STUDIED IN THIS UNIT

Fallacies of Relevance
(Module 34)

1. APPEAL TO FORCE
   a. appeal to fear
2. AD HOMINEM
   a. tu quoque
   b. bad seed
   c. faulty motives
   d. poisoning the well
   e. guilt by association
3. APPEAL TO IGNORANCE
   a. innuendo
4. APPEAL TO SYMPATHY
5. APPEAL TO POPULAR SENTIMENT
   a. appeal to popular people
6. APPEAL TO AUTHORITY
   a. appeal to tradition
   b. appeal to patriotism
   c. appeal to titles
   d. jargon
   e. appeal to progress
7. DIVERSION

Inductive Fallacies
(Module 35)

8. SLIPPERY SLOPE
   a. Domino Theory
9. ATTACKING A STRAW MAN
10. FALSE DILEMMA
    a. the bureaucratic fallacy
11. FALSE ANALOGY
12. FALSE CAUSE
    a. appeal to superstition
    b. the gambler's fallacy
    c. *post hoc, ergo propter hoc*
13. HASTY GENERALIZATION

Fallacies of Presumption
(Module 36)

14. COMPOSITION
15. DIVISION
16. EQUIVOCATION
17. AMBIGUITY
18. SLANTING
19. BEGGING THE QUESTION
    a. circular reasoning
    b. question-begging characterizations
    c. victory by definition
    d. leading questions
    e. alleged certainty
20. COMPLEX QUESTIONS

Formal Fallacies
(Module 37)

21. SPECIAL PLEADING
22. AFFIRMING THE CONSEQUENT
23. DENYING THE ANTECEDENT
24. AFFIRMING A DISJUNCT
25. MALDISTRIBUTED MIDDLE
26. UNEQUAL DISTRIBUTION
27. UNEQUAL NEGATION

MODULE 33

ABOUT FALLACIES

33.1 IN GENERAL

Generally speaking, a **fallacy** is an error in reasoning. *Informal* fallacies are errors in argument form or content which, when they occur, constitute sufficient grounds for rejecting that argument. We may say that informal fallacies are committed when arguments lead or direct us to an unjustifiable conclusion; one that is misleading or misdirected. Such arguments should not persuade or convince us.

Formal fallacies are committed when an argument that appears plausible actually violates a formal rule of inference.

These definitions might not satisfy you. You might well object that they're too loose, and you'd have a point. I suspect that all informal fallacies can be regarded as specimens of a formal fallacious type. For example, we will study below the fallacies of relevance. These are classified as informal fallacies. However, there is a sense in which we can regard them as formal fallacies, since they do fail to satisfy formal rules of inference. For example, in a given conditional argument, where 'P' and 'Q' stand for discrete statements and there are two premises, we might have an argument like the following:

If P then Q;
P.
Therefore, L (where 'L' can stand in for any other statement).

This is a fallacy of relevance, and it is a formal fallacy. All the informal fallacies of relevance resemble this form (though they need not be conditionals). So, what makes a fallacy informal in our sense?

One major distinction is that the informal fallacies emphasize a certain history and tradition whereas study of formal fallacies is more concerned with technique. Most of the informal fallacies you will study here have their origins in the works of Aristotle who catalogued them in his *De Sophisticis Elenchis* ("On Sophistical Refutations").

Perhaps we can say that an informal fallacy is a fallacy with a particular history or biography, or, specifically, it is a fallacy that is described in a certain tradition. That tradition may describe some of the psychological features of our minds that have led us to be persuaded by such faulty arguments down through the ages. That last phrase is striking—'down through the ages'—because in asserting it we are suggesting that despite the passage of a thousand years, the withering away of empires, and the passing of countless cultures, human beings still fall victim to many of the same fallacious modes of thought. If this is true it is a remarkable statement about human psychology.

Perhaps another reason for the longstanding tendency for people to fall into these fallacy-types is that some of the fallacies often resemble, in a distorted way, good argument forms ("rules of inference"). This is especially true of the formal fallacies.

Logic has a long tradition, and a major part of that tradition has been the use of names and mnemonic devices to help recall valid argument forms and detect invalid or unsound argument forms. Much of this work was done by the scholastics and the monasteries of Medieval Europe, and a large portion of this work was devoted to aids of recognition for fallacies in arguments. Consequently, the names of fallacies are more often than not given in Latin.

The names for the fallacies are provided in English and you may use the English names if you prefer, but sometimes we have some good reasons to use the Latin names instead. For one thing, the Latin terms make it clear that a *technical* violation of good argument form has occurred. People tend to be very defensive about their arguments, and they often attach a great deal of personal esteem to their beliefs. A major point of training in logic is to get us to focus on the issues instead of quirks of personality. Can you see how hard it would be to stick to the issue and proceed in a rational, cool-headed way if we point out to our fellow disputant that he's committed a fallacy of "appealing to ignorance"? How much more diplomatic it is to use the Latin terminology, *argumentum ad ignorantiam,* in place of the offensive-sounding English here! In using the appropriate technical terminology, we make clear that we're saying only that our friend has a technical flaw in his argument and, as a result, our friend may actually correct his argument and at the same time remain our friend.

The point is that in describing a fallacy in this way, we are not making an appeal to jargon; we are instead restricting our criticism to a technical flaw in the argument.

The fact that the Latin names for fallacies are a technical terminology is an important one. Consider, for example, that many legal and medical terms are given in Latin—*habeas corpus,* for example. The translation of *habeas corpus* would look silly or be misleading; the term literally means "to have the body." Someone untrained in law might think that had something to do with murder trials when actually *habeas corpus* is a writ requiring a person's presence in court. The Latin helps to keep the term restricted to a technical meaning. *Habeas corpus* is a name for a particular kind of writ.

Some names of fallacies also have unfortunate English translations. By virtue of tradition, the fallacy *petitio principii*, which is very difficult to pronounce for most English-speaking students, has been badly translated as "begging the question" even though when the fallacy is committed, rarely is any question present whatever, nor is anyone begging.

Moreover, even though it is Latin, you will find *habeas corpus* in English dictionaries. You will also find *ad hominem* there (the name of a particular kind of fallacy described in this section) as well as the Latin names of many other fallacies. This is an indication that to be a well-educated person you should know these Latin names because they're also part of good English vocabulary. Well-educated persons will recognize these terms most of the time and understand that you're attempting to say something technical about an argument when you use them. A judge in a court of law, for example, will be almost as likely to recognize the logical term *ad hominem* as she would the legal term *habeas corpus*. As you can imagine, the legal system relies very heavily on good form in argument.

A good exercise is to go through newspapers and magazines at home and look for fallacies of the kind described in these pages. This assignment has certain real advantages. First, it is an active, outgoing way to study the fallacies. One often learns better by doing than by reading textbooks. Second, this is a way of taking logic out of the classroom and applying it in the "real world" that will be the testing ground for any lessons in logic that are of lasting value.

Here are some tips about where to look for fallacies. A good source is the tabloids of the sort sold typically in supermarkets. If you can stomach all the stories about three-headed babies and abductions by extraterrestrials, you'll surely find plenty of hasty generalizations and false cause fallacies thereabouts. Journals and magazines that are highly partisan and devoted to a pressure-group or a single-issue cause are also good sources. For example, *Ms.* magazine on the one hand, and *Playboy* on the other, can often get carried away with their own fervor. Sometimes editorial critical acumen gets dropped in favor of enthusiasms over special issues. Local newspapers are also good sources. Look to the editorials and the letters to the editor.

Don't rule out the national news journals. In an effort to find just a few more fallacies for exercises in this book, I bought out the supermarket's supply of sensational tabloids and turned up nothing I could use. Only when I turned to the *Newsweek* that I purchased for my own reading did I find a "howler" of a fallacy in the opinion column, an article by a professor of English who should have known better.

Pay attention to what's being said, listen and keep an open mind. One should learn from arguments and disagreements. Contrast this with the competition of a debating society where the main object is winning the contest at all costs. The object in logic is communication and discovery. Whoever wins the argument is less important than what there is to be learned. You shouldn't cultivate the study of fallacies as an adversary would.

The ability to recognize and identify informal fallacies is a special talent that must be treated with care and respect. Some novice logic students are like hunters who are poorly trained on how to select their targets. But our presumption is that something is to be learned in the give-and-take of argument. "Blasting away" with charges of fallacies is every bit as misdirected as the hunter who shoots scattershot into the woods.

Beware of "the false charge of fallacy." The temptation is to find a fallacy lurking behind every argument. Later on, when students of logic finally come to realize that not all arguments contain fallacies, the temptation is to find fallacies *only* in arguments with which they strongly disagree. The difficulty is putting one's prejudices aside and seeing fallacies wherever they occur, and *only* when they do. Charges of fallacious reasoning always need to be justified.

Imagine a supporter of Stephen A. Douglas who is listening to the Lincoln-Douglas debates in 1858. First of all, you should know that Lincoln and Douglas were not far apart on the issue of slavery in their debate for the Republican nomination for the U.S. Senate seat. However, human nature being what it is, the temptation (if you invest time, work, and energy in a favored candidate's campaign) is to see only good in your candidate's speeches, and little good in your opponent's orations. Can you imagine how a Douglas supporter might criticize these famous lines delivered by Abraham Lincoln in the debates?

> "A house divided against itself cannot stand. I believe that a government cannot endure permanently half slave and half free."

As a Douglas supporter, you might accuse Lincoln of a false analogy, that a nation ought not to be compared to a structure; or you might accuse Lincoln of attacking a straw man, of suggesting that Stephen A. Douglas believes that a government should endure permanently half slave, half free (which he did not). But given the point of Lincoln's speech, such accusations really miss the mark. Logic demands a commitment to fairness, impartiality, and objectivity, and these are qualities very difficult to cultivate.

Now, let's update the example. In 1980, I was asked by local supporters of third-party candidate John Anderson to score the Anderson-Reagan televised debates. The Anderson people knew I was not a supporter of either candidate. I judged the debates on strictly logical points: Did the candidate answer the questions put to him? Were his answers relevant? Did he commit any fallacies in answering the questions? When the debate concluded, I was asked for my judgment and I said (to some surprise) that I thought Ronald Reagan had won that debate. That also seemed to be the consensus of the professional commentators who had judged the debate (I like to think) on similar logical grounds. My conclusions may not have dissuaded the Anderson supporters from their support of their candidate because substantive as well as logical considerations are relevant in such matters, but both they and I learned a greater respect for the abilities of the man who eventually won the presidency. That respect remained with me later on even when I sharply disagreed with many of the president's policies.

One obstacle to such objectivity in argument is the fact that we often attach a certain amount of self-esteem to arguments and conclusions we endorse. Instead of saying, "I believe *that* disarmament is the only rational policy for the United States to pursue," we tend to say, "I believe *in* disarmament," the latter expression signalling a greater tenacity of belief than the former. Down South, I've heard an even more peculiar expression: "I believe *on* the word of my friend." That suggests an even *greater* tenacity of belief. Anyone who uses an expression such as "I believe *in* . . ." or "I believe *on* . . ." is not likely to be susceptible to dissuasion. Consider anyone who says, "I believe *that* . . ." in comparison to someone else who says, "I believe *in*. . . ." If we invest self-esteem in our beliefs and arguments, we might well regard a criticism of our logic or evidence as a personal attack, and that is the wrong way to approach an argument.

Three lessons to keep in mind, then, before we proceed are: beware of the false charge of fallacy; keep an open mind and *listen* to what's being said; and don't attach your self-esteem to a favorite argument or belief.

In Modules 34 through 37 a list and a discussion of common fallacies follow. Good Hunting!

MODULE 34

FALLACIES OF RELEVANCE

Non-sequiturs

Non-sequiturs are inferences that do not follow from adduced premises or evidence. The term *non-sequitur* is a synonym for fallacy and literally means, "does not follow." The fallacies in this section demonstrate most vividly how a conclusion can stray from its premises or evidence.

PART 1: IRRELEVANT CONCLUSION (*Ignoratio Elenchi*)

The fallacy of Irrelevant Conclusion involves arguments that have gone off the track. The implicit fallacy is one of diverting the argument from the point at issue. There are innumerable ways of doing this. In fact, such fallacies occur so commonly that there are standard ways of cataloguing most instances of it. These are discussed below.

One can commit fallacies of relevance in the most outrageous ways. One very general form of the fallacy of Irrelevant Conclusion is answering a question

A deductive argument is **valid,** if and only if, if all the premises are true, then the conclusion must be true.

with a question. The problem is that a question is never an argument. Sometimes, when one responds with a question, she creates the impression that some sort of argument is underway when really none is.

A favorite sin against logic is the expression we hear so frequently: "Who's to say?" There probably isn't a philosophy teacher alive who hasn't at one time or another been chagrined by students who use this expression in class. It's so common, in fact, that one colleague printed T-shirts with the logo "Who's to say?" as a constant warning to any unsuspecting student that he or she had better beware. What's wrong with this expression? Again, it creates the deception that a real argument is underway. Perhaps the deception also consists in a presumption that somebody ought to be in authority to answer the question; since the suggestion is that no such authority exists, no answer is a proper one.

For example, suppose I argue that Republicans can expect to control the presidency for a long time because the Democratic party is in disarray. You answer, "Who's to say?" You haven't challenged my argument except to presume the impossibility of any good answer. That's the fallacy here; it doesn't address the reasons I gave. (My standard reply to "Who's to say?" is a terse, "Well, me—for one—on the basis of the reasons I already gave.")

But the use of the expression isn't just limited to freshman college students. The following is an excerpt from a prize winning essay published in *The International Lawyer.* The subject is international terrorism:

> "Faced with such a ruthless foe, the would-be reformer comes finally to his own crisis of conscience: the anguished and irreversible choice between giving up his worthy ends or resorting to unworthy means. Confronted by such a savage Hobson's choice, who is to say which is the better, which is the worse?"

Consider each of the following fallacies in this section as species of the general fallacy, *Irrelevant Conclusion.* When a fallacy of irrelevance occurs that does not tightly fit any of the following specific forms, classify it as the fallacy of Irrelevant Conclusion.

The following are some standard fallacies of relevance.

1. Appeal to Force (*argumentum ad baculum*)

One commits the fallacy of an **Appeal to Force** whenever he or she employs force or the threat of force or violence to bring about the acceptance of one's position.

This fallacy constitutes an irrelevant appeal because such "argument" is not designed to convince us of the reasonableness of adopting a belief or course of action, but rather to compel us to accept the conclusion regardless of whether we believe it to be true or prudent.

DEFINITION: **An *Appeal to Force* fallacy occurs whenever one irrelevantly appeals to force or threat of force to win an argument.**

EXAMPLE: **Conversation in a sports tavern:**

FIRST MAN: No way the Yanks will take the pennant this year. They ain't got the pitching.

SECOND MAN: The Yanks *are* gonna take the pennant. Here's why: if you say another defeatist word about them, I'll push your teeth down your throat.

COMMENT: **The second man has let his temper rule his reason. The first man made a reasonable point; the second man replied in an unreasonable fashion which in no way detracts from the former's good point.**

It should be noted that fallacies can be very effective on audiences, and there is no clearer example of that effectiveness than an argument which appeals to force. For example, the Spanish Grand Inquisitor, Tomas de Torquemada, was exceedingly effective in purging heresies from the Church in the fifteenth century, but he did so by trying and executing over 2,000 "heretics" instead of pleading the Divine authority and reasonableness of the edicts of the Church, thereby seeking voluntary acceptance.

It is sometimes argued that rationales for capital punishment in general are based on such fallacious appeals to force. It is a diverting thought that perhaps the theory of nuclear deterrence in the modern world is similarly based on reasoning which proceeds from this fallacy.

Consider one more example, taken from a popular song for children:

"You'd better watch out,
You'd better not cry,
You'd better not pout,
I'm telling you why:
Santa Claus is coming to town.
He knows when you are sleeping;
He knows when you're awake;
He knows if you've been bad or good
So be good for goodness sake."

a. Appeal to Fear (*argumentum ad metum*). This fallacy may be thought of as a close relative to the fallacy of Appeal to Force. The fallacy of *Appeal to Fear* is committed if and when fear is employed to gain acceptance of a view. We commonly call this "using fear tactics."

Militant anti-smoking group blamed after . . .
TRICK CIGAR BLOWS MAN'S HEAD OFF
'He took two puffs and then—BOOM!' says restaurant manager

By Henry Weber

. . . Georges Pressoa, 48, has just finished a meal with his wife and two friends when a middle-aged woman approached the table and offered him a large, foil-wrapped cigar.

"Georges smoked two cigars of his own before dinner and there were so many complaints that we thought the woman was sent by the management to apologize," said Laura Pressoa, the man's grieving widow.

"He sniffed the cigar she gave him and licked the end.

"He lit it with a wooden match and took two or three puffs.

"Then it went off—bang!"

The deafening explosion was so strong that it literally rocked the restaurant, said witnesses. Pressoa's wife and friends miraculously escaped injury.

Authorities said the blast actually vaporized Pressoa's head. . . .

Suspected in the man's death is the small but militant anti-smoking group, Stop Smokers Now!

The group has claimed responsibility in the fire-bombings of several tobacco shops in the past six months.

Police think that attacks on individuals like Pressoa are just a logical step in the group's war against smoking in public.

"SSN is an underground group and extremely elusive," said Sgt. Mendes. "But I promise you we will stop them before another innocent smoker is killed. . . ." —*Weekly World News,* January 19, 1988

Here is an exceptionally direct application of the maxim, "When all else fails, appeal to force." We may certainly doubt the homicide investigator's claim that such attacks are the "logical step" in the group's "war against smoking in public." The problem is that some people *do* see appeals to force as a "logical step." No one doubts that deadly exploding cigars are a highly persuasive means of getting one to quit smoking, but this example shows us just how irrational an Appeal to Force can be.

An example of the use of this fallacy may be seen in politics when demagogues play on the fears of the electorate to effect their own ends. In "red-baiting," for example, some politicians may divert the argument from the merits of a bill or policy by suggesting that such a policy may "play into the hands of the Communists."

Appeal to Fear

How do we distinguish a fallacious Appeal to Force from a fallacious Appeal to Fear? The distinction is not exact, and often the same argument can be at the same time one and the other. However, we can keep this criterion in mind: an Appeal to Force (*ad baculum*) involves either force or the immediate threat of force (where a victim's quite natural response may be fear), whereas an *appeal to fear* (*ad metum*) invokes irrational, often deep-set, fears and anxieties such as xenophobia, the fear of the unknown, or fear of what's new or different. The arguer who commits this fallacy knows how to effectively bring out the fears which will persuade her audience to her point of view.

Below is an excerpt from a sermon by the Puritan preacher and philosopher, Jonathan Edwards, who indulges a classic argument from fear by preaching "hellfire" to induce his congregation to seek salvation. This is his description of hell:

> . . . how dismal it will be, when you are under these racking torments, to know assuredly that you never, never shall be delivered from them; to have no hope: when you shall wish that you might be turned into nothing, but shall have no hope of it; when you shall wish that you might be turned into a toad or a serpent, but shall have no hope of it; when you would rejoice, if you might but have any relief, after you shall have endured these torments millions of ages, but shall have no hope of it; when after you shall have worn out the age of the sun, moon, and stars, in your dolorous groans and lamentations, without any rest day or night, or one minute's ease, yet you shall have no hope of ever being delivered; when you shall have worn out a thousand more such ages, yet you shall have no hope, but shall know that you are not one whit nearer to the end of your torments; but that still there are the same groans, the same shrieks, the same doleful cries, incessantly to be made by you, and that the smoke of your torment shall still ascent up forever and ever; and that your souls, which shall have been agitated with the wrath of God all this while, yet will still exist to bear more wrath; your bodies, which will have been burning and roasting all this while in these glowing flames, yet shall not have been consumed, but will remain to roast through an eternity yet, which will not have been at all shortened by what shall have been past. —"The Eternity of Hell Torments," Sermon XI, Vol. IV, New York: Leavitt & Allen, 1843, p. 278.

We can readily see how irrational this *appeal to fear* is, whatever one's theology may be, once we consider such apparent inconsistencies involved in the argument as the distinction between night and day even after we are said to have 'worn out the age of the sun, moon, and stars'.

2. *Ad Hominem* ("against the man")

An *Ad Hominem* fallacy is committed if one directs his remarks against the person instead of the issue. Name-calling is the most common form this fallacy takes, but it can take other forms as well (as in the case of the Greek king executing the messenger who was the bearer of bad news). One can commit this fallacy if he refuses to consider his opponent's argument on its merits alone, and instead attacks his opponent on the grounds of her beliefs, her prejudices, her motives, or inconsistency between her teaching and practice. Remember that this is a fallacy of *relevance* and it is never proper to re-direct the discussion from premises or evidence presented.

The point is this: it is never proper, from the standpoint of logic, to say, "Consider the source." Logic, instead, directs us to consider the argument.

DEFINITION: **An *Ad Hominem* fallacy occurs whenever one attacks the *person* instead of addressing the relevant *issue*.**

EXAMPLE:

SON: Even Dr. Lukasc, my political science professor, gives good reasons for believing the new treaty to be beneficial to both the United States and the Soviet Union. For one thing, it eases Cold War tensions.

FATHER: Not true! Lukasc is a confirmed glutton, a stinking whoreson, a panderer of drivel and mental hooey, and a Hungarian to boot.

COMMENT: **Father never addresses the issue at hand: whether the treaty is beneficial to both nations. Instead he attacks his opponent by calling him names.**

Here is an example of an Ad Hominem argument, which stands as plain name-calling, an article on the Meese Commission Report on Pornography, from *Rolling Stone* magazine:

> "Predictably, given the Reagan administration's history of political appointments, the Meese commission was no modern Plato's *Symposium*. Some of these people would have trouble getting a job taste-testing cat food. And the majority were as unbiased as East German high-dive judges."

Some sub-versions of this fallacious argument form are considered below:

a. *Tu Quoque* (the "you're another" fallacy). The term *tu quoque* might well remind us of a quarrel between children, but the fallacy is committed by adults more often than one would like to consider. Specifically, the fallacy is most commonly attributed to those who accuse opponents of inconsistency between beliefs and circumstances, or teaching and practice. Often it is relevant to point out such inconsistencies; it is not, however, relevant to do so if the goal is simply to score points against one's opponent in a discussion rather than determining the truth of what has been said.

For example, an editorial by the Soviet news agency, *Tass,* reprimanded a New Jersey school district for prohibiting the students from wearing gloves like those worn by rock singer, Michael Jackson. The editorial maintained that the school district should encourage students to emulate Jackson because he is a fine role model (religious, never takes tobacco or alcohol, etc.). If one were to reply, in response to this editorial, that the argument isn't worth a wooden nickel because "no school in the Soviet Union would ever permit students there to wear Jackson gloves," the rejoinder would simply miss the point. If one wanted to argue the merits of the case, one might defend dress codes in schools, for example, on other grounds, but to point out an inconsistency between practice and belief is an argument that is simply out of bounds in this case, and hence fallacious.

Newscaster Tom Brokaw had an exclusive interview with Soviet Premier Mikhail Gorbachev. "Premier Gorbachev," he asked, "what has the Soviet Union done to demonstrate its concern for the rights of dissenters to emigrate from the Soviet Union?" Gorbachev committed a *tu quoque* when he replied, "And what has the United States done to demonstrate its concern for the rights of Americans to economic security?"

However, not all *tu quoques* are fallacious. Consider the parent who tells his son or daughter: "It's too late for me to quit. I started smoking when I was young. I can't quit; I've tried. But don't you follow my example." We usually type all instances of "Don't do as I do, do as I say" as *tu quoques,* but sometimes this advice should be heeded because the object at hand isn't just to win the argument but to instruct the listener to an important truth. In those cases, no fallacy is committed.

b. Bad Seed. One commits the *Bad Seed* fallacy when she attacks her opponent on the basis of his heritage, lineage, or family's past history. An article in *The National Enquirer* claimed that Ronald Reagan was a distant cousin of Cuban guerrilla-warfare expert, Che Guevara. If one were to dismiss what Ronald Reagan had to say, simply because he may have been Guevara's distant relative, then one has fallen into this fallacious line of thinking.

Likewise, the person who dismisses his opponent by saying, "Consider the source," may be guilty of this fallacy, even if the opponent has a reputation for saying foolish or incorrect things. An editorialist once dismissed the views of National Security Advisor Walt Whitman Rostow, calling him, "Wrong-Way Rostow," and then noting that Rostow had never been right on any matter of international policy before. The argument is fallacious because it is of this form:

> "Rostow has never been right in his life; therefore, he must be wrong this time as well."

The conclusion clearly doesn't follow from its premise.

Exceptions: we may consider some examples of *non*-fallacious appeals to "consider the source." Usually, such arguments are fallacious because they "visit the sins of the father upon the son." But sometimes we may be justified in suspecting a source who has been consistently wrong or deceitful. These are judgment calls, for we all know the story of the "boy who cried 'wolf'." When the wolf actually appeared, no one believed him—but this time, they should have!

Yet, there is a saying: "Fool me once, your fault. Fool me twice, my fault!" Consider this hypothetical exchange:

MARY: I promise I won't go out with anybody else. Why, we're engaged!

TOM: You told me exactly the same thing yesterday, but I saw you with Albert last night.

MARY: That was yesterday; this is today.

Should you believe Mary this time?

c. Faulty Motives. The fallacy of *Faulty Motives* is committed if one diverts the argument from the issue by attacking the motives of the person who is making the case. For example, someone may impugn the motives of the *Tass* editorialist who criticized the American school district by claiming that *Tass* merely wanted to create discord in American schools and didn't care a whit about Michael Jackson as a role model. However, even if that were true, it is irrelevant. What is relevant is whether the New Jersey school superintendent was wrong in making this restriction.

One most often sees this sort of fallacy committed by those who suspect that no one would undertake a worthy project unless "there was something in it for him." For example, consider this possible conversation:

WIFE: Did you see our neighbor on television? He was making a plea for contributions to support Public Television. Maybe we should give something.

HUSBAND: Naw, Vanderbilt just does this to get on television.

Consider one more example: One often sees the fallacy of *faulty motives* in politics. A local referendum is set up on floating a bond for a new municipal Convention Center.

VOTER 1: That Convention Center sounds like a good idea; it might bring some new jobs and new businesses to the region.

VOTER 2: You've got to be kidding! I know for a fact that Councilman Jones stands to make lots of bucks if that Convention Center is built. He owns a hotel!

Now, it may be true that Jones does own a hotel and that he stands to profit by the passage of the referendum. But human beings are complicated creatures, and it may also be true that Councilman Jones is genuinely interested in his community's prosperity and believes the Convention Center will generate new jobs. One would be at fault if he or she attributed the worst of motives to Jones when other explanations are possible (we call the attribution of the worst motive "cynicism"). In fact, people rarely act from a single, unequivocal motive. People are made up of a range of interconnected motives—a psychologist might call this a "constellation" of motives. Selecting the worst possible motive is irrational and fallacious.

d. Poisoning the Well. One commits an *Ad Hominem* fallacy if he "poisons the well before another can drink." That is to say, the fallacy is committed by those who prejudice their audiences in advance of their opponent's presentation of his case.

Consider this example from the 1976 vice-presidential debates between Walter F. Mondale and Robert Dole:

> MONDALE: I think Senator Dole has richly deserved his reputation as a hatchet man tonight.

Senator Mondale made this statement about mid-way in the debate with Senator Dole as part of the continuing argument between them. In fact, Dole *had* made some harsh remarks which tended to divert the arguments from the issues. Mondale would have been justified in simply pointing out that Dole was diverting the argument; instead, he chose some harsh rhetoric of his own, imputing to Dole a reputation as a "hatchet man." The key to seeing the fallacy here is to focus on the word 'reputation'. By using that term, perhaps Mondale is suggesting that we can expect even more abusive language from Dole. That prompts the audience to regard with suspicion everything that Senator Dole may have to say from this point on.

Here is another example: Often, in a court of law, a shrewd prosecuting attorney may make references to a defendant's bad character, knowing that the defense attorney will object, and that her objection will be upheld by the judge. But the prosecutor may do it anyway, knowing that the remark, even if stricken from the record, may slant a jury's verdict and cast doubt on the defendant's testimony. This is an illicit technique, both logically and legally; but, of course, it works (unless the jurors took this course in logic!).

e. Guilt by Association. *Guilt by Association* was a mainstay of the Senator Joseph McCarthy investigation into alleged Communist infiltration of the U.S. State Department of the 1950s. McCarthy, along with the House Committee on Un-American Activities, built cases against individuals by compiling a list of "subversive" organizations and investigating associations of government personnel (as well as movie actors and writers and others) with organizations on the list. The investigation even went so far as to compel witnesses to give names of those associated with individuals who had ties with those organizations; under duress, witnesses gave the names of their friends and relatives to avoid charges of perjury and other penalties. In many cases, individuals lost their jobs or were "blacklisted" from movie careers. Among the many actors blacklisted in the movie industry were Zero Mostel and Charlie Chaplin.

Once again, the fallacy is committed because one is not concerned about debating the issue (presumably, whether an individual in the State Department is betraying his country), but is instead content with attacking the individual by virtue of the friends he keeps.

3. Appeal to Ignorance (*argumentum ad ignorantiam*)

An **Appeal to Ignorance** is committed if and when one argues that the absence of evidence for a conclusion counts as reason to believe the truth of the conclusion. This technique of argument is fallacious because it shifts the burden of proof from those affirming the case to those who remain to be convinced of the correctness of the case. It is the sort of bad thinking that occurs when someone challenges you to "prove that I'm wrong."

DEFINITION: **An *Appeal to Ignorance* fallacy occurs whenever one uses an absence of evidence for a claim as though it were evidence *for* the claim.**

EXAMPLE:

OWNER OF A METAPHYSICAL BOOKSTORE: Certainly we have excellent reason to believe in "out of body" experiences. Scientists have been studying this issue for years and have been unable to provide one shred of evidence that it isn't so!

COMMENT: **Usually, one who makes a controversial claim has the responsibility for producing evidence to defend it. Here, the bookstore owner argues only that scientists haven't been able to disprove the phenomenon.**

A fine example of Appeal to Ignorance, as well as an explanation of why the argument is fallacious, occurs in Graham Greene's novel, *The End of the Affair.* The discussion is between two characters, Maurice Bendrix and Henry Miles. The subject is life after death:

> "Do you believe in survival, Bendrix?"
>
> "If you mean personal survival, no."
>
> "One can't disprove it, Bendrix."
>
> "It's almost impossible to disprove anything. I write a story. How can you prove that the events in it never happened, that the characters aren't real? Listen. I met a man on the Common today with three legs."
>
> "How terrible," Henry said seriously. "An abortion?"

"And they were covered with fish scales."

"You're joking."

"But prove I am, Henry. You can't disprove my story any more than I can. . . ."

a. Innuendo. *Innuendo* is a particularly vicious specimen of the form, Appeal to Ignorance. *Innuendo* is an oblique charge against another's character or circumstances, made without any evidence, and accomplished by the sleight-of-hand of making the audience feel foolish or stupid for "not seeing the obvious." Thus, it reverses the burden of proof.

To see how vicious *innuendo* is, suppose a teacher has taken a particular disliking for one student and, in order to hurt her reputation, announces to his class: "I see Ms. Braithwaite isn't here today. If you only knew what she does outside of class time . . ."

It should be clear that *innuendo* commits two fallacies: both *Ad Hominem* (because it usually constitutes an indirect attack on someone's character) and Appeal to Ignorance (because it shifts the burden of proof). However, we shall classify the fallacy of *innuendo* as Appeal to Ignorance because of the underhanded way it destroys reputations by reversing the rules of evidence. Still, we must take care to remember Augustus de Morgan's famous word of caution that there is no single classification of the ways that arguments can go wrong, so it would be entirely proper in most cases to classify *innuendo* as an *Ad Hominem* fallacy as well.

Consider this further example of *innuendo* (taken once again from the 1976 Mondale-Dole Vice Presidential Debates):

DOLE: I couldn't understand why Governor Carter was in *Playboy* magazine. But he was. We'll give him the bunny vote.

Governor Carter, who later won the 1976 election for president, had consented to a text interview for *Playboy* early in the campaign. Senator Dole here has committed a fallacy of Appeal to Ignorance through *innuendo* as well as a fallacy of *guilt by association* because Dole was aware of the reputation the magazine has for the general public.

Many fallacious arguments commit two or more fallacies at the same time. It's not at all surprising to find the same argument legitimately tagged both an Appeal to Sympathy and an *appeal to patriotism;* another, a Hasty Generalization and a False Cause. Usually, we find reason to describe the argument more accurately one way than another, but often the same argument is equally guilty of two or more fallacies.

Innuendo: A "Double-Decker" Fallacy

Sometimes a single argument may commit a bundle of fallacies. This shouldn't be too surprising, for we see that an argument may go wrong in nearly an infinity of ways since a fallacy is a deviation from the rules of inference, and we have only a small, finite set of such proper rules of thinking.

***Innuendo* is such a package of fallacies. Anyone who uses this technique of rhetoric does so willfully and with malice. It is a species of deceit. Moreover, *innuendo* crosses the invisible line between fallacy forms, and will often violate strictures against arguments *Ad Hominem* (because *innuendo* is malicious) and Appeal to Ignorance (because *innuendo* is the deliberate disregard for established standards of evidence). Here's an example:**

Conservative columnist George Will discovered an outrageous and self-serving attack on the person of the failed nominee for the U.S. Supreme Court, Judge Robert Bork. Senator Howell Heflin opposed Bork's nomination, but found he had to justify his actions to his predominantly conservative constituents in Alabama. George Will learned that Heflin wrote a letter to his constituents claiming that, "The history of his [Bork's] life and his present lifestyle indicated a fondness for the unusual, the unconventional and the strange."

As Will rightfully argued, this was a personal smear on Bork's character. The wording of the letter suggests (but does not offer a shred of evidence) that Bork had a socially unacceptable lifestyle. This is an attack, an *Ad Hominem*. But that's not all. Heflin understood that in genteel circles, and native Alabamans may well consider themselves genteel in this way, one never directly states that another engages in deviant sexual behavior, one only *hints* at it. Heflin's constituents, on reading the letter, would undoubtedly read this message as an attribution of sexual misconduct on Bork's part, but would recognize that these things are never openly discussed in polite company and so would not be likely to look for evidence.

In this way, *innuendo* is a damning accusation that places the burden of proof on the audience to discredit the claim and offers no supporting evidence of its own. Since no one likes to think he doesn't know what's *really* going on for fear of embarrassment and being tagged naive, *innuendo* is also a form of Appeal to Ignorance.

4. Appeal to Sympathy (*argumentum ad misericordiam*)

A fallacy of **Appeal to Sympathy** is committed whenever one diverts the issue by making a plea for sympathy or pity. Sometimes this fallacy is subtly employed just by appealing to the sympathies of listeners, such as a politician who appeals for votes on the basis of having once resided in the community ("I'm one of you").

DEFINITION: An *Appeal to Sympathy* fallacy occurs whenever one makes an appeal to sympathy or pity in a way that is irrelevant to the issue.

EXAMPLE:

SENATOR FOGBOTTOM: Did you know that fifteen years ago my wife Frieda lived in Scranton? That's one good reason to vote for me because we're good plain-thinking folk, just like you.

COMMENT: Senator Fogbottom doesn't show that he has a platform that will appeal to the voters. He merely appeals to their sympathies, that he and his wife are just like they are. [In fact, Mrs. Fogbottom once privately remarked while she resided in Scranton that the army "ought to nuke the place and start all over."] Notice that sometimes and under some circumstances people can legitimately argue, "I'm one of you." For example, a politician may legitimately appeal for votes on the grounds that she shares many of the same interests as the constituents; however, what's wrong with Fogbottom's appeal here (apart from insincerity) is that he's attempting just to create a *feeling* of comradeship in place of reasons for voting for him. Fogbottom is condescending. We can be assured that he and his audience have little in common.

Any state police barrack can relate a hundred stories of Appeal to Sympathy excuses that drivers give for speeding on the highway. One of the best of them came from an overweight man who explained: "I know I was speeding, officer; but, you see, I'm trying to lose weight—doctor's orders. So I drive real fast, and that makes me scared of getting caught. When I get scared, I burn calories." Was the officer right in letting the speeder off with a warning? [The highway patrolman explained, "When I hear a yarn that good, I just gotta let 'em go."]

It is important to recognize, however, that not all appeals for mercy or sympathy are irrelevant from a logical, or even from a legal, point of view. Certainly, such pleas are extremely relevant after a verdict has been reached in a trial and sentence is being pronounced. At these times, pleas for clemency may be heard; in these circumstances, such considerations about whether the defendant is a first offender, or whether her incarceration will represent an undue hardship on her family, are doubtlessly relevant.

But such appeals for compassion may also be relevant outside the courtroom (such as when the suffering of the poor is taken into consideration during budget hearings in the House of Representatives on the appropriations for social welfare programs).

In one of the exercises in *Introduction to Logic,** the classic textbook by Irving M. Copi, the author tags the following as an example of a fallacious Appeal to Sympathy:

> "Mr. Scrooge, my husband certainly deserves a raise in pay. I can hardly manage to feed the children on what you have been paying him. And our youngest child, Tim, needs an operation if he is ever to walk without crutches."

Aside from the fact that we don't want to come down on the same side as Ebenezer Scrooge in this discussion (and that would be a fallacious *Ad Hominem* way of thinking), nevertheless, decent wages for Bob Cratchit may well be justified by the expenses he faces in caring for Tiny Tim. In fact, Bob Cratchit's whole work performance may even be affected detrimentally by his poor diet and his worry about Tiny Tim's fragile health. Much of the substance of this argument turns on what wage Cratchit *deserves* and we can debate whether he deserves a living wage according to certain principles of justice. But somehow, Cratchit's work performance doesn't seem to be as *relevant* to this discussion as does the simple plea for compassion.

Is this a non-fallacious Appeal to Sympathy? What do you think?

5. Appeal to Popular Sentiment (*argumentum ad populum*)

The **Appeal to Popular Sentiment** fallacy is committed if and when one makes an appeal for or a reference to popular support, instead of discussing the argument on its merits (no matter how popular or unpopular the conclusion may be).

This is sometimes called the "democratic fallacy" because of the way it seeks popular support for an issue, but that tag is misleading since no one seriously thinks that democratic majority vote guarantees the truth; at most, it invites us to debate what the truth is. If that were not so, we would not find the following story so amusing: A little boy and girl kept a rabbit for a pet. But they wanted to discover the sex of the rabbit so they could name it accordingly. Yet, no adult could tell them how to find out, so they voted on it.

*Irving M. Copi, *Introduction to Logic,* New York: Macmillan Publishing Co., 1986, seventh edition, p. 106.

DEFINITION: **An *Appeal to Popular Sentiment* fallacy occurs whenever one makes an appeal to popular sentiments in a way that is irrelevant to the issue.**

EXAMPLE:

SALESPERSON: Sure it's a well-made car. More people drive the Yellow Citrus than any other car made in America!

COMMENT: **The salesperson neglected to point out that nobody pays more in car repairs than Yellow Citrus owners. People buy the car because it has fins.**

a. Appeal to Popular People. This is a very familiar form of the Appeal to Popular Sentiment fallacy to television viewers where commercials are concerned. The makers of the commercials hope we won't think through the logic of the argument presented because if we do it looks rather ludicrous. The appeal is one to popular entertainers or role models. The suggestion is: if I can act the way popular person X acts, then I'll be popular (or famous or rich) just as person X is. We see that's utterly irrelevant, especially where breakfast habits are concerned.

A recent corn flakes commercial depicts actors reading a book of corn flakes eaters. One says: "Did you know that Bruce Jenner eats corn flakes?" Another adds: "So do the Beach Boys." A third says: "So do the Grateful Dead."

The suggested conclusion, which is absurd when carefully spelled out, is that if you eat corn flakes, then you'll be just like these people too.

6. Appeal to Authority (*argumentum ad verecundiam*)

If one re-directs the discussion from the worthiness of the argument or the truth of its premises to a stand on ceremony, tradition, or authority, then he has committed a fallacy of **Appeal to Authority.**

Note that the Latin, *verecundiam,* is better translated as an appeal to *modesty* than as an appeal to authority. Perhaps this is because when one commits a fallacious Appeal to Authority, he is really making a claim to false modesty: "Neither my views nor yours count for beans when compared with the word of this mighty authority (and it just so happens that he agrees with me!)."

Not all appeals to authority are fallacious appeals, since it may well be the case that an authority in the field under discussion may have something especially worthwhile to say about the truth of the argument under discussion. A *fallacious* Appeal to Authority is made when the authority appealed to is clearly speaking outside the area of his expertise, or when the appeal is made in order to squash the debate instead of sincerely searching for the truth. There is no place in logic for such "one-upmanship."

DEFINITION: An *Appeal to Authority* fallacy occurs whenever one makes an appeal to authority to support her case without considering the *reasons* the authority presents in favor of the matter.*

EXAMPLE:

PHILOSOPHER: Believe it! I personally heard the great philosopher William Quirk say that the job market for philosophers is worse than it's ever been.

COMMENT: **While Quirk is a great philosopher, he may not know much about markets. Such pronouncements are intimidating because Quirk is greatly respected in his field but, without hearing his reasons, we can't say whether or not he has made an accurate appraisal of the current conditions in the philosophy job market.**

a. Appeal to Tradition. Tradition is a very important part of our comprehensive worldview. Traditions can give us stability and a sense of meaning and place; they can also offer us normally reliable "rules of thumb" for conduct. But it is specious to assume that tradition must always be in the right. It is a central thrust of philosophy that cherished beliefs and attitudes must be subjected to close scrutiny, the better to ascertain their truth (when they are indeed true), the better for us to discover what we believe may be in fact erroneous (when we are in error). This is how advancements in knowledge proceed, and it is the cornerstone of wisdom.

Here is a popular hymn. While it is not, strictly speaking, an argument because it is a kind of poetry, nevertheless (with a little imagination) we can discern an implicit argument in it:

Give me that old time religion,
Give me that old time religion,
Give me that old time religion.
It's good enough for me.

b. Appeal to Patriotism. Closely related to *appeal to tradition* is *Appeal to Patriotism.* Certainly, on the face of it, there is nothing whatever wrong with a strong patriotic feeling, given the proper setting. Fourth of July celebrations, the thrill of your nation's flag being raised at the gold medal ceremonies at the

*This definition represents something of a departure from conventional accounts of the fallacy of Appeal to Authority which many logicians believe occurs only when the authority cited is not legitimate, or when the arguer fails to note legitimate disagreements between legitimate authorities on an issue. However, reasons are the stuff of logical arguments. When we consult a legitimate physician about our health or listen to a meteorologist about the weather, we should still be more interested in why she says what she says than in the fact that an authority said it.

Olympics, and touring the monuments and the buildings of state in Washington, D.C., are all part of the wholesome instinct for community that we all share. However, propagandists can skillfully manipulate that wholesome good feeling and turn it into something that sets groups of people against each other and destroys the sense of community in the name of patriotism. On such occasions, appeals to patriotism lose their moral rationale for being and must be regarded with suspicion.

We have all heard the appeal, "My country right or wrong." (Cardinal Spellman, once the Archbishop of New York, made this statement famous.) The statement takes on the context of an argument as sometimes happens when one uses it to justify any action of the government (U.S. involvement in Vietnam, in Cardinal Spellman's case). To see how irrelevant this appeal is to the issue, consider that Cardinal Spellman was actually quoting somebody else, and out of context at that (see the fallacy of "Slanting" in the next module). To understand why appeals to patriotism can be fallacious, consider the full quotation in context:

> "Our country, right or wrong. When right, to be kept right; when wrong, to be put right." —Carl Schurz

c. Appeal to Titles. A clear sub-category of the fallacy of Appeal to Authority, the *Appeal to Title*, is included here to give the reader a more-rounded appreciation of the subtle ways that appeals to authority can work.

There is some evidence that a physician's authority as a well-educated specialist actually assists in the recovery of her patient. Yet a patient sometimes has a justifiable interest in knowing why his doctor has prescribed a certain course of treatment. Doctors are not omniscient, and certain details of a patient's medical history can escape the physician. For example, you might be allergic to penicillin, and so you want to check in advance that the hypodermic that your doctor is about to inject is free of the drug. Or your physician might direct you for minor surgery, and you want a second opinion on the need for the surgery. Perhaps a less drastic recourse is possible. But if you make such a polite challenge to your doctor, and she replies, "I'm the doctor. I know what's good for you," it's probably time to get another doctor.

In our meritocratic society, titles tend to proliferate. Among the alphabet soup of titles, there are Ph.D.s, M.D.s, R.N.s, D.D.S.s and D.D.s. In fact the list is enormous. There are "sanitation engineers" and "household engineers" (see "Appeal to Jargon"). The size of the list should give us pause. More often than not, such a title is a hard-earned recognition of perseverance and work that has earned the bearer some recognition of specialized knowledge. But no title is a mark of infallibility, and in a democratic society, no special homage should be due when an argument is on the line and you have a worthy point to make. Stick to your guns, but listen to the authority's side of the argument; never, however, tolerate a dismissal out of hand.

d. Jargon. If experts have titles, they also have a specialized jargon. Sometimes, in a given field of study, a technical term says what has to be said in the clearest possible way. That is the ideal. Sometimes, however, technical terms are actually used to cloud the issue instead of illuminate it. When that happens in the context of persuasion (such as in the purchase of a product), the fallacy of an *Appeal to Jargon* has been committed.

When does technical language become jargon? Here are a few rules of thumb: (1) when it says in "ten" sentences what can better be said in one; (2) when it obstructs communication; and (3) when it unnecessarily employs metaphorical language or "sesquipedalian" (literally, "sixty-foot long") words instead of familiar terms.

The following passage from *The Social System* by the sociologist Talcott Parsons amply exemplifies these sins (at least according to fellow sociologist, C. Wright Mills, whose translation follows):

An element of a shared symbolic system which serves as a criterion or standard for selection among the alternatives of orientation which are instrinsically open in a situation may be called a value. . . . But from this motivational orientation aspect of the totality of action it is, in view of the role of symbolic systems, necessary to distinguish a "value-orientation" aspect. This aspect concerns, not the meaning of the expected state of affairs to the actor in terms of his gratification-deprivation balance but the content of the selective standards themselves. The concept of value-orientations in this sense is thus the logical device for formulating one central aspect of the articulation of cultural traditions into the action system.

It follows from the derivation of normative orientation and the role of values in action as stated above, that all values involve what may be called a social reference. . . . It is inherent in an action system that action is, to use one phrase, "normatively oriented." This follows, as was shown, from the concept of expectations and its place in action theory, especially in the "active" phase in which the actor pursues goals. Expectations, then, in combination with the "double contingency" of the process of interaction as it has been called, create a crucially imperative problem of order. Two aspects of this problem of order may in turn be distinguished, order in the symbolic systems which make communication possible, and order in the mutuality of motivational orientation to the normative aspect of expectations, the "Hobbesian" problem of order.

The problem of order, and thus of the nature of the integration of stable systems of social interaction, that is, of social structure, thus focuses on the integration of the motivation of actors with the normative cultural standards which integrate the action system, in our context interpersonally. These standards are, in the terms used in the preceding chapter, patterns of value-orientation, and as such are a particularly crucial part of the cultural tradition of the social system.

Translation: "People often share standards and expect one another to stick to them. In so far as they do, their society may be orderly. (end of translation)" —C. Wright Mills

A fallacy of an *appeal to jargon* occurs when a listener is intimidated into accepting what someone says on the basis of fancy-sounding terminology, thus imputing to the speaker an authority that he may or may not really have. The only way you can tell the difference is by asking what the terms mean.

The physicist, Robert Oppenheimer, once urged his colleagues to give clear explanations when he said: "If you can't explain it to a five-year-old, you don't understand it." Of course, such advice becomes more challenging when the subject under discussion is quantum mechanics or Kant's *Critique of Pure Reason*, but keep in mind that Oppenheimer made the remark with respect to nuclear physics. How much you can explain to a five-year-old depends on the aptitude of the child, and you may well not get finished until the child has grown into his thirties, but we can still sympathize with Oppenheimer's apparent impatience with jargon.

There is an amusing aphorism which directs us to "eschew obfuscation." If you can understand what that means, it's good advice. Sometimes we like to use big words or technical terms to "show off" what we know at cocktail parties, but that's poor logical and rhetorical form, and you're also likely to be very unpopular at the party. Forget "sesquipedalian" words; say what needs to be said using the clearest and shortest possible words that turn the trick. Remember that logic emphasizes clarity and preciseness.

e. Appeal to Progress. If *appeals to tradition* can be fallacious, then so can their opposite: *Appeals to Progress*. Is progress always and in itself a justification for a course of action? We may disagree about what constitutes progress (as do many environmentalists when they oppose construction projects). But *appeals to progress* can represent wholesale rejection of traditions for no very good reason. As such, we may also call this "the liberal fallacy" in the sense that one who appeals to it shows little sensitivity to tradition whatsoever.

For example, consider that Art Deco was once regarded as a common and base architectural style. Many of the Art Deco buildings in the United States were torn down to make way for re-development. In time, however, Art Deco came to be re-appraised in the art world as a noble example of American artistic tradition, and the loss of many fine examples of the style is now lamented.

This fallacy is committed when one argues for tearing down a building solely on the grounds that "you've got to make way for progress." When city councils nowadays give no second thought to paving over sites of one of the first examples of a McDonald's fast food restaurant or a Putt-Putt miniature golf course, they are displaying a sort of insensitivity to their heritage that results when appeals to progress are made without proper regard.

7. Diversion

The fallacy of **Diversion** (also called "Evasion") occurs whenever one allows an argument to be shifted from the issue in an attempt by the opponent to evade or confuse the issue.

DEFINITION: The fallacy of *Diversion* occurs whenever one evades the obvious conclusions of an argument by changing the subject.

EXAMPLE:

FELIX: Oscar, you really must start picking up after yourself. You dropped your racing sheet in the oven, you spilled beer in the fish tank, you left your dirty socks in the dishwasher, and, to top it all off, you left your bedroom slippers on the quiche in the refrigerator.

OSCAR: Hey, Felix! Look at this: the Metropolitan Opera Company is performing "Aida" on public television tonight.

FELIX: Let me see that television guide!

COMMENT: **Felix has made a reasonable argument to which Oscar can make no reasonable reply, so Oscar diverts the issue hoping Felix will forget the point.**

Diversion is an odd sort of fallacy in that the "fallacy" is committed by whoever lets the argument get side-tracked in the first place, as well as by whoever does the side-tracking. In other words, you are responsible for keeping your argument on course. Your opponent may be the one who shifts ground, whether by inadvertence or by design, but you are the one to press home your conclusions. Thus, the fault of allowing an argument to be diverted rests equally with you, the arguer.

Diversion is a special type of fallacy of relevance—one of the most general sorts of the fallacy of *Irrelevant Conclusion*, and so it will be the last of the fallacies of relevance to be considered here. It is special for its oddity; it is a fallacy of omission as well as one of commission. Both you and your opponent share the blame.

If one answers a question with a question in order to divert the argument from the issue at hand, one is attempting diversion. If you, the arguer, permit the stratagem to succeed, then you are just as responsible for clouding the waters.

Here is an example of an attempt at Diversion. The following is taken from Volume I of *The History of the Inquisition*. This is an excerpt from a "handbook" or "manual" for training Church inquisitors written by a Grand Inquisitor of the Middle Ages, Bernard Gui, to show aspiring inquisitors "the quibbles and tergiversations for which they must be prepared when dealing with those who shrank from boldly denying their faith." We see that the accused heretic makes several efforts to divert the Inquisitor's questions from their course (the accused presumably will not tell outright lies, but rather tries to evade and confuse), but the Inquisitor, properly trained in logic, steadfastly refuses to allow his gaze to be diverted from its single-minded (and chilling) purpose:

INQUISITOR: I know your tricks. What the members of your sect believe you hold to be that which a Christian should believe. But we waste time in this fencing. Say simply, Do you believe in one God the Father, and the Son, and the Holy Ghost?

ACCUSED: I believe.

INQUISITOR: Do you believe in Christ, born of the Virgin, suffered, risen, and ascended to heaven?

ACCUSED: (Briskly) I believe.

INQUISITOR: Do you believe the bread and wine in the mass performed by the priests to be changed into the body and blood of Christ by divine virtue?

ACCUSED: Ought I not to believe this?

INQUISITOR: I don't ask if you ought to believe, but if you do believe.

ACCUSED: I believe whatever you and other good doctors order me to believe.

INQUISITOR: Those good doctors are the masters of your sect; if I accord with them you believe with me; if not, not.

ACCUSED: I willingly believe with you if you teach what is good to me.

INQUISITOR: You consider it good to you if I teach what your other masters teach. Say, then, do you believe the body of our Lord Jesus Christ to be in the altar?

ACCUSED: (Promptly) I believe.

INQUISITOR: You know that a body is there, and that all bodies are of our Lord. I ask whether the body there is of the Lord who was born of the Virgin, hung on the cross, arose from the dead, ascended, etc.?

ACCUSED: And you, sir, do you not believe it?

INQUISITOR: I believe it wholly.

ACCUSED: I believe likewise.

INQUISITOR: You believe that I believe it, which is not what I ask, but whether you believe it.

ACCUSED: If you wish to interpret all that I say otherwise than simply and plainly, then I don't know what to day. I am a simple and ignorant man. Pray don't catch me in my words.

INQUISITOR: If you are simple, answer simply, without evasions.

ACCUSED: Willingly.

INQUISITOR: Will you swear that you have never learned anything contrary to the faith which we hold to be true?

ACCUSED: (Growing pale) If I ought to swear, I will willingly swear.

INQUISITOR: I don't ask whether you ought, but whether you will swear.

ACCUSED: If you order me to swear, I will swear.

INQUISITOR: I don't force you to swear, because as you believe oaths to be unlawful, you will transfer the sin to me who forced you; but if you will swear, I will hear it.

ACCUSED: Why should I swear if you do not order me to?

INQUISITOR: So that you may remove the suspicion of being a heretic.

ACCUSED: Sir, I do not know how, unless you teach me.

INQUISITOR: If I had to swear, I would raise my hand and spread my fingers and say, "So help me God, I have never learned heresy or believed what is contrary to the true faith."

[Bernard Gui continues:]

"Then trembling as if he cannot repeat the form, he will stumble along as though speaking for himself or for another, so that there is not an absolute form of oath and yet he may be thought to have sworn. If the words are there, they are so turned around that he does not swear and yet appears to have sworn. Or he converts the oath into a form of prayer, as 'God help me that I am not a heretic or the like'; and when asked whether he had sworn, he will say: 'Did you not hear me swear?' And when further hard pressed he will appeal, saying 'Sir, if I have done amiss in aught, I will willingly bear the penance, only help me to avoid the infamy of which I am accused through malice and without fault of mine'. But a vigorous inquisitor must not allow himself to be worked on in this way, but proceed firmly till he makes these people confess their error, or at least publicly abjure heresy, so that if they are subsequently found to have sworn falsely, he can, without further hearing, abandon them to the secular arm."

EXERCISES

Exercise 1

DIRECTIONS: *Consider Bernard Gui's argument in the passage from* The History of the Inquisition. *Do you detect any fallacies other than Diversion on the part of either the 'Inquisitor' or the 'Accused'? (For example, do you suspect the presence of an inherent Appeal to Force behind the Inquisitor's questioning?) In your opinion, is the Inquisitor really interested in unearthing the truth in his cross-examination of the accused, or has he already presumed the accused to be guilty? How might he ask questions that can elicit a reasonable and honest reply from the accused heretic?*

Exercise 2: Elementary Exercises

DIRECTIONS (Read Carefully): *Most of the arguments that follow contain fallacies of the sort we just studied (some commit no fallacy at all). These are textbook examples; the fallacies are meant to be readily identifiable. (1) Find the fallacies in the following arguments (or parts of arguments) and enter the correct name in the space provided. (2) Give the most specific name you can for the fallacy (e.g.,* tu quoque *instead of* Ad Hominem *when appropriate). (3) If the argument commits no fallacy, write "No Fallacy" in the space provided.* Beware of the False Charge of Fallacy!

A List of Fallacies Studied in This Section

Appeal to Force
  appeal to fear
Ad Hominem
  tu quoque
  bad seed
  faulty motives
  poisoning the well
  guilt by association
Appeal to Ignorance
  innuendo
Appeal to Sympathy
Appeal to Popular Sentiment
  popular people
Appeal to Authority
  tradition
  patriotism
  titles
  jargon
  progress
Diversion

Name That Fallacy!

______ 1. "I know I didn't take any tests for this course, Professor Kingsley. I know that I did none of the homework, that I missed almost all the classes, but if I don't get a 'B' in this class, I'll flunk out of school. So Professor, you really ought to give me a 'B' for the course."

______ 2. "The Bible says it, I believe it, and that settles it!" —a bumper sticker observed in North Carolina

______ 3. Voodoo zombie powder has been found to have significant medical applications since a pharmaceutical company changed the powder's name to "tetradotoxin."

______ 4. "Buy 'BaldRemove.' Our company was taken to court for mail fraud, but, in the opinion of the judge, since no consumers had filed complaints, we won the case. 'BaldRemove' really works!"

______ 5. MANAGER: I asked you to see me to discuss your attitude toward your job. You're always late for work, you spend company time at the water cooler, and you make personal phone calls.

EMPLOYEE: If I may say so, sir, that's a nice tie. Yellow is a "power color," you know.

MANAGER: No, I didn't know that. Thank you.

______ 6. "Oh, yeah—telepathy is real enough. That's what Professor Peachfuzz (Ph.D., D.D.S., M.L.S.) said."

______ 7. "Yes, you should believe it. If you don't, I'll rearrange your face."

______ 8. "I'm not going to vote for him for mayor. He's as ugly as Jabba the Hut!"

______ 9. " 'Political refugees,' my foot! The only reason the Vietnamese came to this country was to take our jobs away."

______ 10. "The Packers ought to be in the thick of things this year. They've improved their defensive line, and that's always been their weakness."

______ 11. "I'm not saying I know this for sure, but I heard that the dean got his job because he knew where all the bodies were buried."

______ 12. TEACHER: Which is the correct answer to the second problem in this exercise? How many say it's an Appeal to Authority fallacy? All of you? Then it must be!

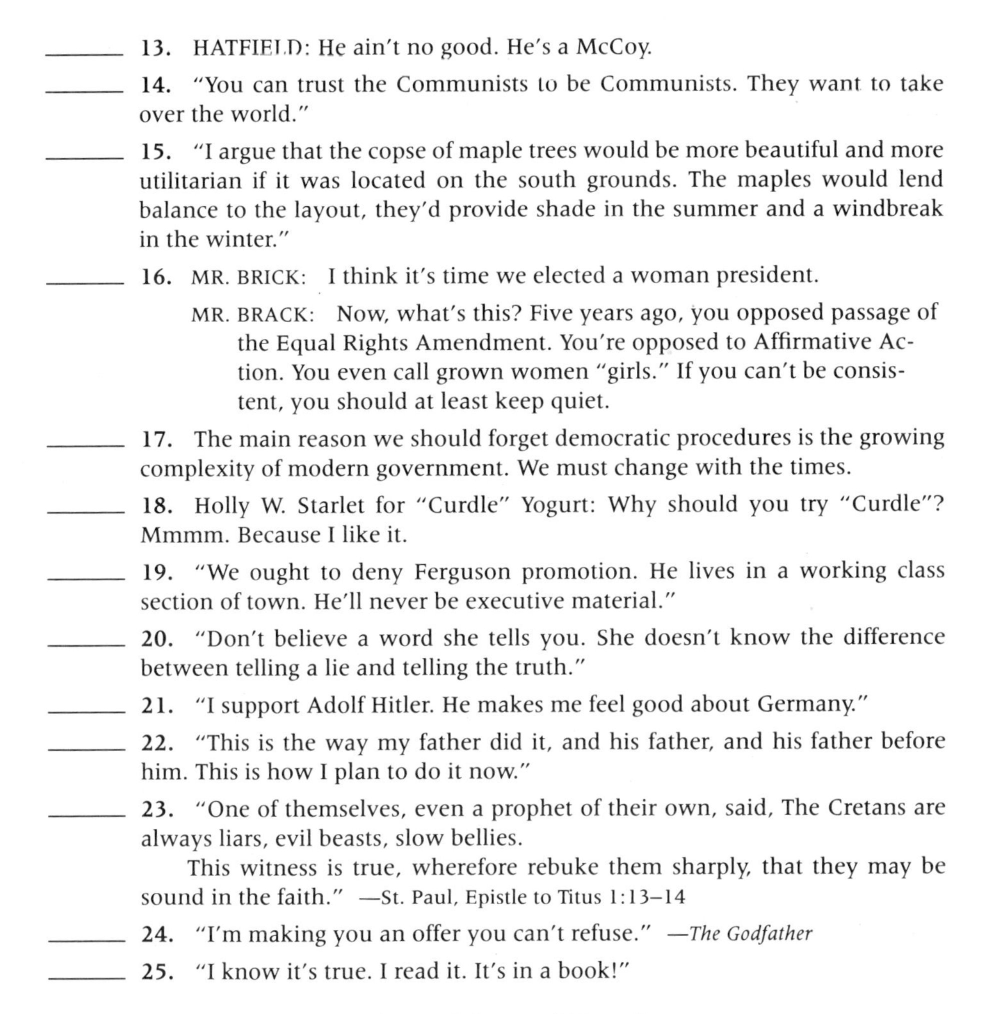

_______ **13.** HATFIELD: He ain't no good. He's a McCoy.

_______ **14.** "You can trust the Communists to be Communists. They want to take over the world."

_______ **15.** "I argue that the copse of maple trees would be more beautiful and more utilitarian if it was located on the south grounds. The maples would lend balance to the layout, they'd provide shade in the summer and a windbreak in the winter."

_______ **16.** MR. BRICK: I think it's time we elected a woman president.

MR. BRACK: Now, what's this? Five years ago, you opposed passage of the Equal Rights Amendment. You're opposed to Affirmative Action. You even call grown women "girls." If you can't be consistent, you should at least keep quiet.

_______ **17.** The main reason we should forget democratic procedures is the growing complexity of modern government. We must change with the times.

_______ **18.** Holly W. Starlet for "Curdle" Yogurt: Why should you try "Curdle"? Mmmm. Because I like it.

_______ **19.** "We ought to deny Ferguson promotion. He lives in a working class section of town. He'll never be executive material."

_______ **20.** "Don't believe a word she tells you. She doesn't know the difference between telling a lie and telling the truth."

_______ **21.** "I support Adolf Hitler. He makes me feel good about Germany."

_______ **22.** "This is the way my father did it, and his father, and his father before him. This is how I plan to do it now."

_______ **23.** "One of themselves, even a prophet of their own, said, The Cretans are always liars, evil beasts, slow bellies.

This witness is true, wherefore rebuke them sharply, that they may be sound in the faith." —St. Paul, Epistle to Titus 1:13–14

_______ **24.** "I'm making you an offer you can't refuse." —*The Godfather*

_______ **25.** "I know it's true. I read it. It's in a book!"

Exercise 3: Advanced Exercises

DIRECTIONS (Read Carefully): *Most of the arguments (or passages that contain arguments)* that follow contain fallacies of the sort we just studied (some commit no fallacy at all). These are "real-life" examples; they're taken from editorials, advertisements, and letters. Because they are taken from real life, they may prove to be more difficult than the "textbook" examples you worked in Exercise 2. (1) Find the fallacies in the following arguments (or parts of arguments) and enter the correct name in the space provided. (2) Give the most specific name you can for the fallacy (e.g.,* tu quoque *instead of* Ad Hominem *when appropriate). (3) If the argument commits no fallacy, write "No Fallacy" in the space provided.* Beware of the False Charge of Fallacy!

*Note: Some of these passages are, strictly speaking, reports, but they contain arguments attributed to others or they lead into arguments. For the purposes of this exercise, examine the arguments contained in the reports for fallacies.

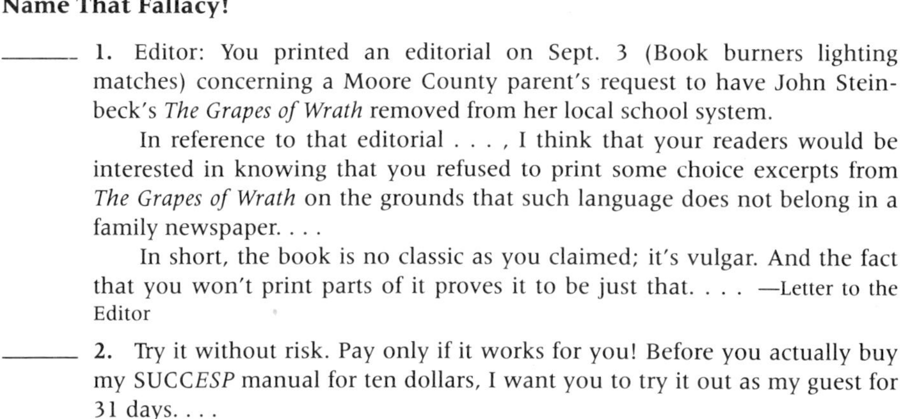

Name That Fallacy!

______ 1. Editor: You printed an editorial on Sept. 3 (Book burners lighting matches) concerning a Moore County parent's request to have John Steinbeck's *The Grapes of Wrath* removed from her local school system.

In reference to that editorial . . . , I think that your readers would be interested in knowing that you refused to print some choice excerpts from *The Grapes of Wrath* on the grounds that such language does not belong in a family newspaper. . . .

In short, the book is no classic as you claimed; it's vulgar. And the fact that you won't print parts of it proves it to be just that. . . . —Letter to the Editor

______ 2. Try it without risk. Pay only if it works for you! Before you actually buy my SUCC*ESP* manual for ten dollars, I want you to try it out as my guest for 31 days. . . .

Try this awesome technique at home, at work—*anywhere*! Convince yourself beyond the shadow of a doubt that SUCC*ESP* really works. That you can conquer anyone! —Advertisement

______ 3. Editor: I buy fish from the Hieronymus Brothers since they use only the *Morning Star* to wrap their product. All has gone well (good reading, great-tasting fish) until recently when the *Morning Star* has started to go downhill; that is, the deletion of *Andy Capp* and the addition of *Zippy*.

The paper has gotten so bad that if *Andy Capp* is not restored I will have to procure my fish and paper elsewhere. —Letter to the Editor

______ 4. This product was developed by Mrs. Ernest Borgnine and many of the top stars in Hollywood absolutely swear by it. In fact, Ruta Lee, Brenda Vaccaro, Connie Stevens, Maureen Dean, Jed Allen, Buck Trent and Debbie Reynolds are just a few of the important celebrities who use and enthusiastically endorse this product. —Advertisement for TOVA Corp.

______ 5. Representative Charlie Rose hints that he has some horrible information about his opponent, Tommy Harrelson. "It would curl the paint off a pickup fender," he says. "Of course," says the Congressman piously, "I am not going to engage in a mudslinging campaign." —Editorial, *Wilmington Morning Star*

______ 6. If Rose has persuasive evidence of wrongdoing, he should share it with the voters; Harrelson and the Republicans certainly asked for it and the voters deserve it.

If Rose does *not* have persuasive evidence of wrongdoing, he should keep his mouth shut. Otherwise, he runs the risk of getting [former President Richard] Nixon's endorsement. —from the same editorial

______ 7. In a pastoral letter issued Thursday, Bishop Louis E. Gelineau said that during October, priests would not have to ask him before forgiving women, their relatives, doctors and others who want to repent following "the sin of abortion." He said the unusual move was an attempt to highlight the church's "Respect Life" month.

But Mary Ann Sorrentino, the excommunicated director of Rhode Island Planned Parenthood, said Friday that Gelineau's gesture was a transparent political move to influence a vote on a proposed anti-abortion amendment to the state constitution. —*Associated Press*, "Absolution extended to include abortion"

_______ **8.** Editor: This is in support of the young men of Delta Tau Delta fraternity of UNCW who are undeserving of the abuse they have received on publishing their calendar, *The Girls of UNCW.*

One is misinformed to refer to a fraternity as a frat. The abbreviation "frat" connotes *Animal House* images to the public. The fraternities of UNCW and the whole Greek system are enthusiastically supported by the administration.

This does not mean that we do not from time to time make mistakes, for we are, after all, still young and are here to learn. When we make mistakes, we have to bear responsibility for our actions. . . . —Letter to the Editor

_______ **9.** The religious controversy swirling around Tammy Bakker isn't the first she's ever been in—during Tammy's childhood her mother was kicked out of their church in a messy divorce scandal, then Tammy herself caused an uproar when she fell in love with the church pastor's son! —"Tammy Bakker—a Fanatical Holy Roller Who Got Booted Out of Bible College," *National Enquirer,* July 21, 1987, p. 1

_______ **10.** The proponent of a new "facial toning" technique being used in Whiteville and Wilmington beauty salons claims he'll make you look years younger by electrically tightening facial muscles.

Federal authorities say the process is a fraud that uses a medical device for unapproved purposes. . . .

"They haven't approved it, but they haven't disapproved it either," McWhorter [who promotes the machine] said. "They haven't said I can't use it." He added that he's been told by the distributors of the machine that it meets all safety requirements. —"Feds say electric 'facial toning' is a fraud," *Wilmington Morning Star,* October 2, 1986

_______ **11.** All great cultures of the past have perished only because the originally creative race died out from blood poisoning. . . .

The man who misjudges and disregards the racial laws actually forfeits the happiness that seems destined to be his. He thwarts the triumphal march of the best race and hence also the precondition for all human progress, and remains, in consequence, burdened with all the sensibility of man, in the animal realm of helpless misery. —Adolf Hitler, *Mein Kampf*

_______ **12.** Bloom and a few kindred spirits are resisting the triumph of relativism and intellectual egalitarianism. To the modern mind, those are related moral imperatives. Relativism is considered a requirement for a free society because the only modern sin is intolerance, and intolerance results from denying that all 'values' are of equal dignity. —Letter to the Editor

_______ **13.** Scorning chronological order, I began with Kierkegaard and Sartre, then moved quickly to Spinoza, Hume, Kafka and Camus. I was not bored, as I had feared I might be; rather, I found myself fascinated by the alacrity with which these great minds unflinchingly attacked morality, art, ethics, life and death. I remember my reaction to a typically luminous observation of Kierkegaard's: "Such a relation which relates to its own self (that is to say, a self) must either have constituted itself or have been constituted by another." The concept brought tears to my eyes. —Woody Allen, "My Philosophy," *Getting Even,* Random House, 1971, p. 27

MODULE 35

INDUCTIVE FALLACIES

PART 2: IS THE ARGUMENT SOUND?

Inductive fallacies are those which usually occur in inductive arguments. Inductive fallacies often challenge the *soundness* of a given argument; they cast doubt on the truth of one or more of the reasons given in support of the conclusion.

Other times, the presence of an inductive fallacy may indicate a *shortage of evidence* needed to warrant the likelihood of the conclusion, or (as is sometimes the case) lots of evidence is offered in the argument, but it is not the *right kind* of evidence. Acquaintance with these types of fallacies may help us to know how much evidence is *enough* evidence, and whether it is the right kind of evidence, to make the argument work.

Before you study the inductive fallacies, you might do well to recall the distinction between *inductive* and *deductive* arguments. An inductive argument is an argument which offers evidence in support of its conclusion and which claims a degree of likelihood or probability for its conclusion; a deductive argument is one which claims that its conclusion *must* follow from its premises. Review Module 8 in Unit One, "Deductive and Inductive Arguments."

An **inductive** argument is **sound**, if and only if, the evidence is sufficient or representative enough to warrant the likelihood of the conclusion, and the premises are true.

8. Slippery Slope

A fallacy of **Slippery Slope** is committed when one affirms an unjustifiable "chain reaction" of causes or reasons which, if it is allowed to continue, leads irrevocably to disaster. This has been colorfully called by the Arabs, "the fallacy of the Camel's Nose Under the Tent." If the camel is allowed to poke its nose ever so slightly under the tent, then, before you know it, you'll find a camel in your tent.

DEFINITION: **The Fallacy of *Slippery Slope* occurs whenever one unjustifiably concludes an "avalanche" will be the result of an unconnected or loosely-connected series of events.**

EXAMPLE:

HIGH SCHOOL PRINCIPAL: Dancing is the work of the devil! Let young people listen and dance to rock music and I tell you that sooner or later they will turn against their parents and authority, they'll take to fornication and pornography, and they'll grow up to be prostitutes or procurers. No good can come of it, so I have cancelled all future high school dances.

COMMENT: **The principal has confused the issue. Rock music and dancing can be a healthy and normal outlet for the energies of young people (and maybe older people too). Perhaps the principal has observed that dancing to rock music is an outlet for sexual energy and has falsely concluded that it must lead down the road to *illicit* sexuality (among other things).**

A Slippery Slope argument is a "before-you-know-it" type of argument. A series of innocent events are linked together in such a way that it seems plausible that if this string continues, it can only lead to disaster. For example, someone might argue that if young people are allowed to drink beer, that it will surely lead them down the road to hard liquor, then marijuana and cocaine, and "before you know it" they'll be addicted to heroin!

To see just how silly this type of argument is, consider just what is faulty in our "heroin addiction" example. The assumption runs through the argument that all of these things—beer, liquor, marijuana, cocaine, etc.—have significant properties in common. It isn't just that they are all addictive substances (even if they are merely psychologically addictive), the supposition is that they are both hazardous and habit-forming. We might parody the argument by pushing back the chain of events until even the most innocuous drugs or beverages seem to propel us down that shady road to perdition. Why start the sequence with beer and liquor? Let's try caffeinated soft drinks or aspirin: "Don't let your son or daughter start on cola or aspirin because that can only lead to the consumption of beer and liquor. It will only be a matter of time before they then start on marijuana and cocaine, and then, before you know it, they'll be addicted to heroin!"

Does this seem silly? Of course it does. But the salient point to note is that what is hazardous and habit-forming about these early entries in the sequence can occur only if they're *abused* by the taker. But the difference between aspirin and coke, beer and other alcoholic beverages on the one hand, and marijuana, cocaine, and heroin on the other is not a difference of *degree* so much as it is a difference in *kind.* Wine, beer, and liquor should be consumed in moderation. Cola beverages are no different in that they must be taken with care. But the act of dropping acid, smoking dope, snorting coke, or shooting up heroin constitutes an abuse in and of itself. One may not drink to get drunk, but it makes little sense to smoke grass if one doesn't intend to get high.

We shall see that Slippery Slope arguments bear some features in common with arguments from analogy (see "False Analogy"). The schematic form of the argument is something like this:

> **Events or items A, B, C, and D, have some property x in common.**
>
> **So, if A, B, and C are allowed, then they lead irrevocably to disaster D.**

The way to challenge the "slippery sloper" is to question how true the assumption is that A, B, C, and D have property x in common. Perhaps it is a weak supposition; perhaps they do have x in common, but some to less degree than others; perhaps (as in our addiction example) there is a difference in kind between the items cited so that the differences between them count more than the properties they have in common. These are some strategies on how to proceed against arguments of this very spurious sort.

Consider the following argument for capital punishment:

> **"When people begin to believe that organized society is unwilling or unable to impose upon criminal offenders the punishment they 'deserve,' then there are sown the seeds of anarchy—of self-help, vigilante justice, and lynch law."**

You can see that this counts as a Slippery Slope argument, but can you say just why it is specious?

It is important to note that not all arguments of the slippery slope variety are necessarily fallacious. It is not a very rigorous type of argumentation, but sometimes tracing a sequence along a chain of events can help us hypothesize an outcome. But the most we can get is a good working hypothesis, and that must be tested according to more rigorous rules of evidence and argumentation.

Next follows a prominent example of a type of slippery slope reasoning that has become so pervasive in foreign policy that it may deserve a heading of its own.

a. The Domino Theory. The *Domino Theory* is a name given to a dogma of recent foreign policy. It is a paradigm of slippery slope argumentation (i.e., a *species* of Slippery Slope,) and we may define the *domino theory* as any Slippery Slope argument specifically applied to foreign policy. The *domino theory* originated in the United States in the 1950s as a model of the alleged plan of the Soviet block to achieve world domination. Its central assumption is that the Communist nations are "monolithic" in their intent to cause subversion, unrest, and revolution in the non-Marxist nations of the world.

The *domino theory* was most tellingly applied to the war in Southeast Asia. It was argued that if Vietnam fell to the Communists, then all of Southeast Asia would soon follow. Now, this remains to be seen. Since the U.S. withdrawal from Vietnam, Cambodia and Laos did install Marxist governments, but their integrity was considerably weakened by the war since their strategic positions offered supply lines for North Vietnamese troops. Even so, these nations proved to be so unlike each other, that conflicts have broken out between them, and Vietnam recently invaded Kampuchea (as Cambodia was known), deposing the totalitarian Marxist regime of Pol Pot. Given the tensions between Vietnam, Cambodia, China, and Laos, there is hardly substance for the monolithic conception of Marxist activity which is the underlying assumption of the *domino theory.*

Moreover, Thailand and Indonesia, two other nations in Southeast Asia, show no sign of "going Communist" and do not appear to be in any danger of invasion.

In spite of the dubious premises of the *domino theory* as a tenet of American foreign policy, it nevertheless seems to have reared its head again in our relations with Central America. Given the lesson of the Vietnam War, one tends to wonder if *domino theory* is something of a "self-fulfilling prophecy." In using counterinsurgency techniques to stem the tide of communism, is the United States in fact alienating the nations neighboring the site of the conflict, or is there really some substance to the scenario of "falling dominoes"?

9. Attacking a Straw Man

The fallacy of **Attacking a Straw Man** occurs when and if a debater *fabricates* the position of his opponent, and then attacks *it* instead of the opponent's real position on the issue. When in training for combat, the ancients and the medievals used their swords against dummies made out of straw in order to better get the "feel" of their weapons, and to improve their coordination. The practice remains with us today in military boot camp, as well as in football training sessions where weighted bags are used for tackling. But, it is one thing to use one's weapons and skills against a straw man; it is yet another to go into combat and face a real opponent.

A news story carried an account of congressional opposition to President Reagan's program to aid the Contras in Nicaragua. Both houses of Congress held a majority of Democrats. We may imagine that many congressmen had sincere and thoughtful reservations about such military aid, but a spokesman for the White House, Marlin Fitzwater, attacked their stand by arguing that the Democrats "think surrender is the best way to achieve peace. We disagree." This

DEFINITION: **The fallacy of *Attacking a Straw Man* occurs whenever a weak, oversimplified, or implausible characterization of an opponent's point of view is attacked in place of the real (and probably more formidable) argument.**

EXAMPLE:

My own view on religion is that of Lucretius. I regard it as a disease born of fear and as a source of untold misery to the human race. I cannot, however, deny that it has made some contributions to civilization. It helped in the early days to fix the calendar, and it caused Egyptian priests to chronicle eclipses with such care that in time they became able to predict them. These two services I am prepared to acknowledge, but I do not know of any others." —Bertrand Russell, "Has Religion Made Useful Contributions to Civilization?"

COMMENT: **Bertrand Russell was certainly one of the great philosophers of all time but, by characterizing religion as having made only *two* contributions to civilization (and exceedingly weak ones at that!), Russell has oversimplified the issue. At minimum, we can reply that religion has also made great contributions to art and philosophy.**

comment badly oversimplifies the position taken by congressional opponents, such as Senator Christopher J. Dodd, who opposed the proposed aid because a Central American peace agreement was under negotiation. The White House response ignores the fact that many Republican members of Congress agreed with Dodd. Marlin Fitzwater was guilty of "attacking a straw man."

Here's another example. When Senator Edward Kennedy's son Patrick developed cancer, the senator became acutely aware of the catastrophe that serious illness could mean to families without means to pay for medical treatment. He visited the hospital ward where many other children had illnesses as serious as his son's, but their parents could not afford the kind of care that Kennedy, a millionaire, could provide for Patrick. As a result, he proposed legislation on the Hill for "National Catastrophic Health Insurance" to provide the means to care for those who cannot afford to pay for adequate treatment. The bill was met with fierce opposition led, in part, by Senator Jesse Helms, who accused Kennedy of advocating "socialism."* The bill was defeated.

10. False Dilemma

Informally speaking, a dilemma occurs when one is faced with a very limited number of alternatives, all of them unsavory. The fallacy of **False Dilemma** occurs whenever one attempts to bring a premature end to a debate by declaring

*Notice that this is also an appeal to fear.

DEFINITION: **One commits the fallacy of *False Dilemma* when posing a restrictive set of undesirable alternatives when other legitimate alternatives may be possible.**

EXAMPLE:

CITIZEN: I don't think I want to vote Republican *or* Democratic for president this year. I don't like either of the candidates.

PARTISAN: Then, obviously, you don't want to vote.

COMMENT: **This dilemma is false since the citizen obviously wants to vote even though she doesn't want to vote for either of the candidates the two major political parties have put up for election this year. To get her to vote the way he wants her to, the partisan argues in a fashion that precludes consideration of a third-party candidate.**

a dilemma when none exists. Other alternatives may be possible, or other courses of action can be pursued.

One common form of False Dilemma is exemplified in the slogan (attributed to Leon Trotsky), "If you're not with us, you're against us." This slogan is often invoked by political pressure groups to compel your support. Against this sort of self-serving argument, consider the cool reasonableness of this excerpt from John F. Kennedy's Inaugural Address:

> "To those new states whom we welcome to the ranks of the free, we pledge our word that one form of colonial control shall not have passed away merely to be replaced by a far more iron tyranny. We shall not always expect to find them supporting our view. But we shall always hope to find them supporting their own freedom . . ."

Another example can be seen in the debate over the creation of a national minimum drinking age. One group argued that the drinking age should *not* be raised to 21 because young people in America are expected to discharge some weighty responsibilities, such as voting and conscription, at the younger age of 18. So, with the responsibilities should come the corresponding privileges of adulthood.

Others argued plausibly that the variability of minimum drinking ages from state to state promoted irresponsible drinking and driving, since many young adults would take long drives to neighboring states that had lower minimum age requirements to purchase alcoholic beverages.

In the midst of the controversy, one newspaper editorialized that while it had been for a time convinced by the first argument, the death toll on the highways that has resulted from young people crossing state lines to buy alcohol had convinced them of the need for the national minimum age of 21. Has the newspaper posed for itself a dilemma that could be resolved by taking into consideration yet another alternative that would satisfy both arguments? What could the further alternative be that would spike the apparent dilemma?

a. The Bureaucratic Fallacy. The *Bureaucratic* fallacy is a sub-category of False Dilemma. You're all familiar with this logical boner. Suppose you want to substitute an advanced logic course for your introduction to mathematics requirement. You've already read the very textbook your math teacher is supplying for the introductory course. You're concerned that you'll find the course boring, a waste of tuition money, and you haven't time to fit in all the courses you really need into your schedule. You might suppose that you have adequate cause to

False Dilemma/Straw Man

Here's an example of a passage which commits (at least) two fallacies: a False Dilemma and the fallacy of Attacking a Straw Man. The following is taken from *The Suicide of the West* by conservative writer James Burnham. Here is an excerpt from his characterization of the differences between American liberalism and conservatism:

Elements comprising the doctrinal dimension of the liberal syndrome:

THE LIBERAL: Human nature is changing and plastic, with an infinite potential for progressive development, and no innate obstacles to the realization of the good society of peace, justice, freedom and well-being.

One possible set of contrasting non-liberal elements:

THE CONSERVATIVE: Human nature exhibits constant as well as changing attributes. It is at least partially defective or corrupt intrinsically, and thus limited in its potential for progressive development; in particular, incapable of realizing the good society of peace, justice, freedom and well-being.

Burnham goes on to list eighteen other ideological differences between the two points of view. Why is this fallacious? The reason is that he is slanting the outcome to be favorable to the "non-liberal" position. Characteristically, he does the following:

The "liberal" position is stated as a bald exaggeration; for example: The liberal believes 'extreme generalization ABC' is true. Then he contrasts the non-liberal who believes that it's not the case that 'extreme generalization ABC' is true, which is surely the more reasonable position. Thus, Burnham commits a Straw Man fallacy by oversimplifying his opponent's position, and a False Dilemma by making it seem no alternative is left to the liberal but to believe 'ABC.'

substitute the course in "Advanced Logic and Recursive Function Theory." However, when you present your case to the university administrator, she says: "If I make an exception for you, I'll have to make an exception for everybody."

Apparently, when the administrator argues this way, she is illogically posing this dilemma: "Either I make an exception for everyone or for no one." Once the dilemma is spelled out, we can clearly see its absurdity: it is odd to describe something done for everyone as an *exception*.

When faced with such obstacles, logically speaking, you should correct the administrator, and reply: "No, you should make exceptions for everybody in like circumstances." If the logic does you no good here, take the math course, get your 'A', and somehow try to fit in the advanced logic course later.

This sub-type of fallacy admits of infinite variations, given your profession or circumstances. The fallacy may be committed by union representatives, bank vice presidents, locker room attendants, corporate managers—in short, wherever bureaucracy exists and more concern is paid to the letter of the law than the reason for it. Such fallacies occur whenever someone, putatively in authority, asserts: "If I make an exception for you. . . ."

An anecdote may help illustrate how this fallacy is capable of variations. An elderly man walked into a bank in his hometown. He had an account with this bank for fifty years and he knew all the employees as neighbors or friends. On this occasion, he approached a teller who mentioned that she hadn't seen him in a while.

CALIFANO: Still, it's nice to see you again, Mr. Murphy.

MURPHY: It's good to see you too, Mrs. Califano.

CALIFANO: How's Mrs. Murphy?

MURPHY: Why, she's fine, thank you.

CALIFANO: What can I do for you today?

MURPHY: Well, I'd like to cash this check.

CALIFANO: Certainly, Mr. Murphy, but may I see some identification please?

When Mr. Murphy asks the teller why she needs to see *his* identification, she replies: "If I have to make an exception for you. . . ."

11. False Analogy

Strictly speaking, all arguments from analogy are risky from a logical point of view. Analogies make useful hypotheses but not good, tight arguments unless the argument from analogy is also accompanied by good, solid, relevant evidence. When that evidence is lacking the argument from analogy results in a **False Analogy.**

Every argument from analogy presupposes that the analogue has sufficient properties in common with the thing to which it is compared to allow one to hypothesize that it possesses this further feature which the other has. As such, loose analogies can be defeated by noting points of *dis-analogy*—i.e., properties

DEFINITION: **A fallacy of *False Analogy* occurs whenever one draws a conclusion on the basis of a questionable comparison or without offering sufficient evidence for the truth of the comparison.**

EXAMPLE:

"The crime rate in the slums is indeed higher than elsewhere; but so is the death rate in hospitals. Slums are no more the 'causes' of crime, than hospitals are of death; they are the locations of crime, as hospitals are of death."
—Ernest Van De Haag, "The Need for Capital Punishment"

COMMENT: **This is a questionable comparison. The author compares slums to hospitals and the location of crime to the location of death, and then concludes that the poverty of the slums has no more to do with the causes of crime than hospitals have to do with causing death. One significant dis-analogy is that one usually *chooses* to seek hospital care whereas one does not usually choose to live in the environs of poverty.***

that one has that the other clearly does not. It is worth noting that such points of dis-analogy will always be present, or else the items compared are one and the same thing. (The philosopher, Gottfried Wilhelm Leibniz, proposed the famous Law of Identity which states that A and B are identical whenever they have all their properties in common.) But the whole point of an argument from analogy is to compare two discrete things which have at least some shared properties. If they turn out to be identical, there is no point to the analogy; if they are in fact discrete entities, then there are always points of dis-analogy, and these constitute grounds for defeating the argument.

The argument from analogy is typically of this form:

A has properties x, y, and z;
B has properties x, y, and z;
A has the additional property q;
Therefore, B has the additional property q.

As a rule, however, this does not follow unless additional justification is given. But precisely the kind of additional justification needed is an entirely different method of argument.

Nevertheless, analogies can be helpful in that they can give us hints to relationships that we hadn't suspected existed prior to making the analogy. In

*Notice that this is not strictly speaking a False Cause fallacy because the author is using an analogy to *dispute* the existence of a cause, whereas one usually employs a False Cause fallacy in the context of finding a cause where none exists.

other words, they can serve as good grounds for making hypotheses. However, additional justification (in the form of good evidence) is always required.

Furthermore, some analogies are just plainly better than others. When you find a howler of an analogy, you are entitled to make the charge of False Analogy; it is here that you should stress the points of dis-analogy. When you encounter a very tight analogy, it is always worth pointing out the inherent shortcomings of the method, recognizing as you may the particular merits of this instance.

Consider another example of False Analogy:

During the "Iranscam" controversy in the Reagan administration, Special Presidential Aide Patrick Buchanan compared Colonel Oliver North to "the heroes of the Underground Railroad." Both, he argued, acted out of conscience, both acted illegally, and both worked to help oppressed people. But the Underground Railroad was set up prior to the Civil War to aid blacks escape from slavery. It was non-violent, whereas the money Colonel North raised from an arms sale to Iran was, putatively, used to buy weapons for the Contras in Nicaragua.

12. False Cause

One commits the fallacy of **False Cause** when hastily characterizing one event or set of circumstances as the *cause* of another event or set of circumstances. Often the mistake is made for effect or for self-serving purposes. We may find it convenient to try to explain why things go wrong by appealing to our own prejudices or grudges. Conversely, we might seek to explain why things go right by appealing to pet theories or beliefs. In such circumstances, one may be too quick to assign causes which reflect one's own favor or dislike.

DEFINITION: **A fallacy of *False Cause* occurs whenever one argues for a particular causal connection on the basis of insufficient evidence or when other explanations are plausible that fit the facts.**

EXAMPLE:

"Admission to Harvard is based on how complicated your parents' tax returns are. After all, look at the financial records of nearly everyone admitted to Harvard, and you'll find out their parents fill out lots of complicated tax returns."

COMMENT: **The person who is making this argument assumes that Harvard admits students according to the talent or wealth of their parents. Harvard admits students on the basis of their own merit. If their parents have complicated tax returns, that may be because the cost of a college education entitles them to special deductions.**

One can see why thinking fallaciously in this way can be dangerous as well as addle-brained. When crops failed in the Massachusetts Bay colony during the seventeenth century, for example, people came to blame the poor harvest on supernatural forces. In Salem, women were put to death as witches; they were said to be the cause of the crop failures.

It should be noted here that (as Bertrand Russell once said) "philosophers and men of science have got going on cause, and it has not anything like the vitality it used to have . . ." When scientists speak among themselves, they are not likely to claim (for example) that cigarette smoking *causes* cancer; they would say instead that there is a high statistical correlation between sustained cigarette smoking and the development of some sorts of cancerous tumors. This is neither excessive caution nor professional neurosis. The reason that scientists speak this way is their awareness of the problems with claims about causation made by old theorists. Scientists and natural philosophers of old thought that they could assemble a compendium of causes and their related effects—sort of an encyclopedia of causation—and with that in hand, they could uncover the basic secrets of the universe from which they could deduce all corresponding actions and behaviors.

Philosophers and scientists have got away from that sort of alchemy, for the world has proven to be far more complicated than that view of the universe would allow. Modern day philosophers and scientists, in their humility, do not make such claims for themselves.

But we do, as a matter of fact, speak of causes and effects on an everyday basis. When we do, it is more important than ever to exercise the utmost caution to avoid attributing unfounded causal relationships to our own inclinations or superstitions.

Here is an example:

John Molloy, author of *The Dress for Success Book* and *The Women's Dress for Success Book,* advances the thesis that "clothing makes the (wo-)man." It is Molloy's contention that there is a fairly direct correlation between success in business and a standard of dress. Is Molloy guilty of a False Cause? What if he is indeed right, and there is such a correlation between success and executives who wear three-piece navy blue, beige, and grey suits (or women who wear conservative blazers and skirts with few frills)? Then is the fallacy Molloy's, or is it on the part of the business "establishment," or perhaps it is on the part of both since the publication of Molloy's book may contribute to the trend? What do you think?

a. **Appeal to Superstition.** The reason *Appeal to Superstition* is an instance of False Cause should be apparent enough. Superstition is the catch term we use for beliefs or attitudes that are inconsistent with either the body of scientific knowledge or with the method of science.

Sports have become fertile ground for the new superstitions. While Americans are less likely to blame bad happenings on the coincidence that we happened to walk under a ladder that day, or failed to knock on wood, some football players will not go on the field of play unless they are wearing their "lucky socks" (which, for some reason, are never supposed to get washed). When Mike Schmidt strikes out, many fans tend to blame themselves for having "thought bad thoughts" when the mighty Michael Jack Schmidt came to bat.

A word on astrology: Unfortunately, astrology has made something of a comeback in recent years. Horoscopes may be a good illustration of self-fulfilling prophecies. If you put any stock in them, ask yourself this question: Do you tend to remember the ones that "come true," and forget the ones that don't? My horoscope for Labor Day predicted that I would have a very relaxing afternoon. Instead, I find myself working hard at the computer keyboard, writing this essay on fallacies and superstitions.

b. The Gambler's Fallacy. A relative of *Ad Superstitionem* is the *Gambler's Fallacy.* The fallacy here occurs when one fails to understand properly the meaning of statistical frequency. Statistics has become the new science of superstition in America, rather decisively replacing the "old science" of astrology. The *gambler's fallacy* is committed when one distorts probabilities into the supposition that the chances of event X occurring are *maturing* on the rather dubious premise that after numerous opportunities, event X hasn't yet occurred—so it's bound to happen sooner or later! Thus, if Dave Winfield is (so far) a lifetime 300 hitter, and he has fanned (struck out) for the last sixteen consecutive times at bat, the gambler may fallaciously conclude that this time, Winfield is "due" for a hit.

Here's another variation. A gambler of this author's acquaintance has a theory that a baseball team is "given only so many runs" in a given week. If yesterday's Atlanta Braves game was a blowout, and the Braves scored eight runs, the gambler-friend will conclude that today's game is certainly going to be a low-scorer. If today the Braves are playing a new team, he'll check to see if the new team (e.g., Cincinnati Reds) has been scoring low this week. If it has, then he'll conclude "they're due," and bet on Cincinnati. This is a good way to go broke! In fact, the more logical you are, the less likely you will be to find gambling an attractive proposition. At the very least, you will become wary of "having a system."

c. Post Hoc Fallacy. *Post hoc ergo propter hoc* translates as "This before that, therefore this because of that": One can quite innocently fall into the fallacy of False Cause—not because the ascription of a cause in the case follows from pet theories or disfavored sources, but because one simply observes a temporal connection between two events and then falsely assumes that the earlier event is the cause of the later. Surely, if there were a set of criteria for causation, one such criterion would be temporal priority ("cause A comes before effect B in time"). But there would have to be other criteria as well.

Some such criteria that have found favor among past philosophers and scientists include spatial proximity ("cause A is close to effect B in space") and constant conjunction ("whenever A appears, then B appears"). So, it is not sufficient to assume that since an event immediately preceded another, the former was the cause of the latter.

All *post hoc* fallacies are species of the fallacy of False Cause, but not all False Cause fallacies are *post hoc.* When the criteria of spatial proximity or constant conjunction are violated, these are instances of the more general False Cause fallacy. *Post hoc* fallacies are violations of the criterion of temporary priority.

Consider this fine example of *post hoc* (taken from Raymond Smullyan's book, *This Book Needs No Title*):

Story of the Fleas

Once upon a time there was a man. This man had a dog. This dog had fleas. The fleas infected the entire household. So the man had to get rid of them. At first he tried killing them individually with a fly swatter. This proved highly inefficient. Then he tried a flea swatter. This was also inefficient. Then he suddenly recalled, "There is such a thing as science! Science is efficient! With modern American equipment, I should have no trouble at all!" So he purchased a can of toxic material guaranteed to "kill all the fleas," and he sprayed the entire house. Sure enough, after three days all the fleas were dead. So he joyously exclaimed, "This flea spray is marvelous! This flea spray is *efficient*!"

But the man was wrong. The flea spray was actually totally inefficient. What really happened was this: Although the spray was inefficient, it was highly odoriferous. Hence he had to open all the windows and doors to ventilate. As a result, all the cold air came in, and the poor fleas all caught cold and died.

13. Hasty Generalization

The fallacy of **Hasty Generalization** may be charged whenever one generalizes to a conclusion on the basis of scanty evidence, a sample that is obviously too small to warrant the generalization, or a failure to maintain tight controls over the sample to prevent the selection of the sample from biasing the results.

Many of the inductive fallacies we've already considered are species of the genus, Hasty Generalization. For example, many instances of the fallacy of False Cause may be properly construed as instances of Hasty Generalization as well. However, it is better form to search for the more specific fallacy, and Hasty Generalization is appropriately one of the more general forms of *Inductive Fallacy.*

It is also useful to realize that the Hasty Generalization can be made on the basis of either (1) too small a sample, or (2) an unrepresentative sample. These are common problems in statistics and in gathering data in science. Is a poll of 1,500 people sufficient to form an accurate judgment about popular opinion on national elections? Many pollsters think so, provided that the poll is based on a wide and carefully selected sample of polling districts. But is the sample truly representative? Suppose a voter in a highly conservative, Republican district in California is exit-polled, answers questions to the effect that he is a conservative Republican in a high income bracket, and then claims he voted for the highly liberal Democratic candidate. Even a few such voters in a few such districts can throw off the results of a poll. When cancer researchers perform tests on labo-

DEFINITION: **A fallacy of *Hasty Generalization* occurs whenever one collects a sample of relevant evidence in support of a conclusion but the sample is just too small or unrepresentative to warrant the force of the conclusion.**

EXAMPLE:

"The latest poll in Minnesota shows that Walter Mondale will win that state decisively. On the basis of that poll, I conclude that Mondale has an excellent chance of winning the election for president."

COMMENT: **Minnesota was the *only* state Mondale won in the 1984 presidential election (he was the "favorite son"). The sample was too small and too unrepresentative of the population at large to justify the conclusion.**

ratory rats and conclude that substance 'X' leads to a high incidence of cancer, we can legitimately question if the environment and anatomy of a rat is sufficiently representative of a human's environment and anatomy. But the fact of the matter is that pollsters and cancer researchers really are concerned about such problems of representation and sample size, and they take great pains to attempt to correct for such factors. You should strive to emulate the care and attention they pay to such matters when you come to inductive judgments of your own.

This fallacy is typically of the form:

I see that A has properties x, y, and z.
Therefore, I conclude that *all* A's have properties x, y, and z.

Consider this example. "We haven't had a hurricane around here in thirty years, so it's not likely that we'll get one this year." This is the reverse of the *gambler's fallacy,* but it's still a fallacy. Without tighter statistical controls, and better meteorological evidence, it doesn't follow that a hurricane is unlikely this year any more than the converse follows: that since we haven't had one in so long, we're due.

EXERCISES

Exercise 1: Elementary Exercises

DIRECTIONS (*Read Carefully*): *Most of the arguments that follow contain fallacies of the sort we just studied (some commit no fallacy at all). These are textbook examples—the fallacies are meant to be readily identifiable. Once you get some practice with these, go on to Exercise 2. (1) Find the fallacies in the following arguments (or parts of arguments) and enter the correct name in the space provided. (2) Give the most specific name you can for the fallacy (e.g.,* post hoc *instead of "False Cause" when appropriate). (3) If the argument commits no fallacy, write "No Fallacy" in the space provided.* Beware of the False Charge of Fallacy!

A List of Fallacies Studied in This Section

Slippery Slope
  the domino theory
Attacking a Straw Man
False Dilemma
  the bureaucratic fallacy
False Analogy
False Cause
  appeal to superstition
  the gambler's fallacy
  post hoc fallacy
Hasty Generalization

Name That Fallacy!

______ 1. SAMMY: This day is different from all the other times I asked you to marry me.

SHERRY: Why do you say that, creep?

SAMMY: My horoscope says so.

______ 2. When preparing a hollandaise sauce, add the melted butter very slowly, in droplets, to the egg yolk mixture while beating. If care isn't taken to do this procedure especially slowly, the sauce may well curdle.

______ 3. The United States has won the gold medal in basketball in every Olympics since the event was introduced as part of the games. The United States will certainly win it this year.

______ 4. TICK: I've never believed the Warren Commission Report that Lee Harvey Oswald acted alone in the assassination of John F. Kennedy.

TACK: Richard Nixon was present in Dallas the day of the assassination, did you know that?

TICK: No! Say, do you think . . . ?

______ 5. The "head and shoulders" formation in a graph of Wall Street activities usually signals a crash is in the offing. We see the formation developing. A crash is imminent.

______ 6. The ship of state needs a firm hand at the helm. That's why I say we should appoint a dictator to replace the president of the United States.

______ 7. HAN SOLO: Hey Luke! You can do the job if anybody can. The Force is with you!

______ 8. "Why do I say that Islam is a violent religion? The Koran teaches that any one who dies under the banner of Mohammed in combat in a holy war will enter paradise, that's why!"

______ 9. "I know why there's a hole in the ozone layer. It's because we're sending up so many spaceships, that's why. They're poking holes in it."

______ 10. TERRORIST SYMPATHIZER: Terrorism is the "poor man's" form of warfare. If we outright condemn terrorism against innocent people, then freedom fighters will have no effective way to make themselves heard.

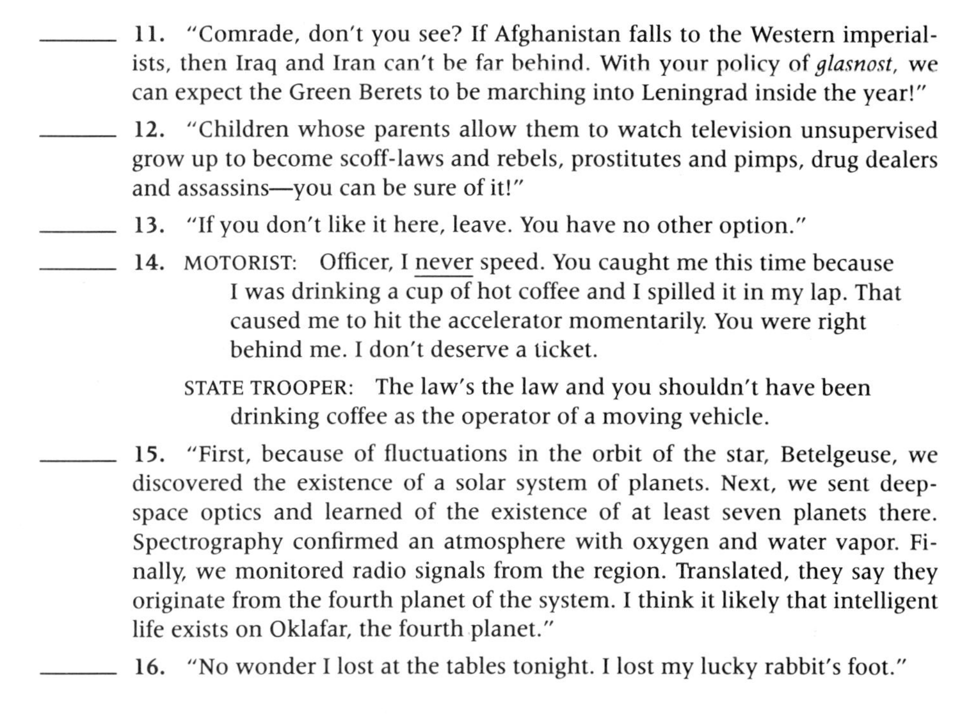

______ **11.** "Comrade, don't you see? If Afghanistan falls to the Western imperialists, then Iraq and Iran can't be far behind. With your policy of *glasnost*, we can expect the Green Berets to be marching into Leningrad inside the year!"

______ **12.** "Children whose parents allow them to watch television unsupervised grow up to become scoff-laws and rebels, prostitutes and pimps, drug dealers and assassins—you can be sure of it!"

______ **13.** "If you don't like it here, leave. You have no other option."

______ **14.** MOTORIST: Officer, I never speed. You caught me this time because I was drinking a cup of hot coffee and I spilled it in my lap. That caused me to hit the accelerator momentarily. You were right behind me. I don't deserve a ticket.

STATE TROOPER: The law's the law and you shouldn't have been drinking coffee as the operator of a moving vehicle.

______ **15.** "First, because of fluctuations in the orbit of the star, Betelgeuse, we discovered the existence of a solar system of planets. Next, we sent deep-space optics and learned of the existence of at least seven planets there. Spectrography confirmed an atmosphere with oxygen and water vapor. Finally, we monitored radio signals from the region. Translated, they say they originate from the fourth planet of the system. I think it likely that intelligent life exists on Oklafar, the fourth planet."

______ **16.** "No wonder I lost at the tables tonight. I lost my lucky rabbit's foot."

Exercise 2: Advanced Exercises

DIRECTIONS (*Read Carefully*): *Most of the arguments (or passages which contain arguments)* given below contain fallacies of the sort we just studied (some commit no fallacy at all). These are "real-life" examples—they're taken from editorials, advertisements, and letters. Because they are taken from real life, they may prove to be more difficult than the "textbook" examples you worked in Exercise 1. (1) Find the fallacies in the following arguments (or parts of arguments) and enter the correct name in the space provided. (2) Give the most specific name you can for the fallacy (e.g.,* post hoc *instead of "False Cause" when appropriate). (3) If the argument commits no fallacy, write "No Fallacy" in the space provided.* Beware of the False Charge of Fallacy!

Name That Fallacy!

______ **1.** I would sincerely hope that the students and faculty on this campus take this threat to our constitutional freedoms as real. If Wooten and others like him can claim that gays and lesbians have no rights on this campus, then the next phases will be, "If you are black, female, non-Christian, un-American or un-conservative, then you have no rights." —"The Tar Heel Forum," *The Daily Tar Heel*

______ **2.** Can animals have forewarnings of disaster? In many instances, it appears they can. The case of Josef Becker is a chilling illustration.

*Note: Some of these passages are, strictly speaking, reports, but they contain arguments attributed to others or they lead into arguments. For the purposes of this exercise, examine the arguments contained in the reports for fallacies.

Becker was enjoying a drink at the local inn in Saarlouis, West Germany, when his faithful Alsatian, Strulli, suddenly became agitated.

The dog was doing everything possible to attract attention. . . .

Tired of this losing battle and seeing that he would never enjoy his drink in peace, Becker left the inn at two minutes to 5. Two minutes later, to a deafening crash, the inn collapsed on its occupants, killing nine and injuring over 20 people.

Investigators later discovered that builders excavating next door had damaged the inn's foundations. Nobody suspected it—but the dog knew something was wrong. —"Dog Forces Man to Leave Pub—Then It Collapses." *National Enquirer,* July 28, 1987

______ **3.** These people believe they are patriots. But pure ideals and burning patriotism do not always justify the actions taken in their name. No doubt there are Oliver Norths around the Ayatollah Khomeini and Daniel Ortega, as there were around Josef Stalin and Adolf Hitler.

No, that isn't to compare North's actions with those of boot lickers in such dictatorships, but it is to point out that zeal without restraint is highly useful to tyrants, who generally entrust the running of their domains to platoons of energetic young patriots eager to follow orders. —Editorial, *The Morning Star,* July 16, 1987

______ **4.** The State of New York, with what passes for wisdom in this lock-step era, seems to be in the process of passing a law to require its citizens to bind themselves to automobiles when traveling.

If that's not "Big Brother," then this isn't 1984 and Orwell's been reincarnated as "Boy George." —Jim Fain, ". . . As New York Tries to Strap Them In," *Wilmington Morning Star,* June 20, 1984

______ **5.** Pursuing that rationale [for laws making seat belts mandatory] would require first of all making it a felony to smoke. Tobacco is certifiably the No. 1 health hazard in the country, producing lung cancer, heart attacks, emphysema and assorted horrors.

Then we would need to enact legislation making it a crime to ingest more than a Spartan minimum of cholesterol a week, to say nothing of the calories that cause overweight.

Policing might be a problem, but that seldom deters devotees of legislated behavior. We could set up weight stations, like they have for trucks, and it would be easy to check cholesterol levels at periodic inspections in clinics, such as many busy-body states now require for autos. —Jim Fain, ". . . As New York Tries to Strap Them In," *Wilmington Morning Star,* June 20, 1984

______ **6.** Recent reports that the Amberly project will be approved by the Board of Aldermen on July 28 are indeed disconcerting. Orange County, Chapel Hill and OWASA representatives have clearly and publicly opposed any development in the University Lake watershed pending an in-depth study of the issue. However, Carboro officials seem ready to possibly compromise our local water quality without consideration of the long-term consequences. Despite the fact that Amberly plans remain unavailable for review, the aldermen appear determined to push this project to reality without thorough analysis. I cannot understand the need for you to proceed with such haste. To act without full information and understanding is certainly not in the best interest of your citizenry.

I urge you to not approve of any development in the University Lake watershed until the OWASA study can be completed. Your urgency to allow Amberly to proceed cannot be more important than our water sources of the future. —An Open Letter to Carrboro's Mayor, *The Chapel Hill Newspaper*, July 26, 1987

______ 7. But you might want a few moments alone with a onetime celebrant of the funky and the far out.

Like the student who called me "an old liberal" when I was only 37, these former followers of the aberrant and the outlandish are easy to spot. They've been rewarded with executive privileges in the corporations they once denounced. And with positions of power in a government that they called fascist. I've also seen them in the academic offices they once occupied by force. —"My Turn," "A State of Incivility," *Newsweek*, February 8, 1988, p. 10

M O D U L E 36

FALLACIES OF PRESUMPTION

PART 3: FAULTY PRESUPPOSITIONS

A fallacy of presumption draws a conclusion on the basis of a faulty presupposition. We may define a presupposition as an assumption made in advance of inquiry. Sometimes we make up our minds about what a sentence means without reading it carefully to find out that other interpretations are possible and those alternate interpretations might just "yank the rug out" from under our conclusion. Such is the case with the fallacies of grammatical structure, such as "Equivocation" or "Ambiguity." Other times we automatically assume the truth of the conclusion and use that assumption (usually in a roundabout way) as a reason for believing the conclusion. When that happens, we are said to be "Begging the Question."

14. Composition

The fallacy of **Composition** is committed on the condition that what is asserted of every part is assumed to hold true for the whole as well. It is a commonplace observation that the individual parts of a thing may contain properties which the composite whole does not. Artists and scientists are well aware, for example, that colors are such properties. A pointillistic painting is made up of a series of

---

A deductive argument is **sound**, if and only if, if all the premises are true, then the conclusion must be true, and all the premises are in fact true.

DEFINITION: **A fallacy of *Composition* occurs whenever one argues that the properties of each member of a class must hold true for all members of the class taken as a whole. (This is a fallacy of generalization going from "each part" to "the whole.")**

EXAMPLE:

"The Graham Chapman movie *Brazil* must be spectacular because I've read there are so many spectacular scenes in it."

COMMENT: **One of the repeated criticisms of the move is that, despite very many superb individual scenes, the movie fails to hang together as a whole.**

colored dots (say, blue and yellow dots), but it does not follow that the completed painting will appear blue and yellow to the observer—in fact, it would appear green.

Here is a simple and clear example of the fallacy of Composition. One sports fan talks to another:

LEFTY: Every member of the college basketball team is ready to play.

RED: In that case, I'm going to bet on the home team to win because the *whole* team must be ready to play.

We can see why this is a fallacy. True, each of the team's players may be potentially good performers and each of them may be ready to play, but it doesn't follow that the *whole* team is ready. Every sports team needs a great deal of unselfish teamwork, and if the players are "hot dogs" or individual glory seekers, no matter how ready each of them is, the team may well fall apart on the court.

15. Division

The fallacy of **Division** occurs whenever the properties of the whole are assumed to hold true for each of its parts. As such, it is the reverse of the fallacy of Composition, but it remains a fallacious way of thinking for much the same reason that Composition signals errant thinking.

Consider again our example of colors. Scientists recognize that macro-objects (i.e., objects of sight) are invariably colored: battleships are grey, chestnut trees are green in summer, etc. But it does not follow from this that their constituent atomic particles possess the same colors. In fact, we are taught that atomic particles are not colored at all, for color is a phenomenon of the visible strata.

Let's return to our basketball example again. After losing his shirt for betting on the home team, Red decides he's learned how important teamwork is for a

DEFINITION: **A fallacy of *Division* occurs whenever one argues that the properties of the whole class must hold true for each of the members of the class taken individually. (This is a fallacy of "instantiation" going from "the whole" to "each part.")**

EXAMPLE:

"That was the best logic class I ever had. Gomer was a student in that class, so Gomer must be one of the best logic students I ever had."

COMMENT: **Gomer was the only student who failed.**

successful basketball team. Lefty tells him that the visiting team is noted for its teamwork and so this time he bets the house on the visitors. Unfortunately for Red, he's thinking badly again. In the meantime, the home team has learned a measure of cooperation, and while the visitors have great team play, no players on the bench are ready to play a varsity game. Two of the visiting team's starters foul out. With only unprepared players to replace them, the home team wins and Red loses his house. Red is victim to the fallacy of Division.

The fallacy of Division may also be called, "the fallacy of the aggregate." Let's imagine a hypothetical college campus that has a parking problem even though the administration repeatedly makes the claim that they provided an adequate number of spaces in the aggregate for the semester's enrollment. The point is that parking spaces may be sufficient *in the aggregate,* but insufficient in particular instances. For example, the free spaces may be at the other end of the campus from where most classes are in attendance. Most students quite naturally try to park close to their classroom building. The result is traffic and congestion at the high demand areas, making it difficult for a student trapped in traffic to get out of the congested lots to drive to the remote parts of the campus where parking spaces may be found. In this respect, parking is inadequate under particular circumstances.

The fallacy of Division teaches us important lessons. Some economic theories base the notion of the "common good" on aggregate measures of utility, such as the Gross National Product (GNP). However, even if the GNP rises, this may reflect only the fact that wealthier people are getting better off while the poorest members of society's condition may be worsening. While millionaires buy luxury goods in a rush, many poor people may become homeless. It doesn't follow that if the whole plight improves, so does each segment of the whole.

16. Equivocation

The fallacy of **Equivocation** occurs whenever the sense of a word or phrase is shifted in the course of an argument to take on a new meaning, thereby creating the false impression that the reason for the conclusion is really a good reason, when it is semantically irrelevant.

DEFINITION: **A fallacy of *Equivocation* occurs whenever one employs a term which has more than one meaning and then, in the course of an argument, shifts meanings to reach an erroneous or irrelevant conclusion.**

EXAMPLE:

PROFESSOR SMITH: I cannot recommend this student for the honors program. Her evaluations are right on the median for all the candidates invited to apply. 'Median' means "in the middle;" so does 'mediocre'. I oppose her admission because she's a mediocre student.

COMMENT: **Professor Smith has fallaciously equivocated on the expression 'in the middle'. *Median* and *mediocre* do indeed mean "in the middle," but mediocre has negative connotations ("nothing exceptional") whereas *median* means that there were as many evaluations above this student as there were below. Since we're informed that the students applied to the program by special invitation, we can expect that this is already an exceptional lot. To say that this student is at the median in such a group is to indicate the likelihood that she's a very gifted student indeed.**

The fallacy of Equivocation can be a source of humor, precisely because it often jars us when we hear it. Will Rogers once quipped, "I belong to no organized political party—I am a Democrat." The equivocation, of course, is on what we ordinarily mean by the phrase *organized political party.* To take another example with an even more obvious equivocation, consider this from James Thurber: "I have always thought of a dog lover as a dog that was in love with another dog."

On the other hand, Equivocation can become a source of profound confusion, and may lead even sophisticated thinkers into dangerous waters.

In his book, *Metaphysics,* philosopher Richard Taylor once argued the rather startling thesis that a single object can well be in two different places at the same time, and he gave as an example the occurrence of an earthquake which strikes two distant cities at the same time (let's say, Los Angeles and Peking). Geologists may point out that the same fault line extends from California to China. Taylor goes on to argue that it's the same earthquake in the following way:

> "Now some would want to insist here that more than one object is involved in this example, or even that an earthquake is no proper object at all. But it is an object in every significant way—it has a location in space and time, has interesting properties, such as the property of being destructive, and might even have a name, as hurricanes, for example, usually do. Nor is it to suggest that, in the example given, more than one such object is involved. The people in the different towns could say, rightly, that they suffered from the same earthquake."

The argument takes on increased plausibility when one realizes that Taylor has carefully defined 'object', so that 'earthquake' (which we more commonly think of as an occurrence than as an object) fits the definition. If we grant him that term, mightn't Taylor have yet equivocated on an equally important term: 'place'? We might concede that the earthquake that struck Los Angeles and Peking was the same object, but disagree that it was in two different places at the same time, since the fault line of the earthquake was in the same place (underneath the earth's crust) all along. The earthquake wasn't here and there; it was all along here the whole time.

Consider another example of the fallacy of Equivocation, which *Rolling Stone* magazine correctly attributes to the Meese Commission Report on Pornography: ". . . look at this fallacy of equivocation: '[The commission] concludes that organized crime . . . exerts substantial influence over the obscenity industry. . . . All major distributors and producers of obscene material are highly organized and carry out illegal activities.' It's organized, and it's crime, so it's Organized Crime."

Consider this example (typical of restroom graffiti):

God is love.
Love is blind.
Ray Charles is blind
Therefore, Ray Charles is God.

We can see two possible equivocations: one on the word 'love'; another on the word 'blind'. Perhaps what this example should teach us is distrust of the logic in restroom graffiti.

By the way, former President Jimmy Carter committed a fallacy with his use of the word 'median' during his unsuccessful campaign for re-election; lamenting that poverty was widespread in the United States because "fifty percent of all families in America earn less than the median income."

17. Ambiguity

An **Ambiguous** sentence (also called "an amphibole") is a sentence which can be construed to yield two or more distinct *senses* or *interpretations*. The fallacy is committed when one of these separate meanings is shifted to lend false support to the conclusion, even when really it is irrelevant. One may find it easier to keep Equivocation and Ambiguity distinct by recalling that Equivocation involves shifts of meanings in words or phrases, whereas Ambiguity involves shifts of meaning in whole sentences.

Another way of looking at the distinction between Equivocation and Ambiguity would be to describe 'Equivocations' as "fallacies of *term* ambiguity" and 'ambiguity' (or 'amphibole') as "fallacies of *grammar* ambiguity." The fallacy of Ambiguity in this sense occurs in the misuse of a sentence which employs a grammatical construction with two meanings in the course of an argument, one meaning which appears to lend support to a conclusion, and another, usually the intended meaning, which does not.

DEFINITION: **A fallacy of *Ambiguity* occurs when a sentence which contains two or more distinct interpretations in such a way in the course of an argument that an erroneous or irrelevant conclusion is obtained.**

EXAMPLE:

Chaerephon, a friend of Socrates in his youth, once asked the Oracle at Delphi if Socrates were the wisest of men. The oracle replied, "There is none wiser." Chaerephon concluded that his friend had great wisdom indeed.

COMMENT: **This is a fallacy on the interpretation of the oracle's statement due to the structure of the sentence. Interestingly, Socrates probably understood that the oracle was posing a riddle, for he later embarked on a life-long exploration into the meaning of what the oracle said. Chaerephon's interpretation was the obvious interpretation, but the oracle may have meant something else entirely: that *no* humans have any wisdom whatsoever—Socrates included!**

Consider the following sentence (from linguist and philosopher Noam Chomsky):

"The police were ordered to stop drinking after midnight."

There are five distinct meanings possible for this sentence on any given reading. Can you name all five? Do you see how so ambiguous a sentence can lead to a false conclusion? If we read the sentence superficially, we might presume that the first interpretation that comes to mind is the correct one, when others equally plausible are possible. Suppose we conclude that the police have been in the habit of drinking during the night shift, when really the statement was intended to make us realize that the police will arrest *us* if we're caught drinking after the clock strikes twelve.

Sometimes ambiguous sentences are used cleverly to upstage an opponent. During the McCarthy hearings, the House Committee on UnAmerican Activities was charged with the responsibility of rooting out subversion in government. Their tactics were so harsh and unforgiving that one unknown reporter was inspired to re-name the committee, "the House *UnAmerican Activities* Committee"—suggesting that what the committee was doing was itself un-American. No one quite knows who this inspired reporter was, but the committee members made themselves look foolish by adopting the locution interchangeably with the correct name in their own speeches and press releases.

A famous ambiguity was found by anti-war activists during the Vietnam era. Some of them noticed that selective service questionnaires asked the question, "Do you advocate the overthrow of the United States government by force or violence?" Some responded by selecting the lesser of the two evils. "Force," they replied.

Ambiguous sentences create problems in communication. Consider this bridge sign, typical of such states as Pennsylvania:

"BRIDGE FREEZES BEFORE ROAD SURFACE."

Probably no other road sign has caused as much confusion as this one. What does it mean? Does it mean that the superstructure of the bridge freezes before the road surface of the bridge does? Some reflection (which should be unnecessary for a road sign) helps us to distinguish between a spatial and a temporal sense of 'before'. Once we do that, we see that the temporal sense is meant. What the Department of Transportation means to say is that the bridge road surface freezes (in time) before the land road surface, so be careful. But carelessness might cause us to presume that the sign is warning us of a few feet of frozen road surface immediately before the bridge. We could conclude, dangerously, that the bridge itself is safe enough; that what we need to worry about is the road surface. That sort of presumption could get us into an accident.

The state of Georgia resolves this confusion with a sign that expresses a simple and unambiguous eloquence:

"BRIDGE MAY ICE IN WINTER."

Here is another example of an ambiguous sentence [taken from *The New Yorker* magazine]:

We Don't Want to Hear About it Department

[*From the Newark Star-Ledger*]

The number of women with infants in the labor force has grown dramatically in the last decade, up from 31 percent in 1976.

This is ambiguous because we can detect at least two meanings to the sentence: first, that the number of *women* in the labor force with infants has grown dramatically; and secondly, the number of women with *infants* in the labor force has grown dramatically. Presumably, the first sense is intended. Can you see what sort of fallacious inference might be drawn from assuming the statement can have only one meaning when yet another equally plausible meaning is possible?

18. Slanting

The fallacy of **Slanting** (also known as "the Contextual Fallacy") is committed whenever a phrase or passage is pulled out of context to make a point which is entirely different from the point originally made. We saw an example of this already, when Cardinal Spellman declared, "My country right or wrong." He was really quoting out of context from a longer statement; the original has a meaning just the reverse of what he intended.

DEFINITION: A fallacy of *Slanting* occurs whenever one takes a sentence or passage out of context and then accents or emphasizes a feature of the sentence or passage in such a way as to allow an erroneous or irrelevant conclusion.

EXAMPLE: Gabby reads this story in a supermarket tabloid:

Woman Executed for Witchcraft

SALEM, MA. Prudence Howland, age 49, was tried, convicted and executed for the practice of witchcraft in this small and ordinarily quiet community.

Witnesses testified that they had overheard Ms. Howland "summon up the devil" and others said they'd observed the accused "engaged in mysterious and unholy rites."

Gabby concludes that witches are still being executed in Salem.

COMMENT: The story has no dateline. Readers naturally presume that newspapers report on current events (that's what 'news' means). That presumption may prompt the reader to conclude that trials for witchcraft still go on in Salem. The event in question probably took place nearly three hundred years ago.

Consider another example, a visual one this time. An *Associated Press* photograph depicted a man throwing his arms in the air during the shelling of Beirut, Lebanon. The caption as it appeared in at least one newspaper read: "Beirut slaughter. A man shouts, 'God is great', as shelling kills 12." Taken at face value, that seems like an awfully arrogant thing for the man to say. At least that was the judgment of one of my classes when I showed them the photograph. It seemed like the man was praising God for bringing this rain of destruction.

However, let us put the caption in a slightly different context. "Allah Akbar" is sometimes loosely translated as "God is great;" a better translation is "God be praised." Perhaps that was what the man was really saying. In that case, his words may not have been in arrogance at all; he may have simply been praying to his Maker, which is very understandable when howitzer shells are raining down, killing and maiming all those around. In fact, "Allah Akbar" is the standard salutation to most Muslim prayers.

19. Begging the Question (*petitio principii*)

The fallacy of **Begging the Question** is committed if and only if the speaker, in the course of his argument, assumes the very thing he is trying to prove. Thus, the conclusion that the speaker is arguing to gets assumed among the premises for the conclusion. The schematic form of the argument is as follows:

Premise 1

.

.

.

Premise n: a

Therefore, A.

One "begs the question" by illicitly using the rule of inference known as "Repetition;" it is normally proper to repeat any of one's premises at any time in the course of an argument. However, when one goes from an assumed premise directly to an identical conclusion via Repetition, then this very special type of fallacy is committed.

Because the fallacy of Begging the Question does indeed follow a rule of inference, it is both a deductive argument form and always valid. However, the

DEFINITION: **One commits the fallacy of *Begging the Question* whenever in the course of argument one assumes the very thing she is trying to prove.**

EXAMPLE: **[Here is an exchange of dialogue between the characters Willis Murdick and Holly from the novel *O-Zone* by Paul Theroux:]**

"Don't you believe it," he said. "There's danger there—it all looks different on the ground!"

"Except Willis never goes on the ground," Holly said.

"That's the reason why!" Murdick said.

COMMENT: **The character Willis Murdick is arguing to the conclusion that danger exists on the ground despite its deceptively tranquil appearance. Holly points out to a third character that Willis has never been there to find out. Murdick gives as his reason the assumption he already made—that there's danger there.**

Sometimes one slips into the fallacy of Begging the Question because one already has his mind made up about the answer in advance of evidence or inquiry. The author of this novel nicely demonstrates how prejudice and preconceived notions can trap us into making such unjustified assumptions. When those preconceived notions are stubbornly held, one may even resort to this sort of illogic.

important point to notice is that when one begs the question, the soundness of the argument is always suspect—it casts the truth of the premises in doubt. For this reason, Begging the Question is a fallacious argument form.

Begging the Question can be committed either flagrantly or very subtly. When it is committed flagrantly, humor or confusion is the likely outcome. For example, imagine a counterfeiter who has tried to pass a phony twenty dollar bill which she printed only that morning. The cashier points out the bill is a fake because the ink rubbed off on his hands. The counterfeiter replies:

> **"No, it is the real thing, and I can prove it. Read what it says in the left hand corner of the bill."**
>
> **The cashier reads, "This note is legal tender for all debts, public and private."**
>
> **"There!" exclaims the counterfeiter. "I told you it was genuine."**

The counterfeiter just begged the question, and if you were the cashier and you didn't have your logical senses about you, she may have even confused you enough to make good her escape.

When one commits this fallacy subtly, it can be very, very subtle indeed. Philosophers are still debating, for example, whether St. Anselm begged the question in his ontological argument for the existence of God when he defined God as "that than which nothing greater can be conceived." From this premise, Anselm argued that it is greater to exist in reality than not. From this, he concluded that God must truly exist in reality.

Even the consummate logician, Bertrand Russell, puzzled over this one. In his *Autobiography* (for which he received the Nobel Prize for Literature), Russell tells how as a graduate student at Cambridge, he was walking down the road, carrying a tin of tobacco, lost in thought. Suddenly, the young Russell stopped in the middle of the street, tossed his tin of tobacco in the air and exclaimed, "Eureka! The Ontological Argument is sound!"

Here is yet another example of Begging the Question, as described by an article in *Rolling Stone* magazine on the 1986 Meese Commission Report on Pornography:

> "Having assembled this blue-ribbon panel, the attorney general's office instructed the members to 'determine the nature, extent, and impact on society of pornography in the United States, and to make specific recommendations to the Attorney General concerning more effective ways in which the spread of pornography can be contained.' In the study of logic, this is called *petitio principii,* or 'begging the question.' The commission was charged with discovering whether pornography is spreading and whether it's harmful, but the phrase 'ways in which the spread of pornography could be contained' presupposes both the spread and the harm."

Below are several variations of the way that one can find himself begging the question.

a. Circular Reasoning. All question-begging arguments are circular in one respect or another. In fact, many concepts are so interrelated that the philosopher W. V. Quine calls these, "close families of terms." For example, look up the word 'truth' in the dictionary. You may find this definition: "that which accords with fact or reality." Now, look up the word 'reality' and you may find: "the quality of being true to life; fact." Look up 'fact' and you find: "a thing that has actually happened or is really true." These terms are so abstract and so close-knit that we can't do much better than this in describing them, and there is nothing wrong with that at all. These are circular definitions, but they are not circular in any way that should overly concern us.

An argument is fallaciously circular, however, when the breadth of the circle is just too tight. Consider, for example, the following quotation: "It's not that Southerners are more violent; it's just that we got more people need killin'." Despite the macabre humor implicit here (a humor that fades fast when we imagine these words in the mouth of a Klansman), what's wrong with the logic of the statement is that it answers the charge of needless killing with the bald rejoinder that the violence is needed. There is no intervening evidence or further argument to support the flat insistence of the conclusion; neither is it likely that further evidence is possible to support such a distasteful conclusion. The argument describes a complete circle, and as such is wholly fallacious.

b. Question-Begging Characterizations. These fallacies involve circular reasoning no less than our previous examples, but there is a special quality to this species of fallacy since it both commits the informal fallacy of *Ad Hominem* and the fallacy of Begging the Question. This fallacy is committed when if by employing certain characterizations of an issue or person, one has implicitly evaluated the issue or the person in a way that, if true, automatically wins the argument.

Consider this hypothetical news account: "The statewide Democratic party convened a caucus today in the hopes of selecting a new candidate for the office of United States senator." The news story is descriptive, non-argumentative, and acceptably unbiased, as a news story should be. But suppose a rival tabloid newspaper engages in "yellow journalism" and headlines the account in this manner: "Democratic Party Bosses Seek Puppet to Buy State's U.S. Senate Seat." It is apparent how the use of certain epithets can completely alter the meaning of the report and utterly destroy its objectivity.

In effect, one commits this species of question-begging by slanting the terms in the argument. If the piece began as a description, the slanting turned it argumentative, and fallaciously so. The use of emotive language also adds to the slanting; keeping in mind our hypothetical news story, recall this quotation from Harry Truman: "When a leader is in the Democratic party he's a boss; when he's in the Republican party he's a leader."

c. Victory by Definition. A question-begging definition ensures victory in a debate since any possible counterexample or rebuttal is ruled out in the definition, or by revisions to the definition. Former U.N. Ambassador Jeane Kirkpatrick's definition of autocratic and totalitarian states may be guilty of this sort of fallacy. She argues in her publications that we can recognize autocratic states as non-democratic nations that are friendly to the United States whereas unfriendly states are totalitarian. When one argues that the Shah's Iran was totalitarian rather than autocratic because of the way that he employed state terror in order to secure his foothold in Iran, Kirkpatrick replies that by her definition the Shah's Iran was autocratic since he was friendly to the United States. Short of the unlikely production of evidence to the effect that the Shah was not friendly to this nation, it is thereby impossible to argue the distinction.

One may also commit this fallacy by defining some things out of existence. President Gerald Ford once argued that terrorist acts couldn't occur in the United States because we simply wouldn't describe such acts as terrorism. Ford was not being facetious. He was merely pointing out that it is the State Department's policy never to recognize an act of terrorism in the borders of the United States so as not to give valuable publicity to the terrorists. They would be called "criminals" instead.

d. Alleged Certainty. One more way in which one might beg the question is to cast a point of view or an opinion in language that puts it beyond reproach. This is an attempt to persuade the listener that the issue is already settled and her opposition to it can only demonstrate her naivete. Thus, our arguer might contend, "Everyone knows that El Salvador is an evil dictatorship." Or he might say, "Don't be so naive. The Communists want you to think that Nicaragua isn't under their control."

Each of the arguments commit one or more other fallacies (e.g., Appeal to Popularity, *Ad Hominem,* etc.) but what makes them distinctive here is that they appeal to a level of certitude that just isn't possible for issues that are far from settled.

In a debate between Governor Jim Hunt and Senator Jesse Helms of North Carolina, Hunt argued against the cost-effectiveness of the proposed "Star Wars" scheme for a space-based weapons system. Helms replied: "I don't want to be harsh on you, but you don't know what you're talking about. This is the first I've heard that it's not going to work."

By replying as he did, Senator Helms was guilty of more than *Ad Hominem* attacks on his opponent (though he was committing those as well). Through his bombast, Helms was posturing feigned certainty, making his opponent look ill-informed without ever offering any argument or evidence in his own defense.

20. Complex Questions

One commits the fallacy of **Complex Questions** when and if one asks two or more questions in a fashion that makes it appear that only one question has been asked, or when the question is asked in such a way that the answer is presupposed in the question.

Questions should elicit information; they shouldn't cut off or complicate questions the way that Complex Questions do. Notice why asking complex questions is considered a fallacy; it is precisely because no information is really being sought. The questioner is actually formulating a hidden argument in the guise of a question.

For example, consider this question:

> **"Are you now, have you ever been, or have you ever had any sympathies with, Communists or Fellow Travelers?"**

This is the sort of question that the House Committee on Un-American Activities used to pose to people called to testify before Congress. Notice that it calls for a "Yes-or-No" type of answer, but that the answer might be more complicated than that. For example, suppose you never belonged to the Communist Party and never had any intention of joining. However, you realize that you *do* have a certain amount of sympathy for what Premier Gorbachev is trying to accomplish with his program of *glasnost*. How do you honestly answer that question? Certainly, not with a "Yes" or a "No." There are actually three questions in one here, but you can't even answer with a "No, to the first question; No, to the second; and Yes, to the third," because the Committee might take your third answer to mean that you are sympathetic to communism, when really you have only some sympathy for a program started by the current premier of the Soviet Union. We may think of such a question as a verbal attempt at entrapment.

To see how a complex question can prejudge an issue in the form of a question, turn to the passage by Bernard Gui from *The History of the Inquisition* at the end of the section entitled, "The Fallacies of Relevance" (pp. 438–439).

a. Leading Questions. *Leading Questions* are familiar devices to lawyers, judges, and fans of courtroom dramas. The sinister prosecuting attorney browbeats the defendant with this question: "Have you given up your criminal ways?" The dashing defense attorney leaps out of his seat, exclaiming, "Objection, your honor! The prosecution is leading the witness." The judge nods approvingly, bangs her gavel, and proclaims: "Objection sustained."

Leading questions are Complex Questions that are also forms of Begging the Question. These are questions that presuppose the possible range of answers in advance of what the respondent has to say. The possible range of answers is limited by the design of the question. In other words, more than one question is being asked, so no straightforward answer is appropriate.

In our example above, the prosecutor is really asking two separate questions. The first one is: "Have you ever engaged in criminal activities before?" The second implicit question assumes a "Yes" answer to the first implicit question and goes on to inquire whether the defendant has continued his criminal behavior. Such questions cause hesitation in a respondent, and such hesitation can look incriminating in the eyes of a jury.

The abuse to the truth caused by *leading questions* need not be restricted to a court of law, however. In the debate between Governor Jim Hunt and Senator Jesse Helms of North Carolina, Hunt asked Helms to join him in a pledge to accept no more out-of-state campaign contributions in their U.S. Senate contents. Helms declined, saying: "Are you having trouble raising out-of-state money?" This was a leading question, since Hunt had already established his moral opposition to such fund-raising. If Hunt answered "Yes," then his opposition to the fund-raising looked like sour grapes; if he answered "No," then his answer would cast him in the role of doing something he morally opposed.

EXERCISES

Exercise 1: Elementary Exercises

DIRECTIONS (*Read Carefully*): *Most of the arguments below contain fallacies of the sort we just studied (some commit no fallacy at all). These are textbook examples; the fallacies are meant to be readily identifiable. Once you get some practice with these, go on to Exercise 2. (1) Find the fallacies in the following arguments (or parts of arguments) and enter the correct name in the space provided. (2) If the argument commits no fallacy, write "No Fallacy" in the space provided.* Beware of the False Charge of Fallacy!

A List of Fallacies Studied in This Section

Equivocation	Begging the Question
Ambiguity	circular reasoning
Complex Questions	question-begging characterizations
leading questions	victory by definition
Slanting	alleged certainty
Composition	
Division	

Name That Fallacy!

_______ 1. MUTT: The police force must be riddled with corruption.

JEFF: Why do you say that?

MUTT: Because they have an entire division called "the Vice-Squad."

_______ 2. MAH: The blowup of the Zapruder film of the assassination of President Kennedy conclusively proves that he was shot by a second gunman on the grassy knoll.

JONG: All I see are blotches.

MAH: But this blotch looks just like a rifle.

JONG: Yeah, I see what you mean.

_______ 3. News caption for a story on a bill sponsored by the president to raise taxes on the incomes of the wealthiest ten percent of Americans: "PRESIDENT WANTS TO HIKE TAXES."

_______ 4. MOE: The dean said that no one can withdraw from classes on a holiday.

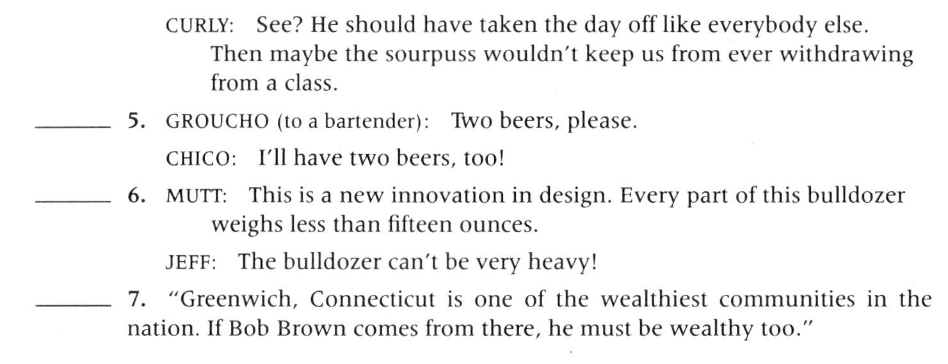

CURLY: See? He should have taken the day off like everybody else. Then maybe the sourpuss wouldn't keep us from ever withdrawing from a class.

_______ 5. GROUCHO (to a bartender): Two beers, please.

CHICO: I'll have two beers, too!

_______ 6. MUTT: This is a new innovation in design. Every part of this bulldozer weighs less than fifteen ounces.

JEFF: The bulldozer can't be very heavy!

_______ 7. "Greenwich, Connecticut is one of the wealthiest communities in the nation. If Bob Brown comes from there, he must be wealthy too."

Exercise 2: Advanced Exercises

DIRECTIONS (*Read Carefully*). *Most of the arguments (or passages which contain arguments)* given below contain fallacies of the sort we just studied; some commit no fallacy at all. These are "real-life" examples—they're taken from editorials, advertisements, and letters. Because they are taken from real life, they may prove to be more difficult than the "textbook" examples you worked in Exercise 1. (1) Find the fallacies in the following arguments (or parts of arguments) and enter the correct name in the space provided. (2) If the argument commits no fallacy, write "No Fallacy" in the space provided.* Beware of the False Charge of Fallacy!

Name That Fallacy!

_______ 1. Sign in university lounge:

> NO SMOKING
> FOOD OR DRINK

Upon reading the sign, Barney thought it was okay to bring his cheese-steak sandwich in since he wasn't trying to smoke it.

_______ 2. *Be bloody, bold, and resolute; laugh to scorn,*
The power of man, for none of woman born
Shall harm Macbeth. . . .
But yet I'll make assurance double sure,
And take a bond of fate. . . .
Macbeth shall never vanquish'd be until
Great Birnam wood to high Dunsinane hill
Shall come against him.

—William Shakespeare, *Macbeth*, act 4, sc. 1.

[Note: The fallacy is on the part of the character Macbeth, not on Shakespeare. On the basis of this prophecy, he made an unwarranted assump-

*Note: Some of these passages, are, strictly speaking, reports, but they contain arguments attributed to others or they lead into arguments. For the purposes of this exercise, examine the arguments contained in the reports for fallacies.

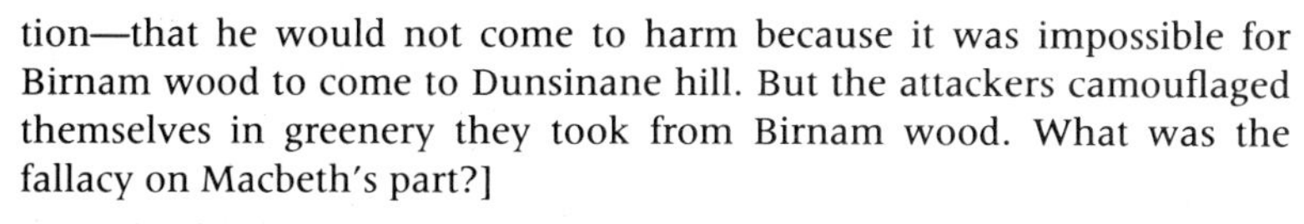

tion—that he would not come to harm because it was impossible for Birnam wood to come to Dunsinane hill. But the attackers camouflaged themselves in greenery they took from Birnam wood. What was the fallacy on Macbeth's part?]

_______ **3.** "The fact is that AIDS is transmitted by sex. This is important because you're sleeping with every person your partner has slept with over the last ten years." —Actress Elizabeth McGovern in a Public Service television announcement sponsored by the Red Cross.

_______ **4.** STUDENT: "You wrote a comment on my essay test that I think was insulting and uncalled for."

INSTRUCTOR: "What was the comment?"

STUDENT: "You wrote, 'The only good thing on this test was the last page.' I checked, and the last page of the examination booklet was blank."

INSTRUCTOR: "I meant the last page *you wrote*."

[Identify who committed the fallacy, if there is one, as well as what it is.]

_______ **5.** "We hold these truths to be self-evident, that all men are created equal, and that they are endowed by their Creator with certain unalienable rights: among these are life, liberty and the pursuit of happiness." —Thomas Jefferson, *The Declaration of Independence.*

M O D U L E 37

FORMAL FALLACIES

PART 4: FORMAL INCONSISTENCIES

21. Special Pleading

The fallacy of **Special Pleading** is a distinctive type of formal inconsistency that occurs if and when two statements, mutually contradictory, are advanced by the same speaker or writer.

Here is an example of formal inconsistency taken from a newspaper horoscope cited by *New Yorker* magazine:

> **How's That Again? Department**
>
> [Horoscope in the Detroit Free Press]
>
> ARIES (March 21–April 19): Defer to someone who is more experienced than you. Don't take a backseat to anyone.

We see that this horoscope is not only an appeal to superstition; it is also logically contradictory. Such advice is not only difficult to follow—it is *impossible* to follow. Since logical contradictions baffle people, many people sometimes are fooled into believing there is profundity there. This is a favorite technique of con-men and charlatans. Since you can't understand a contradiction, you may

---

A deductive argument is **valid**, if and only if, if all the premises are true, then the conclusion must be true.

be tricked into believing the inability to understand what is being said is your deficiency, when the real deficiency is in the argument or advice itself. A clear-thinking person will recognize such nonsense for what it is.

It is likely that many of the beliefs we have, were we to examine them, would reveal inconsistencies if we were to spell them out on paper. Perhaps we believe that "a stitch in time saves nine," and we use the proverb whenever we want to do a job right on the spot. Other times, though, we may feel a bit like procrastinating; then we may invoke another proverb, "Haste makes waste." There is nothing formally inconsistent about the two proverbs (although there may be a slight tension between the two), but using one to do whatever it is you want to do right now, and the other to put off 'til tomorrow whatever you don't want to do may constitute a case of special pleading. Special Pleading is an inconsistency of special interests. One adopts a principle of action, follows through on it when it suits him, and falls down on the job when it doesn't suit him. When a parent tells a child, "Don't do what I do; do what I say!" the parent may well be committing the fallacy of Special Pleading.

DEFINITION: A fallacy of *Special Pleading* occurs whenever one argues for a special exemption from inconsistency on the basis of one's own advantage or interest.

EXAMPLE:

AIDE: Congressman, you promised your voters that you would oppose any pay increase for House members until we got the budget under control, but you just voted for a special pay raise for yourself.

CONGRESSMAN: When I made that promise, I didn't intend it to include *me*.

COMMENT: The Congressman had campaigned against *all* pay raises.

The following is an example of Special Pleading taken from *Ms.* magazine:

> "We don't practice discrimination in reverse. No, we hire more women because women, for the most part, work harder than men and are simply better in their jobs—less posturing, less walking *around* the problem. . . . we just get in and get it done."

To see why this is Special Pleading, turn the example around and substitute the word 'men' for women and the word 'women' for men. If a male executive used that argument to defend the practice of hiring only men, would that sound like discrimination to you?

Here is a rather blatant example of Special Pleading, taken from *Barnotes,* a publication for those in the legal profession, published by the North Carolina Bar Association. The topic is capital punishment.

> A death sentence in appropriate cases is not "murder" by a society indifferent to the value of human life. It in fact confirms society's value for human life. To fail to impose death for the most outrageous and aggravated murders would reflect just how lightly the human life of the victim was considered by the community.

One is reminded of the story of the lawyer who said he had to cheat his clients in order to make an "honest" living.

22. Affirming the Consequent

The fallacy of **Affirming the Consequent** occurs whenever an argument is of the following form: Given any two discrete statements (P and Q), the fallacy is committed if we assert:

If P then Q;
Q;
Therefore, P.

Consider the following example: Someone might argue that if the citizens of the United States were living in a totalitarian society as sophisticated as that portrayed in George Orwell's book, *1984*, no one would know it because thought control would be too pervasive to allow so critical a judgment of one's society. "We don't know, as a matter of fact, that we're in an Orwellian thought-controlled totalitarian state. Therefore, that only proves my point: this is '1984'."

If you reflect on this argument for a moment, you will see why this is muddled thinking: Maybe we don't *recognize* this as a totalitarian society because it *isn't*.

Pointing out that the argument is of the form of the fallacy of Affirming the Consequent is sufficient to dismiss the argument altogether, but we might also recall that the argument about Orwell is also a specimen of *argumentum ad ignorantiam*. The fact that we don't know that this is an Orwellian dystopia is not reason to believe that it is.

23. Denying the Antecedent

A **Denying the Antecedent** fallacy is committed if one asserts:

If P then Q;
It is not the case that P;
Therefore, Q (or not Q).

For example, if one thinks illogically, he might argue: "If you want to become a successful businesswoman, then you must dress well. You obviously have no interest in entering business. Therefore, you're sure to be a sloppy dresser."

24. Affirming a Disjunct [We might also call this: "The Fallacy of the Christmas Tree Lights."]

Given any two discrete statements (P and Q), the fallacy of **Affirming a Disjunct** is of the following form:

Either P or Q;
P;
Therefore, not Q.

Anyone who has ever tried testing Christmas tree lights wired in series will know why this is a fallacy. When such lights are in series, if any one of them is burned out, none of the bulbs will light. Suppose you suspect it is the red bulb at the end of the string. You test the bulb and find it is burned out. You might be making a bad mistake if you went ahead and strung the wires over the tree, because there is a very good chance that at least one other bulb is burned out as well. The argument looks like this: "Either the red ('R') or the green ('G') or the blue ('B') or the yellow ('Y') bulb is burned out." You know, after testing, that 'R' is burned out. It is a fallacy, however, to conclude that no other bulb is defective—or that any other bulb is! Thus, either 'R' or 'G' or 'B' or 'Y' is burned out; 'R' is burned out; but no other conclusion necessarily follows.

25. Maldistributed Middle

Use of the **Maldistributed Middle** is one of the three formal ways that a categorical syllogism can go wrong. It violates the condition that the middle-term in a syllogism must be distributed exactly once. Thus, the fallacy is committed (for example) if one argues that all battleships are grey; moreover, some dinghies are grey as well. Therefore, some dinghies are battleships.

26. Unequal Distribution

Unequal Distribution is the second of three formal ways that a syllogism can go wrong. It breaks the rule that no end-term may be distributed only once. For example, the following argument is fallacious in this way:

"All battleships are warships; this destroyer is not a battleship. Therefore, no destroyer is a warship."

In standard syllogistic form, this becomes:

All battleships are warships.
Not all destroyers are battleships.

No destroyers are warships.

As you can see by the rules of distribution, each of the end-terms is distributed once, thus violating the Three Rules (see Module 18) on both accounts.

Notice that you need to find just one end-term that is distributed only once in order to detect a violation of the Second Rule.

27. Unequal Negation

Unequal Negation is the third way a syllogism can exhibit faulty thinking. It violates the rule that states that the number of negative premises must be equal to the number of negative statements in the conclusion. An example of Unequal Negation: "No destroyers are submersible; no submarines are destroyers. Therefore, no submarines are submersible."

EXERCISES

DIRECTIONS (Read Carefully): *Find the fallacies in the following arguments (or parts of arguments) and enter the correct name in the space provided. If the argument commits no fallacy, write "No Fallacy" in the space provided.* Beware of the False Charge of Fallacy! *Note: Some passages may contain two or more formal fallacies. Identify as many as you can.*

______ 1. Fascism denies, in democracy, the absurd conventional untruth of political equality, dressed out in the garb of collective irresponsibility, and the myth of "happiness" and infinite progress. But if democracy may be conceived in diverse forms—that is to say, taking democracy to mean a state of society in which the populace are not reduced to impotence in the State—Fascism may write itself down as "an organized, centralized and authoritative democracy." —Benito Mussolini, "The Political and Social Doctrine of Fascism"

______ 2. "I never forget a face, but in your case I'll make an exception." —Groucho Marx

______ 3. LAUREL: Everytime we have a Democratic president, we have a war.

HARDY: Jimmy Carter was a president. We had no war while he was in office.

LAUREL: He's the exception that proves the rule!

______ 4. The automatic camera being tested by the city of Pasadena, as well as its photo radar, would be considerably less useful if traffic lights were properly synchronized. I believe that the principle attraction of these devices is their revenue-generating potential. The problem for us drivers is that the city gets more return on its "investment" by not synchronizing the lights.

______ 5. If lupins have colorful clusters and not all azaleas have colorful clusters, then azaleas are lupins.

______ 6. All lungfish are Dipnoi; Dipnoi have lungs as well as gills. Therefore, everything that has lungs as well as fins is a lungfish.

______ 7. Halberds are weapons. Thus, some axes are halberds because some weapons are axes.

_______ **8.** If someone is predestined to hell, he won't realize it. Ellen doesn't have the faintest idea that she's going to hell. Therefore, she must be!

_______ **9.** SALLY: According to the student handbook, if a student misses more than six classes during a term, then that student fails the course.

JOHN: But you missed most of the classes in logic last term.

SALLY: That doesn't count. I'm the former student body president, and I made the rule.

M O D U L E 38

PROPAGANDA

38.1 PROPAGANDA DEFINED

Shakespeare reminds us that the devil can quote scripture to his own purposes. A superficial knowledge of the fallacies can be used in a cynical way by commercial interests and governments to manipulate public opinion. Notice that this is rhetoric, not logic. Logic strictly speaking should aim at the truth. Rhetoric aims at clear communication, but sometimes rhetoric can be used to manipulate opinion without regard for the truth.

A real danger exists that we will become so skilled at rhetoric that we deliberately violate the strictures of logic and direct our sights only at success in argument. When masses of people are persuaded of views that are argued for with suave disregard for the truth or logic, we call these techniques **propaganda.** Propaganda comes in many forms, and we must always be on the lookout for it. Sometimes propaganda is the instrument of governments and politicians;

> *"All propaganda has to be popular and has to adopt its spiritual level to the perception of the least intelligent of those towards whom it intends to direct itself."*
>
> —ADOLF HITLER, *Mein Kampf*

A deductive argument is **valid**, if and only if, if all the premises are true, then the conclusion must be true.

other times it may be the instrument of businesses and advertising agencies. Madison Avenue may attempt to persuade us to buy products we don't really need for prices made exorbitant by the very costs of advertising. Politicians and pressure groups may attempt to get us to support policies that serve only special interests instead of the general good. A major benefit from the study of logic is the development of a critical ability to recognize and debunk propaganda.

The dictionary defines propaganda as "the systematic propagation of a given doctrine or allegations reflecting its views and interests."* This raises a crucial difference between logical argument and propagandistic technique. For the student of logic, the only interest should be in discovering the truth. The propagandist is more concerned with promoting her own point of view even if it's at the expense of the truth.

Where did the word 'propaganda' come from? Interestingly, it got its start in religion. The word 'propaganda' derives from an arm of the Catholic Church by the same name. 'Propaganda' is the short name for the *Sacra Congregatio de Propaganda Fide*—the "Society for the Propagation of the Faith." This congregation of the Catholic Church once went under the name, "the Inquisition." Thus, the word 'propaganda' has long-standing negative connotations.

Some analysts of propaganda like to argue that propaganda is value-neutral; it can be used to mislead, deceive, or manipulate, but sometimes it can be used for worthy ends and purposes as well. That point of view confuses ends and means—the ends or purposes of indulging in propaganda may be good and honorable (such as "winning the war") but the *means* used to effect the ends can be unworthy (such as portraying "the yellow race" as brutal and cowardly). William Hummel and Keith Huntress, who wrote an influential book on the subject, take the "value-neutral" stance on the subject of propaganda, but even they concede that 'propaganda' ". . . is a sinister word. The average American, hearing it, is reminded of spies and secret police, of cynical reporters and biased magazines, of lobbies and special privilege and lies."† Hummel and Huntress concede too much. The word 'propaganda' does indeed have negative connotations. Since at least World War II, 'propaganda' has had the suggestion of moral censure built into it.

However, these analysts argue that "propaganda can be used for a good cause as well as for a bad one"‡ and certainly we can agree that's true. The Red Cross can use propaganda to promote its blood drives, the government can use propaganda to sell war bonds, and so on, but this misses the point. Advertising can be informative or it can be "propaganda for a product." When it is informative, it doesn't need to use spurious logical techniques to persuade us to buy a product—we buy the product because we have a need for it and we are well-

***The American Heritage Dictionary of the English Language*, Boston: Houghton-Mifflin, 1970, p. 1048.

†William Hummel and Keith Huntress, *The Analysis of Propaganda*, New York: The Dryden Press, 1949, p. 1.

‡*Ibid.*, p. 2.

informed consumers. But when advertising is "propaganda for a product," it may sell us a product for which we have little or no need, or on false pretenses.

Hummel and Huntress contrast German and American news accounts about the Battle of the Bulge to support their contention. They note that German newspapers

> "cited tremendous Allied losses, and claimed that the front would be stabilized, that German arms would yet be victorious. American newspapers reported the Battle of the Bulge as a limited German success, headlined Allied heroism, and minimized the possibilities of further German exploitation of the break-through."

They continue:

> "In the event the American newspapers were right, but you must note that the news stories of the two countries were trying to do the same thing—inform readers about some important events and give them confidence in the armies of their countries. The American stories were the more accurate, but the purpose was the same."*

This simply isn't true. If a newspaper is committed to reporting the news accurately, it is not indulging in propaganda. The editorial page may carry a point of view, but objectivity is the watchword on the front pages where the news is reported. I have often read the [conservative] *Los Angeles Times* and the [liberal] *New York Times* accounts of the same news events, and they are almost identical, paragraph by paragraph. This suggests that a standard of journalistic objectivity does indeed exist.† If such standards for accuracy are rigorously followed, no accusation of propaganda is justified.

38.2 EXAMPLES OF PROPAGANDA

Political Propaganda

Propaganda may use a combination of lies and deliberate illogic to persuade. Let's consider some classic examples.

The Nazis were masters of propaganda. The Germans of the 1920s and 1930s were a people demoralized by a crushing armistice at the conclusion of World War I and by oppressive inflation brought on during a global depression. The Nazis capitalized on these conditions and came to power by appealing to

**Ibid.*, p. 2.

†The main difference that I detect between the *New York Times* and the *Los Angeles Times* in their front-page news coverage is that the *Los Angeles Times* tends to use more adjectives per sentence than does its East Coast competitor. Perhaps that accounts for why the *Los Angeles Times* tends to be a *thicker* newspaper.

the worst in human nature. Deliberately appealing to the fears and prejudices of the nation, the Nazis mounted a campaign which irrationally blamed hard times on a religious minority whose customs and practices were different from those of the majority of the people. They sought scapegoats for Germany's woes and rode to power on irrelevant and hateful appeals to fear. The eventual result was World War II and the murder of Jews in what we've come to call "the Holocaust." Interestingly, many top scientists and philosophers who were skilled in logic saw through the propaganda and successfully escaped to Britain and the United States where some of them played major roles in helping the Allies win the war.*

Similarly, propagandists can make good use of warm images and enthralling music to create a good feeling about the subject matter. This is a much more subtle (and, I think, more effective) technique than a crude appeal to fear. This approach relies on a fallacious appeal to sympathy to work its magic, and one of the first to master the technique was filmmaker and photographer Leni Riefenstahl.

Riefenstahl was already a filmmaker of some note when she was approached by Adolf Hitler and the Nazi Party in 1933 to do a "documentary" of the upcoming Nuremberg Party rally. She decided to direct the film as though it were a commercial project instead of the overt propaganda originally conceived by the film's planners. The result was one of the most powerful and effective pieces of propaganda ever attempted—a film that stands to this day as a benchmark of innovation in both film and propaganda technique.

A film of sounds and images, it utilizes no narrative overlay whatsoever. After a series of rather clumsy title cards (Riefenstahl confessed to being poor at presenting titles), the film makes a dramatic switch in tone and presentation. The viewer is transported through billowing cloud banks as though he were soaring through the air like an eagle. Soft strains of martial music play in the background—the German national anthem and the official anthem of the Nazi Party, "The Horst Wessel Song."

Suddenly, a majestic and powerful plane bursts through the clouds bearing Adolf Hitler to the cheering masses below. The plane lands in Nuremberg, and Hitler makes a dramatic exit to throngs of admirers shouting "Heil Hitler!" with arms outstretched in the Nazi salute. Hitler is transported by a motorcade through the streets of Nuremberg, lined with cheering onlookers, en route to the site of the rally, a large convention hall filled to capacity and decorated with gigantic red banners bedecked with swastikas. This is a film so well executed that even now one might find its overpowering charm hard to resist. It was a film designed to make every German citizen proud to be a German! Although a fallacious Appeal to Sympathy—*argumentum ad misericordiam*—that was precisely its fatal attraction. With hindsight, we recognize the lie behind the glowing promise.

*For example, Albert Einstein, Enrico Fermi, Karl Popper, and Rudolf Carnap.

By the way, every president of the United States since John F. Kennedy has re-enacted the entrance at the Nuremberg airport in "The Triumph of the Will" when emerging from Air Force One. To be sure, they saw the film too.*

Propaganda in Cinema and Television

'Propaganda' has become a term of disfavor since Joseph Goebbels called himself the Nazi "Minister of Propaganda." Never again can a propagandist be so candid about the nature of his work and expect to be listened to. Consequently, we should not expect to find propaganda openly packaged as such. We'll be told that what we're reading or watching is "news" or "entertainment"—never "propaganda."

Two major sources of contemporary propaganda are the motion picture industry and television. Consider some recent examples of movies that have propagandistic undertones:

Red Dawn. On the surface, the movie *Red Dawn* (1984) looks like a harmless and juvenile adventure tale about what might happen during "the next war." However, a closer look reveals a very cynical design and purpose. The storyline is rather absurd: war breaks out between the United States and "the Communist Bloc." The Soviet Union attempts a land invasion of the United States but gets bogged down in heavy resistance in Alaska. On the other hand, the Communist Cubans and Nicaraguans have much more success; they send in squads of paratroopers who occupy the entire Midwest of the United States.

Any critical thinking focused on this movie will amply show its absurdity (tiny Nicaragua successfully invades the United States?), but some reflection will reveal its obvious message: a Communist Nicaragua is a serious threat to U.S. national security. The film was released at the height of the U.S. debate over whether to aid the Contras in Nicaragua.

Is this really propaganda? Consider two giveaways: first, the opening credits reveal that *Red Dawn* is "a Valkyrie Production"—'Valkyrie' is a mythical Norse name for the gods who hover over battlefields—a name that has distinct Wagnerian overtones.† Second, the film begins with stirring martial music and images of being transported through swirling clouds—an opening that is almost identical, frame-by-frame, to Leni Riefenstahl's "The Triumph of the Will." This film has so many political associations that it strains our belief in coincidence.

Rambo: First Blood, Part II. Another film with right-wing associations is Sylvester Stallone's, *Rambo: First Blood, Part II* (1985). This is a film that height-

*A point of interest: Leni Riefenstahl is still very much respected for her work. Recently, she published a book of photographs of Marilyn Monroe.

†According to William L. Shirer in his book, *The Rise and Fall of the Third Reich,* Hitler used to say, "Whoever wants to understand National Socialist Germany must know Wagner." (New York: Simon and Schuster, 1960, p. 101). *The Valkyrie* was the name of an opera by German composer Richard Wagner, composed in 1857, and was based on the ancient Germanic legends.

ens the apprehensions of Americans about M.I.A.s and P.O.W.s in Vietnam. Stallone plays "Rambo," an alienated "super-soldier" who returns to Vietnam fully a decade after the end of the war to rescue Americans who are still captives there.

We discern its propagandistic effect from critical reflection on what the film shows us. For example, the Vietnamese soldiers are depicted as wearing World War II Japanese army uniforms. This is especially inappropriate because Vietnam (as the French colony called "Southeast Asia") was subject to harsh Japanese occupation during World War II. The Vietnamese were on the other side in that war! Either a fallacious appeal to fear is intended by the filmmakers, or the properties department of the film studio had a special on Japanese uniforms during the filming of *Rambo*.

The "Russian advisor," without whom we are led to believe the Vietnamese would be helpless bunglers, does indeed wear a Russian uniform (even if the insignia are wrong for his mission), but he smokes his cigarettes by holding them with the thumb and forefinger in the manner of World War II Nazi soldiers. Russians customarily smoke cigarettes by holding them between the third and fourth fingers of the hand. This misrepresentation either shows callous disregard for or ignorance of the facts, or the intent to associate the Russians with Nazis (their tragic nemesis in World War II) in a deliberate and fallacious appeal to fear and prejudice (*argumentum ad metum*).

Hollywood isn't exclusively conservative in its politics and many films are made by the industry that have liberal political overtones. Interestingly, the fallacies appealed to in liberal propaganda are usually different from those featuring conservative propaganda. Here are a couple of candidates.

The China Syndrome. *The China Syndrome* (1979) is a film about a major accident at a commercial nuclear power plant. The film's propagandistic intent was obscured by a remarkable coincidence: the nuclear meltdown at Three Mile Island occurred within weeks of *The China Syndrome*'s release. Thus, the film was hailed as prophetic.

'China Syndrome' is a term in the nuclear industry used to describe the hypothetical result of a total meltdown of the fissionable core of a nuclear power plant; theoretically, the super-heated core will melt through its housing and continue melting through the crust of the earth. Some scientists predict tongue-in-cheek that, with nothing to stop it, the hot core will continue melting all the way through to the other side of the earth—all the way to China (hence, the origin of the term).

Viewed apart from the incident at Three Mile Island, *The China Syndrome* appears to make a fallacious appeal to fear. However, because an incident very much like the one described in the film actually did occur, the implicit appeal to fear seems to be amply justified: nuclear power plants have proven that they can be exceedingly dangerous.

Nevertheless, I want to argue that the film still has a propagandistic effect. True, nuclear power is potentially more dangerous than anyone originally imagined. Still, *The China Syndrome* presents the construction of nuclear power plants as an industrial *conspiracy.* Those responsible for the construction and maintenance of the plant are portrayed as greedy and irresponsible people who built and maintained this plant despite knowing the dangers it presented to the community.

In fact, most of the people who build and operate nuclear power plants are not like that at all. The nuclear engineers that I know have always regarded commercial nuclear power as a potential boon to the consumer and the community. Since the accidents at Three Mile Island and Chernobyl, these scientists have been concerned about finding ways to make nuclear power genuinely safe and affordable. They may or may not succeed, but they are not the monsters the film portrayed them to be.

In what ways was *The China Syndrome* propaganda? Examine the way the film attacked the character and motives of the people who ran the nuclear power plant, even allowing for an artwork's license to portray characters that are not well-rounded. The film used *Ad Hominem* attacks and appealed to Faulty Motives to make its case; in that respect, the film is propagandistic in its presentation.

The Last Emperor. A film by Italian director Bernardo Bertolucci, *The Last Emperor* (1987) is an award-winning look at the recent history of China. China in the twentieth century is revealed to us through the eyes of the last emperor, who was kept a virtual prisoner in "the Forbidden City"—the emperor's palatial grounds—from childhood. China is shown to be so torn apart by factions and colonial powers that by the end of the film, we come to regard the imposition of Communist rule as a welcome relief. To be fair to Bertolucci, the film is at least mildly critical of the Chinese Communists at one point, during the "Cultural Revolution."

At the film's conclusion, the emperor (who has undergone "re-education") visits the Forbidden City as a tourist and sees the palace from an entirely fresh point of view. After meeting a young boy dressed in the uniform of the Young Communist League, he returns to a section of the throne room, now roped off, and locates a "cricket box" which he had placed there as a boy many years before. The cricket is a symbol of luck. He gives the box to the boy who opens it and discovers the cricket, alive after all this time.

I interpret this episode of the film to represent a passing of the scepter of power from the old China to the new. The matter of the transference of power by itself, of course, would be an uninteresting fact of life in China: the Chinese Communists obviously did come to power. What this scene seems to represent, however, is an assignment of legitimacy to Chinese Communist rule; the boy as a member of the Young Communist League is portrayed in that role as the *rightful heir* to the powers of the emperor.

We may argue with some justice that *The Last Emperor* oversimplifies this segment of history.* In the film, we are so exposed to the abuses of the old ways of Chinese rule that the Communists seem to represent the only alternative. This

*Especially given recent events, such as the slaughter in Tiananmen Square.

may not have been true—perhaps China could have given up the imperial rule in favor of a kind of democracy. If we can make this case, then the film may be viewed as a type of "left-wing" propaganda which commits a fallacy of False Dilemma. Other alternatives may have been possible and preferable to the outcome portrayed in the film.

Television. Television sitcoms and melodramas may also indulge in political propaganda. Consider, for example, how labor leaders are usually portrayed on television as almost always criminal bosses or thugs. One such show that's in re-runs is the detective series from the 1970s, "Hart to Hart." Recall the storyline: a wealthy and handsome married couple use their independence, skill, and intelligence to fight crime. They are the "corporate heroes," the front-line of capitalism. They are a loving couple, happy and carefree, but they choose to devote their energies to combating the enemies of law and order. Every now and then, they employ their cunning and expertise against corrupt union bosses who are portrayed as evil and avaricious. What corporation wouldn't be delighted to sponsor such a show?

Even so innocuous a program as the old "Star Trek" series occasionally ran episodes that were very obviously justifications for American involvement in the Vietnam War. One such program portrayed the noble Federation and the evil "Klingons" as competing for possession of a planet in a strategic quadrant of the galaxy by supporting the backward inhabitants in a "brush-fire war."

Though they have changed somewhat in recent years, sitcoms used to present highly idealized representations of American family life that created role models for families that were almost impossible to emulate. Nowadays, many sitcoms portray adults as hopeless bunglers who would be far better off if they would only listen to the wise counsel of their own children. Perhaps the recent success of some sitcoms like "The Cosby Show" and "Roseanne" testify to a change in national consciousness. In those programs, the *adults* tend to get the funny lines.

Some of the portrayals of American life in television and in the movies may have unintended (and unwanted!) propagandistic effects. Consider, for example, how much violence is portrayed in film and television and imagine the image of American society that someone living in Kuwait would form as a result of watching imported detective and police dramas that display such violence. They would certainly imagine American cities as horribly dangerous, and violence as an American way of life.

Have you ever considered that music videos may be a sort of propaganda? What do you see when you watch "MTV" for example? Do the videos have strong sexual overtones? Do they portray sex as healthy, or do they often titillate us by conveying the suggestion that sex is forbidden or obscene? How do such videos depict women? Do they make them out to be creatures of fashion? Do they show beautiful women standing in the background? Have you seen videos that parade lines of beautiful women who are dressed and made-up to look exactly alike? Consider the video for Robert Palmer's "Simply Irresistible" where a chorus line of women are made up to look like identical mannequins, or Prince's "Batdance" video which shows a troupe of identical "Vicky Vales" wearing T-shirts that read: "All this, and brains too!"

Consider the lyrics of some music in this decade: "I'm a material girl in a material world." "Girls just wanna have fun." What are these songs saying? Have you ever thought about it?

Some social critics have tagged a recent phenomenon in current music as "Fascist Rock." "*We will, we will* rock *you!*" are the driving lyrics to a stirring song by "Queen" in the early 1980s. Perhaps this song is innocent enough, but I've been to some sporting events where the public address system played this song, the audience joined in, and the crowd's fervor seemed to take on a sharper, ominous, more vitriolic intensity. More recently, "Guns N Roses" demands "Police and niggers, get out of my way" in some especially offensive lyrics to the song, "One in a Million." Rock music, unfortunately, lends itself to this sort of treatment very well.*

Commercial Propaganda

No doubt, we could devote whole volumes to the subject of commercial advertising as propaganda, and, indeed, many books are dedicated to the subject. However, I will consider only a few particularly noteworthy cases.

A few years ago, a major insurance company ran a series of television commercials designed to get you to buy life insurance. The commercial showed a typical American family in a tranquil and happy domestic scene—a handsome father, a beautiful mother and their young child. Suddenly, the frame freezes and the handsome father disappears from the picture.

"If you were suddenly 'out of the picture,' " intones a narrator, "what would happen to them?" The widowed mother and her daughter look despondent, worried, frightened for the future. You begin to worry whether you've adequately provided for your family. But cheer up. If you took out a policy with "XYZ" life insurance, everything is rosy! Cut to a slow-motion scene of happy mother and daughter running carefree through a field of flowers.

This was a blunt piece of propaganda. Its appeal to the emotional—to your sense of guilt, fear and pity—are relentless and obvious. The commercial described here enjoyed some brief success but was soon removed from the air (for whatever reason) never to be seen again.

Another commercial that appeared in years past and was removed, apparently for its unintended consequences, was a commercial by a major holding company advertising its many diversified branches. The commercial showed familiar product after familiar product and identified each household name as

*Compare these songs to the pounding cadences of "The Horst Wessel Song," the official anthem of the Nazi party, composed by the son of a Protestant minister who turned his back on his family and the ministry to fight for the Nazi cause in its early years. Before he was killed, supposedly by Communists, in 1930, Horst Wessel composed these words for his song: "Raise high the flags! Stand rank on rank together./Storm troopers march with steady, quiet tread. . . ."

an acquisition of this major company. "We're Beatrice," boasted the commercial's narrator.

Upon seeing this commercial for the first time, however, I was at a loss to see what Beatrice Company was trying to get us to buy or believe. Perhaps the commercial was intended to be a wholesale ad for its various branches. But it didn't have that effect on me. As a matter of fact, I came to regard the commercial as highly intimidating. My attitude was: "You mean one company owns *all* that?" I saw the Beatrice commercial as an unabashed boast; they were bragging about their *power.* I can't say for sure whether that was the commercial's original intent, but perhaps it had this unforeseen result because the ad soon disappeared from the air.

On the other hand, what if the intimidation were intended? Perhaps the suggestion they wished to implant was something like, "Nothing succeeds like success." We're supposed to buy their subsidiaries' products because we want to be on the winning side—we're intimidated by their sheer size and power to want to buy their goods. If that was the intent, then recognizing it constitutes a sufficient antidote to the commercial's subliminal message. A little bit of critical thinking goes a long way.

Here's an example of a television commercial that is seen by viewers in the Western United States. We see some cowboys in a pickup truck watching a Western movie in a drive-in theatre. They run out of Pace® Picante Sauce and one of the cowboys offers a jar of his own. One of them reads the label and exclaims, "This sauce is made in *New York City*!" Suddenly, the characters in the movie stop their gunfight and exclaim, "NEW YORK CITY!" A tough looking hombre comes on the screen and moans, "That really chaps my hide!"

As an experiment, I went into several well-stocked supermarkets locally to look for a picante sauce made in New York. I couldn't find any. Then I began to reflect on what the real message of the commercial might be. The company was trying to sell its product, not on the basis of its quality, taste, or price, but just because it was not produced in that iniquitous metropolis of the East Coast. In other words, it was playing on pure prejudice.

People just let commercials wash over them. Commercials are most dangerous when their message is subliminal; much of the time, we have little difficulty in identifying the intended message, whether it's "Buy Care-Free Cola!" or "Save at Infidelity Bank!" That is a minimal amount of critical thinking, but that's not all that's required of us. Even a little reflection will deflect the greatest effect of the subliminal message; if we're conscious of what they're trying to sell us or to get us to believe, we've taken the first major step to avoid being conned.*

A final word about advertising: Commercial advertising doesn't need to be propagandistic to be effective. Consider some of the ads that are designed to appeal to educated audiences—commercials by some computer companies or commercials for oil companies that sponsor major plays or symphonies on television. These ads are as often appeals to intellect as to desire. "You have a need

*Remember that annoying commercial that concluded, "Offer will expire by midnight tonight!" and we'd see it every night for years!

for a product that does this and that? Ours does it well and at a price you can afford." "M__________ Corporation is working to insure that oil reserves will be adequate to supply the nation's needs well into the next century." Even if you don't agree with them, such companies typically give good reasons for their claims. I've had discussions with executives and scientists from such computer and oil companies who believed these messages. Sometimes they showed me where my misgivings were unfounded; other times they listened to me with respect and attention. Such commercials are invitations to discussion and argumentation.

Sometimes public relations statements can create the appearance of concern and commitment when the reality is very different. Consider the following example.

Shortly after the catastrophic oil spill in Prince William Sound, Alaska, Exxon Corporation ran the following full-page advertisement in many of the major newspapers in the United States. This accurate replica is taken from the *New York Times:*

> **An Open Letter to the Public**
>
> **On March 24, in the early morning hours, a disastrous accident happened in the waters of Prince William Sound, Alaska. By now you all know that our tanker, the Exxon Valdez, hit a submerged reef and lost 240,000 barrels of oil into the waters of the Sound.**
>
> **We believe that Exxon has moved swiftly and competently to minimize the effect this oil will have on the environment, fish and other wildlife. Further, I hope that you know we have already committed several hundred people to work on the cleanup. We also will meet our obligations to all those who have suffered damage from the spill.**
>
> **Finally, and most importantly, I want to tell you how very sorry I am that this accident took place. We at Exxon are especially sympathetic to the residents of Valdez and the people of the State of Alaska. We cannot, of course, undo what has been done. But I can assure you that since March 24, the accident has been receiving our full attention and will continue to do so.**
>
> **L. G. Rawl**
> **Chairman**

As you see, Exxon claimed that it "moved swiftly and competently" to deal with the problem. However, the *Los Angeles Times* editorialized on May 6, 1989 that Exxon's first clean-up plan was "found to be inadequate by federal officials" and that its second plan "seems to retreat even from the commitment of the first plan, proposing 3,400 workers compared with 4,000, and calculating 364 miles of contaminated beach when Alaska officials count double that amount."

The editorial went on to cite Alaskan Secretary of the Interior as saying that the effort was "make-work for appearance's sake." Moreover, the *Los Angeles Times* chided Exxon for noting that any costs incurred in the clean-up "not covered by insurance will be tax deductible." The editorial concluded: ". . . Exxon seemed to treat the disaster as just one of those things. Yes, the oil spill was a bad one, the firm said in a recent statement, 'but accidents can happen to anyone.' The arrogance of the company seems to be exceeded only by its insensitivity to the dimensions of the disaster and the concern of the people throughout the country for the Alaskan environment and wildlife."

Other Types of Propaganda

So far we've looked at two types of propaganda: that with a political purpose and that with commercial intent. Yet any amount of reflection will reveal that the attempt to persuade mass audiences on any subject of controversy is susceptible to a propagandistic treatment.

As mentioned before, the word 'propaganda' derives from the official Latin name of the Catholic Church's "Inquisition." It follows then that one very fertile ground for propaganda is religious conversion. This should be obvious, but for some reason many people seem to be reluctant to draw such conclusions or to apply the same standards of logical propriety to religious issues. I would argue that logical caution is as much in order here as it is in politics.

One of the fertile grounds for logically bamboozling the public is television evangelism. When the preacher makes an appeal for donations, be on your guard. One North Carolina preacher made this appeal on television: "Send your donation of ten, twenty, a hundred dollars, or whatever you can afford. In return for your donation, the ministry will send you this holy, holy dime, blessed by the Lord Jesus Christ, and as long as you have this holy, holy dime you will never be without money in your pocket." The tautology is delightful (of course you will never be without money—you have the dime), but perhaps we can detect a fallacious Equivocation here because what the audience will wishfully believe is that with this "magic charm" they will never have need for money, and that is just superstition.

Of course, maybe we should distinguish between borderline "con-artistry" and propaganda. Religious leaders make propagandistic appeals most clearly when they demand your uncritical belief on penalty of damnation or exclusion from the group. It's in these cases that fallacious appeals to fear come into play. However, we should note that these are *abuses* of religious sentiment rather than anything intrinsic to religious appeal. Moreover, we should also recognize that from a historical point of view logic and the appeal to reason took great strides under the patronage of many religious organizations. The attempt to prove the existence of God, for example, did at least as much to perfect systems of logical analysis as it did to advance belief in basic religious tenets.

Another area of controversy that is susceptible to propagandistic treatment is the subject of sexual politics, a subject now very current in the United States. Any area where lines of allegiance are sharply drawn is ripe for propagandistic conversion and too often the basis for reasoned discussion may fall victim to a pressure-group sponsored party line. Perhaps both sides of the abortion controversy exemplify this point but we may see the opposing positions as logical *contraries*, where both positions can be false but it is not the case that both positions can be true. The careful, reasoned stance of the independent thinker may actually help to dispel the controversy.

38.3 RECOGNIZING PROPAGANDA

Be cautious of films, books, or shows with political, commercial, or religious topicality.

Remember the words of Samuel Johnson, that patriotism "is the last refuge of the scoundrel." *Any* film, *any* book, or *any* paid political announcement designed to "make us feel good about America" will likely be asking us to swallow a bitter pill whether it's inflation, a new tax program, a rationing, or an attempt to drum up support for an intended war. In any case, such deliberate emotional appeals necessarily evade the issues present in any political campaign or political argument. That's why we should view them with suspicion.

Look for fallacies in the mass appeals related to religion, politics, or commercialism. If you find such fallacies, and if you judge that the people who are trying to persuade you have used fallacies cynically (i.e., knowing that they're committing fallacies but doing so anyway), then you are reasonably justified in assuming that you have encountered propaganda.

Propaganda may appeal to fallacious argument or it may just engage in deliberate lying. Adolf Hitler said that people "will more easily fall victim to a big lie than to a small one." He observed that the way to get people to swallow your lie is to repeat it over and over so many times that it becomes familiar. Then, no matter how outrageous the lie, people will come to believe it. We call this technique "indoctrination."

Most propaganda results in a distortion of the truth in some way or other. It may even present a conclusion that is true (for example, that the United States was justified in World War II) but use deceptive or illicit means to persuade you of that. When propagandists lie to you outright, they may create the illusion of justifying themselves by appeals to fallacious argument or they may just go on repeating the lie. Polish schoolchildren, for example, were openly taught in history classes that the massacre of the Polish army officers in the Katyn Forest in World War II was done by Jews. This absurd and vicious bit of indoctrination can be easily exposed once you make the slightest attempt to think critically. The Jews and Poles fought side-by-side against the Nazis in the Warsaw ghetto. They stood to gain nothing by killing each other, and the decimated Jews were hardly in any position to massacre anyone. Who did stand to gain by the massacre of the Polish officers at the conclusion of World War II? Only with the new policy of *glasnost* did the Soviets finally come to admit to the Katyn Forest massacre.

The lesson is just this: don't believe everything you read or hear. Investigate. Think critically. Be on the lookout for fallacies or indoctrination. These are the ways a free people have of making themselves immune to propaganda.

38.4 THE TECHNIQUES OF PROPAGANDA

A final word on the techniques of propaganda is in order. Ever since Leni Riefenstahl made the film, "The Triumph of the Will," propaganda has relied increasingly on visual media. Rarely does propaganda take the form of conventional arguments with stated premises and a conclusion anymore. Instead, we are more likely to have our attention grabbed by a photograph in a newspaper or magazine or a visual image on television or a movie. "Waving the bloody shirt" is a time-honored propaganda appeal used to stir up crowds, but modern media can extend the range of this technique to mass audiences. A photograph of outraged elderly people can be used to dissuade the public from adopting a voter's initiative on insurance provisions, the image of an intimidating-looking black man can be used in a racist appeal to elect a presidential candidate, and so on.

We may ask if any fallacies can be committed where there is no conventional argument. My answer to this is simply, "Yes, because non-conventional argument forms are appealed to." A visual image can serve as a premise in an argument where certain conventions of symbols and structure are recognized. Such visual images work much as ellipsis or enthymemes do in ordinary linguistic argument. In effect, such techniques can be used in a presidential candidate's campaign ads roughly as follows: "Vote for ____________ for president. [The opponent's] revolving-door policy on crime allowed this man to go free. He subsequently murdered a housewife in Maryland." [Show photograph of a frightening-looking black man.] Why vote for ____________ for president?—That's why."

The appeal to fear is decidedly present, but no words conveyed that message; a visual image did.

This point about visual images constituting something akin to premises in non-conventional argument is a controversial one, hardly accepted by all logicians. I reserve an extended discussion of the issue for the next unit. (See Module 45, "Arguments and Visual Images.")

EXERCISES

1. A class project: Write to the chief editor of your local newspaper and request a statement of the newspaper's editorial policy. What criteria does the paper use to report the news objectively?

Next, ask yourself these questions: Does the newspaper adhere to its statement of criteria? Does the newspaper slant the news to appeal to the political and religious views of its patrons or advertisers? Does it look to the sensational?

2. A class project: Write to a television news department and request a statement of the station's editorial policy. Repeat the steps in exercise 1, making the necessary changes.

3. Write a model editorial policy for an imaginary newspaper, news magazine, or television news department.

4. Go see a movie that you suspect will indulge in conservative political propaganda. Identify any fallacies or techniques used to persuade you of a political point of view.

5. Go see a movie that you suspect will indulge in liberal political propaganda. Identify any fallacies or techniques used to persuade you of a political point of view.

6. Identify a fallacious appeal in an advertisement taken from radio, television, or print media.

EXERCISES FOR UNIT FOUR

DIRECTIONS (*Read Carefully*): *(1) Find the fallacies in the following arguments (or parts of arguments); (2) identify the person who has committed the fallacy; (3) give the most specific name you can for the fallacy (e.g.,* Post Hoc *instead of "False Cause" when appropriate); (4) answer each of the multiple choice problems. Caution: Some of the passages may contain no fallacy at all.* Beware of the False Charge of Fallacy! ***A suggestion:*** *Select some of these debates that interest you the most and identify the fallacies in those. Especially select those debates you might actually have witnessed or read about. You should work enough of the exercises to satisfy yourself that you can recognize the fallacies.**

*Each of the passages contained in this set of exercises is taken from actual transcripts or journalistic accounts of the debates. Most of them are taken from the *New York Times* transcripts. I assure the reader that I did my best to present the passages *in context,* and with approximately as many opportunities for fallacy-hunting from one candidate to another. Despite some urging to present the candidates in a "better light," I have resisted the application of cosmetics.

Be aware that reading a transcript is very different from hearing a debate. Sometimes *reading* the words of our favorite politicians can be a sobering experience—a transcript will reproduce the stutters, the stammers and the grammatical errors that go largely unnoticed over the television airwaves.

Name That Fallacy!

(From the 1988 elections: the Bush-Dukakis Presidential Debates and the Bentsen-Quayle Vice Presidential Debates; from the 1984 elections: the Mondale-Reagan Presidential Debates, the Ferraro-Bush Vice Presidential Debates, and from the Hunt-Helms Debates for the U.S. Senate from North Carolina):

The 1988 Bush/Dukakis First Debate

BUSH: You see, last year in the primary he [i.e., Dukakis] expressed his passion. He said, "I am a strong liberal Democrat"—August '87. Then he said, "I am a card-carrying member of the A.C.L.U." That is what he said. He is out there on—out of the mainstream. He is very passionate.

______ 1. The fallacy committed is best characterized as one of:
(A) Appeal to Force
(B) Irrelevant Appeal
(C) Leading Question
(D) *Ad Hominem*
(E) No fallacy at all

BUSH: But I don't agree with a lot of the—most of the positions of the A.C.L.U. I simply don't want to see the ratings on movies—I don't want my 10-year-old grandchild to go into an X-rated movie. I like those ratings systems. I don't think they're right to try to take away the tax exemption from the Catholic Church. I don't want to see the kiddie pornographic laws repealed. I don't want to see "under God" come out from our currency. Now these are all positions of the A.C.L.U., and I don't agree with them.

______ 2. The fallacy committed is best characterized as one of:
(A) Attacking a Straw Man
(B) Begging the Question
(C) Hasty Generalization
(D) *Ad Hominem*
(E) No fallacy at all

DUKAKIS: Well, when it comes to ridicule, George, you win a gold medal; I think we can agree on that.

______ 3. The fallacy committed is best characterized as one of:
(A) Attacking a Straw Man
(B) *Ad Hominem*
(C) Composition
(D) Division
(E) No fallacy at all

BUSH: Wouldn't it be nice to be perfect. Wouldn't it be nice to be the ice-man so you never make a mistake.

______ 4. What fallacy was committed?
(A) Appeal to Authority
(B) Appeal to Sympathy
(C) *Ad Hominem*

(D) Appeal to Force
(E) No fallacy at all

BUSH: Yes, I think it's fine to do business with them, but I don't want to see us exporting our highly sensitive national security-oriented technology to the Soviet Union. I don't want to see us making unilateral cuts in our strategic system, while we are negotiating with them, and so I'm encouraged by what I see when I talk to Mr. Gor—what I hear when I talk to Mr. Gorbachev and Mr. Schevardnadze. But can they pull it off?

DUKAKIS: Nobody is suggesting that we unilaterally disarm, or somehow reduce our strength. Of course not. What we're talking about is a combination of a strong and effective nuclear deterrent, strong, well-equipped, well-trained, well-maintained conventional forces and at the same time, a willingness to move forward steadily, thoughtfully, cautiously.

______ 5. The fallacy is committed by:
(A) George Bush
(B) Michael Dukakis
(C) Hagar the Horrible
(D) Wierd Al Yankovic
(E) None of the above

______ 6. The fallacy committed is best characterized as one of:
(A) Appeal to Popular Sentiment
(B) Appeal to Ignorance
(C) Leading Question
(D) Attacking a Straw Man
(E) No fallacy at all

BUSH: Is this the time to unleash our one-liner? That answer was about as clear as Boston Harbor. Now let me —let me —let me —let me —let me help the Governor. There's so many things there I don't quite know where to begin.

______ 7. What fallacy was committed?
(A) Appeal to Ignorance
(B) Appeal to Sympathy
(C) *Ad Hominem*
(D) Hasty Generalization
(E) No fallacy at all

BUSH: I talked in New Orleans about a gentler and kinder nation. And I have made specific proposals on education and the environment, on ethics, on energy, and on how we do better in battling crime in our country.

______ 8. What fallacy was committed?
(A) Composition
(B) Division
(C) Slippery Slope
(D) Begging the Question
(E) No fallacy at all

DUKAKIS: You know my friends, my parents came to this country as—as immigrants, like millions and millions of Americans before them and since, seeking opportunity, seeking the American Dream. They made sure their sons understood that this was the greatest country in the world, that those of us especially that were immigrants had a special responsibility to give something back to the country that had opened up its arms to our parents and given so much to them. I believe in the American Dream.

_____ 9. What fallacy was committed?
(A) Appeal to Patriotism
(B) Slippery Slope
(C) Appeal to Sympathy
(D) Both (A) and (C)
(E) No fallacy at all

The 1988 Quayle/Bentsen Debates

QUAYLE: The one thing he [i.e., Senator Bentsen] tried to point out about Governor Dukakis is that he cut taxes. The fact of the matter is, Senator Bentsen, he's raised taxes five times. He just raised taxes this last year. And that's why a lot of people refer to him as Tax Hike Mike. That's why they refer to the state of Massachusetts as Taxachusetts. Because every time there's a problem, the liberal Governor from Massachusetts raises taxes.

_____ 10. What fallacy was committed?
(A) Innuendo
(B) False Dilemma
(C) *Ad Hominem*
(D) False Cause
(E) No fallacy at all

BENTSEN: This Administration came in and put in a James Watt, an Ann Gorsuch. Now that's the Bonnie and Clyde, really, in environmental protection.

_____ 11. What fallacy was committed?
(A) Begging the Question
(B) Irrelevant Appeal
(C) False Analogy
(D) Equivocation
(E) No fallacy at all

QUAYLE: Senator Bentsen talks about recapturing the foreign markets. Well, I'll tell you one way that we're not going to recapture the foreign markets, and that is if in fact we have another Jimmy Carter grain embargo. Jimmy—Jimmy Carter—Jimmy Carter grain embargo—Jimmy Carter [*sic*] grain embargo set the American farmer back.

_____ 12. What fallacy was committed?
(A) False Analogy
(B) Irrelevant Appeal
(C) Ambiguity

(D) Equivocation
(E) No fallacy at all

QUAYLE: That's what his [i.e., Dukakis]—that's what he and his Harvard buddies think of the American farmer. . . .

______ 13. The fallacy committed is best characterized as one of:
(A) Appeal to Popular Sentiment
(B) Appeal to Ignorance
(C) Leading Question
(D) Guilt by Association
(E) No fallacy at all

Q.: Senator, I want to take you back to the question that I asked you earlier about what would happen if you were to take over in an emergency, and what you would do first and why. You said you'd say a prayer and you said something about a meeting. What would you do next?

QUAYLE: I don't believe that it's proper for me to get into the specifics of a hypothetical situation like that. The situation is that if I was called upon to serve as the president of this country or the responsibilities of this country, would I be capable and qualified to do that.

______ 14. The fallacy is committed by:
(A) Lloyd Bentsen
(B) Daniel Quayle
(C) The Questioner
(D) ZZ Top
(E) None of the above

______ 15. The fallacy committed is best characterized as one of:
(A) Evasion
(B) Appeal to Ignorance
(C) Leading Question
(D) Attacking a Straw Man
(E) No fallacy at all

QUAYLE: I have as much experience in the Congress as Jack Kennedy did when he sought the presidency. I will be prepared to deal with the people in the Bush Administration if that unfortunate event [i.e., the death of the president] would ever occur.

BENTSEN: Senator, I served with Jack Kennedy. I knew Jack Kennedy. Jack Kennedy was a friend of mine. Senator, you're no Jack Kennedy.

______ 16. The fallacy is committed by:
(A) Lloyd Bentsen
(B) Daniel Quayle
(C) The Questioner
(D) Madonna
(E) None of the above

______ 17. The fallacy committed is best characterized as one of:
(A) *Ad Hominem*
(B) Appeal to Ignorance

(C) Leading Question
(D) Attacking a Straw Man
(E) No fallacy at all

The 1988 Bush/Dukakis Second Debate

Q.: Governor, if Kitty Dukakis were raped and murdered, would you favor an irrevocable death penalty for the killer?

DUKAKIS: No, I don't, Bernard, and I think you know that I've opposed the death penalty during all of my life.

______ **18.** The fallacy is committed by:
(A) Michael Dukakis
(B) George Bush
(C) The Questioner
(D) Cyndi Lauper
(E) None of the above

______ **19.** The fallacy committed is best characterized as one of:
(A) *Ad Hominem*
(B) Appeal to Ignorance
(C) Leading Question
(D) Attacking a Straw Man
(E) No fallacy at all

BUSH: No. Look, I'll tell you—now let me tell you something about that. And Barbara and I were sitting there before that Democratic convention and we saw the Governor and his son on television the night before, and his family, and his mother who was there. And we're—I'm saying to Barbara, "You know, we've always kept family as a bit of an oasis for us." You all know me. And we've held it back a little. But we use that as a role model. The way he took understandable pride in his heritage. What his family means to him. And we've got a strong family and we watched that and we said, "Hey, we've got to unleash the Bush kids." And so I would say that the concept of Dukakis' family has my great respect. And I'd say that—I don't know if that's kind or not. It's just an objective statement.

______ **20.** The fallacy committed is best characterized as one of:
(A) *Ad Hominem*
(B) Appeal to Ignorance
(C) Appeal to Sympathy
(D) Hasty Generalization
(E) False Cause

DUKAKIS: Well, Margaret, we've heard it again tonight and I'm not surprised—the labels. I guess the Vice President called me a liberal two or three times. Said I was coming from the left. . . . And I would hope that tonight and in the course of the rest of this campaign we can have our good solid disagreements on issues. There's nothing the matter with that, but let's stop labeling each other and let's get to the heart of the matter, which is the future of this country.

______ **21.** What fallacy was committed?
(A) Appeal to Authority
(B) Appeal to Sympathy
(C) *Ad Hominem*
(D) Appeal to Force
(E) No fallacy at all

DUKAKIS: Mr. Bush says we're going to put the IRS on every taxpayer; that's not what we're going to do. I'm for the taxpayer bill of rights. And I think it's unconscionable that we should be talking, or thinking about, imposing new taxes on average Americans when there are billions out there, over $100 billion in taxes owed, that aren't being paid. . . .

BUSH: . . . I have been all for the taxpayer's bill of rights all along. And this idea of unleashing a whole bunch, an army, a conventional force army of IRS agents into everybody's kitchen. I mean he's against most defense matters and now he wants to get an army of IRS auditors going out there. I oppose that.

______ **22.** The fallacy is committed by:
(A) Michael Dukakis
(B) George Bush
(C) The Questioner
(D) Winston Churchill
(E) None of the above

______ **23.** The fallacy committed is best characterized as one of:
(A) *Ad Hominem*
(B) Appeal to Ignorance
(C) Leading Question
(D) Attacking a Straw Man
(E) No fallacy at all

Q.: Governor, you won the first debate on intellect and yet you lost it on heart.

BUSH: Just a minute.

Q.: You'll get your turn.

DUKAKIS: I don't know if the Vice President agrees with that.

Q.: The American people admire your performance but didn't seem to like you much. Now, Ronald Reagan has found his personal warmth to be a tremendous political asset. Do you think a president has to be likable to be an effective leader?

______ **24.** The fallacy is committed by:
(A) Michael Dukakis
(B) George Bush
(C) The Questioner
(D) Joe Piscopo
(E) None of the above

______ **25.** The fallacy committed is best characterized as one of:
(A) Appeal to Force
(B) Appeal to Ignorance

(C) Leading Question
(D) False Cause
(E) No fallacy at all

The 1984 Mondale/Reagan Presidential Debates

Q.: Mr. Mondale, two related questions on the crucial issue of Central America. You and the Democratic party have said that the only policy toward the horrendous civil wars in Central America should be on the economic developments and negotiations with, perhaps, a quarantine of Marxist Nicaragua. Do you believe that these answers would in any way solve the bitter conflicts there? Do you really believe that there is no need to resort to force at all? Are not these solutions to Central America's problems simply again too weak and too late?

MONDALE: I believe that the question oversimplifies the difficulties of what we must do in Central America.

______ **26.** The fallacy is committed by:
(A) Walter Mondale
(B) Ronald Reagan
(C) The Questioner
(D) None of the above

______ **27.** The fallacy committed is best characterized as one of:
(A) Appeal to Force
(B) Irrelevant Appeal
(C) Leading Question
(D) No fallacy at all

(From the closing statement, the second presidential debate):

REAGAN: The question before comes down to this: do you want to see America return to the policies of weakness of the last four years, or do we want to go forward marching together as a nation of strength and that's going to continue to be strong?

______ **28.** The fallacy committed is best characterized as one of:
(A) Appeal to Force
(B) False Dilemma
(C) Guilt by Association
(D) No fallacy at all

(From the second presidential debate):

MONDALE: A minute ago, the President quoted Cicero, I believe. I want to quote somebody a little closer home, Harry Truman. He said the buck stops here. We just heard the President's answer for the problems at the barracks in Lebanon where 241 Marines were killed. . . . Good intentions, I grant, but it takes more than that.

______ **29.** The fallacy committed can be best described as:
(A) Appeal to Authority
(B) Appeal to Sympathy

(C) Appeal to Ignorance
(D) No fallacy

(From the second presidential debate):

MONDALE: Mr. President, I accept your commitment to peace, but I want you to accept my commitment to a strong national defense. . . . I have proposed a budget which would increase our nation's strength by, in real terms, by double that of the Soviet Union.

______ **30.** The fallacy committed can be best characterized as:
(A) Appeal to Force
(B) Attacking a Straw Man
(C) Slippery Slope
(D) No fallacy

(From the second presidential debate):

MONDALE (to Reagan): Your definition of national strength is to throw money at the Defense Department. My definition of national strength is to make certain that a dollar spent buys us a dollar's worth of defense.

______ **31.** The fallacy committed?
(A) Equivocation
(B) Attacking a Straw Man
(C) Ambiguity
(D) No fallacy

(From the second presidential debate):

MONDALE: These high interest rates, real rates, that have doubled under this Administration, have had the same effect on Mexico and so on, and the cost of repaying these debts is so enormous that it results in massive unemployment, hardship, and heartache. And that drives our friends to the north . . . up into our region, and we need to end those deficits as well.

REAGAN: Well, my rebuttal is I've heard the national debt blamed for a lot of things, but not for illegal immigration across our border, and it has nothing to do with it.

______ **32.** The person who committed the fallacy is:
(A) Walter Mondale
(B) Ronald Reagan
(C) No fallacy was committed

______ **33.** The fallacy committed?
(A) Appeal to Sympathy
(B) Appeal to Ridicule (*Ad Hominem*)
(C) False Analogy
(D) No fallacy

(From the first presidential debates):

REAGAN: I know there was [*sic*] weeks and weeks of testimony before a Senate committee. There were medical authorities, there were religious, there were clerics there, everyone talking about this matter, of pro-life. And at the end of all of that, not one shred of evidence was introduced that the unborn child was not alive. . . . So this has been my feeling about abortion, that we have a problem now to determine. And all the evidence so far comes down on the side of the unborn child being a living human being.

______ **34.** The fallacy committed?
(A) Appeal to Ignorance
(B) Begging the Question
(C) Both A and B
(D) No fallacy

The 1984 Bush/Ferraro Vice-Presidential Debates

BUSH: [Walter Mondale] also made a reference that troubled me very much, Mr. Boyd. He started talking about my chauffeur, and you know, I'm driven to work by the Secret Service—so is Mrs. Ferraro—so is Mr. Mondale—they protected his life for four years and now they've done a beautiful job for Barbara and mine [*sic*]. They saved the life of the President of the United States. I think that was a cheap shot—telling the American people to try to divide class—rich and poor.

______ **35.** The fallacy committed by George Bush?
(A) Appeal to Fear
(B) Irrelevant Appeal
(C) Faulty Motives
(D) No fallacy

BUSH: But let me help you with the difference, Mrs. Ferraro, between Iran and the embassy in Lebanon. . . . We went to Lebanon to give peace a chance, to stop the bombing of civilians in Beirut, to remove 13,000 terrorists from Lebanon—and we did. We saw the formation of a government of reconciliation and for somebody to suggest, as our two opponents have, that these men died in shame—they better not tell the parents of those young marines. They gave peace a chance.

FERRARO: . . . please don't categorize my answers either. Leave the interpretation of my answers to the American people who are watching this debate. And let me say further that no one has ever said that those young men who were killed through the negligence of this Administration and others ever died in shame.

______ **36.** There are several things going on in this exchange: it is emotional and heated, and both Bush and Ferraro have indulged emotional excesses. But there is one particular glaring fallacy made by one of the two debaters. Who committed the more serious fallacy?
(A) George Bush
(B) Geraldine Ferraro

______ **37.** What is the fallacy?
(A) Appeal to Fear

(B) Appeal to Patriotism
(C) Appeal to Sympathy
(D) Attacking a Straw Man

BUSH: Almost every place you can point, contrary to Mr. Mondale's—I gotta be careful—but contrary of how he goes around just saying everything bad. If somebody sees a silver lining, he finds a big black cloud out there. Whine on harvest moon! I mean, there's a lot going on, a lotta opportunity.

_______ **38.** What is the fallacy?
(A) *Ad Hominem*
(B) Attacking a Straw Man
(C) Both A and B
(D) None of the above

The 1984 First Helms/Hunt Debate for U.S. Senator from North Carolina

HUNT (to Helms): [Isn't it true that] taxes paid for your junkets, since these trips were paid for by the tax breaks you pushed through Congress for right-wing millionaires?

HELMS: You haven't got your facts straight. I am the Chairman of the Foreign Relations Committee. I've actually been saving money, developing the anti-communist position around the world. . . .

HUNT: You didn't answer the question. [Repeats the question.] Have you disclosed the costs of your trips to the American public?

HELMS: Certainly, I'll answer your question. You identify all these millionaires. You explain it to me.

_______ **39.** Note: Both candidates have committed fallacies in this exchange. Which fallacies were committed by Jesse Helms?
(A) Diversion
(B) Appeal to Ignorance
(C) Irrelevant Appeal
(D) All of the above
(E) None of the above

_______ **40.** Which fallacies were committed by Jim Hunt?
(A) Leading Question
(B) False Dilemma
(C) Special Pleading
(D) All of the above
(E) None of the above

(From the first Helms/Hunt Debate):

"Mr. Helms portrayed Mr. Hunt as ill-informed on national issues, and several times tried to link Mr. Hunt's positions with those of Senator Edward M. Kennedy, the Massachusetts Democrat who is not a favorite among North Carolina conservatives. At one point he read recent statements in which Mr. Hunt had offered mild praise for Mr. Kennedy and Walter F. Mondale, the Democratic presidential nominee."

_______ **41.** What is the best characterization of the fallacy committed by Helms?
(A) Appeal to Patriotism
(B) *Ad Hominem*
(C) Guilt by Association
(D) Both A and B
(E) Both B and C

(From the first Helms/Hunt Debate):

HELMS: "If North Carolina loses its first chairman of the Senate Agriculture Committee in 149 years, you could kiss the tobacco program goodbye."

_______ **42.** What best characterizes the fallacy committed here?
(A) Special Pleading
(B) Slippery Slope
(C) Appeal to Fear
(D) All of the above
(E) None of the above

(From the first Helms/Hunt Debate—the *Raleigh News & Observer* account):

"Hunt noted a number of instances in which Helms had voted against or spoken against Reagan's policies on Central America, citing specifically his support of right-wing Salvadoran leader Roberto d'Aubuisson over that country's president, Jose Napolean Duarte, who is supported by the Reagan Administration.

" 'What you are accusing me of is being strongly anti-communist,' Helms replied, adding, 'Yes sir, I plead guilty to being opposed to communism and I plead guilty to doing everything to stop it.' "

_______ **43.** What best describes the fallacy on Jim Hunt?
(A) *Ad Hominem* (Circumstantial)
(B) Guilt by Association
(C) *Post Hoc, Ergo Propter Hoc*
(D) All of the above
(E) None of the above

_______ **44.** What best describes the fallacy committed by Jesse Helms?
(A) Diversion
(B) Appeal to Patriotism
(C) Appeal to Fear
(D) All of the above
(E) None of the above

(From the second Helms/Hunt Debate):

HELMS (the closing statement): Jim Hunt is a Mondale liberal and is ashamed of it. I am a Reagan conservative and I'm proud of it. If you're for Ronald Reagan, vote for me.

_______ **45.** What is the fallacy?
(A) *Ad Hominem*

(B) Guilt by Association
(C) Appeal to Popular Sentiment
(D) All of the above
(E) None of the above

(From a television commercial for Jesse Helms):

"I love it [referring to a new car], and I can afford it. But I'm going to wait until after the elections to buy it. Because if Jim Hunt and Walter Mondale are elected, then I'll be spending that $157 a month in higher taxes."

______ **46.** What is the fallacy?
(A) Attacking a Straw Man
(B) Guilt by Association
(C) Division
(D) All of the above
(E) None of the above

(From a campaign speech by Ronald Reagan):

"Have you ever noticed that the depressed areas of this country are represented by Democratic congressmen? No wonder they're depressed areas!"

______ **47.** What best describes the fallacy?
(A) Appeal to Popular Sentiment
(B) False Cause
(C) Appeal to Authority
(D) *Post Hoc*
(E) No fallacy

RECOMMENDED READING FOR UNIT FOUR

Aristotle. *On Sophistical Refutations on Coming-to-Be and Passing-Away.* Translated by E. S. Forster. Cambridge, MA: Harvard University Press, 1955.

Bentham, Jeremy. *Bentham's Handbook of Political Fallacies.* Edited by Harold A. Larrabee. New York: Thomas Y. Crowell Co., 1952.

UNIT FIVE

CREATIVE THINKING

MODULE 39

AN INTRODUCTION TO CREATIVE THINKING

39.1 IN GENERAL

Careful, goal-directed thinking is almost indispensable to any sort of problem solving. Approaching any problem, we would do well to heed the philosopher René Descartes' advice to break up the problem into manageable components; to take care to leave nothing out in one's appraisal of a problem; and to do careful and comprehensive reviews. This is always good advice, and it exemplifies the kind of critical thinking we've been doing throughout most of the book. But—and this is a crucial reservation—critical thinking is far from the whole story where problem solving of any kind is underway, whether the problem is a deduction in logic, the formulation of a scientific hypothesis, or merely a way to get your stubborn lawn mower started.

What else is needed? One needs to think *creatively* as well as critically. A computer can solve most—but not all—of the problems in this book, but what separates the skilled logician from the finest and most advanced computer is the ability to find entirely new approaches to the conception of a problem. In other words, effective problem solving requires *creative* as well as *critical* thought.

Creative-thinking books are the latest craze, especially in executive boardrooms, advertising, and marketing seminars. Sometimes they offer productive ideas. However, such books are rather one-dimensional and highly subjective. They're one-dimensional, not because they're geared to the concerns of business, but because they lack a corresponding stress on *critical* thinking, which is the true complement of creative thinking. These books tend to be highly subjective because they stress one person's way of thinking creatively without any evidence that what works for one person can in any way be generalized to others.

We will attempt to explode an old dogma about the sharp division between critical and creative thinking. Surely the two are separate phenomena, but one cannot think well without at least holding a share in both.

Nowhere is the critical/creative duality of thought more in evidence than in the propagation of a new theory of science. The great scientists and philosophers of science recognize the indispensability of creativity to the scientific enterprise, but display an utter inability to account for it. Here's what Albert Einstein had to say:

> ***"The most beautiful thing we can experience is the mysterious. It is the source of all true art and science."***

When the philosopher Karl Popper attempts to account for the element of creativity in the formulation of scientific hypotheses, he is likewise consigned to contemplating the ineffable, remarking that we must rely on "myth and inspiration."

However, we can surely say something more about how the process of creativity works. My contention is that we need to focus on the similarities between creativity and criticism in good thinking (whereas the usual approach is to stress their dissimilarities) without attempting the bogus reduction of induction into a subspecies of deduction.*

In doing so, we encounter many problems, few of which we can deal with effectively in this place. For one thing, despite many forays into the subject from a scientific point of view, especially in the body of research done by professional psychologists, little evidence exists that any one *routine* or *program* can make a person more creative in her thinking. In addition, judging by a large sample of competing textbooks and courses on the subject, the evidence is far from conclusive that any one is superior to any other in the results it produces. However, we may be able to shift to more productive ground here if we consider that the attempt to find a creativity-producing *routine* is ill-conceived from the start. The most important factor in the development of a creativity which complements critical thinking is the simple recognition that the two *are* complementary. We need to *infuse* creativity into crucial points along the critical thinking spectrum.

*Some of what I have to say here bears some resemblance to what C. S. Peirce calls *abduction*—his account of the process of discovery in science—which Peirce regards as a thought process separate from both induction and deduction but which serves as an aid to them both. Unlike Popper and Einstein, Peirce did not think that discovery in science was in any way "mysterious" but, unfortunately, his account of abduction is itself notoriously obscure. Consequently, I will refrain from using that term and talk instead about the element of creativity in inductive and deductive thought.

In summation, I submit that we need to accept three propositions:

1. That creativity is possible
2. That to some degree, creativity can be learned and
3. That creative thinking is a natural complement to critical thinking

39.2 THINKING IN A RUT

One way that creativity and criticism complement each other is that their combined effect on the process of problem solving helps us to avoid "thinking in a rut." The plodding thinker is one who suffers from a "hardening of the categories"—he needs to be able to break out of the routine and the conventional. One who thinks in a rut is unable to think critically about what he's missing, and the needed element is assuredly a liberal dose of creativity. Take the following case, which is based on a true account of the operations of a police force in a modestly sized midwestern town.

The veteran police chief in this town (call him "Captain First") presided over the police force for many years. He came up through the ranks and applied the time-tested techniques of his predecessors. For one thing, he continued the old practice of having policemen walk a beat. Patrol cars were used as backup, or as the means to get policemen to the location of a disturbance quickly, but the backbone of his organization was the foot patrolman. The crime rate in this town was low and always had been.

Upon Captain First's retirement, the town council hired a replacement (call him "Captain Second") who was highly trained in the latest techniques of criminology, and he vowed to bring the small police force "into the twentieth century." His reorganization of the police force included taking patrolmen off the beat. He doubled the number of police cars and required his police to patrol the streets in their vehicles, two policemen to a car.

Gradually, but perceptibly, the town's crime rate increased, and citizens were alarmed. So great was the increase over time that the town council called Captain Second to account. He argued, with some plausibility, that the town's increased rate of crime reflected a nationwide trend. He contended that the rise in the incidence of crime would have happened eventually in any case, and that the town was lucky to have someone with his expertise at such a critical moment. He proceeded to lay out his recommendations that the town council increase the budget expenditure on law enforcement, and double the number of police and materiel. The council was persuaded, and complied with his request.

Now, with twice the patrolmen and twice the patrol cars, the crime rate in the town *did* decrease to well below its peak, but never to the level prior to the appointment of the new chief of police. Captain Second offered the reduction in crime as the vindication of his expertise, and the town council, recognizing that their town was once again a safe place to live, rewarded the chief with an increase in salary.

As time went on, however, the crime rate began to increase once more. Called to account, Captain Second made similar recommendations about larger budget expenditures. The town council balked at the price tag and took the expedient of firing Captain Second and replacing him with someone of the "old school" (call him "Captain Third"). Captain Third re-instated the old practice of having most of the patrolmen walk the beat, but he kept the patrol cars in reserve. The crime rate dropped almost immediately, returning to the levels that approximated the crime rate under Captain First.

What do we make of this story? I suggest that the hotshot new police chief, Captain Second, who had vowed to bring the town "into the twentieth century," was victim to a kind of "thinking in a rut." His solution to every problem was to apply a "technological fix."

The moral of the story is this: we are too often taught to approach a problem methodically—applying familiar routines to new problems—without asking ourselves if a new problem requires new or different ways of thinking. When we go at a problem in the same way, over and over, without getting results, we're guilty of thinking in a rut. A major part of creative thinking is to liberate ourselves from routine procedures in problem solving or finding an explanation.

Consider the fact that a good computer program for playing chess has built-in limitations which severely constrain the skill of the computer to play chess well. The very best such programs developed at places like M.I.T. and Carnegie-Mellon University are unable to defeat chess masters with any regularity. We can discern some very good reasons for this.

In a nutshell, we may say that the main reason that computer chess programs are badly matched when they confront chess masters is that such programs follow routines rigorously. Within the confines of those routines, they rarely make tactical mistakes (for example, in an exchange, they rarely fail to take a piece that is higher rated). However, they are often terribly guilty of strategic errors.

When a chess program confronts a problem in a position on a chess board, strictly speaking its approach is mere process of elimination. It turns up every avenue to see what will happen. This is time consuming, even given a sophisticated computer's rapid ability to calculate.

A chess-playing human, on the other hand, can see right away which pawns are blocked, which pieces can be moved, and so forth, in a single glance. The chess player considers overall *patterns,* not single moves of single pieces. That gives the human chess player a huge advantage both in terms of time saved and overall strategy.

The chess master is one who can combine both tremendous insight into overall strategies and an appreciation of arrangements of patterns with a finely tuned attention to detail.

To avoid thinking in a rut, we need to mimic the chess master, not the computer chess program. We need to try out new ideas, devise new strategies, and expand our vision of the sets of possible patterns which constitute the background position behind the problem. In what follows, we will attempt to emulate the chess master's strengths in the solution of problems which are more practical or more important than a mere game of chess.

39.3 INDUCTION AND CREATIVE THINKING

We need a new approach to induction. Induction is not justifiably presented as an analogue to deductive thought. We make too much of probability calculations and not enough of how to assess evidence and generate hypotheses. We need to emphasize the latter. Surely, we are better served reflecting on the paradigm of the scientific method in order to learn something about induction. To this end, Module 40 looks at scientific method, a vast subject that can only serve as an introduction to inductive method in science, but an introduction that is necessary to obtain a well-rounded appreciation of inductive and creative thought.

Next, let's turn to Module 41 on practical problem solving. This module presupposes that we've done some of the work in the modules that precede it so that we know how to go about working methodically, but on the whole Module 41 stands on its own. Our interest here is to follow some hints (admittedly highly subjective) about how to go about solving a problem. Most of these tips are purely common sense.

If we learn a good general and practical approach to scientific thinking, then we also need a good general and practical approach to generating hypotheses (which is one of the first steps to applying the method of science). To this purpose, Module 42 gives tips on how to form hypotheses and how to generate them.

In the final module, we consider a broader thesis: arguments and visual images. The section is a defense of the contention that visual images can serve as (something like) assertions in the formation of argument. In a wider sense, perhaps the module will also serve as a study of how we can use observation in a way to aid our thinking.

MODULE 40

SCIENCE AS A METHOD OF INDUCTION

1. Observation
2. Hypothesis
3. Experimentation
4. Theory

40.1 IN GENERAL

Science is a method for solving most kinds of problems. It is not a collection of facts. It is a *method.*

Scientists don't pretend to be able to solve every kind of problem using this method, but a wide array of them fall under their scrutiny. We know from our readings, our past education, and our general knowledge that some of the problems that scientists attempt to solve include questions about the age of the earth, the solar system and the universe, the identity of the basic particle of matter, whether life exists on Mars, and the causes and cures for disease. But a good, scientific approach can help us solve more common problems. For example:

An **inductive** argument is **sound**, if and only if, the evidence is sufficient or representative enough to warrant the likelihood of the conclusion, and the premises are true.

"Why is Dr. Cook an easy grader and why is Dr. Baker such a hard grader?" "Is there a sound reason why the cafeteria consistently overcooks steak while undercooking chicken?" "If I buy a hundred lottery tickets, will I win the lottery?" "Why do football players grow so much larger than the rest of us?" "Why is Bret attracted to Iris and not to me?" "How can Annie eat so much and stay so thin?"

Some of these problems can be solved by a bit of inquiry and a good dose of rigorous method. It also helps to study the sciences that take a generalized interest in such questions—sociology, biology, psychology, and physics. But the answers that science textbooks give us merely reflect the repeated and painstaking application of the kind of methods we are going to study here. Finding the answers takes detective work, and we should learn how to do that detective work for ourselves.

We're about to look at science as a method of induction. Although we've talked about induction in other Units,* most of this book has been dedicated to the study of deduction. In fact, as we shall see, the two modes of thinking frequently work together, but they are different enough that we should take a long, hard look at the methods of induction and the best place to find that is in the study of the scientific method.

Philosophers disagree about what induction is and how it works. Many treat induction as though it were a spinoff of deductive thought. However, it's widely agreed that the operations of the Scientific Method—making observations, assessing evidence, and generating hypotheses–-are the best showcases for inductive thinking in action and we shall argue that creativity is an indispensable element to such thinking.

This module will follow the basic suggestions of the philosopher Karl Popper on scientific methodology with occasional glimpses of other theories on the subject. Popper's method is particularly accessible and it is a good, respected and reflective exposition of the sorts of uncertainties that typify scientific inquiry.

Also included throughout this Unit are some simple problem-solving approaches which are little more than common sense, but may well prove useful especially to the college student who is just beginning to learn techniques of research and problem solving in college.

40.2 AN INTRODUCTION TO THE SCIENTIFIC METHOD

When you were in high school, you probably learned a version of the Scientific Method as a sequence of steps which proceeded in the following order:

1. Observation
2. Hypothesis
3. Experimentation
4. Theory

*For example, see Module 8, "Deductive vs. Inductive Arguments," in Unit One.

To solve a problem we first need to take a good look around us. If we're observant, we may be able to formulate sound guesses ("hypotheses") about the solution to our problem. The next step is to put our hypotheses to the test and evaluate our data. If we do that well, we can then confidently devise a theory which provides our solution and finally arrive at a set of constant rules ("laws") that we can apply to get new solutions to similar problems in the future.

What you may not have learned is that this is mere *shorthand* for how science works. It is by no means the final word on the matter and, as we shall see in this Unit, as it stands it is somewhat misleading, inaccurate, and controversial. We may even refer to these four steps as the "prototype" version of the Scientific Method because it is only a launching point for further discussion. The four steps provide something of a "static" model of the operations of science, but later we'll come to regard the Scientific Method in a more dynamic way than the prototype allows. For the sake of simplicity, however, the prototype gives us a good place to start.

The most important thing we can learn from this early version of the Scientific Method is that it's a *method*. Science is not a *collection of facts;* it is not a compendium of knowledge nor an encyclopedia of everything that humanity has learned over the course of history. Rather, it is a procedure. It does not yield certainty; at most, it offers justification or assurance. The Scientific Method is the highest refinement of the logic of induction: as such, it renders judgments that are justified or corroborated, making them more or less worthy of belief. In other words, science is a method which arrives at conclusions that are likely or probable.

We may begin our discussion of science with an account of what observation is. By all accounts, we mean *sensory* observation: we start with the data we collect from our senses. Other sorts of experience are possible—artistic, sexual, or religious experience, for example—but even though sciences like psychology may make these an *object* of inquiry, they are not the kinds of experience needed for scientific procedure.

To be a good inductive thinker (i.e., to be a good scientist) you must be first a good observer. That sounds easier than it is. Few of us have the ability to see what's really there without adding layers of obstructive beliefs and interpretation. In fact, interpretation might be an indispensable aid to observation, but then we need to be able to interpret what we see objectively, without bias.

Moreover, we must be able to notice what other people do not. When Galileo was a young boy, he saw something in church that many other people must have seen many times before him. These old Romanesque churches had large oil lamps suspended by long chains from the ceiling. The church sexton was in charge of keeping the lamps fueled and lit. To get to the lamps, the sexton had to walk along a parapet and snag the lamp with a long hook. He would then fill the cup inside the lamp with oil, ignite the wick, close the lamp, and let it swing back in place. Galileo noticed one day that when the lamp swung after the sexton released it, it swung in ever-decreasing arcs until its motion was hardly noticeable. (See Figure 1.) However, the young scientist *timed* the swing of each arc and discovered that, while the arcs decreased in distance covered,

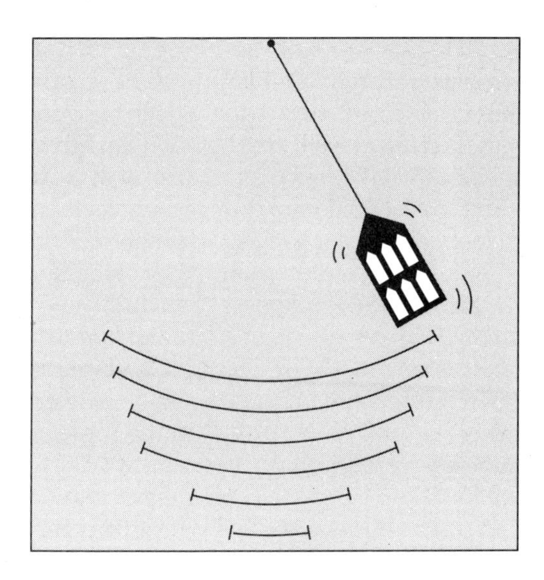

FIGURE 1

the time elapsed to cover the full swing remained exactly the same. From this observation, Galileo was able to discover the principles behind the pendulum, which served as the basis for the construction of the analog clock.

Next, the scientist must frame a **hypothesis.** The hypothesis consists of a prediction that if certain events will occur, another event will follow them. As such, the hypothesis often involves a *prediction.** Moreover, it is a hypothetical: it should be framed as a material conditional—"*If* event x occurs, *then* so will event y." The prediction contained in the hypothetical is subject to testing.

Testing, or "**experimentation,**" may well be one of the most controversial areas of the Scientific Method. What makes a good test? How much does an experiment "prove" (if it proves anything at all), and how well does it establish (or refute) the hypothesis under consideration?

When a hypothesis is corroborated, confirmed, verified (or otherwise "established") by experimentation, the result is embedded in a "**theory**;" which may in turn yield a set of "laws of nature" (or just "scientific laws"). Scientific laws express generalizations about certain uniform regularities in nature, generalizations that have been carefully and repeatedly confirmed by testing. For example, consider Newton's Laws for theoretical physics: "Every action has an equal and opposite reaction. An object in motion tends to stay in motion." Such laws are useful in laying the groundwork for scientific explanations and predictions.

Even though "law" is sometimes given as a fifth and last step of the "prototype" Scientific Method, most scientists and philosophers of science nowadays accord more status to the 'theory'. That fact represents a more recent stage of

*Sometimes hypotheses may be about the past—for example, hypotheses in the fields of archeology or anthropology.

development in the Scientific Method. At one time, scientists conferred the status of certainty on *laws of nature* (and many scientists and philosophers still do). Such laws represented "fixed points" that were established beyond dispute and many thinkers took refuge in the comfort that such certainties offered. In Newtonian physics, scientific laws were the "stuff" of scientific knowledge, the "facts" which could fill a universal encyclopedia. However, with the introduction of Einstein's Theory, all that changed. Newton's Laws were called into question, the resting point of absolute certainty gave way to cautious assurance and theory supplanted law as the focal point of scientific endeavor.

Sometimes laymen speak of "proving" scientific theories (as in "The theory of evolution hasn't yet been *proven* to be a fact"). However, this way of speaking about theories reveals a terrible misconception about science: no longer do most scientists regard science as capable of "proof" in the sense of reaching conclusions that are immune to revision. The enterprise of science is founded upon uncertainties. These "foundations" are always subject to revision in the light of further evidence. The "theory" has come to be regarded as the highest court of appeal in science. We once spoke of "Newton's Laws," but we now speak of "Einstein's Theory" which supplanted Newton's findings. Much of Newton's work has been discarded, but Einstein's theory has survived ingenious and concerted attempts at refutation. Darwin's "Theory of Evolution" has come under attack by many Creationists on the grounds that Darwin's theory is just that—a "theory;" presumably the fact that it is a theory makes it less worthy of faith. But such arguments display a lack of sophistication about what may well be the most well-confirmed body of knowledge in the history of science.

What is a theory? Consider this discussion from the great philosopher of science, Carl Hempel:

> Theories are usually introduced when previous study of a class of phenomena has revealed a system of uniformities that can be expressed in the form of empirical laws. Theories then seek to explain those regularities and, generally, to afford a deeper and more accurate understanding of the phenomena in question. To this end, a theory construes those phenomena as manifestations of entities and processes that lie behind or beneath them, as it were. These are assumed to be governed by characteristic theoretical laws, or theoretical principles, by means of which the theory then explains the empirical uniformities that have been previously discovered, and usually also predicts "new" regularities of similar kinds.*

Theories, therefore, serve two functions: *explanation* and (sometimes) *prediction*. As such, theories are dynamic; they are not merely after-the-fact expositions of what happened when we tested a hypothesis. Instead, they seek to enlarge the repository of knowledge by speculating on the consequences that follow given the theory. Theories do this by proposing grounds for new predictions as hypotheses.

*Carl G. Hempel, *Philosophy of Natural Science*. Foundations of Philosophy Series, Elizabeth and Monroe Beardsley, eds., Englewood Cliffs, NJ: Prentice-Hall, Inc., 1966, p. 70.

This brings us to a final realization, that the "prototype" version of the Scientific Method is seriously deficient in many ways. It is deficient because in its original version it presupposes that the search for natural laws is the central task of science; in fact, theory-building is science's most ambitious aim. Moreover, it supposes that science is built up of a collection of certainties. We know instead that all of science is tentative and subject to revision in the light of further evidence. Finally, the "prototype" lays out a series of steps in the method that is presumed to be a strict, methodical sequence; in fact, the steps in the process of science are interactive: all observation, for example, is at least somewhat interpretive, certainly selective, and is based on prior theoretical conceptions of what the world is like. Hypotheses may not be sequentially prior to theories; they may actually follow them.

The enterprise of science is evolutionary and changing. The greatest scientific discoveries have actually been directed at the "establishment" of science. Science is not generally a respecter of tradition and propriety; it thrives in the atmosphere of the radical and new. It relies on the greatest creative instincts of human beings: their ability to see the world and the universe anew.

40.3 THE HISTORICAL ORIGINS OF THE SCIENTIFIC METHOD

What is the origin of the prototypical Scientific Method and from what did modern science's emphasis on method evolve? We can't hope to give an exhaustive answer to that question here, but we can sketch a few important details. Certainly we could go all the way back to ancient times to find our answers and a complete account would demand that we do so. Most of the history of philosophy can be read as an account of the origins and development of science. However, a significant change in the way philosophers and scientists asked questions occurred in Europe in the seventeenth century. It was a change that many historians were later to call "the Epistemological Turn."*

Prior to that time, what we ordinarily think of as the subject matter of science was then called "natural philosophy." Natural philosophy was largely speculative and concerned itself with essences and the hidden natures of things in the world. With the Epistemological Turn, however, some philosophers began to shift their attention away from essences and natures to questions about knowledge ('epistemology' is the study of knowledge) and how we know what we know.

Most historians of science cite the influence of three major figures in the seventeenth century on the origins of modern science: Francis Bacon in England, Galileo Galilei in Italy, and René Descartes in France. Each of these men sought to liberate science from the past in a special way. They saw nature as an object of study apart from spiritual or human influences and subject to laws and regularities that could be an independent focus of study.

*The American philosopher, Robert Paul Wolff, is among them.

Of the three, Galileo may well have been the greatest contributor to practical science, especially for his experiments relating to the study of gravity. Galileo set his sights on the Scholastic (or Aristotelean) methods of induction and sought a new way to express physical relationships through mathematical formulae (for example, $g = 32$ ft./sec^2).

Francis Bacon also assailed the Aristotelean concepts of induction. He believed that the Scholastic thinkers were too easily disposed to hasty generalization and he sought to purge scientific thinking from biases and presuppositions. According to one writer, the "two principle features of Bacon's new method were an emphasis on gradual, progressive inductions, and a method of exclusion."* However, Bacon made few if any practical contributions in scientific research.

René Descartes may well have made the most significant contribution to the development of modern science (for the purposes of our study) for his distinctive new emphasis on *method.* It is Descartes that we credit with the inception of the Epistemological Turn. Moreover, he was not only a fine theoretician, but an accomplished scientific researcher as well.

In his book *The Discourse on Method,* Descartes laid down a set of rules for thinking well. He believed that these rules for the mind would enable anyone to discover any truth whatever. The four rules he proposed are as follows:

1. Never accept anything as true unless one recognizes it to be evidently so.
2. Divide all problems into as many parts as possible in order to arrive at an easy solution.
3. Think in an orderly manner, starting with the simplest and most understandable and then gradually proceeding to the more complex.
4. Always make enumerations so complete and reviews so general that one can be certain nothing is omitted.

Recognize those rules? Probably not, though they will surely strike the reader as good advice when tackling any problem. However, some thought will reveal that these rules are the ancestor of our Scientific Method. In fact, the full title of Descartes' work is *The Discourse on the Method of Rightly Conducting the Reason and Seeking Truth in the Sciences.* Moreover, Descartes appended three scientific studies—in optics, meteorology, and geometry. Descartes added these three items of research, in his own words, "to show how successful this method has been in my hands."

Notice something else about Descartes' rules: they offer valuable hints at problem solving, but they are the sorts of hints that prove to be most valuable when solving deductive or mathematical problems. If you are working a prob-

*John Losee, *A Historical Introduction to the Philosophy of Science,* London: Oxford University Press, 1972, p. 64. See Losee's book for a comprehensive discussion of the roles of Galileo, Descartes, Bacon, and Newton in the formulation of the Scientific Method.

lem in direct derivation,* for example, you'd benefit greatly from following Descartes' advice. You'd want to start from axioms or given assumptions, divide your problem into parts starting from the simplest and working to the more complex, and reviewing your work to make sure it was thorough and that you made no errors in the process. However, that's not the way modern science works—at least, not in all of its operations. Descartes' method was a promising start, but it was only a crude beginning in the evolution of the method.

The origins of the evolution of the Scientific Method may be seen as an interplay between the work of Descartes, Galileo, and Bacon. Descartes stressed the deductive approach to science. His work proceeded from general principles (which he saw as certain and immovable) to more specific propositions. In a very rough way, we owe the concepts of hypothesis and theory to Descartes. However, it was Francis Bacon who stressed inductive procedures, proceeding from specifics to the most general propositions, and Galileo taught us more about the value of experimentation and applied science.

The next evolution in the Scientific Method came from the British and the school of philosophy known as "Empiricism." John Locke, Bishop George Berkeley, and David Hume added to the theorizing of Bacon, Descartes, and Galileo in a way that now placed direct, sensory experience under central focus. Thanks to the Empiricists, we add the indispensability of observation and sense experience to the brew of the method of science. Experimentation now took a firm hold in the world of experience and observation was deemed to be the starting point of scientific endeavor.

From this basis, the physics of Isaac Newton got its start and with Newton we truly entered the modern age of science. Newton's work was as much a reaction to Descartes as Descartes' own work was a reaction to Aristotelean scholasticism. Newton sought a blend of inductive and deductive reasoning in science—a bridge between the rationalism of Descartes and the empiricism of Locke and Hume. The "prototype" Scientific Method, as we have described it in previous pages, had now taken full shape.

40.4 INDUCTION AND DEDUCTION WORK TOGETHER

In previous modules, we defined the differences between deductive and inductive arguments according to two features: the sort of *reasons* given in support of the conclusion, and the *conclusion force* of the argument. We said that deductive arguments convey the tone of necessity in drawing the conclusion from the premises; also all of the information or evidence is already contained in some respect in the premises. It is an argument that moves from assumed or given premises to a conclusion that follows with decisive force.

An inductive argument, we know, is much the reverse of a deductive in that the argument moves from a collection of evidence to a conclusion and that

*See Module 29, "Direct Derivations," in Unit Three.

evidence serves to confer a degree of likelihood or acceptability upon the conclusion.

Normally, both deductive *and* inductive thinking works together in science (a lesson taught to us chiefly by Isaac Newton)—and in fact in all problem solving. To see how this works, let's take a simple example from everyday life.

Suppose a friend asks you to work on his car. The engine will not crank. You consult a car repair manual and learn that the failure of an engine to turn over will be due to one of four possible conditions, and those only: a bad battery, a faulty solenoid, a faulty starter, or faulty connections between them. Your friend assures you that the battery is "brand new" and nothing is wrong with it, so you rule that out. You proceed with this process of elimination (which is a deductive task) and test the solenoid, finding nothing wrong. You check out the electrical connections and find nothing wrong there either. Therefore, you conclude (by deductive exclusion) that the problem must be the starter, and so you replace it.*

When the starter is replaced, nothing happens: the engine still will not crank. Our deductions were correct as far as the evidence goes, but there must be something the matter with our induction. Possibly the test on the solenoid was misleading; possibly the new starter is faulty and we need to test it; possibly we overlooked a bad electrical connection. Then your friends concedes he never tested the battery. You get a battery tester and, sure enough, the "brand new" battery is dead after all.

What was wrong with your reasoning? Only that you didn't sufficiently test the battery; you took your friend's word on its reliability. That assumption led you down the wrong track and you may have replaced an expensive starter unnecessarily. On the other hand, maybe not; the starter may have been faulty *too* and that may have caused the battery to run dead.

We see, therefore, how the improper collection of evidence and the weighing of "inductive" likelihoods can carry even the best deductive reasoning to an unsound conclusion. The best science (and the best home repair work) is a sound combination of both valid deductive reasoning and good empirical testing. Science works hand in hand with both types of reasoning.

40.5 KARL POPPER AND THE "PROBLEM OF INDUCTION"

One difficult but important consideration remains. What I have been calling "inductive reasoning" thus far has been misleading, but intentionally. By 'induction', I mean any sort of reasoning that carries us to the likelihood of a conclu-

*This is *deductive* thinking because we are given only four possible causes. If we can rule out three of them then the fourth *must* be the problem. The *inductive* reasoning concerns the testing of the individual parts; those tests may be faulty, so the conclusions about the mechanical or electrical integrity of the parts are not certain.

sion on the basis of some system of relevant evidence. However, the word 'induction' has a more common and traditional usage, and that concept has come under sustained and possibly irrefutable objections.

Traditionally, induction has meant that a collection of empirical instances serve to *support* the likelihood or probability of a conclusion. The key word here is 'support'. One danger of looking for support for any theory (or hypothesis) is that you see only what you want to see. Consider, for example, that you have been assigned a term paper for history class and your topic is whether the American Revolution was necessary to achieve independence from Britain. You're predisposed to say it was and so you begin your research by digging out whatever quote supports your thesis. You find many authoritative sources that do agree, but you find one particularly troublesome one. This source (let's call her "Professor Iconoclast") argues with great plausibility and cogency that if the American Revolution had not occurred, the American colonies would soon after have attained a type of independence comparable to that of Canada in modern times (never mind, for now, your knowledge of the merits of that particular argument). Let's say that you are so astonished by this argument and so impressed by its force that you cannot find a reasonable reply to it, so you take your notes from that source and then conveniently "lose" them. You proceed to take all the supporting sources, write a paper around them and then hand it in. What happens?

You get a low grade. Your teacher commends your style, perhaps, and your diligence, but writes: "What about Professor Iconoclast's argument? For an acceptable grade on this paper, you should at least have done enough research to discover it. For a good grade, you should have addressed it."

Is that a familiar experience? If you are honest with yourself, you will acknowledge the justice of your teacher's criticism, but the larger point is that no amount of supporting evidence alone will ever suffice to make a hypothesis acceptable. What warrants your acceptance of a theory should be a sustained criticism of the thesis you favor.

Philosopher David Hume identified much of the same problem with the procedures of scientific evidence. He considered that all inductive instances have an underlying presupposition that the future will resemble the past. For example, "every day of my conscious experience I am aware that the sun has risen; therefore, the sun will rise tomorrow." The fact that the sun has risen every day of my experience (and upon reliable reports every day of the experience of every individual I have ever known or read about) does not offer one shred of evidence for the underlying thesis that the future will resemble the past. Therefore, there is *no reason whatever* to believe on the basis of this evidence that the sun will rise tomorrow!

That is a startling conclusion, but it is a powerful one; one more often ignored, even by scientists and philosophers of science, because otherwise, they believe, the method of science is without any credence—a plainly absurd conclusion. However, one modern philosopher of science has taken Hume's "Problem of Induction" (as it is called) to heart, philosopher Karl Popper.

Popper agrees with Hume. He agrees that no set of supporting evidence can ever substantiate or even warrant a scientific conclusion in the way of positive support. As such, he rejects induction altogether, or at least what I've called here the "traditional" account of induction.

Popper contends that all science is the process of criticism. He notes an "asymmetry" between *proof* and *disproof,* claiming that no amount of evidence can ever *prove* a hypothesis or theory but, theoretically at least, contrary evidence can *disprove* a hypothesis or theory by *falsifying* it.

More will be said of Popper's method later, but let's take one recent example to see what he means. In April 1989, two chemists at the University of Utah, B. Stanley Pons and Martin Fleischmann, claimed a discovery that would revolutionize physics: the development of "cold" fusion in a "test tube." All the conventional thinking said it was impossible and the scientific establishment was in an uproar. It was exactly the kind of scenario that Karl Popper would have relished. For one thing, if they were right, these two Utah scientists would have turned modern science on its ear; the attempt alone would have been enough to endear them in Popper's heart forever.

Immediately, more urbane and tradition-bound scientists scoffed at the two. From many quarters, they received rejection out of hand. Some prominent scientists disregarded the experiments even before they reviewed the data. Others, however, proceeded more methodically—more in a manner in accord with Popper's theory.

Scientists in Japan and Switzerland attempted to *replicate* (i.e., to repeat accurately) the Pons-Fleischmann experiment without success. Harwell Laboratory in Britain was also unable to replicate the experiment. Sustained research from a collaboration between Brookhaven National Laboratory and Yale University was also (so far) unable to find any substance to the claims made by Pons and Fleischmann. This is the process of criticism that Popper hailed as the mainstay of science. The attempt at falsification is the heart and soul of the scientific enterprise. Skeptical scientists continue to study the results in an effort to falsify what Pons and Fleischmann claim.

Then the two Utah scientists did something that would greatly displease a philosopher who ascribes to Popper's theory. They announced that they withheld some of the details of their experiments from public purview, presumably pending further funding of their research. This renewed the furor and this time with much more justice. Pons even announced that the failures to replicate their experiments were due to the fact that the attempts did not exactly follow their procedures, but then failed to say what those additional procedures should be.

In so doing, Pons and Fleischmann were reducing the empirical testability of their claim—the chance to falsify it—and thereby asking the scientific community (and the funding agencies) to take their claims on faith. In effect, they brought the process of scientific criticism to a complete halt.

As of this writing, the scientific establishment and the press in general are now highly cynical of the Pons-Fleischmann "discovery" and it is a justified cynicism, since the two apparently haven't followed sound procedure. Some laboratories, however, seem to have made some unusual discoveries of their own in attempting to replicate the Pons-Fleischmann results that say some new but different things about how nuclear fusion works, and that may prove to be important news indeed. In all, the attempt to shake up the establishment in "its dogmatic slumber" must be viewed with encouragement from a Popperian point of view, but the refusal of the two Utah scientists to provide enough detail to allow for proper attempts at falsification (according to Popper, the proper procedure of science) must be disdained.*

In the coming modules, we will look at the philosophy of science of Karl Popper in more detail, as well as some of the further evolutions of the Scientific Method in practice. In particular, we will study some recent speculation on the nature of observation, testing, and the evaluation of data. In all, the accent of the overall Unit will be on the processes of *creative* thinking.

EXERCISES

Recall the discussion about the young Galileo and his observation about the swing of the church lamp. Galileo noticed that while the arcs diminished in length, the elapsed time it took for the lamp to complete an arc did not. From this, Galileo derived the idea of the pendulum, which in turn was the idea behind the first clock.

If the clock was not yet invented, how did Galileo time the swing of the lamp?†

*For an account of the Pons-Fleischmann controversy, see the article "Fusion or Illusion?" in *Time*, Vol. 133, No. 19, May 8, 1989, pp. 72–89. Note that the withholding of details on procedure by Pons and Fleischmann may be due to patent complications.

†I am indebted to Charles Dyke of Temple University for the idea of this exercise.

MODULE 41

CREATIVITY, OBSERVATION, AND PROBLEM SOLVING*

1. **Observation**
2. Hypothesis
3. Experimentation
4. Theory

41.1 IN GENERAL

Suppose your compact disc player is broken; you push the "power" button and nothing happens. What do you do? Pack it up and ship it to an "authorized repair center"? Unless the repair is under warranty, you can expect a steep bill in return; if the unit is under warranty, you can expect a delay of a month or six weeks or more before you get the disc player returned. There's got to be a better way.

Often, there is. Fix it yourself.

An **inductive** argument is **sound**, if and only if, the evidence is sufficient or representative enough to warrant the likelihood of the conclusion, and the premises are true.

*Special thanks to Dr. Gerald Shinn of the University of North Carolina at Wilmington who inspired many of the suggestions contained in this module.

In this module, we'll look at some tips at how to approach some simple solutions to apparently formidable problems. These tips may strike you as little more than common sense; and they are. But the first step in the solution of any problem is a determination to solve the problem yourself. Behind these common sense tips is the heart and soul of scientific inquiry. The key to creative thinking is inspiration and determination. The hope is that by reading this module, you'll find some of that inspiration and methodical persistence needed to solve any problem.

41.2 GETTING THE PICTURE

When attempting to solve a problem, whether it's repairing your automobile or finding the solution to a logical deduction, the first rule is to take a look at the dimensions of the problem. Check it out.

To see how important it is to recognize a variety of different perspectives in problem solving, let's consider an analogy in art. If we're familiar with the study of paintings which are two-dimensional, we may find it hard to analyze a sculpture, which is *three*-dimensional. We may notice the composition and texture of a painting and try to apply that analysis to the study of a sculpture. That is the wrong way to look at it. Consider a familiar example: Auguste Rodin's *The Thinker.* The proper way to view that sculpture, as any other, is to walk around it and view it from different perspectives. As we circle the sculpture, we come to appreciate it more.

Analogously, sometimes we need to walk in circles around a problem we're analyzing. We try a solution from this angle and if it doesn't work, we need to try another solution from a different angle. That is frequently true whether we're doing laboratory research for physics or repairing our household electrical wiring. The "tried and true" and "the conventional wisdom" may not always be the best guides to a solution, especially if the problem is an unfamiliar one, or an especially difficult one.

Recall that the first step in the "prototype" of the Scientific Method is observation. In order to be really good problem solvers, we first need to be good observers. In other words, take a good look. Maybe you can *see* what the problem is or what the solution may be.

Let's go back to our earlier example of the broken compact disc player. Once you've found the resolve to try to repair the unit by yourself and you've reserved an adequate amount of time to tackle the problem, the next step is to take a good, hard, long look at what might be wrong. Is it plugged in? How obvious a question, but sometimes we need to be taught to see the obvious! Ask any appliance repair person how often he is called in to fix an appliance only to find there's nothing wrong with it; the owner just forgot to plug it in again.

But let's suppose that our problem is a bit more severe than that. What's the next step? Make sure the disc player is *unplugged* to prevent electric shock, get a screwdriver, and take off the cover.

Auguste Rodin's *The Thinker* (1904)

"What??? But that's high technology! That's got a *laser*!" you say, and that's certainly true. But, chances are, it won't be the laser itself that's out of commission, but a wiring connection or a minor component. You can sometimes see that by just opening the unit. Behold! Some dust and lint have gotten inside and fouled a connection! A simple repair after all.

Why don't people take this advice and scout out the dimensions of the problem before shipping it off to an expert? Why don't more people take a look for themselves first? This is a psychological and sociological question, but we may be justified in hypothesizing an answer. Perhaps their timidity is based on a fear that the problem is just so complicated that they'll only make things

A word of caution: Be careful that removing the cover won't void the warranty if so specified. However, we should be a bit cynical about warranties since they exist as much to absolve companies of liability as to hold them to their responsibilities. For my own part, I ignore such prohibitions because I'm usually confident that I can fix it myself (and if I can't, I still insist that the company honor the warranty—they usually do).

More importantly, be careful of hazardous conditions. Some components of electrical appliances, such as some television sets, may store up powerful charges even when unplugged. Be careful not to touch exposed parts with unprotected hands. Best of all, buy or borrow an appliance repair manual and read it. Some very good ones are on the market, and they will inform you when and where to be especially cautious.

worse. Sometimes that does happen,* but the satisfaction of finding your own solution outweighs the occasional mishaps and contributes to your self-confidence and esteem.

But getting in and "getting down and dirty" is a step removed from just *looking.* If you try to take on a major problem by yourself without proper training or experience, you may well get in over your head. But looking hardly ever gets you into trouble; it's just a matter of taking the small amount of time and *thinking* about what you see.

To my mind, this is the scientific outlook *par excellence.* Put aside the complex formulae and the intricate laboratory apparatus; the scientist is one who can look at the world afresh. As Albert Einstein said, "The whole of science is nothing more than a refinement of everyday thinking." The scientific attitude is marked by curiosity and wonder and a determination to think about what we see. As one science fiction writer has observed, the scientifically minded person is one who never takes for granted what she sees. Another person might take the automatic doors in a supermarket for granted as just another everyday experience; the scientifically minded person is the one who marvels at the experience and wonders how it works.

The ancient Greeks were among the first scientifically minded people and their skill at "natural philosophy" (as it was called) resulted in many important discoveries and advances. It's a common misconception that people of the world uniformly believed that the earth was flat until the voyages of Columbus. In fact, the average ancient Greek knew better. Many of the old Greek city-states were made up of traders and seafarers, so knowledge of the curvature of the earth was a well-known fact to most Greeks. How did they come to this remarkable revelation? By taking seriously what they saw, and by thinking about it.

It was a simple observation. The ancient Greeks watched their ships come into port. The first thing they could see was the ship's masthead.

*When replacing a friend's fuel pump in her automobile, I inserted it improperly and that caused a piece on the crankshaft to snap off; a repair that would have saved my friend about $100 ended up costing me nearly $400. It happens.

Soon the mainsail came into view.

Finally, as the boat came closer, the observer on the shore could make out the whole front of the boat.

How could this phenomenon be accounted for? "Well," the ancient Greeks reasoned, "the three sightings of the boat are exactly analogous to the view of a horse cart coming over the top of a large hill. Perhaps the earth is round: that would account for the different perspectives of the boat."

Three centuries before the birth of Christ, the curator of the library at Alexandria, a Greek named "Eratosthenes," carried this well-known observation one step further. He noticed the differing lengths of shadows at different seasons of the year at two different locations: Alexandria, Egypt and the city of Syene (now Aswan), far south in Africa. He reasoned that this was further evidence for the curvature of the earth. By taking measurements of the shadows in the two locations during the two seasons, he was able to calculate the circumference of the earth. His calculations were accurate to within the equivalent of 225 miles!

What was obvious and commonplace for the ancient Greeks became lost in ignorance to succeeding generations of the world. The great library at Alexandria was destroyed (we know of Eratosthenes' work by secondhand accounts) because the knowledge contained therein was deemed to be dangerous and heretical. Surely, many people in other places in later times observed the same phenomena as the Greek seafarers and Eratosthenes, but they were unable (or forbidden) to think about what these observations might mean.

We may sometimes neglect to check out the dimensions of a problem, to refuse to think about what we see, because we're taught or indoctrinated *not* to see.

41.3 AVOIDING PRESUPPOSITIONS

We talked about the dangers of presuppositions in Module 40 in this Unit when we discussed "thinking in a rut;" recall the story of the midwestern police chief who knew only technological solutions to combating crime and how he was unable to see through the trap of thinking just one way about a problem. We may say that the police chief was trapped by his own "presuppositions."

A **presupposition** is an assumption made in advance of inquiry. The most dangerous of these are the systematic or ideological presuppositions that lock us into seeing the world around us through one main perspective when others may be more appropriate.

Freeing ourselves from perspectives which we take for granted in problem solving is a difficult task. First, we need to recognize that we *have* a presupposition which is directing our thinking. Then we need to evaluate it and see if it's an appropriate way to think about a problem. Sometimes, no matter how methodical we are, if we see a problem from only one narrow perspective, we may never find the solution.

How do we find out that we hold presuppositions if these are assumptions made in advance of inquiry? We probably don't even know that they're there. One time-honored and effective technique that philosophers use is called "the thought-experiment." The thought-experiment is to philosophy what the laboratory is to physics. The object of the thought-experiment is to take us out of parochial or narrow outlooks and widen our thinking on a subject. Thought-experiments are a way of doing the "ground-clearing" for hidden presuppositions.

Sometimes the job of ground-clearing for presuppositions can hold surprises for us; it may reveal that we have prejudices or biases that we didn't know we had. For example, recall this riddle of the women's movement from the 1970s:

A father and his young son were traveling along a dark road late at night. Because of poor visibility, the car veered off the road and struck a tree. The father was killed and the boy was badly injured. An ambulance rushed the boy to the nearest hospital. The attending surgeon looked at the boy and exclaimed: "I can't operate on him! This boy is my son!"

How is that possible?

The answer is obvious to most of us nowadays: the surgeon was the boy's mother. But only a decade ago, sex roles were so rigidly implanted in the thinking of most Americans that few could solve this simple riddle. We tended to *presuppose* that the surgeon had to be a man. Members of the women's movement used this thought-experiment to reveal to many well-meaning and even sympathetic men and women that they had systematic presuppositions about rigid sex roles that needed to be reconsidered.

Uncovering presuppositions need not be so dramatic or revealing. Even in solving the simplest problems, we sometimes let presuppositions block our way to obvious solutions. For example, maybe we know that our car is overdue for a tune-up and that thought has been worrying us for a while. Suddenly, the car chugs to a halt. Immediate inference: it needs to be tuned. That presupposition may interfere with our systematic solution to the problem. Before you run out and buy a tune-up kit, check the fuel gauge. Maybe you ran out of gas!

EXERCISES

DIRECTIONS: *Examine the following five black pieces. The object is to arrange them into an arrow. Can you solve the problem?*

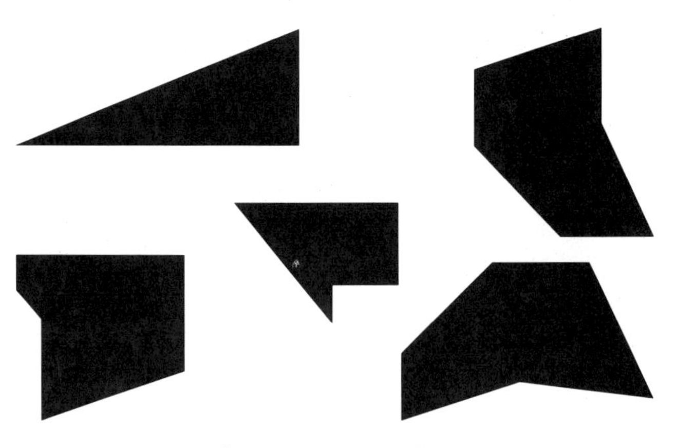

The "Arrow Puzzle"

As you might expect, the puzzle is designed to test your ability to discover hidden presuppositions which may interfere with the solution of the problem. The solution to the puzzle is shown at the end of this module.

41.4 OBSERVATION AND INTERPRETATION

Many philosophers now believe that observation is inextricably bound with interpretation. While this viewpoint is still somewhat controversial, it is gaining wider acceptance among philosophers of science. In opposition is the doctrine of the logical positivists who maintained that observation statements—precise and irreducible statements about the objects of sensation—could be devised in a way that would make them independent of interpretation.

The issue is a complicated one but we can consider an example which will help to make the matter more understandable. The example is a brief, one-act play about the reliability of eyewitness reports, staged for a legal studies classroom, but only the teacher and the actors know it's put on.

A male and a female student interrupt a class and the teacher's lecture by quarreling as they come into the room late. They exchange some unusual lines (the male student says to the female, "I don't care, you're still my *husband*"). Then the female student pulls out a banana and "shoots" the male student, who falls to the floor. While all of this is happening, the teacher holds up a series of unusual articles without comment—an orange, a dog biscuit, a red ball. When the "shooting" is over, the actors take their seats and the instructor then asks the class to write down exactly what they saw and heard.

The results are highly instructive; in classes in which I have put on such an unannounced play, few of the written statements come close to what actually happened (and what is recorded in the "script" originally handed out to the actors). They generally do record the "shooting" with the banana, but many students will identify it as a gun (failing to realize, I expect, that I want a *literal* account of what happened).

Most students *interpret* and *anticipate* what's said and done according to familiar categories. For example, when the male student says to the female, "You're my husband," a majority of students report that "You're my *wife*" was the actual utterance.

Only on one occasion in my experience of performing this play did one student give a completely accurate account of what happened. I learned later that he was an air traffic controller—a *trained* observer. The point of this story is that we should *train* ourselves to be accurate observers—that it *does* take training. Further, we must do our best to detach ourselves from emotions, evaluations, and presuppositions that may interfere with how objectively we see the world and what happens around us.

41.5 RESOURCES

The main thrust of this module so far has been on self-reliance; the emphasis has been on checking things out for ourselves and seeking our own solutions. We've considered a few cases where a failure of imagination can lead to conditions that interfere with the reliability of our own processes of observation, but by just being aware of those obstacles we may be better able to overcome them.

However, to become effective problem solvers, we must very often rely on the observations of others. None of us has the requisite time or resources to make all our observations in person, even though such firsthand observations are usually more desirable. If we want to know, for example, the chief export of the African nation of Togo or we want to know how many people attended the Woodstock Music Festival in 1969, we rely on the testimony of those who have studied the exports of West Africa or reporters who were present at Woodstock. That information is available to us in a good library.

Of course, this advice will likely strike the reader as commonplace, since the average college student spends a good portion of his college education doing research in a library. By now, you are very likely familiar with the periodical sections of the library, the card catalogue (and computer retrieval systems, if your college is fortunate enough to have one), the inter-library loan systems, and the cross-index to periodicals, newspapers, and specialized subjects. (If not, make an appointment with your school or community library for a tour of the library and its facilities. Most libraries offer regular tours and seminars.)

Perhaps you'll agree, however, that what most college students lack is enough hands-on experience, firsthand observation, and first-person interviews. For this reason, I will devote the remainder of this module to a discussion of that need, and some suggestions about how to get such experience.

The Detective and the Clues

What is tantalizing about the proverbial task of finding the needle in the haystack is that you are assured the needle is there. The space is restricted, the object is unique, it must be possible to find it. And no doubt, with enough care and patience it could be found with just a pair of eyes and a pair of hands. Of course, if the hay were to be packed in small cubes and a large magnet brought to bear as the contents of each were spread out, then the task would be made much easier—it would almost turn into a game.

This fairy tale problem has its analogue in the researcher's hunt for the facts. The probability is great that any one fact he wants is in some printed work and the work in some library. To find it he has to find the right cube of hay—the book or periodical—and then use his magnetic intelligence to draw out the needlelike fact. —Jacques Barzun and Henry F. Graff, *The Modern Researcher*

Ask Somebody Who Knows!

I have one simple piece of advice: *ask somebody who knows.* That advice, like all good advice, is obvious but, as we've learned in the pages of this module, sometimes the obvious is the hardest thing to see. Many times we confront a problem and the answer to the problem readily awaits us in the library. We go there straightaway, find just the right index, just the right citation, and the needed source awaiting us on the shelf or in a readily available periodical. That's the first step in doing college research (or most other research, for that matter). But time is precious; college students (and others as well) are usually facing a fast-approaching deadline. How do you get the facts you need as quickly as possible? By asking.

Perhaps the first person to ask when you've failed to turn up the information you need is your instructor. Very likely, she will be pleased you've taken such an interest in your subject. She has many years of training and experience in doing research and will prove to be the very best source on how to research an assignment, especially in her own field. If she can't tell you the answer or where to look, then ask her to refer you to someone who can.

Perhaps the next step is to consult the research librarian. That's his job. He knows the holdings of the library and also has a great deal of experience in tracking down obscure and difficult-to-find information. I have always found research librarians to be a great help; the more obscure the information, the greater the challenge, the more effort they expend in finding the facts you need. On some occasions, research librarians have succeeded in turning up some arcane information I requested even after I've given up hope of finding it.

Every student should make a practice of touring the great libraries of the world whenever possible. Some libraries are more receptive to guests from out of town than others, but I have never been turned away and in some cases (in the Institute for Medieval Studies in Montreal and in the Special Collections Division of the Library of Congress, to name two) I have succeeded in finding answers to my questions with eager and painstaking assistance of the research librarian. It's always good to familiarize yourself with the card catalogues and holdings of the libraries you use most often so you can do most of the research yourself, but the first thing to do when you're at an unfamiliar library is to ask for assistance. If you're polite and patient, you will obtain the help you need more often than not.

A next step: conduct first-person interviews. Every college and university and nearly every community has a treasure-trove of experts on a wide variety of subjects. Many universities will boast some of the foremost experts in their fields. Consult them. If you've been assigned a class project to do a marketing report for a hypothetical product, for example, contact a local marketing firm and ask them if they'll look over your work. If you need to find out the content of a bill in the legislature for your political science class, contact your representative's office; they're usually very happy to provide such assistance. If you have trouble applying a formula for aerodynamics in physics, you might try talking to a local engineer and asking her help.

These business people, politicians, and professionals probably already support your college or university financially and in many other ways. Few will be reluctant to help out individual students who politely and patiently ask their assistance; they'll often be pleased you asked. But remember: their time is valuable (possibly beyond your imagination). You must be exceedingly polite and you must not try their patience. Remember also that you are asking them to do for free the very services that they perform for fees or salaries, so show them the utmost respect and courtesy.

We can extend this idea beyond the confines of your community as well: use the telephone. In a class I once taught in Business Ethics, I asked a student to find out if a particular case study we were reading had yet been resolved in court. I thought the student would spend weary hours in the library. Instead, he called the airline that was the subject of the study; they weren't much help directly (the airline was the subject of the litigation) but they directed my student to the person who was the litigant. To the student's surprise, the litigant was very forthcoming, explaining the details of the case from his point of view and also noting what success he had had so far in court and informing the student what had yet to be resolved.

The telephone is the most remarkable and rapid source of information we have; most business people and professionals are well aware of this. If you can't be on the scene to make the observations you need, call someone who *is* there. The caveat, to be polite and patient, should be underlined and recited when making such phone calls. If someone is reluctant to help you, don't push. Make sure you know what you want to ask and be brief; know what you're talking about before you make the call. Above all, be considerate. A whole other set of rules for courtesy apply when you're on the telephone since you are contacting a person where he or she lives or works and every telephone call constitutes an intrusion into the work and life of the person called.

If you've been successful in obtaining information in an interview or a telephone call, cite the persons you talked to (if they're willing). Give the date and the time of the interview and be careful to report exactly what they did say.

The scientist, scholar, detective, and reporter share many attributes in common: a commitment to discovering the truth, the procedures for finding the truth, and the methods of obtaining evidence and evaluating it. In Modules 44 and 45 of this Unit we will consider other methods for obtaining evidence and evaluating it, a process that is at the heart of scientific research and the basis for good inductive thinking.

EXERCISES

Exercise 1

DIRECTIONS: *Following is a series of projects that the class may undertake individually or in groups. Use the suggestions contained in this module for finding the solutions. Note that* *some are more difficult than others, and some are deceptively difficult. The more difficult projects are marked by a star (*).*

1. Find out Plato's *real* name. ('Plato' is a nickname; he used it as a pen name.)

2. Discover the title of the 1942 popular movie which inspired the recent film *Crocodile Dundee.*

3. Find the source of the quotation, "Truth is truth, and error error. As long as I know the difference, I am not mad."

*4. Find the name of the congressman for whose legendary speech the term 'bunkum' was coined.

5. How did the father of the character Lucy Binet spend his time in prison?

6. Name the only ordained minister who played major league baseball.

*7. Who was the last switch hitter to win the American League "Most Valuable Player" award?

8. Find the derivation of the name 'HAL' in the film *2001: A Space Odyssey.*

*9. Name a superstition that originated and became common in the United States in the twentieth century.

Exercise 2

DIRECTIONS: *Compile a set of problems like the ones in Exercise 1 for the rest of the class to research. The problems should be somewhat arcane, require unconventional modes of research, or help expose hidden presuppositions about the way we approach problems.*

Solution to the Exercise in 41.3

The "Arrow Puzzle"

When you tried to solve the puzzle, did you think that the arrow would be in black?

The solution to the puzzle is obvious once you realize you had a presupposition about how the result would look. Once you tried to free yourself from conventional ways of thinking, you were on the right track. Maybe, after a few false starts, you tried thinking of the pieces three-dimensionally (as I did). I even tried imagining the pieces standing on their sides! That led me to change perspectives and soon I was able to think of the arrow as *outlined* by the black pieces instead of composed of them.

MODULE 42

CREATING HYPOTHESES

1. Observation
2. **Hypothesis**
3. Experimentation
4. Theory

42.1 IN GENERAL

The single most creative step in the Scientific Method may well be the task of framing hypotheses. Hypotheses are often seen as following upon the footheels of observation, but framing a suitable hypothesis may require repeated observations and occasional second looks so the process is really more a synthesis than a sequence.

Hypotheses are usually conditional in form ("if . . . then"), and they usually make a testable prediction. For example, suppose that you have noticed over the years that whenever people lie to you, they display a tendency, however

An **inductive** argument is **sound**, if and only if, the evidence is sufficient or representative enough to warrant the likelihood of the conclusion, and the premises are true.

briefly, to touch their noses. Have you discovered an important fact about body language? To find out, we need a hypothesis to test. We could fashion it to look something like the following:

***Hypothesis:* If I discover any person *X* touching his nose when conveying information to me, then I should check out what *X* is saying to see if *X* is telling the truth.**

Now it's time to put the hypothesis to the test: Suppose that you miss a class and an acquaintance named Sarah tells you that a forthcoming test was postponed one class day. While she's telling you this, you notice that she unobtrusively brushed the tip of her nose with a knuckle. You check it out; the test wasn't really postponed! If you'd listened to her, you would have gone into the test unprepared. [Now it's time to frame a new hypothesis about why she didn't tell you the truth!]

What does this prove? Nothing at all, but you do appear to have a worthy subject for investigation. Notice that no number (no matter how large) of positive confirmations will ever establish our hypothesis about body language. Now that you have a handle on a suitable hypothesis, you can begin testing in earnest. The object should be to look for *falsifying* cases—and you will find them if you look hard enough. Yet, you won't want to give up such a juicy theory-in-the-making about body language, so as the negative instances pile in, you find you can adjust and fine tune your hypotheses even more, make them more precise, make their explanatory power greater, and account for the apparent falsifications; then you're on your way to an interesting theory. [More about testing a hypothesis in the next module.]

But where did such an intriguing idea about the connection between body language and lying come from in the first place?

42.2 BRAINSTORMING

The statement "You can't teach creativity" is itself a hypothesis and it's probably false, but like most false hypotheses, we have something to learn from it. Surely, we look in vain if we're trying to find a recipe (or an algorithm—an "effective decision procedure") for creative thinking. Such straitjackets are precisely the reverse of the creative process. Nor should we try to find ways of thinking creatively that can be universalized to all people or under all circumstances. But can we say nothing worthwhile about the subject at all?

Notice that asking how we come up with ideas for scientific hypotheses is exactly like the question successful writers get asked so frequently: "Where do you get your ideas from?" No question gets more tiresome than this one for the professional writer and possibly that's because the writer rarely thinks about where the ideas flow from; they're just there when she needs them! So writers tend to shrug off the question with a curt reply: "from the Muse" or something else irreverent.

This question so distressed Stephen King, the writer of horror stories, that he wrote a horror story based on it! The story is called "The Ballad of the Flexible Bullet"* and it has as its underlying premise the idea of a fiction writer who has a little elf-like creature (called a "Fornit") in his typewriter to whom he feeds pieces of cookies and peanut butter (gumming up the works from time to time) and in return the Fornit brings him good luck for his writing.

Well, this is certainly a *creative* answer to the question of where ideas come from (and probably a very appropriate sort of answer, given the subject), but surely we can come up with more constructive answers, even if they're only tentative ones.

In Module 41 we learned that the process of observation is what gets the creative juices flowing. Observation provides the sort of range of information we need to get going. But, as we also saw in that module, sometimes the most obvious solutions are the most difficult to see. A wide range of information is just the starting point; surely, the more informed we are, the better are our chances of generating new ideas. Where do we go from there?

Perhaps the very first step is something we call "brainstorming." The dictionary defines a 'brainstorm' as "a sudden clever, whimsical, or foolish idea."† That's a revealing definition because it assumes so much. First, it suggests that many new ideas are "sudden." In what sense is that true? Certainly, many good ideas come to us "in a flash," but that description conceals a great deal of deliberation that went before. In fact, most good ideas come only after a sustained amount of thinking, but it is true that when they come, they seem to come all of a sudden. Second, the ideas (when they come) seem to be "whimsical"—as though they were there all along, or at least they should have been obvious to us all along. Clever ideas are often like that. However, we already know from the module on observation that the obvious is often the hardest thing to see. (Recall Edgar Allen Poe's short story, "The Purloined Letter"—where was the letter? Right in front of you all along!) Finally, the definition suggests that brainstorming may seem "foolish." That's the point that needs discussion here.

In his book *Effective Problem Solving,* Martin Levine says the following about "brainstorming:"

> The first and, perhaps, primary rule is that criticism is not allowed. Take any proposed solution, no matter how far-fetched, and write it down. It is easy to be critical of novelty, especially of other people's suggestions. However, nothing inhibits wide-ranging, flexible, lateral thinking like a critical, judg-

*The story is included in the collection called, *Skeleton Crew,* by Stephen King and published in paperback by Signet.

†*The American Heritage Dictionary of the English Language,* William Morris ed., Boston: Houghton-Mifflin, Co.

> mental atmosphere. Don't judge. Even the crudest ideas are to be respected. The brilliant critic, who immediately sees the flaws in every suggestion, does not belong here. Neither does the bureaucrat who, comfortable with his routine, is afraid of change. One must feel free to be "lateral," to search for strange and wonderful ideas.*

The irony here is that you've been taking a course in logic or "critical thinking" and now, all of a sudden, you've been asked to suspend your critical faculties—but just for the moment. Perhaps the best way to see the point being made here is to imagine you're a writer of fiction, like Stephen King. Every writer at one time or another faces the phenomenon known as "the terror of the blank page." How do you get to develop an idea from its first and crudest incarnations to its polished and sophisticated form? The answer is by writing "a rough draft." The rough draft is precisely that—it's *rough*. It's crude. It's filled with misspelled words and inaccuracies. The "bureaucrat" can never be a fine writer (or creative thinker) because the bureaucrat can never get past those crude first steps that every good writer must take to get to something new. The good writer is often the person who can momentarily suspend critical judgment to get to a working prospectus. Criticism does indeed play an indispensable role, but that comes later. For now, the objective is to do something different. Not everything that is different is good (consider, for example, the sorts of religious cults that pop up from time to time on the West Coast). However, when this kind of inspiration is combined with a critical faculty, the result may be a new business entrepreneurship, a revolution in art, or a breakthrough in philosophy or science.

Every new idea must begin with a "rough draft." The temptation is to apply the most stringent critical standards to it before it has a chance to get off the ground. However, a certain amount of tenacity is required, a determination to revise and refine your initial idea until it can stand up to the justifiable criticisms that it will require, and (most important of all) you must then seek out the most devastating sorts of criticisms that you can find. We must make a distinction between two sorts of criticism: *cynicism* on the one hand and *constructive criticism* on the other. Some people reject all criticism, but they can never hope to come up with a worthy idea; the best ideas are precisely the ones that can be honed in the face of criticism. Other people abandon any idea at the first sign of criticism, and many otherwise worthy ideas are completely lost precisely that way. The object is to show some persistence and loyalty to an idea until the approach to the solution of a problem you have in mind meets such sustained criticism (or evidence, or testing) that it can no longer stand up to the assault.

So, suppose you have what you think is a genuinely good idea for a new fiction novel. So very many people do have such ideas. In his book, *You Can't Go Home Again,* Thomas Wolfe tells stories of ordinary people he meets who have wonderful ideas for stories but have no hope of putting them into prose.

*Martin Levine, *Effective Problem Solving*, Englewood Cliffs, NJ: Prentice-Hall, Inc., p. 83.

Wolfe does it for them. That's a magnificent revelation to any aspiring writer: the difference between an *aspiring* writer and the *successful* writer is the ability to take a sparkling idea and work it into the strict discipline of written prose. The moral of the story is this: take your good ideas, afford them enough respect to give them the discipline they're due, and pound them into shape. That's the whole story behind creating constructive hypotheses. Neither the person who gives up at the first sign of difficulties nor the person who holds to an idea in its original form despite the well-meaning criticisms of others will ever produce an original idea that's worthwhile.

Roughly, we need these things to go from a brainstorm to a constructive hypothesis: first, we need a new idea, however crude (some suggestions about how to do this will follow), and then we need both the dedication to the idea to see it through hard times and the self-esteem to hear criticisms to hone it into a testable shape. The best way to do this is to seek a wide range of information and be willing to *rearrange* that information into a shape that will help you solve your problem. The uninformed person is rarely the creative person.

Second, the way to rearrange your information is to step outside your problem and look at it as if you're facing it for the first time. Imagine that you're a child (or an extraterrestrial) who is looking at the world for the first time. Discard all dogmas and "common sense." Presumably, the problem is a problem precisely because the old routines haven't worked. If they did, you wouldn't have a problem. Don't be a bureaucrat. Don't allow yourself to be trapped in a rut. Find a new way out. And don't be shouted down by people who object that you're not "going by the book." The "book" is very likely why you have a problem to begin with. Write your "rough draft" and write it all the way through. That's when self-criticism comes into play. If you're a good thinker, then you will likely be your own best critic: then and only then, give it your best, sustained attack. If you are intellectually honest (and able) you will anticipate your opponents' criticisms, and your work will be better for it.

When you seek out criticism, seek out criticism first from people you think may be sympathetic to what you're trying to do. This advice is obvious, but it needs to be engraved in your thinking. These people will be the ones who will try to make your ideas more refined and immune to sustained attack. They will want to see you succeed but you surely can't do it if you don't pay some heed to what they say. Here's a good rule of thumb: if two or more of your "sympathetic" critics find a common ground of criticism, give your idea more thought. Granted: you may have an idea that's so new, so different, that even friendly critics have a hard time accepting it. But make sure that's your problem before you go ahead. Remember also that someone once said that a *fanatic* is "someone who redoubles his efforts when he loses sight of his goals." Be prepared to give up your hypothesis when the criticism is both sincere and irresistible.

That brings us to the next stage: make your hypothesis as *testable* as possible. Once you have framed your rough idea in a way that you think can stand up to friendly criticism, re-shape it in a way that gives your opponents a chance to falsify it. If your idea has any staying power at all, it, or some version of it, will survive the assault. What better idea is there than one that can stand up to its harshest critics? Make your critics show some ingenuity. Give them something to aim for.

Now let's say a few words about where some of these ideas we call 'hypotheses' may come from.

42.3 GENERALIZATION

We're taught from such an early age *not* to make generalizations that we might lose sight of the fact that *generally speaking* knowledge is framed in terms of generalizations. The best of these generalizations we call "laws of nature."

Laws of nature are hard to obtain, and no "law" is immune to criticism and contrary evidence. But every "law" begins with a tentative hypothesis.

Some generalizations are so bad, so full of bias and presuppositions, that they don't deserve the wind to speak them aloud. Generalizations about people and races are like this. If we say that "All Northeasterners are rude" or "All Southerners are bigots" or "All Californians are flaky," we are making generalizations that are not worth the time it takes to say the words. There's no possible way we could ever find enough evidence to support such nasty generalizations about people. Philosopher Norman Malcolm tells a story about his teacher Ludwig Wittgenstein. Malcolm had returned to England on leave during World War II and he said something to Wittgenstein about the "character" of the "English people." Wittgenstein was furious and demanded to know if the young Malcolm had "learned anything" from his teaching. Looking for "national" or "racial" characteristics of people reveals the most heinous form of prejudice conceivable.*

On the other hand, where do we start looking for testable hypotheses unless we're willing to start looking for generalizations of one sort or another? The opposite of bigotry is the moral timidity to look for generalizations at all—for example, the refusal to generalize definitions of fallacies to cover religious tracts which express beliefs you share or the refusal to generalize the fallacies to cover political speeches by politicians you support just because you support them. This is just a flat-out form of anti-intellectualism. When we do this, we are in effect "locking" generalizations into compartments so that they are removed from our interests, so they do not affect our personal beliefs or our concerns.

The problem, as we have already suggested, is that many people "shoot down" without any good reason any attempt at all to look for a generalization. This is where some persistence of conviction must necessarily come into play. Follow up on a generalization until you come up with good reasons to think it's groundless or too riddled with exceptions. Even very good generalizations will have some exceptions. In that case, revise them and refine them until you

*The one exception to this, of course, is when one suspects some sort of racial, ethnic, or sexual discrimination is already in place. In such circumstances, looking at the racial, ethnic, or gender memberships of institutions or organizations may have something revealing to say about the levels of discrimination present. But this, like all generalizations, requires caution in light of other possible explanations.

can have more confidence in them; always remember that generalizations will have many borderline cases, that's in the nature of the discipline. But if the generalization is so pitted with falsifications and exceptions that it cannot be sustained except with the most heavy-handed and critically blind interpretations, give it up.

When does a generalization of the sort we're discussing become a "hasty" generalization? When does the process of generalizing from experience become fallacious? To give a rough-and-ready answer, we may say that a generalization harbors the fallacy of Hasty Generalization when a generalized hypothesis is treated like a Law of Nature—when the sweeping generalization we made as a result of brainstorming is accorded the status and durability of a thoroughly tested and time-honored bedrock of science.*

In most of the day-to-day business of science, the art of generating hypotheses is dry, procedural, and obvious. When Thomas Edison was commissioned to invent the electric light bulb, for example, he spent fourteen months in search of a suitable filament. When one material didn't work, he tried another, and so on until he found one that did work. Perhaps this is what he meant when he said: "Genius is one percent inspiration and ninety-nine percent perspiration." So much of the Scientific Method is *methodical*—the painstaking application of the principle of trial-and-error.

On the other hand, Edison's remark also implies that inspiration is indispensable to science. Surely, nowhere is it more important than in the art of generating hypotheses. Sometimes we need to look for bizarre hypotheses and wild coincidences so we have a place to start our investigations. I'm especially fond of a remark made by the protagonist in a popular detective novel: "Coincidence exists but believing in it never did me any good."†

In what follows, we will consider some suggestions about searching for inspiration for hypotheses. Some approaches are more methodical than others, but each has served an important role as a "hypothesis-generator" at one time or another.

> ***. . . every recognition of a truth is preceded by an imaginative preconception of what the truth might be—by hypotheses such as William Whewell first called "happy guesses," until, as if recollecting that he was Master of Trinity, he wrote: "felicitous strokes of inventive talent."***
>
> —PETER MEDAWAR

*For a more detailed treatment of the fallacy of Hasty Generalization, see Module 35, "Inductive Fallacies," in Unit Four.

†Robert B. Parker, *Pale Kings and Princes*, New York: Dell Publishing, 1987.

42.4 ANALOGY

Another very powerful way of conjuring up the basic ingredients for a useful hypothesis is through forming analogies from the unfamiliar to more familiar experience. Analogies are formed upon recognizing that two entities or processes share a set of common features. The presumption is that if both share properties x, y, and z, then they may share other properties in common as well.

Let's consider an easy example of how analogies may help us to think creatively. So much of Einstein's theory of relativity was so abstract that the claim was made for many years that only a handful of scientists understood it (a claim that certainly is no longer true). One of the ways that teachers of physics have come to communicate the ideas behind the theory to students and laypersons is through concrete analogies to more familiar experiences. For example, the common phenomenon of passing trains helps us to explain the relativity of motion to the observer. When two trains pass each other from opposite directions, for a moment the other train looks frozen in space. If your train is at rest and another train begins to move parallel to it, you may be deceived into thinking that your train is moving away from the station, and so on.

Such analogies have been highly useful in expounding upon Einstein's special theory of gravitation. Einstein regarded gravity as a distortion in the "fabric" of space. The word 'fabric' here is a metaphor, and an analogy is made between space and a piece of cloth. Some scientists have extended the analogy in highly productive ways. Carl Sagan, for example, asks us to imagine space as a "sheet of graph paper" made of rubber and if we drop any object of mass on it, it "deforms or puckers" the surface. That gives us a three-dimensional representation of space. The sun may be imagined to be a very weighty ball which creates a deep indentation in the rubber sheet. A smaller and lighter ball may be rolled with sufficient velocity to spin around the edges of the indentation made by our "sun," thus giving us a representation of a solar system. Such a representation has been helpful in theorizing about "black holes," cosmic areas of amazingly high gravitational density from which even light cannot escape. Sagan compares a black hole to a "bottomless pit" creating such a distortion in the fabric of the rubber sheet that if an object were to fall in it might emerge in an entirely different place (and time!). Scientists have given the name 'wormholes' to these theoretical distortions in the fabric (picture the deep indentation that would result; they do seem to resemble something like a "worm's hole") and have speculated that if such wormholes exist, they might one day serve as a kind of "subway" for interstellar travel.*

A word of caution: we don't know for sure that black holes really exist. At this time, they are little more than a subject for scientific speculation based on some observational evidence of apparent distortions in light and radio signals received on earth from deep space. However, the fact that they are subjects of speculation is precisely our point. We're in the business of generating new ideas,

*Carl Sagan, *Cosmos*, New York: Random House, Inc., 1980, p. 242.

new *hypotheses,* and one very productive way of doing that is by means of such analogies. Notice how many analogies have emerged from thinking about space and time in this way: (1) we compare space to a sheet of rubberized graph paper; (2) we imagine the gravitational fields of stars as indentations in the fabric; (3) we imagine black holes as very, very deep indentations in a small space—so deep that we may compare the indentations to holes made by worms; (4) we create an analogy to a "subway" system and imagine the possibility of interstellar travel through the wormholes.

We have no evidence whatsoever that black holes (if they exist at all!) could be used for space travel (and the enormous forces that must build up in them make the idea seem prohibitive), but the result is "food for thought." It is a way to frame new hypotheses that we may someday be in a position to test. That's one way that science and all knowledge can grow.

The problem with *arguments from analogy* (as we mentioned in Unit Four on "Arguments and Fallacies")* is the danger that we may jump to conclusions without evidence or reason simply because we have an attractive analogy. As more evidence rolls in, we may have to revise or discard our analogy. The Greek playwright Aristophanes parodies this problem when he has "Socrates" compare the universe to a pot-bellied stove which heats up and creates "clouds" of steam, thunder and lightning.† (Come to think of it, that's not such a bad analogy after all, even though it was meant and received as an object of ridicule at the time.)

Another danger is that the analogy is so pretty or so well established that we become entranced with it and refuse to discard it even when the weight of further evidence has made it obsolete. Like one of the spinning balls on the rubberized sheet, we may slow down and get "stuck in a rut." We may be so taken up with imagining the universe to be a rubberized graph sheet (or a pot-bellied stove) that we become tempted to discard contrary data and observations, even though the contrary evidence is far more the useful object to focus on. We may therefore refuse to conjure up far more realistic and (perhaps) even more beautiful and elegant analogies for the conditions we're describing just because we're fixed on the old way of looking at things.

Analogies are never more than hypotheses-generators. They do not confirm, corroborate, verify, "prove," or otherwise establish any firm theoretical footing or natural laws. That cannot be done without good, tight, and ample testing and evidence. The elegance or beauty of a good analogy is only a way of getting us to look in places that we might not have thought of otherwise. But analogies are indeed "hypotheses-generators;" they generate new ideas and new ways of looking at problems that can result in testable hypotheses. Never treat the result of analogical thinking as established fact. Any analogy can be "knocked down" just by reflecting on all the possible *dis-analogies.* For example, as useful as the

*See Module 35, "Inductive Fallacies," in Unit Four of this book.

†The forerunner of the traditional "pot-bellied stove" was actually invented by Benjamin Franklin but a modern translator, William Arrowsmith, translates the Greek word for stove that way for its effect on modern audiences. Actually, the old Greek stove was a less suitable model for the convection principle. [Does anybody read these footnotes?]

rubberized graph sheet model of the universe appears to be for modern physics, you'll be sorely disappointed if you point your telescope to a quadrant of the universe where you expect the dotted graph lines to appear. The absence of dotted graph lines in reality in no way affects the viability of that analogy, but it still stands as a point of dis-analogy, however frivolous.

More pertinent are dis-analogies that are so pronounced that they affect the very worth of the hypothesis. Nowadays many scientists and philosophers like to compare thinking to the operations of a computer, an obvious analogy. Before the invention of the computer, thinking was compared to the operations of a telephone switchboard and, before that, to the operations of a telegraph. The dis-analogies are obvious, especially for the older examples, and the effect of clinging too closely to such analogies is that they might "freeze out" important aspects from our consideration. For example, one who is convinced that human thought is very much like the operations of a computer may be less inclined to attach importance to the role of ingenuity and creativity in thinking than someone who does not.

42.5 CAUSE AND EFFECT: MILL'S METHODS

Another way of coming up with ideas for testable hypotheses (and perhaps the most common way) is to observe effects and search for possible causes. For example, suppose you want to find a way to account for the extraordinary economic conditions of the late 1970s. The end of that decade was marked by severe recession and runaway inflation, a combination that many economists had supposed to be incompatible until then. The search for such causes is often impeded by ideological blindness or single-mindedness. For example, an entire school of economists grew up around the thesis that the "stagflation" (as it came to be called) was a consequence of "the welfare state" and sought to remedy the conditions by infusing more free enterprise capitalism into the system. Actually, we can expect the "causes" of the period's stagflation to be much more complicated than that.

How do we proceed? By looking for the events that characterized the era. Stagflation was a new and unprecedented phenomenon, so we can expect that its causes would be unique or unprecedented. The excessive government spending of the time is surely a productive place to start looking, but that runaway spending was not just limited to welfare programs. Defense spending had hit all-time highs as well. If we can accept the possibility of some "lag time" we can look to two other major events in the decade that might also account for the phenomenon. For example, the Arab oil embargo of 1973 had a devastating effect on both the actual workings of the economy and the confidence that people placed in the working of the economy as well. The war in Vietnam had come to an end in 1973 as well with a cease-fire agreement that provided for U.S. withdrawal from Vietnam without any similar agreement from the North Vietnamese. The end of the war on such terms brought about severe economic dislocations, taking us from a prosperous war-tuned economy to a sudden peace-time economy. These events could have contributed to the economic difficulties the nation faced in the years ahead. In fact, we may even find such ideas to be useful hypotheses in accounting for the stagflation four years later.

Whether or not that analysis bears up to criticism, the idea is that it gives us a place to start looking, and that's the role of any hypothesis formation. Finding an event or a phenomenon we need to explain, we quite naturally begin to speculate on its causes. The problem is that the word 'cause' is an antiquated term which has ties to very old-fashioned ways of looking at the world. Medieval alchemists, for example, were in search of a compendium of causes and corresponding effects, sort of an almanac for science, which ascribed a single, unique cause to each and every effect. Science doesn't work that way anymore. As we discussed in Module 35, Section 12 on "False Cause" in Unit Four on "Arguments and Fallacies," many modern scientists are leery of the word 'cause' and some prefer to talk about statistical correlation instead ("Cigarette smoking has a high statistical correlation with incidents of lung cancer," for example, instead of: "Cigarette smoking *causes* cancer").* That caution is not excessive. If we wish to avoid the "book of facts" way of thinking about science and the world that typified alchemy, we can look for causes only in the most circumspect way. We mustn't be misled into searching for certainties, for one thing. The postulation of causes for observed effects is better thought of as a way of generating testable hypotheses, and nothing more. Once we've done our groundwork and have begun our testing program in earnest, we'll likely stop thinking of causes and look to more tentative correlations instead.

However, we have some guidelines for searching for putative causes as testable hypotheses, and some of the best of these have been laid down for us by nineteenth century British philosopher, scientist, and economist John Stuart Mill. These have come to be called, "Mill's Methods of Induction," and they consist of five methods, as follows:

1. The method of agreement
2. The method of difference
3. The joint method of agreement and difference
4. The method of residue
5. The method of concomitant variation

The Method of Agreement

The method of agreement is characterized by searching for characteristics that any set of circumstances have in common. Ideally (but usually unrealistically) we look for the *one* common feature. The best example of this sort of analysis can be found in medicine. If, for example, we encounter a previously undiagnosed disease (let's call it "Logicbookitis") we may begin by looking for the

*See Module 35, "Inductive Fallacies," in Unit Four.

constant presence of a single, new strain of bacteria or virus that each person who contracts the disease has present in his bloodstream. If we detect such a new strain, find it in every person who has the disease and on the cover of every logic book, we may begin to hypothesize that "logicbookitis" is caused by the contact between people and logic books. But that's a hypothesis only, and (as always) it is subject to extensive testing.

The problem with this method is its emphasis on finding a *single* cause that produces the condition; rarely are things so neat in practice. Even a disease can undergo variations and mutations, so that some strains are harmful, some are not; the ones that are harmful may not affect all people the same way so that differing symptoms emerge. Other people seem to be immune to the disease. In fact, the detection of such immunity (which can be overlooked by this method) can offer the very key to curing the disease, since reproduction of those conditions (perhaps reading your logic book with care and attention) can produce the same immunities in those who have been stricken by the disease.

Furthermore, recall our more complicated example about stagflation in the 1970s. A little reflection will prompt us to consider that not one event caused the condition, but a whole series of events. The method of agreement is a useful place to begin, but too much reliance on it can take us astray.

The Method of Difference

The method of difference consists of looking for ways in which the sets of circumstances differ. We look at two cases (at least) in which one exhibits the phenomenon under study and in which another does not. How do they differ? The answer to that question can provide us a clue to a hypothesis-generating "cause."

For example, suppose we know two identical twins, one suffers from agitation and nervousness, the other does not. How can this be? We ask them to monitor their behavior for a few days and find one notable difference: the one twin drinks four or five cups of coffee a day; the other drinks none. We therefore have a difference that can usefully serve as a hypothesis for the condition of the one nervous twin.

Once more, the danger is that we prematurely cut short the possible number of other differences between the two. For example, if we begin to test our twin subjects under controlled conditions, we may find that the coffee ingested by the one is not enough by itself to produce the agitation, but when we take into account the fact that the nervous twin also has weak grades in school (whereas the other does not) and drinks the coffee to stay awake studying at night, we have a more powerful explanation for the condition than we would otherwise have.

The Joint Method of Agreement and Difference

The joint method of agreement and difference does precisely what the name suggests: it uses a combination of both methods to generate a hypothesis about a cause. We look for a respect in which two or more cases agree, and a respect in which they differ.

For example, we have a suspected case of food poisoning. Ten students in the school cafeteria come down with gastroenteritis, whereas the other ten who ate there that day did not. What they all have in common is that they ate in the cafeteria. However, the difference is that ten students ate "mystery meat" that evening, and the rest had "slop on a shingle." We are entitled to speculate that the mystery meat was the cause of the food poisoning.

Once more, we face the same danger of oversimplification as before. More careful examination may also reveal that the meatballs were tainted as well. But that condition may be concealed by the fact that only the students who ate the mystery meat also ate the meatballs, whereas none of the others ate any meatballs. That fact might not come out until the next day, when the cafeteria decides to recycle the meatballs and serve them to the unaffected students. (However, the more prudent of them, the ones who read this book, decided to eat off campus that evening.)

The Method of Residue

The method of residue relies on noticing the conditions before and the conditions after a particular event. You look for what's "left over" after an inventory of circumstances. Notice that this is an example of a sort of rudimentary process of elimination when only a limited set of the possible totality of circumstances is taken into consideration. You can account for the presence of x and y, but z (the *residue*) is a condition that goes unaccounted for by conventional explanations.

This method can be either deductive or inductive. It is deductive when you have limited your range of explanations to a tightly controlled range. For example, suppose you want to find out your dog's weight. You place the bathroom scale on the floor and you plead with Merlin, your dog, to get on it (he won't). You try to push him on (he doesn't cooperate and, anyway, you can't get all four paws on it). So what do you do? First, you take your own weight. Then, you lift up Merlin in your arms, stand on the scale, take a reading, then subtract your weight from the total: presto! You have Merlin's weight (the residue).

Of course, that method isn't foolproof because (as you know) Merlin will be struggling and wiggling the whole time and that could throw off your balance and the scale's reading. But, assuming you have matters under control, and that the scale is accurate, you can now safely conclude you have an accurate reading of your pet's weight.

A more "inductive" use of this method is clear when you have a possibly wide range of explanations, but lack of complete knowledge of them or a lack of imagination limits you to a few probable alternatives. For example, you wake up one morning to a foul odor in the house—a smell of rotten eggs. You had eggs for breakfast yesterday, so you hypothesize at first that you need to take out the garbage. You do that, but the odor lingers. You clean out the refrigerator, but find nothing which can account for the odor. You walk outside, and still the odor of "rotten eggs" is in the air. You know there's a paper manufacturing plant thirty miles away that emits sulfur dioxide in the treatment process, and that gives off an odor very much like that of rotten eggs. You test the direction of the wind and find it's blowing from that general direction. You have thus eliminated other possibilities and the one that remains—that the paper plant is at fault—becomes a good working hypothesis (one good enough to warrant a call to your local health agency for investigation).

However, this is at most a probable conclusion—a hypothesis only. It needs to be checked out. Other possible explanations which may seem less probable to you at the time may still bear examination. (Use your imagination: there may even be a giant, rotten dinosaur egg nearby, but that's not very likely, is it?)

The Method of Concomitant Variation

The method of concomitant variation relies on measuring or approximating variations between two (or more) sets of phenomena. If the variations of one decline or appreciate in proportion to the variations of the other, you have grounds for supposing the presence of a causal relationship between the two.

A clear example of this method may be the observation made by health experts who noted a high correlation between heart disease and high levels of cholesterol. They observed that subjects who maintained high cholesterol diets tended to have much higher rates of heart disease than patients who regularly ingested a low cholesterol diet. That observation served as a hypothesis for further testing under controlled conditions and as a result of that testing we now have a much better understanding of the inception of heart disease.

One writer (I think it was Ayn Rand) once hypothesized a correlation between colorful clothing and economic prosperity. In times of economic hardship, fashions tend to be drab and uniform. In economic booms, people wear bright, colorful, even flashy clothes. This is a fascinating idea, but it demonstrates a certain problem with the method. First of all, Ayn Rand had a tendency to paint with very broad brushstrokes. She may have been thinking about the stereotypes of the Communist Chinese wearing grey tunics and the Soviets wearing wide-lapelled, drab business suits in comparison to the splash of fashions evident in the Western capitalist nations. But notice how much of a stereotype that is: the traditional ceremonial robes of the Chinese and the traditional *peasant* dress of the Russians are highly colorful. And American business executives are notorious for wearing blue, grey, or tan suits (or suit dresses) even if they are made of expensive material.

But, aside from the dangers of stereotyping in this example, you can see that it is a hypothesis that is practically impossible to test. We noted before that the United States underwent a period of economic hardship in the late 1970s. Did fashions change appreciably? Were they less colorful? On hindsight, we may be tempted to say, "Yes," unequivocally. We might think back to the wild fashions of the 1960s and compare them to the more subdued fashions of the next decade. But more careful thought will reveal that to be every bit as much a case of selective perception as the comparison between the Communist Chinese and Americans.

This is a problem regularly faced by sociologists who need to devise ingenious ways of taking such complicated data into account in making an appraisal of such a hypothesis. Such methods are surely prone to severe limitations. We can't set up any careful controls on an experiment that would test for such phenomena so we are very uncertain of the results such surveys may provide (and what counts as "colorful" and what does not is an issue strewn with highly subjective overtones).

This method, and all the others we've considered, point to one very important fact: at most, they generate very uncertain hypotheses which are badly in need of testing before we can place any rational degree of confidence in them. Fortunately, Mill's Methods are interactive and—to a certain degree—self-corrective when used in combinations. Many philosophers and scientists regard such repeated applications of Mill's Methods as the mainstay of day-to-day *methodical* science, where innovation, creativity, and ingenuity is not immediately demanded. The Method of Agreement and the Method of Difference, for example, work so closely together that Mill gave a separate heading to "The Joint Method of Agreement and Difference." When you're checking out the odor of "rotten eggs" as mentioned in an earlier example, the Method of Residue may prompt you to hypothesize that the nearby paper mill is the source of the odor, but a jointly applied Method of Concomitant Variation may prompt you to look for another source as the odor lingers even when the plant is shut down at the end of the day.

42.6 CONCLUSION

We've looked at a number of different ways we can begin to construct hypotheses for further study. These may serve as guidelines for ways of going at the highly creative process of thinking of new avenues for discovering knowledge. Necessarily, this is an incomplete list. We do not have any failsafe recipe for creative thinking. A few of these methods even may be among those used time to time by imaginative writers like Stephen King and Isaac Asimov when they're attempting to come up with ideas for their novels. They are less likely to be used by some avant-garde artists like Jackson Pollock or Andy Warhol; however, we really can't say much about their sources of ingenuity and creativity. That point may be more important than it sounds, for the kinds of systematic guidelines we offer here cannot begin to exhaust the many sources of inspiration for artists and scientists alike. In the final analysis, what matters is that we *get* the ideas in the end, not where they come from. The moral of the story is not to get trapped in thinking you're breaking the rules of logic if you have utterly irrational ways of generating new ideas for hypotheses. What matters is that you have a hypothesis worthy of testing.

Worthy of testing is a key phrase, for no one should take one of these methods and infuse complete and utter faith into their conclusions. They are inductive hypotheses only. To warrant any confidence at all, these hypotheses must now undergo a rigorous program and a demanding discipline. Some of these hypotheses will be more susceptible to testing than others. We can set up more suitable conditions for experimentation with the correlation between cholesterol and heart disease, for example, than we ever could for the charming but rather unscientific hypothesis about the connection between colorful fashions and prosperity. Other hypotheses stand in wait for some imaginative ideas about testing, and some others (as in the case of black holes) await much greater developments in technology.

But the emphasis is on the ultimate possibility of a future test. The worth of the hypothesis is the empirical content it has: is it something that can be borne out by experience? Can we construct any sort of test for it at all? The thing to notice is that creativity isn't limited to the art of building hypotheses. Constructing the proper tests may require the most ingenuity and creativity of all.

EXERCISES

Curing a Virus

You are working with a personal computer on a class project that must be completed this day. Suddenly, your PC is afflicted by a computer "virus." An image of a hungry beast materializes on the screen with the message, "Cookie Monster. Give me a cookie!" Then, it disappears as suddenly as it came.

You work a while longer when the same image and message suddenly flashes on again, longer this time. You're concerned that this is interrupting your work, but you plod ahead, making slow progress when once again the Cookie Monster appears and lingers even longer.

You break out in a cold sweat; you check the clock—only minutes to go, but you're almost done when, with malicious timing, the Cookie Monster appears once more and the image and message begins flashing on and off the screen. It won't go away. You find you can't save the file so you don't dare shut off the computer.

What do you do? Form a hypothesis.

MODULE 43

TESTING AND EXPERIMENTATION

1. Observation
2. Hypothesis
3. **Experimentation**
4. Theory

43.1 IN GENERAL

Once again, recall the "prototype" version of the Scientific Method (see box).

The prototype gives the first step in science as "observation." In other words, take a good look. Maybe you can *see* what the problem is, or what the solution may be.

Next, the prototype tells us to form a testable hypothesis. The art of creating good hypotheses is one of the most creative acts in science. The hard part is coming up with a new idea in the first place. (We've discussed some ways of doing that in the previous module.) But, once we've done that, the act of

---

An **inductive** argument is **sound**, if and only if, the evidence is sufficient or representative enough to warrant the likelihood of the conclusion, and the premises are true.

creating a hypothesis is very forgiving: we don't need to place any great deal of reliance or invest much of our self-esteem in finding a hypothesis to test; the whole idea is to have a place to start. Once we have that, we can then proceed to the next, and perhaps the most important, step in the Scientific Method: testing and experimentation.

If creating hypotheses is the most creative act in science, then coming up with an ingenious and effective test has to be the next most creative act. As we shall see, many wonderful hypotheses subsist in scientific limbo for lack of the ingenuity or technology needed to perform the tests that will corroborate or falsify them.

The key word here is 'falsify'. As before, we will follow the philosophy of Karl Popper in describing the methods of science. Popper's main contribution to the method may well be his insistence on the attempt to find evidence that will *refute* a hypothesis.

What constitutes a good test of a hypothesis? It is important to realize that the answer to that question is every bit as much a matter of debate as the answers to the questions of whether black holes exist or if an opposing force to the force of gravity exists that pushes, however weakly, a falling object "upwards" against the force of gravity (a question that is now being debated in some circles). The point is that the Scientific Method itself is a matter of continuous debate, and properly so.

The conventional wisdom about testing is that it should consist in the collection of data or evidence which *supports* the hypothesis. This is analogous to a student writing a term paper for a history class by compiling a collection of authoritative quotations that support her thesis while ignoring those sources which stand as possible refutations. Of course, that's an oversimplification in our case. Scientists recognize that such a procedure would be every bit as bad for scientific knowledge as the method is for writing a college term paper. Nevertheless, such bad science results occasionally, whether from ignorance of the finer methods of scientific procedure, ignoble motives such as marketplace pressures, or fraud for the sake of personal advancement. We shouldn't make too much of these cases—surely, science is not rampant with fraud—but the fact that some cases exist may tell us something about what else is needed to practice good scientific methods and good inductive thinking.

Consider a famous case of scientific fraud in recent years. In the spring of 1981, a graduate student at Cornell University named Mark Spector and his mentor, the esteemed professor Efraim Racker, announced a remarkable new unified theory of cancer causation based on experiments conducted mainly by Spector. For some time, Racker had believed that an inefficiency in the operation of an enzyme called "ATPase" was distinctive of cancer cells. Spector produced results from a series of ingenious experiments which seemed to confirm his teacher's pet theory. The two were hailed in scientific circles and were expected to become prime candidates for a Nobel Prize. Unfortunately, the experiments were faked and the data was fraudulent.

Efraim Racker so wanted to find a confirmation of his pet theory and had invested so much confidence in his star pupil that he never questioned the results that Mark Spector was getting. In fact, the autoradiograms that Spector produced were forged.

How does the scientific establishment uncover such bad science when it is practiced? The answer can be found in one key word: *replication.* **Replication** is the act of repeating an experiment in an attempt to duplicate the results.* Closely related to replication are peer review and the referee system. These may be thought of as the "checks and balances" system of scientific inquiry in which the scientific community engages in the processes of checking the credentials of experimenters and of copying another's experiment to see if the results check out. Such methods are considered routine in the scientific community and are usually without any suspicion of fraud involved. The idea behind such "checks" is the advancement of knowledge, and when new people work on the same experiments, replication can often yield new insights as well as perfect the methods involved.

When Mark Spector's results were announced, one of Racker's colleagues, a biochemist named Victor Vogt, became fascinated by the experiments. He said later that these results "were so clean and beautiful and convincing that I was seduced into putting my time in on this project."† However, Vogt was puzzled by the erratic results obtained in the replication of the experiments by Spector. Consequently, he assigned his student, Blake Pepinsky, to work with Spector. It was Pepinsky who finally detected the forgery.

What is the moral of the story? According to William Broad and Nicholas Wade, the authors of *Betrayers of the Truth* (a book about fraud in science), replication in science "is a philosophical construct, not an everyday reality."‡ They contend that such replication, the watchdog of good science, is done infrequently and erratically; the discovery of fraud in the Mark Spector case just happened to be a case of good luck. Broad and Wade contend that every case of fraud that comes to public attention "is representative of 100,000 others" which may never come to light.

Broad and Wade don't give any evidence for that supposition, but if it were correct, it would suggest that something is very wrong in the way that science is practiced. The cases of fraud and deceit in science, however frequently they may occur, are even less important than the presumably far greater number of simple, well-intentioned mistakes that go undetected because replication is such a costly and time-consuming practice. If this is true, then we need a better way of going

*The purpose of replication is methodological rather than moral. It weeds out bad science.

†William Broad and Nicholas Wade, *The Betrayers of the Truth: Fraud and Deceit in the Halls of Science,* New York: Simon and Schuster, 1982, p. 66. My account of the facts in the Mark Spector case are taken from this source.

‡*Ibid.*, p. 86. The reader should note that Broad and Wade, who were reporters for the journal *Science,* are more interested in the detective story about fraud in science than they are in a study of Scientific Method, and so they show a tendency to dismiss these issues as mere "philosophical constructs." Their ingenuous grasp of these issues causes them to make a number of mistakes, including their identification of Karl Popper with the School of Logical Positivism (which is manifestly false).

about the discipline of testing and experimentation in order to assure ourselves of the continuing process of refinement and correction in science—one that is more likely to bring such errors to light on the part of intellectually honest experimenters. Karl Popper has proposed such a new way of thinking about experimentation, and we will look at his proposals in the coming pages.

43.2 PROOF AND DISPROOF

Here's a radical proposal (especially for a logic textbook): when discussing science and induction, let's throw away the word *proof.* That remark should raise some eyebrows, but that's often the first step in getting us to do some original and constructive thinking. However, I'm confident that even a little bit of reflection will show that this proposal isn't really so radical after all.

Our subject is "induction" in the widest sense of the word, meaning that our interest lies in rendering judgments based on evidence that is more or less worthy of belief. Induction differs from deduction precisely with regard to the notion of "proof." Only deductive arguments can be conclusively proven; inductive arguments are at most warranted, or made likely, by the available evidence. The two sorts of argumentation differ in their very claims to conclusiveness.

Let's consider an example about how the word 'proof' can interfere with clear thinking about a problem. Every so often, someone introduces legislation on Capitol Hill to require automobile manufacturers to install airbags in new automobiles, and lobbyists from Detroit traditionally mount campaigns to quash such legislation. In one instance a Michigan congressman argued against such a proposal by contending that there is "no proof" that airbags reduce fatalities, so the bill should be defeated (and it was).

Technically, the congressman from Michigan is correct: airbags have not been proven to reduce fatalities. However, the congressman was displaying a terrible ignorance about the nature of proof when he made that remark: airbags have not been proven to reduce fatalities precisely because such proof is absolutely impossible to attain; such is the nature of any argument based on inductive evidence. Asking for such proof is just plain asking for too much. However, ample evidence exists that placing airbags in automobiles greatly increases the *likelihood* that traffic fatalities will decrease, and that fact puts us in another ballpark entirely.

The point is that arguments based on scientific evidence should never suppose that such evidence can ever prove anything. That just isn't what science is about. However (and this may be a surprise), if science cannot *prove* any positive

> *Aristotle could have avoided the mistake of thinking that women have fewer teeth than men by the simple device of asking Mrs. Aristotle to open her mouth.*
>
> —BERTRAND RUSSELL

thesis, it sure can *disprove* a great deal. We may not be able ever to *prove* the Warren Commission's thesis that Lee Harvey Oswald was the sole assassin of John F. Kennedy, for example but we have a clear idea of what would *disprove* the Warren Commission Report—namely, the discovery that the fatal shot came from another gunman on the grassy knoll in Dallas that day.

Karl Popper would have us regard *proof* and *disproof* as "asymmetrical" in the practice of science. In other words, Popper believes that while *proof* is impossible to obtain in science, *disproof* is theoretically possible. Moreover, while we should disdain the search for proof in science, we should always strive to obtain disproof (or "falsifications") for our pet theories and hypotheses.

As science is often practiced, the search for disproof is usually conducted (when it is conducted at all) by the method of replication by a neutral party. However, as we have already seen, we have some reason to suspect that such replication is infrequently or erratically carried out. Karl Popper, on the other hand, doesn't think the search for disproof should fall to an antagonist, or to some neutral "third party," or to some disinterested observer. He believes that searching for falsifications should be the business of the experimenter herself at all stages of the implementation of the experiment.

Let's take a simple example of how Popper's theory of falsification works. Suppose you are the first person to discover that water boils at a temperature of 212° F. You came up with this hypothesis one day in your laboratory in Chicago when you dropped a thermometer in some water you were boiling with a Bunsen burner. The thought occurred to you that maybe water has a set boiling point, and so you set up a series of controlled experiments to see if you will get the same results. You repeat the experiment once, twice, a thousand times, and every time the water comes to a boil, you get a reading of 212° F.

You become excited about your new discovery, so you write an article on your experiment and get it published in a leading scientific journal. That brings your work to the attention of scientists around the world and many of them decide to replicate your experiment to see if they get the same results. Teams of scientists set up shop in New York, Boston, Los Angeles, and London. They perform hundreds of experiments and get your result each time. Your hypothesis is so successful that you create a new system of measuring temperature that you call "the Celsius scale" which sets the boiling point of water at 100° *Centigrade*. The Celsius scale becomes the standard for scientific research.

Have you "proven" your hypothesis? The answer is, "absolutely not." No number of such experiments designed to get that result will ever "prove" that water boils at 212° F. However, you have attained a measure of celebrity in the scientific community and receive invitations to speak all around the world. You accept two of them: one of them in Denver, Colorado; the other, in Quito, Ecuador.

In Denver, you get a shock. You set up your Bunsen burner and, behold! your container of water doesn't boil at 212° F at all! You repeat your experiment many times, and every time you do, the water comes to a boil at a few degrees *below* 212° F. You leave Denver puzzled and embarrassed and get on your scheduled flight for Quito where you set up your experiment all over again and this time your results are even more radically different. This time the water boils at a temperature a few degrees *below* even the Denver results.

Have you *disproven* your original hypothesis? Maybe you've been doing something wrong. Maybe your thermometer is faulty, and so on. But, theoretically speaking, you have got something approaching a practical disproof (Popper admits that an authentic disproof may never be beyond a reasonable doubt, practically speaking, though such disproofs are possible *in theory*). You experiment with new vessels, different thermometers, and so on, but you always get the same reading in Quito, and its different from that you got in Denver, and that's different from the one you got at home in Chicago. For all intents and purposes, you have got a disproof, you realize. Somehow, conditions in Denver and in Quito are different from the conditions in Chicago, New York, and elsewhere.

According to Popper's theory, you have now made a magnificent discovery. The point at which you falsified your original hypothesis is the point at which you truly began to make a valuable contribution to science. You come to realize that what's different about Denver and Quito, as compared to Chicago and New York, is their high altitudes. You can now hypothesize about the boiling point of water with regard to differing conditions of atmospheric pressure. You've learned an invaluable lesson and so, each time you set up an experiment, you no longer search for *supporting* instances; instead, you look for *falsifications.* You vary the experiments in every way you can: whereas before you always experimented with water in open containers, you now try the same experiment with *closed* containers, and so forth. Every time you get an unexpected result, you add volumes more to the existing body of science.

Bryan Magee, a one-time student of Popper's and a popularizer of his theories, has written the following about an experiment much like the one we described above:

> If we had set out to 'verify' our original statement that water boils at 100° Centigrade by accumulating confirming instances of it we should have found no difficulty whatever in accumulating any number of confirming instances we liked, billions and billions of them. But this would not have proved the truth of the statement, nor would it (and this realization may come as a shock) have increased the probability of its being true. Worst of all, our accumulation of confirming instances would of itself never have given us reason to doubt, let alone replace, our original statement, and we should never have progressed beyond it. Our knowledge would not have grown as it has—unless in our search for confirming instances we accidentally hit upon a counter-instance. Such an accident would have been the best thing that could have happened to us. (It is in this sense that so many discoveries in science have been 'accidental'.)*

*Bryan Magee, *Philosophy and the Real World: An Introduction to Karl Popper,* La Salle, IL: Open Court Publishing Co., 1985, pp. 21–22.

This example vividly shows the flaws inherent in any attempt to seek verification (or "proof") for a scientific hypothesis and it also reveals that many shortcomings of the practice of replication in science. Replication is most valuable when it is intended to *disprove* the results of experiments—not necessarily in any antagonistic way, as between professional rivals or enemies, but rather in a disciplined and detached way. The example also shows that the Scientific Method itself is not beyond attempts at refutation; in fact, the method *benefits most* from well-intentioned criticism. That's what aids it in its continuing evolution from the days of Descartes and Bacon, to the work of Popper, and beyond.

However, we should note that Popper is less interested in second- or third-party refutations through the process of replication than he is in assuming the self-critical stance. In other words, the burden of "disproof" is on you. It is up to you as the experimenter to search for falsifications of your *own* pet theories and hypotheses, and it falls to disinterested parties or friendly antagonists to come up with those refutations only secondarily. The object, after all, is the advancement of knowledge, and knowledge advances best in conditions of constant but constructive criticism.

To see why this is so, let's consider one more example. In the last decade, Nobel Prize laureate Linus Pauling proposed the hypothesis that vitamin C was a "cure" for the common cold. Pauling was met with instant skepticism from the scientific community who attacked him for working "outside of his field" (his Nobel Prize was in the field of chemistry, not biology).* To Pauling's embarrassment, he also developed a cult following of health-conscious groupies who took his hypothesis as gospel and spread the idea with the zeal of religious converts.

Since Pauling's critics were mainly unfriendly, they set up experiments to "disprove" his hypothesis with near-decisive results. Pauling's crude experiments were quickly refuted by more professional experiments than he ever undertook for himself. However, one laboratory in Canada, that admitted they shared the general bias against it, got some surprising results when they set out to disprove Pauling's thesis. Nevertheless, the clinicians were sufficiently *self-critical* and open-minded to find a discrepancy in their own results. About that study, *Consumers Union* reported the following:

> One well-controlled study purporting to show the efficacy of vitamin C was published in Canada in September 1972. More than 800 people were involved in the double-blind study which extended over several months. Although the vitamin C group experienced slightly fewer colds and total days of illness, the difference was not statistically significant. The study did show, however, a substantial reduction in the number of days members of the vitamin C group lost from their jobs, because their illness was characterized

*Be skeptical of charges that assert that one is working outside her field of expertise and look at the merits of the argument instead. Linus Pauling, by the way, had done work on alpha-helix protein and on the nature of the chemical bond (i.e., biochemistry).

by a "lower incidence of constitutional symptoms such as chills and severe malaise."*

That particular study may now be obsolete in light of more recent evidence, but whether or not those studies still stand is beside the point. The point is that the members of the laboratory in question admitted to their bias against the Pauling thesis and yet still had the professionalism to be self-critical. Their willingness to acknowledge the results that were contrary to their own preconceived ideas may have helped the science of medicine to make new in-roads in the treatment of the common cold, even though they were disposed to entirely different conclusions than when they began their experiments. According to Popper, that's precisely how important discoveries ought to be made in science. The search for falsification is best conducted when the falsifications are aimed at one's *own* preconceptions and ideas. That's not only good advice but also the essence of the Socratic ideal of the attainment of wisdom.

43.3 UNTESTABLE HYPOTHESES AND "NONSENSE"

We've already mentioned in several places how important it is to come up with *testable* hypotheses. The reasons for that insistence will now become clear in light of what Popper has to say about nonsense and non-science.

Consider the following "thought-experiment" given us by British philosopher Bertrand Russell:

> The world came into existence five minutes ago, complete with memories, histories, museums, relics and fossils.

No doubt, we want to say that the statement is not true, but how can we *show* that it's not true? It looks like a meaningful statement in the sense that we know what the words mean, and it is grammatical, but we find ourselves at a complete loss in trying to *prove* or *disprove* it. In fact, we're also at a loss when it comes to considering what sorts of evidence might support it or, more importantly, what sorts of evidence might count against it. If we rely on carbon-dating (through such methods as "neutron activation analysis") to show the age of the earth, we're still without a disproof because our original premise is that fossils *also* came into existence five minutes ago, and they have complete carbon histories. The trees that came into existence five minutes ago have the rings in the tree trunks intact, philosophers have gray beards, histories came into existence together with the libraries that hold them, and we have vivid memories of our childhood; all of these are irrelevant because they are covered as part of the terms of the thought-experiment. In fact, no evidence at all can possibly count

*The editors of *Consumer Reports, The Medicine Show,* revised edition, Mount Vernon, NY: Consumers Union, 1974, p. 37.

against the proposal. Therefore, Karl Popper would have counted the proposal as literal *nonsense.*

How can this be? We've already said that the statement appears to be meaningful. However, that's deceptive. A little reflection will show that some grammatical sentences are not truly meaningful at all. Consider this example, given us by philosopher-linguist Noam Chomsky: "Green ideas sleep furiously." The sentence is grammatical, but it's not really a statement because it is neither true nor false. Likewise, Popper in his earliest work (such as *The Logic of Scientific Discovery*) would deem Russell's proposal nonsense in the same way: it can be neither true nor false, and it is not a genuine statement.

Let's consider a more elaborate example of a full-fledged theory full of such "nonsense." In the nineteenth century, British astronomer Piazzi Smith constructed a magnificent folly about the Great Pyramid at Giza. Smith and his wife traveled to Egypt where they studied the pyramids, especially the one at Giza, with exhausting care, bringing to bear the state of the art techniques of scientific measurement. In a large volume entitled, *The Great Pyramid: Its Secrets and Mysteries Revealed,* Smith concluded that the pyramids had "sacred origins" and, in addition, he recognized "the extraordinary fact that the pyramid was, in many respects, a structural duplication of the solar system."*

Far-fetched? You bet! But the layman as well as the professional philosopher and scientist must stand in awe at the extraordinary detail that went into the writing of this book. First, Piazzi Smith rejected with powerful arguments the existing interpretations of the systems of measurement that went into the con-

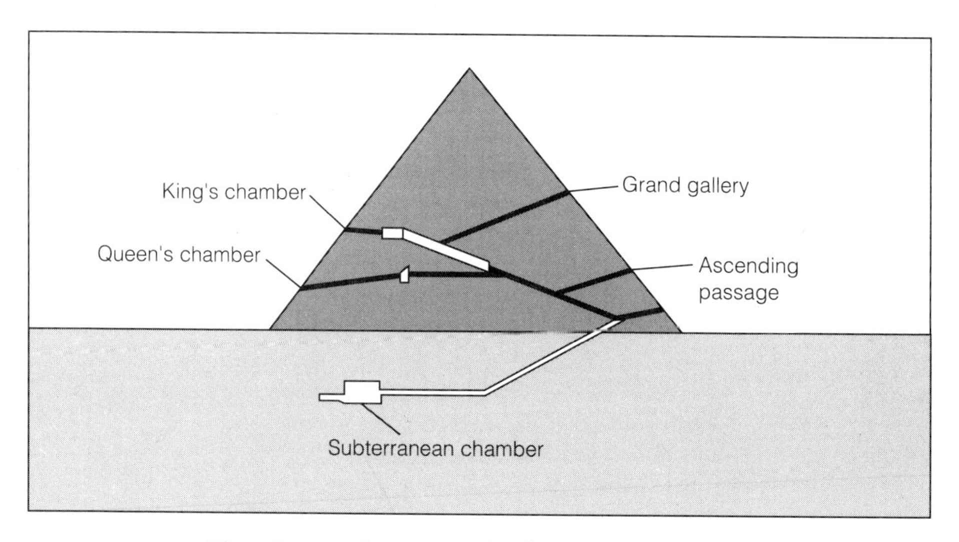

Chambers and Passages in the Pyramid at Giza

*Fatma Turkkan in the foreword to Piazzi Smith's *The Great Pyramid: Its Secrets and Mysteries Revealed,* 4th edition, New York: Bell Publishing Co. Originally published in 1880; reprinted by Bell in 1978, p. iv.

struction of the pyramids, substituting what he regarded as a more suitable unit of measurement which he called, "the pyramid-inch." Then, with a three-fold method which relied on "pure mathematics," "applied mathematics," and "revelations from the Bible," Piazzi Smith drew measurements from the base of the pyramid at Giza and concluded that the construction of the pyramid was actually inspired by God as a prophecy of the coming of Christ. Moreover, Piazzi Smith argued that further revelations, including the fact that "the end is close upon us," were in store for anyone who would study the dimensions of the pyramid. Smith writes: "How close is such an end? The answer to that question must depend solely on the length of the Grand Gallery by measure in Pyramid inches."* Such measurements, Smith believed, tallied with calculations about the length of certain books of prophecy contained in the Bible.

This theory is, of course, pure nonsense. But how do we know that? We are now in a position to give an answer: the thesis is simply incapable of falsification. Confronted with an apparent grounds for refutation, Smith backs off and explains the refutation away by another calculation, or another set of measurements. For example, Piazzi Smith published this book in 1880, and the end of the world has not yet come to be (he gave no precise date). How can Smith account for this fact? He does, because he notes that there is a "passage-floor" leading out of the chamber that can serve as an escape route. What he thinks the Grand Gallery's dimensions divine to us through parallels in the Bible is the inescapable fact that "unexampled days of future trouble" lay in wait.

That's true enough, for two world wars were to follow in the next century. But notice how vague, how wide open to interpretation Piazzi Smith's pronouncements were. We can't imagine what would count for sure as a falsification of his theory. Popper would say this theory about the pyramids lacks solid empirical content; it lacks *testable* hypotheses. The contention that the Great Pyramid's construction was divinely inspired as prophecy is nonsense, pure and simple.

Popper's point is that good science should make itself vulnerable to refutation. If we know what it takes to falsify a theory or a hypothesis, we have a firm footing for science. Such theories and hypotheses have high empirical content precisely because they are testable.

Contrast this folly of Piazzi Smith's with Einstein's Theory of Relativity. When Einstein first propounded his theory at the beginning of the twentieth century, he was careful to spell out a set of testable hypotheses. The general theory turned on a crucial assumption—that the force of gravity could actually bend light. Moreover, Einstein gave a set of equations which calculated with precision just how much a gravitational force could deflect light rays. By doing so, Einstein had revealed his one vulnerability. If an experiment could be constructed that could measure the deflection of light rays and scientists could find no such deflection, then Einstein's theory would have to be dramatically revised or abandoned.

*Piazzi Smith, *The Great Pyramid: Its Secrets and Mysteries Revealed,* 4th edition, New York: Bell Publishing Co., 1978, p. 546.

In 1919, Sir Arthur Eddington and a team of scientists sponsored by the Royal Society of London finally hit on a way to conduct such an experiment that would be the general theory's "acid test." They set up instruments and cameras at two locations along the path of a total eclipse of the sun. When the measurements were taken, they discovered to their amazement that the sun's gravity had indeed deflected light en route to earth and the degree of that reflection was very close to Einstein's own figure within a small margin of error. (We will say more about this experiment in the next section.)

Was Einstein's theory "proven"? No, of course not. But the results were so impressive that Einstein's theory of relativity became widely studied and praised, Einstein himself became famous almost overnight, and his work assumed the prestige and acceptance that had been formerly accorded to the physics of Isaac Newton. You may interpret Einstein's confidence in his own work to be brash. It was risky and perhaps even foolhardy for him to publish the exact figures by which the sun's gravitational field might deflect light, but by Popper's criteria, Einstein was engaged in the true spirit of scientific inquiry: he showed what it would take to prove him wrong.

In the parlance of *The Logic of Scientific Discovery,* we say that Einstein's general theory has been highly *corroborated* by the Royal Society's ingenious experiments, as well as by others. Einstein carefully set out what would disprove his theory and others made repeated attempts to do just that, not only failing to disprove him but also getting results that were very close to Einstein's own calculations. This is the model of inquiry, of testing and experimentation, that we should strive to emulate in our own thinking and research.

A final note: in his later writings, Popper re-assessed calling untestable hypotheses "nonsense" and gave to such hypotheses the new description "non-science." We can discern good reason for this. Many proposals and statements are sufficiently unlike the Piazzi Smith project to be worthy of consideration, even though they can never be disproven. Certain religious and moral tenets may be like that. For example, we may be unable to imagine what sort of evidence would conclusively disprove the existence of God, but belief in God may not be "nonsense." We may do well to believe that "crime doesn't pay," but be unable to say how we could decisively test such a statement. Such proposals and statements are not "nonsensical" in the way the sentence "Green ideas sleep furiously" is nonsense, but they are not matters for scientific investigation precisely because we don't know what evidence would conclusively count against them. Such statements are better regarded as being "outside" the scope of science—hence, the term *non-science* is given to them.

43.4 INGENUITY IN TESTING

Sometimes we confront hypotheses for which we cannot presently devise a test; not all of them are nonsense or non-science. For example, does the third planet around the star Vega support intelligent life? Can wormholes serve as galactic "subways" for future travel to the stars?

We just don't have the technology (or, perhaps, the *ingenuity*) at present to construct proper tests for these hypotheses. We can't train a telescope on the third planet around Vega powerful enough to reveal whether life exists there; we don't even know whether any other stars have *planets*, though some evidence suggests that some do. Likewise, we're not even reasonably sure that black holes and wormholes in space exist, never mind whether they can serve as a future means of travel. Nevertheless, the faith is that someday we will have the technology (or, more importantly, the ingenuity) to construct tests that can refute or corroborate these hypotheses. In any case, we certainly know what it would take to disprove these hypotheses: an absence of any planets around Vega, or the non-existence of black holes. Testing for these, in the present state of our knowledge, awaits someone's ingenuity or else a great leap forward in our technology.

The point, once more, is that creativity plays an indispensable role in devising most tests for new hypotheses. Experimentation itself is an exercise in creative thinking. The experiment by Sir Arthur Eddington which resulted in such high corroboration for Einstein's general theory of relativity is an excellent example of ingenuity and creative thinking.

The outbreak of World War I prevented any serious attempts at testing Einstein's theory but in 1919, one year after the Armistice, Eddington (a Quaker and a pacifist) hit on an idea of putting together an international team of scientists that would work on such a test. Eddington had hoped that such cooperation between scientists of different nationalities would serve as a model for cooperation between the nations themselves.

Eddington and his team learned of an upcoming total eclipse of the sun in the Southern Hemisphere in the spring of that year. He broke up his team into two parts: one would study the eclipse in Brazil, and he would head up the team on Principe Island off the coast of Africa. They learned that the eclipsed sun would be framed by the thick star cluster of Hyades. Such a dense star field for a backdrop was a rare piece of good luck. Eddington began to see how the experiment could proceed. His team and its counterpart in Brazil set up "shadow boxes" to photograph and measure the star field during the eclipse. By overlaying the exposed negatives on top of past negatives of the Hyades star field, they discovered that the positions of the stars in the field were slightly distorted alongside the eclipsed sun. Einstein had predicted a displacement of 1.74 seconds of arc. The Principe team got a reading of 1.61 seconds with a margin of error of 30 seconds and the Brazilian team got a reading of 1.98 seconds with an error of 12 seconds. That evidence was more than enough to provide Einstein with the corroboration necessary to vault him to world fame and his theory to wide acceptance. Einstein's risky prediction had paid off thanks to Eddington's ingenuity in constructing such an appropriate experiment.*

*Jeremy Bernstein, *Einstein*, New York: The Viking Press, 1973, pp. 141–143.

Questions about the existence of life in the Vega system may wait upon great leaps in technology, but sufficient ingenuity (with a bit of luck thrown in) may give us the means and the opportunity to construct adequate tests without having to rely on the development of interstellar space travel. Perhaps we rely too much on the advances in applied science and technology without giving due attention to the creativity that is so important to the art and the science of experimentation.

EXERCISES

1. Have you ever noticed that you can't get a hot cup of coffee on a jet flight? Let's suppose you're a frequent flyer and you've had many cups of coffee while in flight but all of them were no more than tepid at best. What can account for your observation? Possibly flight attendants are trained to make tepid coffee, but that doesn't make much sense. You know that coffee is made with boiling water and you remember that water boils at lower temperatures with increasing altitudes. You hit on the idea that maybe coffee brews at a much lower temperature at such altitudes. However, jet cabins are pressurized to simulate conditions at sea level. If such conditions are genuinely reproduced in flight, then your coffee should be hot.

   How can you set up an experiment to test your hypothesis? Can you vary the conditions in flight to get differing results? Try it out.

2. Read some of the popular books on the market that claim to document cases of abductions of humans by aliens, such as Whitley Streiber's bestselling account in *Communion*, or John G. Fuller's classic account in *The Interrupted Journey*. Notice how well these accounts tally with each other regarding eyewitness reports and descriptions of the aliens and in other details. Can you think of an explanation that might conceivably falsify these reports of alien abductions? How could you test for refutations? Are these accounts non-science or nonsense?

3. Read the science fiction classic, *Solaris*, by Stanislaw Lem. In this novel, the protagonist encounters an ocean on a distant planet and harbors the suspicion—never confirmed nor disconfirmed by the author—that the ocean is both alive and intelligent. Ask yourself how you would go about testing (1) whether the ocean was alive; and (2) whether it was intelligent.

4. Residents of the San Joaquin Valley of central California have noticed that the humidity in this valley of reclaimed desert has increased significantly in recent years. Many attribute the increase in humidity to intensive irrigation over the decades which has allowed the valley to turn green with vegetation. Some believe that with the increase in humidity and the increase in greenery the average rainfall will also increase over the years, changing the climate from desert conditions to temperate.

   What sort of measurements and tests would you set up to check for the truth of this hypothesis? What sorts of things could interfere with this experiment in climatology?

5. Because of increased levels of atmospheric pollution and the resultant destruction of the ozone layer, many scientists predict the onset of something they call the "green-

house effect." Some scientists predict that with diminished protection from the ozone layer, the earth will heat up, arable land will turn into desert, and more droughts and famine will occur. Still others say the opposite will happen: the earth will cool off, arable land will be subject to severe frosts, and a "mini-ice age" will ensue.

What sorts of tests would you set up for the presence of a "greenhouse" effect? Is this a testable hypothesis?

MODULE 44

THEORY AND PROBABILITY: EVALUATING THE EVIDENCE

1. **Observation**
2. **Hypothesis**
3. **Experimentation**
4. **Theory**

44.1 IN GENERAL

Once more, call to mind the "prototype" version of the Scientific Method (see box).

As we have emphasized many times in this Unit, this presentation of the Scientific Method is no more than a "launching point" for discussion. If it were to be taken literally, this prototype would be at the very least misrepresentative of the actual workings of modern science. We have seen how observation actually may presuppose some theoretical interpretations, how hypotheses can follow from theories, that Laws of Nature are nowadays thought to be subordinate to theoretical concerns, and so forth.

An **inductive** argument is **sound**, if and only if, the evidence is sufficient or representative enough to warrant the likelihood of the conclusion, and the premises are true.

Now we will look at a step that has come to be regarded as *intermediate* between (3) experimentation and (4) theory: evaluating the evidence (or "evaluating the data").

To see how important that step is, recall the unmanned Viking lander mission which went to the planet Mars in the year 1976. Viking was a computer-controlled robot survey unit that sampled the soil for microorganisms. Most of the tests it conducted yielded no positive data, but a few did give some mixed results. Hypotheses were laid out, observations were made (even a television camera was aboard), and experiments were run. How do we evaluate the data?

The Viking mission is especially interesting in this context because its mission was to test for the presence of life on Mars. However, we don't have much idea of how extraterrestrial life might behave or look or be composed; our whole idea of life is earth-bound and limited to one course of evolution. Life might be entirely—well, for lack of a better word—*alien* to us on another planet. As a result, when Viking set out to test for the existence of life on the planet Mars, we were predisposed to look for the sorts of conditions that would be analogous to the nurture of life on our own earth. That is understandable, but it is limiting.

The presupposition behind Viking's tests was that if life existed on Mars, it would be enough like microorganic life on earth to be recognizable by familiar procedures used on earth. That was the basic hypothesis of the mission. The tests were accordingly designed to test for features of earth-like microorganisms: respiration, reproduction, and consumption of nutrients. But, these are hugely unfounded assumptions: What *sorts* of nutrients? Water? But couldn't an alien lifeform evolve without a steady dependence on water? These tests represented a point of departure, a place to start our search, but they can't be regarded as conclusive because they can't be regarded as comprehensive. The very fact that the whole emphasis of the mission was to test for *microorganisms* came under serious criticism from some quarters. Carl Sagan, for one, asked the question, "Wouldn't it seem silly if we spent all our time looking for bacteria when at one moment a Martian elephant strolled by and no one could see it?" Because of Sagan's mischievous question, the mission planners conceded to placing a television camera on board and that's why we got those wonderful pictures from the Martian surface.

Another problem was the Viking lander was taking soil samples from only one place on the surface of the planet—a miniscule portion of the entire land mass. If the lander had set down even just a few miles away, it might have gotten radically different results.

Given the experiments conducted by the Viking lander, we may now ask the question: "Does life exist on Mars?" Many of my students have told me that their high school science teachers answered that question with a definitive "No—the Viking lander *proved* that life does not exist on Mars." That answer terribly overstates the results of the experiments and shows a mistaken idea of how science works. The best we can say in evaluating the data received from Viking is: "Based on the limited experiments of one device on no more than one or two meters of the entire planet's surface, the existence of life *as we know it* on the planet Mars seems very *unlikely*." We have added to our theoretical stock of our knowledge of the planets, but very little if anything can be stated conclusively.

Another way of putting this evaluation of the evidence is that "the existence of life as we know it on the planet Mars is very *improbable.*" Notice that we are not forced to choose between the false alternatives that either the experiments *proved* the existence of life there, or that they *disproved* it. Rather, we have a great deal of uncertainty left over and the accurate answer has to do with *degrees* of assurance or acceptability of the hypothesis.

Mathematical Probability Theory was devised to give more precision to estimations of likelihood in such cases of uncertain results, which is by far the usual case in all scientific experimentation. The idea was that we should be able to say in the first case whether an event was either *probable* or *improbable,* as opposed to *proven* or *disproven,* given the evidence. From there, scientists and mathematicians proposed to give numerical values to those probabilities. It's just like those games we played as children: "On a scale of one to ten, what do you think the chances are for the Yankees to win the pennant this year?"

Probability theory begins with the idea of an approachable (but, in practice, unrealizable) limit of 'one'—which stands for certain knowledge. The scale is a set of decimals (or fractions) between '0' and '1'. Thus, the chance of rain tomorrow is .9 (or "90 percent") because of a storm front moving in from the west; this indicates a very high level of probability. Correspondingly, if the chance of the Phillies winning the pennant this year is a mere .2 probability since Mike Schmidt retired, that indicates a very low probability.

In the sections that follow, we will discuss the centrality of theory to the Scientific Method. We will present the promised polished analysis of a *dynamic* model of the Scientific Method. The rest of this unit will be an introduction to the evaluation of data by means of probability theory. That will be given to you at first in a largely uncritical fashion. Probability calculations do have their uses, especially in the gaming tables of Las Vegas or Atlantic City, or for "actuarial tables" of life expectancy for insurance companies. I do not know how pleased insurance underwriters would be to find themselves compared to gambling casinos, but the analogy is nonetheless warranted given the nature of the discipline; in both cases, the "house" is heavily favored. However, in a later section, we will question how useful the method of assigning numerical values to probabilities is when applied to such procedures as evaluating the evidence of the experiments of the Viking mission on the likelihood of the existence of life on Mars.

44.2 THE CENTRALITY OF THEORY IN THE SCIENTIFIC METHOD

We are now at the point where we can abandon the prototype version of the Scientific Method and put in its place a more *dynamic* model of the actual workings of science.

The purpose of this section is to give you a solid basis in inductive reasoning, and the best way to do that is to introduce you to the methods of science. We looked at the prototype version of the Scientific Method in order to give us a place to start. Now, having considered how observation is indispensable to good reasoning; strategies are for generating hypotheses; ingenuity is in testing; and some procedures are for constructing good tests, we have reached a watershed in our discussion: the concept of *theory.*

Many of the discussions we've had so far have shown how the process of observation interacts with the formation of hypotheses and the construction of experiments; how testing may affect our processes of observation; and how creativity plays an important role in each of observation, hypothesis formation, and experimentation. Now it remains to be said that hovering over the entire process is the act of theorizing.

Theory, we have emphasized, lies at the heart of modern conceptions of the Scientific Method. Now we can look at General Theory as the counterpart of Particular Observation and see how they interact. Identification of presuppositions, we have said, is important in making observations and formulating hypotheses. We can now regard presuppositions as theoretical limitations on observation.

I have adapted the following chart, depicting a dynamic model of Scientific Method, from one given by V. James Mannoia in his book, *What is Science?.** My account differs in some respects from Mannoia's—for example, I use Popperian terminology like 'corroboration' and he doesn't—but the central idea is the same: *that the process of science in particular and induction in general has an inherently* cyclic *character.*

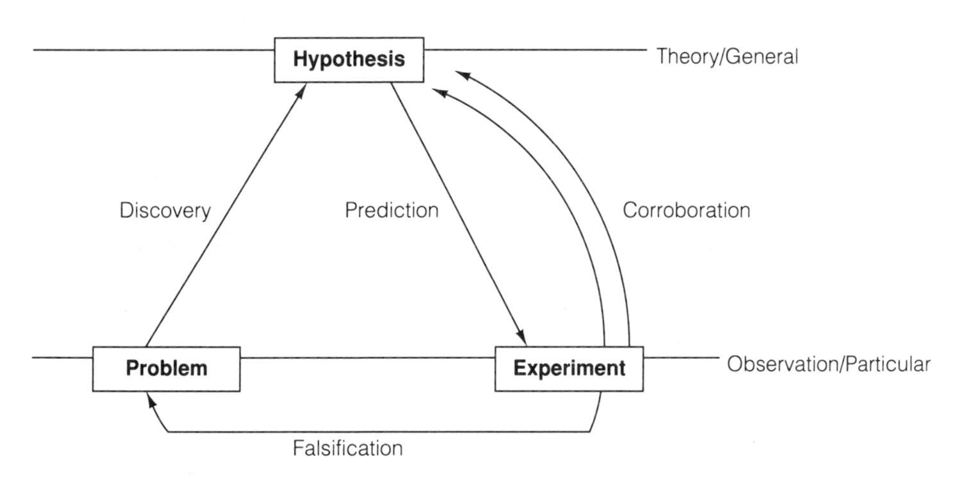

The Scientific Method

*V. James Mannoia, *What is Science?: An Introduction to the Structure and Methodology of Science,* Washington, DC: University Press of America, 1980, p. 23.

You will notice that the chart is divided into two parts: "Theory/General" and "Observation/Particular." On the "Theory/General" line we have *Hypothesis*. The "Observation/Particular" line is shared between *Problem* and *Experiment*. We can regard 'hypothesis' as a kind of *particular* theory which is a unit of a more comprehensive General Theory, such as the "Theory of Relativity." Both *Problem* and *Experiment* rely on particular observations. We recognize a problem by observing our surroundings. We test a hypothesis by looking at data from observable experiments.

The chart moves from problem to hypothesis by means of a process of creative discovery. That is the intuition behind the thesis we proposed about the creative role of hypothesis formation. Next, by means of making predictions based on the hypothesis, we move to experimentation. If our observed results falsify the prediction based on the hypothesis, we then return to formulating the problem stage. If despite repeated attempts we are *unable* to falsify the predictions we based on the hypotheses, we then have some corroboration for the hypothesis as a part of a larger, general theory.

The *cyclic* or *dynamic* model of the Scientific Method, therefore, is much more expressive of the actual workings of science and inductive thought than the static series of steps given in our prototype. Typically, a scientist does not move from one independent step to another. Actually, the process is interactive and full of constant refinement.

What this model does not show is the reaction between general theory and particular observation that many scientists and philosophers of science believe takes place on a regular basis. According to these theorists, every observation we make is "colored" by how we see the world through the filters of a general theory of the world. For example, suppose a poet, a psychiatrist, and a physician see a man standing at the guard rail of a bridge. Each will be disposed to interpret what they see according to their training. The poet may interpret the scene to be that of a reflective and romantic person who looks out upon the world around him. The physician may well be concerned about only the possibility of injury that can come to the man and sees only the hazards that surround him. But the psychiatrist will likely interpret the man's position as a prelude to suicide and as an expression of despair. We know from testimony in courts of law that such eyewitness accounts are notoriously unreliable precisely because they do reflect a tendency to predisposed interpretation. This is what we identified before as the disposition to presupposition. If such suppositions are correct, then the chart given above may even be as subject to revision as was the prototype version of the Scientific Method which we have now abandoned.

This is all we can say about the more theoretical aspects of the Scientific Method in this place, so we turn to one final topic in this series: the way we interpret the data obtained from experimentation. Such evaluations of data may well be seen as theoretical constructs themselves. After all, the data is meaningless unless we have a theory of how to interpret the data. That interpretation is usually done by means of probability theory. A discussion of probability theory, its strengths and limitations, is given below.

44.3 AN INTRODUCTION TO NUMERICAL PROBABILITY THEORY

What is the probability of obtaining a "heads" on the flip of a coin? We have only two possibilities: 'heads' or 'tails', so the probability must be one in two (or $\frac{1}{2}$) (see Figure 1). Likewise, the probability of rolling a six on a single die must be one in six ($\frac{1}{6}$), and the probability of selecting a queen of hearts from a regulation deck of fifty-two cards must be one in fifty-two ($\frac{1}{52}$). This much is obvious. What, however, is the probability of selecting only hearts on two draws from one deck? That complicates matters. In order to account for the added complications of such problems, we need to qualify our interpretations of probability calculations and find the interpretation (or the formula) which is appropriate in each case.

Suppose we ask what the probability of obtaining a heads will be if we flip a coin a second time. You might have an expectation that the previous flip has an effect on the outcome of the second flip, but this is a fact about your psychological expectations, not a fact about the action of a flipping coin. In fact, going by the book, the probability is once again $\frac{1}{2}$.

Suppose, in fact, you are "hot" and you find yourself "on a run" of flipping heads. Suppose you've just flipped seventeen consecutive heads on the same coin, and let's suppose that there's nothing unbalanced with the coin. What is the probability of the next flip of the coin resulting in heads? Strict classical theory tells us that the probability is (you guessed it!) once more $\frac{1}{2}$. "But how can that be?" you might well ask. "Gamblers frequently bet heavily on such runs, and often with a great deal of success. If they bet that way, and they win, is that just luck? Are they being irrational?"

The answer most mathematicians and logicians give is a decisive "yes;" gambling is irrational. However, let's be a bit irreverent for a moment and consider the fact that most mathematicians might be wrong. Is it possible that we could do the successful gambler's rationality more justice by giving the event a differing interpretation from the standard one?

In fact, we have three standard (but by no means exhaustive) interpretations of logical probability calculations: "the classical theory," "the relative frequency theory," and "logical inductive probability." The faith is that one or the other (or some combination of the three) will be more appropriate to the kind of problem we confront, whether the problem is calculating the odds of the next flip of a

FIGURE 1

coin turning up heads after a run of seventeen consecutive flips, or the odds of selecting two hearts on consecutive draws from a deck of cards, or whether the problem is calculating the likelihood of the existence of life on the planet Mars after the vast majority of the Viking lander tests were negative. We have been given three basic interpretations of probability theory to fit such varieties of circumstances and we will now consider each in turn.

The Classical Theory

The basic formula of *classical probability theory* is given as a ratio:

$$P(A) = m/n$$

. . . where 'm' = the number of favorable possibilities, and 'n' = the number of possible outcomes overall.

Let's look at a simple example. If we toss a single die, what is the probability of a *one* turning up?

We have six possible outcomes overall, namely:

Since we have but one favorable case out of a total of six possible outcomes, the ratio is expressed as one in six according to our formula, or just $1/6$.

Let's take another such example: on the single toss of a die, what is the probability of an odd number turning up?

We see that there are three possible favorable outcomes:

Since we've already established that there are six possible outcomes, the probability of an odd number coming up on a single toss of a single die is the ratio of three to six, or $3/6 = 1/2$.

The basic assumption of classical probability theory is that we are provided with an equally possible set of events. This is sometimes called the **Principle of Indifference,** which states that without sufficient reason to do otherwise, we must assume that the assertion of any possible alternatives has an equal initial probability. The problem with the Principle of Indifference is that it may be misleading (except in simple cases, such as the example of the single toss of the die here) to mark off any alternative in advance as equally probable. That determination requires an outside assessment based on knowledge that we just may not happen to have. For example, you may think it equally probable that the sun will rise this morning, just as it has for every morning since time immemorial; however, you didn't know that the sun would supernova last night!

Sometimes, because of additional knowledge, the classical theory which assumes the Principle of Indifference may be inappropriate. From time to time, we may just have good reason to favor one alternative over another (we may know, for example, that the dice are loaded). That certainly changes the probability calculation considerably. The point is that in the absence of other knowledge, we go with the classical theory and assume the equiprobability of outcomes. But a little bit of knowledge can change our calculations dramatically.

How can we reflect that added information in our probability calculations?

The Relative Frequency Theory

One way of showing the relevance of additional input in our calculations is called the *relative frequency theory.* This theory of probability calculation has more application to the actual work done in scientific research. This interpretation of probability theorizes about the relative frequency with which a particular event will turn up in a series of repetitions of some random set of events. Here the formula is given as:

$$P(A) = \frac{m_o}{n_o}$$

. . . where 'm_0' = the number of *observed* favorable outcomes, and 'n_o' = the overall number of *observed* outcomes.

Insurance companies, for example, use a formula like this to set up their actuarial tables for projected life expectancies. What is the probability that a 35-year-old woman will live for another ten years? Suppose we have data that of 1 million 35-year-old women, 990,000 lived another ten years. That would be a ratio of 990,000 over 1 million, or a probability (expressed as a decimal here) of .990 that any given 35-year-old woman will live another ten years.

The problem is that this assumes the probability will hold true over the long run. Insurance companies gamble that this will be so. Suppose, however, that a rare disease is discovered that strikes only women who are in their mid-30s. That would significantly alter the projections and the insurance company that sold life insurance only to women in their mid-30s would be in deep trouble indeed.

However, if the insurance company can stay in business long enough, the faith is that they'll still make money in the long run (provided that a cure is found).

In any case, relative frequency theory really does share many of the presuppositions of the classical theory. For example, suppose we witness a run of 2,000 consecutive rolls of sixes on a single die (that would be incredible indeed!). What are the chances of rolling a six on the next toss?

The relative frequency theory tells us we need to compile all of the total observed rolls of dice in our set of experiences of rolling dice to make a proper probability projection (it's for the long run, you see). Suppose, then, that we have total recall and know that we have personally witnessed 200,000 rolls of dice in our lifetimes. Suppose that in that entire experience we have never witnessed a roll of six before. What is the probability that the next toss of the die will be a six? According to the relative frequency theory, the probability is apparently 2,000 in 200,000 (or one in 200!)—not a good bet!

The professional gambler next to you however wants to bet that the next toss will be a six, and bets heavily. Should you talk her out of it? She calculates the set of outcomes to be 2000 out of 2000 so that there is a lopsided probability that the next toss will be a six. Your calculations are based on a life's experience of tosses, so you regard the toss of a six as highly improbable. Who is right?

Many mathematicians will invoke the Principle of Indifference here, just as they would in classical theory. The chances are once more one in six. Their reasoning is that given an indefinitely large set of observations (in an ideal universe) over an indefinite period of time, the tosses of dice will even themselves out so that they approach a ratio of one in six.

We see then that relative frequency theory really doesn't help us out at all. This theory depends on an indefinitely long run, but that "long run" can be presumed to be very, very, long indeed, completely removing it from the realm of actual experience.

For my own part, I'd probably bet on the six coming up again (but I'd bet only what I could afford to lose).

Logical Inductive Probability

So far, we have confined our attention to problems in *mathematical* probability theory, but the same problems can be rendered as problems in logic. To see how this is so, consider this brief discussion on "logical probability:"

Logical Probability is a logical relation of probability between statements. Consider, for example, the assertion:

"In a toss of a coin, either the result is heads or the result is tails."

The conclusion, "The result will be heads," is a *statement* which asserts a result that has a one in two chance of occurring. The point to notice is that we are now interpreting probability as a relation between *asserted statements,* as opposed to a numerical or mathematical relation. Interpreted in terms of asserted statements, the assertion of degrees of probability can be rendered as *inductive arguments* instead of as deductive calculations.

In the preceding example, we are treating the assertion as a premise offered in support of the conclusion, "The result will be heads." Notice that the argument is deductively *invalid,* but we can render it an *inductive* worth by assigning a degree of acceptability (or probability).

Consider another example: suppose there are fifty students enrolled in your logic class of whom forty are female and ten are male. The class period is just beginning, and students are just about to come through the door. On the basis of logical or inductive probability, I can draw the following conclusion: "The first student who comes through the door will be a female."

Is that a good argument? Rendered with deductive conclusion force (i.e., "The first student who comes through the door *must* be a female"), it would be a terrible argument indeed. However, if I render the argument as an inductive probability ("It is *likely* that the first student who comes through the door will be a female"), the argument becomes much more acceptable. Of course, it's entirely possible that the first student may turn out to be a male, in which case the argument is unsound.

That argument is rather trivial because it holds no obvious importance for the advancement of knowledge. However, the same principles apply to scientific inductions whose assertions of evidence can be interpreted to be premises in arguments of an inductive form.

Consider some problems in the science of genetics. Thanks to the work of Gregor Mendel, we know that some genetic characteristics have a greater likelihood of being inherited by succeeding generations than do others; we call this phenomenon "dominant inheritance." For example, some people with very dark hair exhibit a streak of white (known as the "piebald trait") from a very early age. The chromosome which produces the piebald trait has been shown to be dominant over those chromosomes that do not. If we were to study the inheritance of the piebald trait over a series of generations, we would notice a high probability that offspring will exhibit the trait as well. Suppose a man with a white forelock marries a woman who does not have the piebald trait. The man's father had the trait, and his grandmother also had the trait. The probability is very high that their first child will also share the piebald trait (as high as a 50 percent probability). Thus, the argument may be presented as follows: "If the first generation has a piebald parent and a non-piebald parent, there is a 50/50 chance that their child will have the trait as well. If their offspring has the trait and marries someone who doesn't, then there is a 50/50 chance that their child will have the trait as well. Therefore, the probability is very high that half of the third-generation children will exhibit the trait as well."

This is an inductive argument so we have no guarantees that half of the third-generation children will also be piebald. In fact, it is entirely possible that the trait may "skip" an entire generation (although the odds remain high that *their* children will inherit the trait).

Such studies have important consequences for science even though the inheritance of a white forelock has little more than aesthetic or cosmetic value. Such studies may be generalized to some degree of probability regarding the inheritance of other dominant traits, such as genetic diseases or deformities. Those sorts of studies can have important medical and moral consequences that need to be examined. Likewise, such studies have been generalized in an attempt to extend their results to predictions about the inheritance of intelligence—a subject which is highly controversial.

Some philosophers, such as Rudolf Carnap, have attempted to assign precise numerical values to such inductive probabilities.* Carnap believed that such numerical probability theory was an invaluable step in showing the extent to which hypotheses in science have been confirmed or verified. Later in this section, we will consider serious problems connected with such attempts, but for now it is sufficient to observe that scientific evidence given in the form of asserted statements can be evaluated in terms of numerical probabilities. In fact, each of the formulae we have thus far considered in this module can be given in terms of logical statements as well. For example, we may interpret the "relative frequency formula" in this way, where 'm' is taken to be the number of favorable statements and 'n' = the number of statements of equally possible events overall.

The Probability Calculus

So far, we have looked at fairly straightforward cases of probabilistic calculations (and we found problems even with these!) but suppose we complicated our world of experience even more; suppose we tried calculations with *pairs* of dice (instead of one die) or calculations with mutually exclusive events (such as the probability between a one and a two coming up on a single die). How do we make these calculations?

The answer is the "Probability Calculus." *Probability Calculus* allows us to make calculations based on combinations of formulae under differing descriptions that require differing interpretations. These interpretations are based on concepts in logic that are already familiar to you: conjunction, conditional, exclusive disjunction, and inclusive disjunction. We may want to know the probability of the outcome of a conjunction of events independent of each other occurring, so that the occurrence of one in no way affects the occurrence of the other. (For example, we may want to know thc probability of both you and your friend getting an "A" in a logic course.) For such a calculation, we use the appropriate rule, called "The Conjunction Rule."

What if your class is being graded on a curve so that only 10 percent of you will receive "As" at the end of the course? What is the probability of both you and your friend getting "As" in that event? To make that calculation, we have to adjust the conditions. Perhaps whether you receive the "A" depends on the quality of work of your friend in a team project. In that case, you may want to express your calculation as a material conditional. Or perhaps it's the case that your teacher can assign only one more "A" from an original allotment of ten. What are the odds of one or the other of you or your friend getting an "A" in that event? That calculation must be expressed in terms of an exclusive disjunction.

The Probability Calculus, then, requires us to figure out which of the formulae are appropriate to the sort of calculation we are doing. Here are some examples:

*See Rudolf Carnap, *Logical Foundations of Probability*, Chicago: University of Chicago Press, 1950.

Conjunction. Consider any possible pair of individual events, such as the probability of tossing two consecutive heads on a single coin. This is the interpretation of conjunction that you will recognize as the standard "ampersand" (or '&') in the Sentential Calculus of Unit Three. Here we simply follow the "Conjunction Rule:"

P(A & B) = P (A) × P(B)

Thus, the probability of tossing two consecutive heads is calculated by the probability of tossing the first (i.e., ½) *times* the probability of tossing the second (also ½), so the result is:

P(A & B) = P(½) × P(½)

Therefore, we get the result:

P(A & B) = P(¼)

. . . or a probability of one in four.

Material Conditionals. We can also make calculations of probability based on conditionals. This is the interpretation of conditionals or hypotheticals that you will recognize as the standard "horseshoe" (or '⊃') in the Sentential Calculus of Unit Three. For example, suppose we want to know the probability of drawing two clubs from a standard deck of cards on two consecutive draws. We know that a standard deck of fifty-two cards has thirteen cards in the suit of clubs. Thus, the probability of drawing a club in the first draw is obviously thirteen in fifty-two, or $^{13}/_{52}$ or just ¼. However, if we succeed in drawing a club on the first draw, that changes the conditions for the second draw by reducing the number of clubs in the deck so that the probability for drawing a club on the second draw becomes twelve in fifty-one, or $^{12}/_{51}$. How do we calculate the probability that we will draw two clubs from the deck on consecutive draws?

The answer can be expressed in the form of a material conditional. We say that the probability of drawing one club after another from the deck is the probability of drawing a club on the first draw *times* the probability of drawing a club from the remaining deck *if* the first card is also a club. We express this idea in terms of the "Conditional Probability Rule:"

P(A & B) = P(A) × P(B if A)

Thus the probability of drawing two clubs on consecutive draws from a standard deck is ¼ × ($^{12}/_{51}$ if the first draw is a club) or ¼ × $^{12}/_{51}$ = $^{1}/_{17}$ (or one in seventeen).

Exclusive Disjunction. What if we want to calculate the probability of an outcome based on mutually exclusive events? *Mutual exclusion* is a disjunction in which it is impossible for both events to occur. For example, suppose we want to know the probability of *either* a one or a two occurring on one toss of one die. We express this in terms of the "Mutually Exclusive Disjunction Rule," or:

P(A or B) = P(A) + P(B)

Thus, when confronting the odds of one or the other of mutually exclusive events occurring, we calculate the probability by the *addition* of the two probabilities. (Notice that in each of the other cases, the formulae were based on the *products* of the probabilities in some way or other.)

Let's return to our example. What is the probability of a one or a two turning up on a single toss of die? The probability of rolling a "one" taken by itself is obviously one in six (or $1/6$), and the probability of rolling a "two" is also one in six (or $1/6$.) According to our formula what is the joint probability of the occurrence of one *or the other*? The answer is found by adding the two probabilities together—$1/6 + 1/6 = 2/6$ (or one chance in three).

Inclusive Disjunction. A disjunction is said to be *inclusive* if it is possible for one or the other *or both* to occur; in other words, the occurrence of these events are *not* mutually exclusive. This is the interpretation of disjunction that you will recognize as the standard "wedge" (or '$\vee$') in the Sentential Calculus of Unit Three.

Let's consider an easy example. Suppose you want to know the odds of obtaining either an even number or a six on a single roll of die. Any of the following rolls of the die will satisfy the first disjunct:

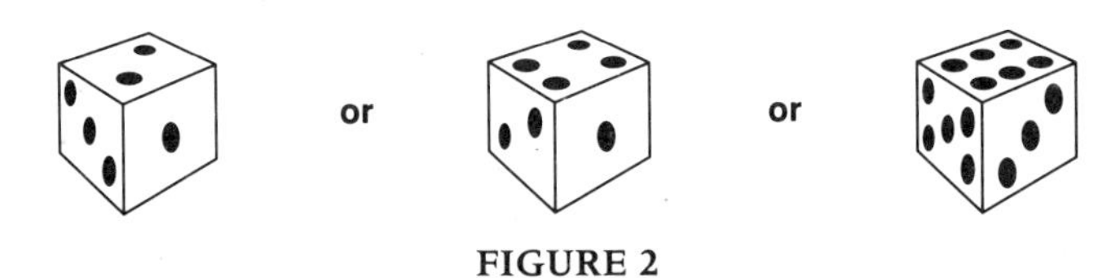

FIGURE 2

But only the toss of a "six" will satisfy the second disjunct. Notice however that if we throw a "six" we have satisfied both disjuncts, whereas if we throw a "two" or a "four" we satisfy only the first of the two disjuncts.

How do we calculate the probability of any of the events occurring as described in the disjuncts? We decide that by following the "Inclusive Probability Rule":

P(A $\vee$ B) = P(A) + P(B) − P(A & B)

The probability of the occurrence of the conditions described in the first disjunct (as illustrated in Figure 2) is three in six (or just ½). In other words, the chances are one in two that an even number will turn up on a single roll of die. We will call that 'P(A)' (or the probability of disjunct 'A').

The probability of the second disjunct (that the roll of die will turn out to be a "six") is obviously one in six (or ⅙).

'P(A & B)' is the probability of the joint occurrence of an even roll and a "six." The "Conjunction Rule" tells us that the probability is determined by the *product* of the two conjuncts, so that P(A & B) = ½ × ⅙. Resolving the equation, we have P(A & B) = 1⁄12.

Applying the formula, we have "½ + ⅙ − 1⁄12."

Resolving the addition first, we have "4⁄6 − 1⁄12."

The result is: P(A ∨ B) = 7⁄12. In other words, the probability of either an even roll or a six occurring is seven chances out of twelve (or about 58 percent).

44.4 SOME DOUBTS ABOUT NUMERICAL PROBABILITIES

Some form of probability theory is appropriate to every rational estimation of likelihood we can make, according to advocates. The presumption is that if we make an estimation of likelihood that one of the major interpretations cannot account for (such as, perhaps, the success of the gambler's bet on the run of seventeen consecutive heads on the flip of the coin) that the claim to probability there is *irrational.* That nicely fits our prevailing sense of morality as well!

But is the label justified? To suppose that it might not be is to throw the *whole* of modern probability theory into doubt—but, why not? Let's do it! We never learn anything by agreement.

The philosopher Karl Popper has raised some serious misgivings about the entire attempt to assign numerical values to probability calculations. This is a broadside, not just against certain interpretations of probability theory (such as Carnap's version of logical probability, the frequency theory, the classical theory, or even the probability calculus in general) but to *all* attempts to gauge probability on calculations based on supporting evidence which has as its purpose to make hypotheses in science as *certain* as possible. Popper regards this whole approach as wrongheaded.

To see why, we must realize that Popper rejects any attempt to prop up a hypothesis by means of positive, supporting evidence; he doesn't think that evidence is ever capable of confirming, establishing, verifying, or proving a hypothesis. Instead he argues that the role of science is to seek refutation—falsifying instances. To an extent that a theory or hypothesis resists serious and sustained attempts at falsification, the theory or hypothesis is said to be *corroborated.* This theory of corroboration sets Popper apart from the philosophers of science of his day. A consequence of the theory is a decisive and insightful idea that all knowledge proceeds out of good, hard, sincere, and sustained criticism.

Popper points out that the degree of corroboration can never be established "by counting the number of the corroborating instances," and he continues: "we cannot define a numerically calculable degree of calculation, but can speak only roughly in terms of positive degrees of corroboration, negative degrees of corroboration, and so forth."*

Why does Popper come to such an astounding conclusion? He argues that a high formal probability is actually correlated with lower empirical testability of a hypothesis. For example, his contention is that no amount of empirical evidence can establish beyond doubt that the probability of "heads" on the next flip of a coin will be ½.

Take another example which Popper gives us: suppose you consult an astrologer who makes a series of predictions, some of which seem to come true. We cannot say that such predictions have any positive degree of corroboration because the formulations of astrologers are typically so ambiguous, general, and cautious. Yet, such predictions may well have a high degree of probability! They have a high degree of probability precisely because they have so little basis for refutation (or falsification) in experience. Thus, Popper writes:

> . . . a higher degree of corroboration will, in general, be combined with a lower degree of probability; which shows not only that we must distinguish sharply between probability (in the sense of the probability calculus) and degree of corroboration or conformation, but also that *the probabilistic theory of induction, or the idea of an inductive probability, is untenable.*†

Those are strong words (and, to some, they appear to mean the demise of all scientific inquiry!), but let's see what they amount to in terms of a concrete and somewhat amusing example.

Skylab was a seventy-seven ton American space station which had serviced manned space missions. On several occasions, the astronauts used the space station then returned to earth. In 1979, NASA ground control noticed a wobble in the orbit of the then unmanned Skylab and recognized that the spacecraft's orbit was degenerating. Ironically, their calculations (*probability* calculations) had led them to suppose that the orbit would remain stable long enough to get the space shuttle program off the ground in a few more years (the shuttle could boost Skylab to a higher and more stable orbit), but those calculations proved to be overly optimistic. Their best guess now was that the large space station would come crashing down to earth within the year.

Nothing this large had ever been placed in orbit before and NASA officials legitimately feared the possibility that large pieces of Skylab would fail to break up in the atmosphere and might crash on population centers. The Battelle Me-

*Karl R. Popper, *The Logic of Scientific Discovery,* New York: Harper & Row, 1968, pp. 267–268. Notice that Popper does concede in an appendix to the 1968 edition that some type of measure of degrees of corroboration may be devised, but he emphasizes that "this measure cannot be a probability."

†*Ibid.*, Appendix vii, p. 363.

morial Center In Columbus, Ohio was commissioned to calculate the probabilities of Skylab debris striking any people when the space station finally broke up. They estimated the odds at 1 in 150 that a piece of Skylab would hit someone, but the chances of any particular person (say, you or me) being hit were 1 in 6 billion! "A person is 3,000 times more likely to be hit by lightning that day than to be hit by a piece of Skylab," a spokesman exclaimed.*

How did NASA determine that remarkable statistic? It was a simple calculation in classical probability theory. To make that calculation, first NASA had to determine what would be the final orbit of Skylab when it began to fall and break up in the earth's atmosphere. They projected the orbit to be something like the Mercator map representation in Figure 3.

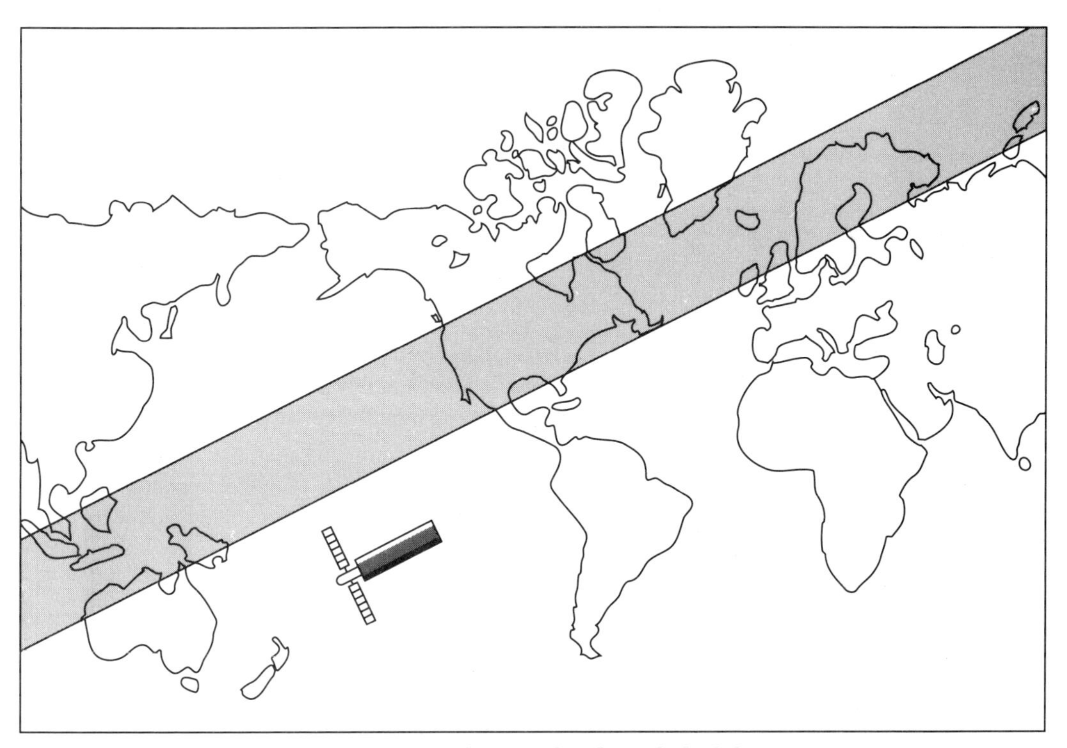

FIGURE 3 **The Final Orbit of Skylab**

Figure 3 is a rough estimation of that orbit; actually, the orbit would look more like a wave on a Mercator projection, but this should be enough to give you the idea. The shaded area represents the path of Skylab on its final approach, about 500 miles in width. NASA experts calculated that some intact Skylab debris weighing as much as 5,000 lbs. could fall in that area. Skylab also passed

*"NASA Doubts Disaster In Break-up of Skylab," *New York Times,* Sunday, May 20, 1979, p. 41.

over a densely populated part of the planet, so concern was generated that Skylab debris could well pose a genuine hazard to the people in its path.

Next, we consult a population density map, as in Figure 4:

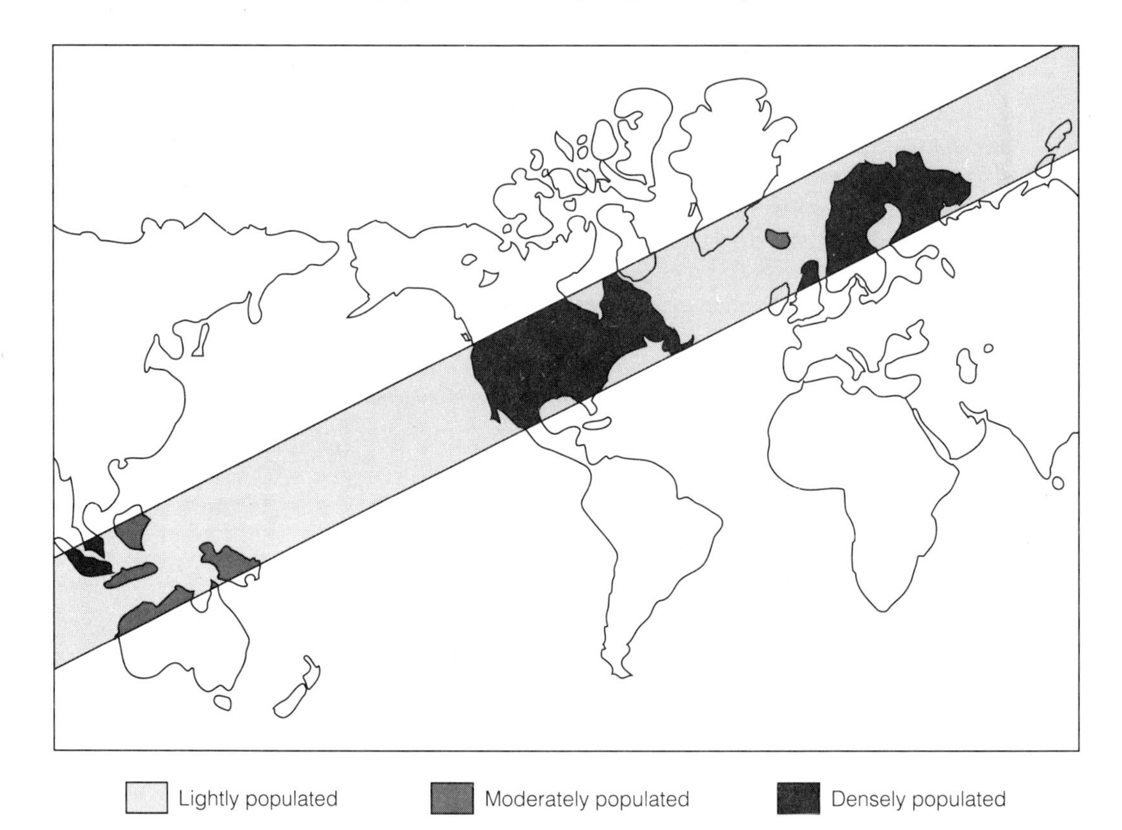

FIGURE 4 Population Density Map

As this map shows, Skylab did indeed pass over some highly populated areas of the earth (including most of the United States). Now, if the average population density over the 500-mile wide area of the entire path of Skylab is calculated, then the odds of some person being bonked on the head by a piece of Skylab are about 1 in 150. (I even saw projections that made the calculation as precise as "1 in 152.") Presumably, that represents the average population density of the area of the path of Skylab in what was expected to be its final orbit. (The original orbit of Skylab actually passed over the greater part of the earth's population—approximately 4 billion people! However, a last-minute adjustment was made in the orbit which carried Skylab over the less-densely populated area represented, albeit somewhat inaccurately, by the map above. The probability calculations were based on the original orbit that swung over the more-densely populated areas.)

However, that figure represents the possibility of anyone at all getting bonked, and people tend to be less worried about the guy in Borneo than they are about the guy next door—or even more pointedly, themselves! So, what were the odds of you being bonked on the head by a piece of Skylab? The answer: 1 in 150 *times* 4 billion = 1 in 600 billion.

Feel safe? We do have the benefit of hindsight; since you're here today, reading this book, you know you weren't hit by a piece of Skylab, so you can be very confident indeed! However, back in 1979, many people did get very anxious about the possibility of falling Skylab debris. These statistics were confusing. "Which is it," they'd ask, "1 in 150, or 1 in 6 billion?" The probability calculations didn't serve to quell anxieties; in many cases, they heightened them.

Then, to the rescue came some dashing individuals! Entrepreneurs began hawking "Skylab" memorabilia. For example, you could buy a bumper sticker that said, "The Skylab is falling! The Skylab is falling!" Or you could buy an "official" Skylab parts bag to collect your debris in. Most ingenious of all, in some cities in the country you could buy an "official" Skylab "protective" helmet to ward off falling debris.

Remarkably, some of the Skylab helmets were made of paper. That didn't matter because the salespeople could demonstrate that the chances of your being bonked on the head by a piece of Skylab were far less than the chances without such protection. How did they figure that?

Well, suppose the vendor had made up and sold 300 protective paper Skylab helmets. If you were one of 300 people who wore the Skylab helmet the day that Skylab fell, you succeeded in individuating yourself even more than the other people who were just blindly compounded in the original calculation comprehending a population of 4 billion in the path of Skylab. That individuating factor allowed you to multiply an additional 300 *times* the 4 billion *times* the 1 in 150 chance of being bonked on the head by a piece of Skylab. In other words, the chances of *you*, wearing the paper Skylab protective helmet, being bonked were a mere 1 in 1.8 *trillion*. That was a difference of 1 in 1.4 trillion of your ever being bonked on the head by a piece of Skylab. Didn't you owe yourself and your family the added margin of protection offered by this ingenious helmet? Isn't that worth the extra three dollars you spent on it?

The fallacy in the calculation is easy to see: why would anyone ever expect that the fact you're wearing a piece of paper on your head should make you less likely to be hit by space debris? But like all of the best hoaxes, this one has a valuable lesson to teach us. For one thing, determining what are and what are not relevant individuating factors in such calculations is highly subjective. Living outside the possible range of the path of Skylab, for example, should count for more protection than wearing a helmet within the possible range of Skylab debris. Even so, pure probability calculations can be blind to such subjective input; in any case, experts present their final figure with an air of authority that represents what many more honest NASA officials had to concede was nothing more than a blind guess. In fact, after a last-minute adjustment of the orbit designed to place Skylab over a less populated area, the debris fell to earth off the coast of Australia (no reports of injuries). The prediction that you were 3,000 times more likely to be struck by lightning that day seems to have been borne out.

However, we should beware of such overconfidence. The former head of the (now defunct) Atomic Energy Commission, Dixie Lee Ray, predicted that the odds of a major accident at a commercial nuclear power plant in the United States were roughly equal to the chances of being bitten by a rattlesnake while crossing a busy intersection in downtown Washington, D.C. She made this pronouncement about one year before the near meltdown of the reactor at Three Mile Island.

Perhaps the moral of the story is: one should be especially cautious of snakebites on Pennsylvania Avenue.

EXERCISES

1. Suppose your logic class is composed of forty women and ten men. What is the probability that the first person to come to class today will be a woman? a man?

2. Calculate the following probabilities (using conventional numerical probability rules)

a. five sixes in a row
b. two tails in a row
c. a one, two, or three on one die
d. three spades in a row (standard deck)
e. heads on all three tosses

3. What is the probability of drawing three hearts in a row from a standard deck of cards if the cards are returned to the deck and shuffled after each draw?

4. What is the probability of drawing three hearts in a row from a standard deck of cards if the cards are not returned to the deck?

5. One day you decide to compare apples and oranges for sweetness and sourness. You have a basket full of thirty oranges and twenty apples mixed together (a total of fifty fruits in all). You have been reliably advised that ten of the oranges are sweet and five of the apples are sour, as shown in the following table:

	sweet	**sour**
oranges	10	20
apples	15	5

You reach into the basket blindfolded.

a. What are the chances that you will select an orange?
b. An apple?
c. A sweet orange?
d. A sour apple?

6. You have two dice. What are the chances of obtaining the following result on one throw?

7. What are the chances of obtaining two aces (as in the drawing) on three consecutive throws of two dice?

8. What is the probability of getting three heads on a single toss of four coins?

9. What is the probability of drawing a royal flush on five cards from a standard deck?

List of Rules in this Module

The Conjunction Rule
P(A & B) = P(A) × P(B)

The Conditional Probability Rule
P(A & B) = P(A) × P(B if A)

The Mutually Exclusive Disjunction Rule
P(A or B) = P(A) + P(B)

The Inclusive Probability Rule
P(A ∨ B) = P(A) + P(B) − P(A & B)

MODULE 45

ARGUMENTS AND VISUAL IMAGES

1. **Observation**
2. Hypothesis
3. Experimentation
4. Theory

45.1 THE VISUAL IMAGE

A presupposition of logic and philosophy is that arguments must be linguistic (i.e., in spoken or written form). The problem with such presuppositions is that they often take on the status of dogma or revealed truth without reason or evidence. In fact, as we shall argue, visualization is an excellent problem solver and hypothesis-generator.

We live in an age in which the visual image is of ever-increasing importance. Visual images are used to manipulate the belief and actions of their intended audiences. Certain conventions of presentation of visual images now exist that

---

An **inductive** argument is **sound**, if and only if, the evidence is sufficient or representative enough to warrant the likelihood of the conclusion, and the premises are true.

allow us to interpret them as analogues to linguistic arguments. These conventions have taken the form in the visual arts: painting, dance, film, and theater.

For example, perhaps the most recognized work by Picasso is his painting, *Guernica*. *Guernica* is now on display in the Prado in Madrid, though the Museum of Modern Art in New York was graced with this masterpiece for many years at the direction of Picasso's will until such time as Spain returned to democracy. The painting exemplifies the aerial bombing of the Spanish town of Guernica by the German war machine during the Spanish Civil War. At the center of the painting is a highly stylized, cubist depiction of a horse caught in a scream of agony. Now, certain conventions have long existed in painting as an art form that sometimes deems the depiction of a horse as a symbol of death. The art connoisseur recognizes these conventions; for her, the cubist horse caught in the throes of death has greater impact than for someone who is not aware of these conventions.

And yet, we must distinguish *signs* and *symbols* from statements and assertions. Symbols may add up to statements but they do not by themselves constitute a statement (just as the letters of the alphabet do not constitute statements though a proper combination of them may do the job). Nevertheless, we sometimes do speak metaphorically about a painting "making a statement."* This talk is loose, but it is a good metaphor. In a sense the entirety of the composition of symbols contained in *Guernica* does make a "statement" of sorts about the Spanish Civil War. Certainly, Picasso wished to assert something about that war; a study of the painting can help us feel the pain and dismay he must have visualized in the painting of *Guernica*.

Art is the domain of the emergent emotions. Arguments should appeal to reason. In this respect the two do indeed differ. However, in fact, arguments *do* sometimes make appeals to emotion. When they do, they are usually fallacious (after all, we did say arguments *ought* to engage our *reason*). In this respect, the conventions which regulate and interpret the presentation of visual images may be subject to much the same sort of criticism as are faulty arguments in logic.

Let's make this qualification right from the start: the conventions which regulate the display and interpretation of visual images do not constitute statements, assertions, or arguments strictly speaking. However, and this is the important point, something very much *like* argument may be going on when visual images are displayed according to recognized conventions used in persuading us to come to a conclusion or a particular course of action. Such attempts at persuasion are commonly found in visual displays in propaganda and commercial advertising. What a shame it would be to suspend our understanding of what constitutes a fallacy when we attempt to criticize what is bogus in such propaganda and advertising! How much more equipped we are to see through visual manipulation of thoughts and emotion when we can bring the standards of critical logic to bear on our examination of propaganda and advertising, even if we must do so in a metaphorical fashion.

*For example, as in the expression, "Making a fashion statement."

Pablo Picasso's *Guernica* (1937)

Consequently, it is my contention that in certain contexts, together with recognized conventions, a visual image can serve as a *premise* in logical argumentation. In some respects, this is a difficult contention to maintain because visual images admit to many ambiguous interpretations. However, I hope to show that certain recognized conventions exist which serve to limit the range of how an image can be interpreted in a given context. Moreover, I contend that they are very often used in just that way to manipulate public opinion or action by making fallacious appeals to the public. We need to be in a position to say that by flashing a photograph on a television screen that arouses fear or sympathy a fallacy of appeal to fear or appeal to sympathy is being committed.

For example, in the 1988 presidential election, George Bush's campaign committee ran a commercial that was highly critical of what it called Michael Dukakis' "revolving door policy" with regard to furloughs for convicts in his state of Massachusetts. It pointed to the case of Willy Horton, a convict who escaped on one of the Governor's furloughs and then raped a woman in Maryland. A photograph of Willy Horton appeared on the screen which portrayed him as highly intimidating and a *black* man.

Many civil rights advocates protested the use of the Willy Horton photograph in that manner. They argued it was an implicit appeal to fear, playing on the racial fears of the public. But how can that be a fallacious appeal to fear if the visual image of Horton cannot serve as a premise in an argument just because it is not a statement in a language? My argument is that the critics of this commercial are justified in their protests and that visual images of this sort *can* provisionally take the place of linguistic statements in the setting of an argument.

Using visual images fallaciously in this way is only one side of the coin; the other is that sometimes a visual image can serve as good supporting evidence in the context of an argument.

To detect the use of visual images in either of these ways, as fallacious appeals or as good supporting evidence, we must train ourselves to be good *observers*. In order to actively use visual images as good supporting evidence, we must also be *creative* in how we employ them. That requires us to study some recent considerations in aesthetics, particularly those which address the "languages" of visual images.

45.2 THE "LANGUAGE" OF VISUAL IMAGERY

In 1968, the film, *2001: A Space Odyssey,* was released to American theaters. Film "auteur" Stanley Kubrick was the director. The movie was loosely based on a series of short stories by science fiction author, Arthur C. Clarke. Kubrick made the story very much his own, and told it from a point of view that even Arthur C. Clarke did not anticipate. Not only was his theme very different from Clarke's, but his manner of presentation was revolutionary for art. Most movies *explain* the plot through dialogue or even narration. In *2001: A Space Odyssey,* Kubrick *revealed* his plot and theme through a series of visual images.

Says one film critic in a book written later about the making of the film:

> Why does *2001* (1968) continue to fascinate us? What qualities bring us back to the film again and again, each time to be reintoxicated by its imagery and its mood, and to discover some new insight or pattern? The answer, I think, has to do with the kind of relationship the film establishes with us. *2001* tells its story—through images rather than words—in a way that is uniquely cinematic, and yet its narrative method is not quite like that of any other film. Though it certainly evokes memories of silent screen technique, its visual vocabulary is not quite that of silent films; if *2001* is not exactly a talking picture, it is certainly not a silent one, as its sound track is crucial in establishing tone and atmosphere and meaning. Sound and image work with, and against, each other to create ironies, parallels, counterpoints, enveloping the film in an aura both of comedy and mystery.
>
> What gives the film its special quality, and what compels us to return to it, over and over, is its fierce reluctance to explain or to interpret its visual and metaphysical mysteries; its profound narrative reticence.*

2001: A Space Odyssey is a film about the evolution of humanity from the earliest beginnings to the very near future. The story is an epic, told with little dialogue and thin characterization. Its only real predecessor is W. D. Griffith's classic 1916 epic, *Intolerance,* which is told (like *2001*) in four parts spanning four ages. *Intolerance* is a silent film, of course, and spawned many of the visual conventions of the cinema that followed.

2001: A Space Odyssey is likewise a major innovation, adapting the conventions of visual imagery from films over the century, incorporating them into the story, uniquely relying on them to expound the story and theme, and introducing very many conventions of its own.

Part One of the film shows primitive ape-humans locked in the struggle for survival, victims to predators and their environment. A mysterious black obelisk is discovered which somehow transforms or advances the evolution of the ape-human into a tool-making creature of intelligence. It is the discovery of the use of tools which marks the crucial evolutionary phase of humanity, according to the film. The ape-humans finally vanquish their predator foes with the use of a bone as a weapon. A victorious ape-human tosses the bone into the air in triumph, where it reaches its apex, spins, and transforms into a spinning space station above the earth. The entire sequence is related without a word of dialogue or even narration.

For my purposes, I will not discuss in much detail the other parts of the film. Part One sets the motif and demands that we follow and interpret it through visual images. That demand, we should understand, is made throughout the rest of the film and, even though dialogue is sometimes present there, it is almost irrelevant to understanding the theme and story. The spinning bone/weapon transformed into a technological marvel tells us all we need to know:

*Arthur C. Clarke, *The Lost Worlds of 2001,* Boston: Gregg Press, 1979, from the introduction by Foster Hirsch, p. v.

advanced technology is merely an extension of that basic, primitive tool. The film will tell us something about the role of tools in the advancement of human evolution.

The spinning bone is a powerful visual image that tells us much. To make the analogy of visual images to language, we need to see that something like a "grammar" is present in the relating of the film to the audience: it is structured; it follows certain rules (conventions); and the success or failure of the film to communicate its concealed message depends intimately on the reliance upon those rules. Communication is the object, but only through the device of visual sequences.

To make this point more compelling, consider one special episode in Part One. Before the discovery of tools, primitive humans are helpless victims to the attack of the panther. In one scene, a panther attacks the humans in their caves; the humans flee. One is apprehended and dragged out of the cave by the predator. The others watch helplessly. The panther is their greatest threat.

At the end of the scene, the panther sits on a high ledge outside the cave. The sun is setting in the distance. The fading light catches the eyes of the panther, and the panther's eyes glow menacingly. I have attempted to capture the scene in Figure 1.

Those who have followed Stanley Kubrick's films over the years know that a key to understanding his motifs lies in watching the eyes of his characters. In this case, an important key to the film is in the observation that the eyes of the panther glow in a menacing fashion. Later, in the second part of the film, Kubrick cashes in on that observation. A space shuttle which resembles a pod makes a journey from the space station we observed earlier and makes its landing approach on the moon. As it sets down on the moon base, the observant

FIGURE 1

FIGURE 2

viewer notices that for a moment the visor in the space pod takes on much the same eerie glow that we saw in the eyes of the panther. I've represented that scene in Figure 2.

The suggestion is that the highly advanced technological tools of modern humans pose to them as great a threat as the panther did to the earliest humans. War and rumors of war fill the radio waves. The fear is ever-present that humans in the early part of the twenty-first century will unleash the horror of nuclear warfare on themselves. The very tools which came to serve humankind now represent their greatest threat to survival.

We could easily make much, much more of this film in illustrating the contention that, given accepted conventions, visual images put together in a structured format can approach or even wholly succeed in mimicking language in the art of communication. Yet, these very conventions surface not only in Stanley Kubrick films but in less explicit ways in so very many aspects of our daily visual experiences as communicated to us through such media as television and films.

For example, the aphorism has it that "a picture is worth a thousand words." Some philosophers (especially aestheticians) have argued at great length that the conventions of the visual arts can be seen to serve as something analogous to languages. Prominent among these is Nelson Goodman who argues as follows:

> Symbols from other systems—gestural, pictorial, diagrammatic, etc.—may be exemplified and otherwise function much as predicates of a language. Such non-linguistic systems, some of them developed before the advent or the acquisition of language, are in constant use. Exemplification of an un-

named property usually amounts to exemplification of a non-verbal symbol for which we have no word or corresponding description.*

The problem is that non-verbal symbols usually do not possess a grammar which unifies them into linguistic assertions. Norman Malcolm relates the story of how his teacher Ludwig Wittgenstein came to be dissuaded of his "picture-theory" of language from *The Tractatus:*

> Wittgenstein and P. Sraffa, a lecturer of economics at Cambridge, argued together a great deal over the ideas of *The Tractatus.* One day [they were riding, I think, in a train] when Wittgenstein was insisting that a proposition and that which it describes must have the same 'logical form', the same 'logical multiplicity'. Sraffa made a gesture, familiar to Neopolitans as meaning something like disgust or contempt, of brushing the underneath of his chin with an outward sweep of the finger-tips of one hand. And he asked, "What is the logical form of *that*?" Sraffa's example produced in Wittgenstein the feeling that there was an absurdity in the insistence that a proposition and what it describes must have the same 'form'. This broke the hold on him of the conception that a proposition must literally be a 'picture' of the reality it describes.†

Can visual images have a grammar (or, alternately, a *logical form*)? Surely such gestures as the one supplied by the Neopolitan economist to Wittgenstein fail on such a test, but must all visual images fail in that way? Clearly not, for works of art do arrange their symbolic images in ways that may conform to something like a grammar in language. The greater context of visual art allows for implicit rules in a system to which symbols must conform if they are to produce meaning. Naturally, such meaning is not without ambiguity (language itself rarely is), but according to some, the greater ambiguity of images in visual art is precisely their virtue.

An artist once told me that she was formerly shy about displaying her art "because it reveals so much." In time, however, she came to realize how little most art patrons understand of what's contained on the canvas. They often fail to fully appreciate the conventions, but more importantly, every work of art is less exacting in the way it displays its meaning than is language. Certainly the contention that every interpretation of the work is legitimate is false; but the sort of looseness, despite the conventions which govern the arrangement of visual art, which any such work inherently contains is what makes visual art so rich. We can't expect the conventions in the arrangement of visual images to be without ambiguity, but equally false is the supposition that such works have no such rules of interpretation, or that they contain no meaning analogous to linguistic meaning at all.

*Nelson Goodman, *Languages of Art*, Indianapolis: Bobbs-Merrill Company, Inc., 1968, p. 57.

†Norman Malcolm, *Ibid.*, p. 58. In a footnote, Malcolm relates that the matter happened somewhat differently according to G. H. von Wright's memory. What Sraffa asked is the question, "What is the *grammar* of that?"

Thus, a range of possible interpretations for, say, a film such as Stanley Kubrick's *2001: A Space Odyssey,* is certainly warranted, but the important point to notice is that it's a very limited range indeed. While several interpretations may stand up to sustained scrutiny, very many others will not do at all. *2001,* for example, cannot properly be interpreted as a xenophobic film; the intelligence that guides the placement of the monolith which assists humans in their preservation and development cannot be made out to be hostile.*

45.3 THE "ARGUMENT" FROM VISUAL IMAGERY

One place in which visual images are arranged in such a fashion as to constitute analogues to assertions in the context of argument is commercial advertising. Another is political propaganda.

The whole point of advertising and propaganda, like one aspect of argument, is to *persuade.* Not all linguistic arguments attempt to persuade (one other function is discovery), but many (and very probably most) do. They attempt to persuade us to accept a conclusion on the basis of reasons presented. Advertising and propaganda which employ either visual imagery exclusively, or a mix of sound, image, and text, can approach argumentation because the whole of their context, their entire reason for being, is to persuade us of something and to act upon it. Such presentations engage our reason to a certain degree; otherwise they would be entirely subliminal (as, I grant, some ads and propaganda are). They are rarely without fallacies as they most often appeal to our emotions in the presentation of images. But many bad linguistic arguments make such appeals: appeals to fear, appeals to force, appeals to sympathy, and appeals to bias. We don't say that there aren't arguments; we say that they're *bad* arguments because they commit fallacies. The same occurs in propaganda and advertising.

Sometimes such arrangements of visual images do assist us, providing reasons in reaching a certain intended conclusion. When photographs of young children with bellies distended from malnutrition accompany ads appealing for financial aid for the relief of the Ethiopian famine, an appeal to sympathy is surely intended, but it is not a fallacious appeal to sympathy. We decide to contribute because we see the horrible effects that the famine is producing, and we see it in no uncertain terms. The depiction of suffering serves as a *reason* for the conclusion that we ought to contribute to this cause, and it's a *worthy* cause.

The point is that we must look to the entire context of the presentation in order to see the relation between visual images and assertions in an argument. To make this point more strongly, let's consider an example of what I contend to be an argument in a mixed-media presentation. The following is a representation of an actual advertisement which appeared in health magazines in recent years. The sponsor is the American Dairy Council.

*I am assuming a lot of familiarity with *2001: A Space Odyssey* here in order to make my point that visual images together with accepted conventions can mimic a grammar. If you haven't seen the movie, rent the video.

Calcium goes to work your whole life to build and maintain strong, healthy bones. Yet, as you get older your bones may begin losing calcium. This can lead to "osteoporosis" causing severe problems later in life. And while it can be caused by a lack of calcium, getting enough calcium in your diet is very easy to do.

One of the most natural ways to replenish your calcium is with dairy foods like milk, yogurt, cottage cheese, and cheese. All you need are three servings a day to meet your recommended dietary calcium allowance of 800 milligrams. And dairy foods offer you a wide variety of delicious and nutritious ways to get a lot of your other essential nutrients.

A public service announcement brought to you by

The American Dairy Council

Notice that as it stands the text on the right is no more than an explanation of the need for calcium in our diet. If we took the parts of the ad out of context, we would surely miss the point. The "photograph" (here represented by a drawing) would be a mere photograph of a pitcher and a glass of milk on a table with a rose in a vase in between. The explanation of the need for calcium would be a mere explanation; but *taken as a whole* and understanding its intent as an advertisement, we can decipher the workings of an argument.

An analogy may be helpful here. Consider that a map is an intermediate case between an uninterpreted picture and a set of conventions. Such maps contain conventional representational techniques as well as a legend. We may think of a visual image that is structured along the lines of accepted conventions in film or photography in much the same way. Both are subject to longstanding conventions of interpretation, and both contain something analogous to a legend.

Arguments are intended to persuade or convince us of the truth of something, or else they are intended to make a discovery. Like argument, this ad is intended to persuade the reader. Arguments are sets of reasons to a conclusion. Like an argument, this ad presents reasons for a conclusion so obvious it's left unstated: "You ought to drink milk and consume dairy products frequently." The reasons include a conventional presentation explaining the need for calcium in our diet; yet, we also discern a more unconventional presentation. The photograph of the table setting is appealing and the rose serves to heighten that appeal. The glass of milk is strategically placed at the forefront of the ad; conventions of visual imagery tell us from past experience that the glass of milk is being offered to us—we should find the idea of drinking milk inviting. Thus, we can say that another reason we ought to drink milk (and consume dairy products in general) is that it's appealing.

Notice that this is for the most part an appeal to our reason as well. Certainly, the explanation of the need for calcium (which dairy products supply) is an appeal to reason for the most part, and not to our emotions. Still, we can also come to regard the attractiveness of the setting in the picture as an appeal to reason in much the same way. The ad is telling us through the picture that getting our minimum daily requirement of calcium can be pleasurable. It doesn't do this through conventional linguistic means; it does so through an art convention, the convention of *exemplifying* the pleasurable nature of the thing under review, much in the way (to borrow a fine example from Nelson Goodman) a swatch of cloth exemplifies the quality and texture of a suit one wishes to have made in a tailor shop.

If we "go by the book," if we follow the traditions and restrictions of most logic books, we'd have to label this as "no argument," but that would do injustice to the apparent intent of the ad. One might object that every argument must have an explicit conclusion, a *stated* conclusion, and no conclusion is stated here. But this objection ignores the fact that many arguments in daily life do not bother to state the conclusion because to do so is to "state the obvious." Recall the definition of the scholastic term 'enthymeme'. An **enthymeme** is a categorical syllogism that is missing a premise *or a conclusion*. Traditional logic then does accept the possibility that one may neglect to state a conclusion if that conclusion is obvious, given the strength of the argument. What we find in this advertisement is something very close to an enthymeme in traditional logic.

Finally, to secure the analogy of this mixed-media advertisement to logical argument, we may consider a *reductio ad absurdum*. If we do not classify the American Dairy Council advertisement as an argument, then we should be unable to use descriptions of fallacies in the criticism of the reasoning of the techniques employed in this advertisement, because fallacious reasoning occurs only where reasoning occurs. Yet, when the ad makes reference to the dangers of insufficient intake of calcium in one's diet by referring to the possibility of the onset of osteoporosis in later life, one may be concerned that the advertisement is making a fallacious appeal to fear (*argumentum ad metum*). Hosts of readers who, by reading the sorts of health magazines in which this ad appeared, may be especially susceptible to such techniques of fallacious reasoning in this regard. But, I'm not sure that such a fallacy can be legitimately ascribed to the American Dairy Council in this case, for the caveats seem reasonable enough. The point to notice is that we are at least legitimate in our *concern* that such techniques of fallacious reasoning may be present. That is a proper critical approach to any ad we read. Yet, without the supposition that an argument is present, we are left without the resources of critical analysis supplied to us by a study of the informal fallacies. Therefore, critical analysis is better served if we treat such advertisements as implicit arguments than if we exclude them as the tradition of logic seems to warrant.

Much the same sort of case can be made with regard to the picture of the milk as well. When the ad displays milk as so appealing, is our reason being engaged or is it our appetites? Placed in context, we see that the ad may confuse our reason and our emotions in a way that goes beyond mere stimulus-response manipulation, and we ought to be wary. Certainly, in reading the ad, we are confronted with the time-honored technique of "the carrot and the stick," but at least *part* of the technique of the ad is an appeal to rational considerations, even if the rational appeal may be confused with the non-rational.

Familiarity with the strictures of logical analysis, then, actually aids us in our ability to read the ad critically. That in itself is a telling reason for treating such ads as arguments, even if they do so in part through such unconventional logic techniques as displaying visual images.

CONCLUSION AND SUMMARY

In summation, we have examined whether visual images can properly serve as (something like) assertions in assembling the building blocks of argument. One reason to credit this proposal as legitimate includes the ever-widening reliance on visual images in techniques of persuasion such as commercial advertising and political propaganda. My contention is that enough of an analogy can be discerned in the properties of visual imagery to allow this conclusion.

One obstacle we faced in making this contention is that visual imagery appears not to follow rules in ways sufficient to convey meaning consistent with the manner that an assertion conveys meaning. Yet, we found that objection to be without merit given the prevalence and wide acknowledgment of conventions which regulate usage of visual images appropriate to transmitting a content to viewers who have even a cursory acquaintance with those conventions. Contemporary aestheticians have advanced this claim with increasing conviction, some of whom even make reference to "languages" of art; whatever the merits of that contention, nevertheless, we find ample reason to suppose that the conventions are well-enough understood to allow us to discern something like a grammar of visual imagery when placed in proper context.

The key to the argument is a contention derived from common sense: we must not cut off a passage or a selection of visual images from their wider relations if we are to understand the new and unique roles they employ in contemporary life. Treating some usages of visual imagery as though they serve as reasons in a wider argument gives us more explanatory and critical power to bring to bear in the criticism of those techniques which employ visual imagery. The need for such analysis is greater now than it has been ever before.

EXERCISES

DIRECTIONS: *In this module, we based our thesis primarily on one example: the Stanley Kubrick film* 2001: A Space Odyssey. *Compile a short list of films you have seen that you think rely on visual images in this way, and describe how the interpretation of the movie depends on specific visual images and conventions of the cinema.*

RECOMMENDED READING

On Problem Solving

Adams, James L. *Conceptual Blockbusting: A Guide to Better Ideas.* 2nd ed. New York: W. W. Norton & Co., 1979.

Barzun, Jacques and Graff, Henry F. *The Modern Researcher.* 4th ed. New York: Harcourt, Brace and Jovanovich, 1985.

de Bono, Edward. *Lateral Thinking: Creativity Step by Step.* New York: Harper & Row, 1970.

Levine, Martin. *Effective Problem Solving.* Englewood Cliffs, NJ: Prentice-Hall, Inc., 1988.

Osborne, Alex F. *Applied Imagination: Principles and Procedures of Creative Problem Solving.* New York: Scribners, 1963.

Paulos, John Allen. *Mathematics and Humor.* Chicago: University of Chicago Press, 1980.

Raudsepp, Eugene, and Hough, George P., Jr. *Creative Growth Games.* New York: Jove Publications, 1977.

Tobin, John, and Feyen, Kathleen. *How to Make Decisions Creatively: A Guide to Decisive Thinking.* Sacramento, CA: Hartnell Publications, 1981.

von Oech, Roger. *A Whack on the Side of the Head.* New York: Warner Books, 1983.

On Visual Images

Goodman, Nelson. *Languages of Art.* Indianapolis: Bobbs-Merrill Company, Inc., 1968.

Gombrich, E. H. *Art and Illusion: A Study in the Psychology of Pictorial Representation.* Princeton: Bollingen Series, 1972.

On the Philosophy of Science

Bernstein, Jeremy. *Einstein.* New York: The Viking Press, 1973.

Broad, William and Wade, Nicholas. *The Betrayers of the Truth: Fraud and Deceit in the Halls of Science.* New York: Simon and Schuster, 1982.

Feyerabend, Paul. *Against Method.* London: Verso, 1975.

Hempel, Carl G. *Philosophy of Natural Science.* Foundations of Philosophy Series, edited by Elizabeth and Monroe Beardsley. Englewood Cliffs, NJ: Prentice-Hall, Inc., 1966.

Keynes, John M. *A Treatise on Probability.* London: Macmillan & Co., 1921.

Kuhn, Thomas S. *The Structure of Scientific Revolutions,* 2nd ed. Chicago: University of Chicago Press, 1970.

Losee, John. *A Historical Introduction to the Philosophy of Science.* London: Oxford University Press, 1972.

Magee, Bryan. *Philosophy and the Real World: An Introduction to Karl Popper.* La Salle, IL: Open Court Publishing Co., 1985.

Popper, Karl R. *The Logic of Scientific Discovery.* New York: Harper & Row, 1968.

Russell, Bertrand. *The ABC of Relativity.* London: George Allen & Unwin, Ltd., 1969.

Answers to Even-Numbered Exercises

UNIT ONE

MODULE 2

Part I, page 27

2. Not a statement (simple interrogative).

4. Asserted statement.

6. Not a statement (simple exclamation).

8. Asserted statement (but false).

10. Asserted statement.

12. Asserted statement (but false).

14. Asserted statement (the context suggests both that Hubert says it and that the speaker believes it).

Part II, pages 27–28

2. Asserted statement.

4. Asserted statement (a complex interrogative—what is asserted is that this is "the only question to be settled now").

6. Not an assertion (complex interrogative, but the questioner is looking for information he does not have).

8. Asserted statements (this is actually a small argument that concludes: we *ought* to pity the meek).

10. Not a statement (simple interrogative).

12. Asserted statement (in fact, it's necessarily true).

14. Not a statement (simple imperative).

16. Asserted statement (Wilde is asserting that he sometimes thinks this way).

18. Asserted statement (a complex imperative—W. C. Fields may well be asserting that he believes this).

20. Asserted statements (complex imperative—you *ought* to plagiarize).

MODULE 3, *pages 34–36*

2. a. **4.** b. **6.** b.

MODULE 4.2, *pages 44–45*

2. Exposition. 4. Argument.

Part I, pages 47–48

2. Argument
4. Argument.
6. Explanation.
8. Explanation.
10. Argument.
12. Explanation.
14. Neither.

Part II, pages 48–50

2. Argument (If his interest wanes, it doesn't mean you're unattractive).
4. Explanation (how to announce your wedding).
6. Explanation (why they make it easy).
8. Argument (we should mind our own business).
10. Explanation (why there's not enough time).
12. Argument (reasons why the community lost a unique person).

MODULE 6

Part I, pages 68–69

2.

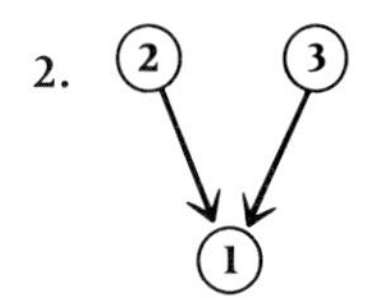

4. (1) The outrage . . . Republican.
(2) First, . . . conservative.
(3) Second, . . . Latinos.
(4) Third, . . . values.

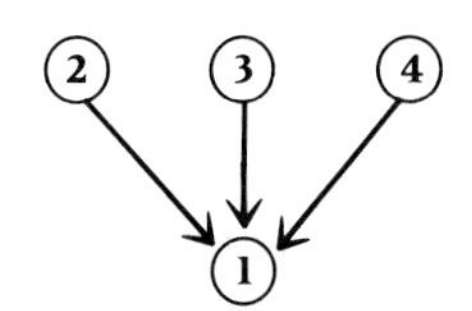

Part II

Exercise 1, pages 69–70

2. (1) = (5) It seems that . . . analogical. (These two sentences contain basically the same statement.)

For
(2) many different senses . . . argument.

Hence
(3) no argument, . . . propositions.

(4) But Holy Scripture . . . fallacy.

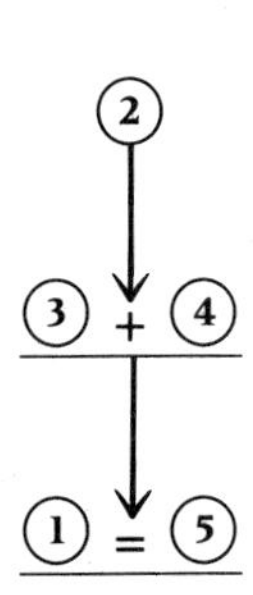

4. (1) Yet . . . much:

   (2) the new . . . mythology.

   (3) Lifting us . . . over.

   (4) He re-established . . . history.

   (5) He transformed . . . memory.

   (6) The energies . . . come.

   (7) Above all . . . humanity.
   So

   (8) the people . . . brother.

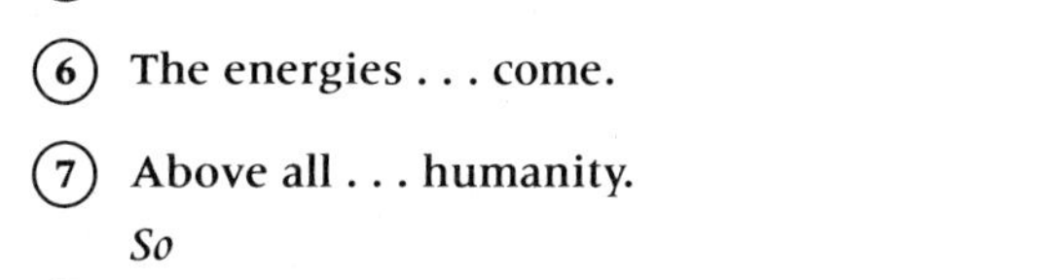

Exercise 2, pages 70–71

The details are left to the reader. (For a hint, look at the solution to Exercise 3.)

Exercise 3, page 71

The conclusion is that conscience makes cowards of us all and makes us lose our resolve for action. (The act Hamlet is contemplating is his own suicide!)

MODULE 7

Exercise 1, page 81

Paragraphs:

[1] Editor: I am . . .

[2] On Sept. 26 . . .

[3] After a few . . .

[4] I am convinced . . .

Asserted Statements for Paragraph 1:

(1) I am a 20-year-old . . . lifestyle
because
(2) I have never . . . clubs.

(3) Due to a recent . . . price.

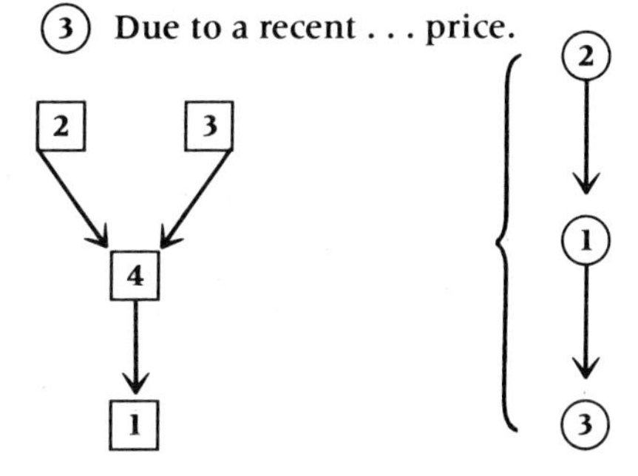

Exercise 2, pages 81–82

1. **Paragraphs:**

[1] Could America's . . .

[2] Economy watchers . . .

[3] The answer, . . .

[4] The similarities . . .

[5] But it is in . . .

[6] Each decade . . .

[7] As the bull market . . .

[8] The '80s have . . .

[9] But this is . . .

[10] Where will it . . .

[11] Unfortunately, . . .

Asserted Statements in Paragraph 3:

(1) The anwer . . . yes.

(2) The more . . . Harding.

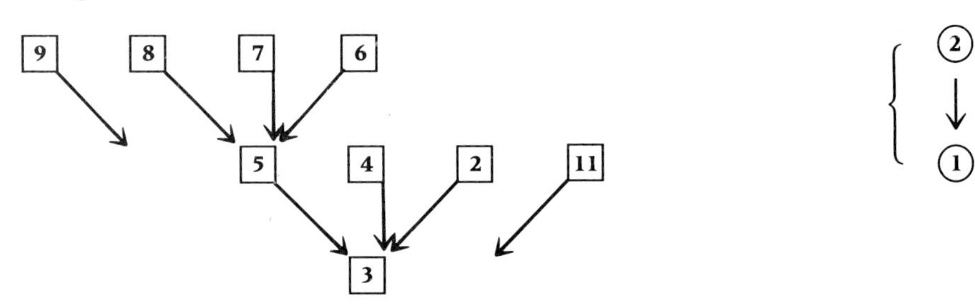

Paragraphs [1] and [10] are comprised of questions, not asserted statements, so we can omit these paragraphs from the diagram.

Exercise 3, pages 83–84

[1] The American Constitution . . .

[2] We are now witnessing

[3] First, there is . . .

[4] Second, there is . . .

[5] And, under pressure . . .

[6] Perhaps the most . . .

[7] But with the . . .

[8] Another serious challenge . . .

[9] The Constitution imposes . . .

[10] We are now confronted . . .

[11] With respect to . . .

[12] So, too, in another . . .

[13] No less disturbing . . .

[14] More serious, . . .

[15] When we reflect . . .

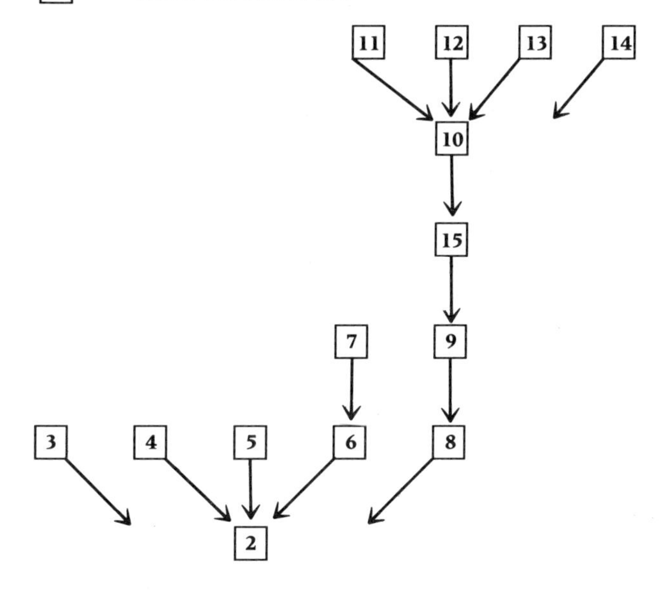

If we had done this diagram as an *assertion* diagram instead of a paragraph diagram, we would have numbered as many as 46 distinct asserted statements and the assertion diagram would have been ponderous. For example, where assertion (1) is the whole of the first paragraph and assertions (2) and (3) together make up paragraph [2], the "root" structure of the assertion diagram would look something like this:

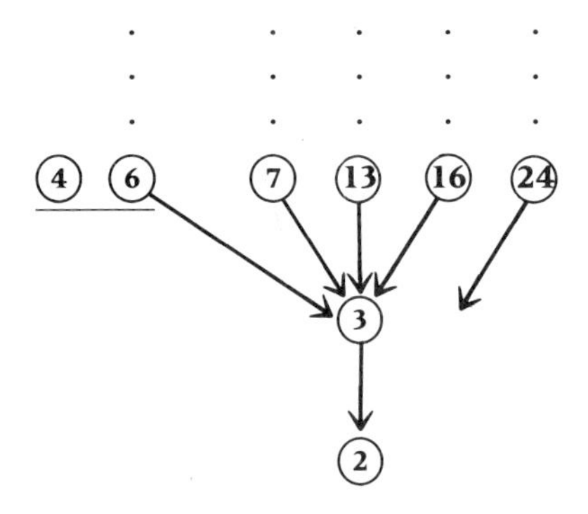

You can imagine the detail needed to complete this assertion diagram! Notice, however, that the *paragraph* diagram does an excellent job capturing the overall structure without all this attention to needless detail.

MODULE 8.1

Part I, page 89

2. Deductive (granting that thirty-five is a very young age to die, it *must* be the case that Mozart did much of his work at an early age).

4. Deductive (but the calculation is wrong: Washington would be *256* years old in 1988. It's still a deductive argument, but a bad one).

6. Deductive (given what we understand to be a "large" state, it must be the case that Altoona is in Pennsylvania. [It is.]).

8. Inductive (circumstantial evidence).

10. Deductive (no other possibility given the definition).

Part II, pages 90–91

2. Deductive (a calculation).

4. Inductive (Kristol "seems" to take glee).

6. Deductive (Spock uses process of elimination).

8. Deductive (the conclusion follows from the instructions).

10. Inductive (the consequences "could be" severe).

MODULE 8.2, *pages 92–93*

2. Inappropriate (assuming the premises are true, it *must* be the case that the Chancellor will cancel classes).

4. Inappropriate (nobody else could have done it. Ngaio Marsh *must* be the killer).

MODULE 9

Exercise 1, page 110

This exercise is left to the reader or classroom discussion.

Exercise 2, page 110

Some suggested answers. (The student should devise answers of his or her own.)

Term	Your suggested meaning	Euphemism or Buzzword?
Example:		
clemency list	*"enemies list"*	*euphemism*
rose garden	"press conference"	euphemism
health club	"abortion clinic"	euphemism
butcher shop	"abortion clinic"	buzzword
mother's milk	"cocaine"	euphemism
cotton candy	"an easy course"	buzzword

Exercise 3, pages 110-111

2. Freud is sometimes criticized for his extensive use of metaphors, especially by analytic philosophers, but here is a case where he uses the metaphors to coin new technical terms that he uses with reasonable consistency and precision.

MODULE 11, *page 127*

Note: These answers are open to some discussion.

2. Apparently contingent (we might use an egg substitute).
4. Apparently contingent (even if it is true).
6. Apparently contingent (when compared to the next example).
8. Apparently contingent (they could have been found elsewhere).
10. Apparently contingent (we might find a "cure").
12. Apparently contingent (entropy is theoretical).
14. Apparently contingent (theoretical).

EXERCISES FOR UNIT ONE

Exercise 1, pages 138–139

2. Not an argument (this is a report).
4. Argument (reason and conclusion).
6. Argument (the report contains Timex's argument, but then *Sports Illustrated* draws its own conclusion).
8. Argument ("this means that . . . ").
10. Argument ("It is now time to start thinking of exploring the stars . . . ").

Exercise 2, pages 139–140

2. Argument (the court "directed a chilly breeze").
4. Explanation (how he did it).
6. Explanation (how children are taught to think of themselves).

Exercise 3, pages 140–141

2. Not an argument (a report).
4. Deductive (process of elimination).

Exercise 4, pages 141–143

2. ①

②

4. ① I oppose . . . law.

② It's . . . many.

③ Seat belts . . . another.

④ As an adult, . . . government.

⑤ It concerns me . . . laws.

⑥ The mandatory . . . Constitution.

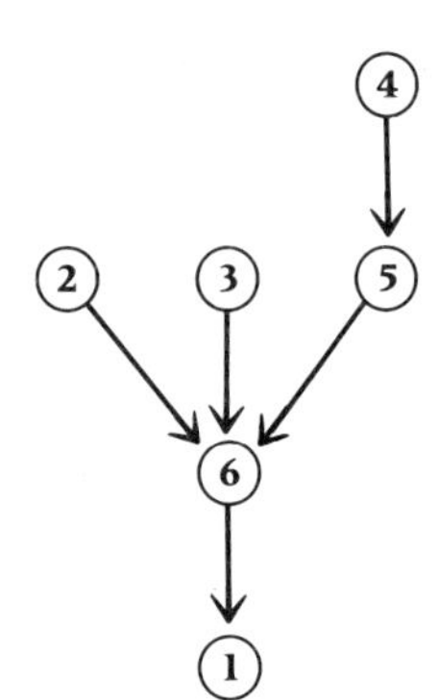

UNIT TWO

MODULE 14

Exercise 3, page 160

Some suggestions: In place of BARBARA, try 'ALASKA' or 'BANANA' (AAA). In place of BAROCO, try 'PLATOON' (AOO). In place of FESTINO, 'ELLIOT' (EIO).

Exercise 4, page 160

Some suggestions: In place of FALEPTON, try 'ASBESTOS' (AEO). In place of FESAPO, 'EMBARGO' (EAO). In place of DARAPTI, 'VANADIUM' (AAI). In place of BRAMANTIP, 'ARACHNID' or 'AQUATIC' (AAI).

MODULE 15, *page 162*

2. contraries
4. subcontraries
6. contradictories

MODULE 16, *pages 166–167*

2. *No* telephone companies are the same. (Universal Negative.)

4. *All* unjust men are nonchalant. (Remains Universal Affirmative—'unjust' and 'nonchalant' must remain intact.)

6. *No* non-aggressors are combatants. (Universal Negative—the subject is *non-*aggressors; we are not saying anything about aggressors.)

8. *Not all* dolphins are playful. (Particular Negative.)

10. *Not all* husbands say it. (Particular Negative.)

MODULE 17

Exercise 1, page 172

2. Ambiguous (does this mean that long-distance credit card calls are an *exception* to the rule? If not, this is a straight out contradiction).

4. Unambiguous (surely, the man *also* has a chow chow, but he brought a lab with him on the hike).

6. Unambiguous in context (the placement of 'only' here may be ambiguous in ordinary usage, but not in this context. Compare with problem 5).

Exercise 2, pages 172–173

Most of Gwen's confusions are warranted by the ambiguous use of quantifiers in this story. See the solutions for Exercise 1 to guide your unambiguous re-write. A few different interpretations of what happened are possible, but make your interpretaton consistent.

MODULE 18, *page 179*

2. Invalid (Rule 1).
4. Invalid (Rule 2).
6. Invalid (Rules 1, 2).
8. Invalid (Rules 1, 2, 3).
10. Invalid (Rules 1, 2).
12. Valid
14. Valid

MODULE 20

Exercise 1, page 202

2.

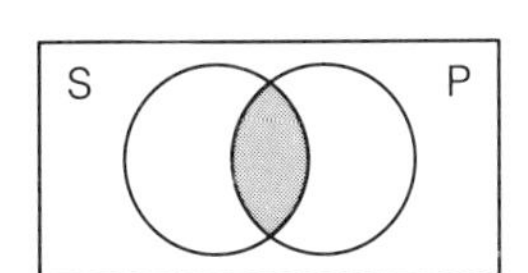

"No S are P."

4. *Note placement of 'P' and 'S.'*

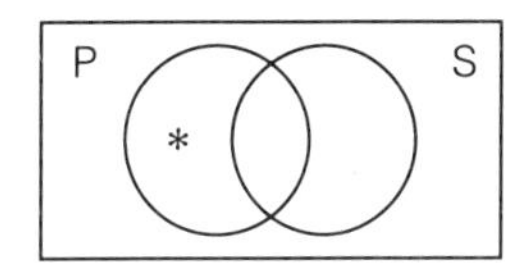

"Not all P are S."

Exercise 2, page 203

2.

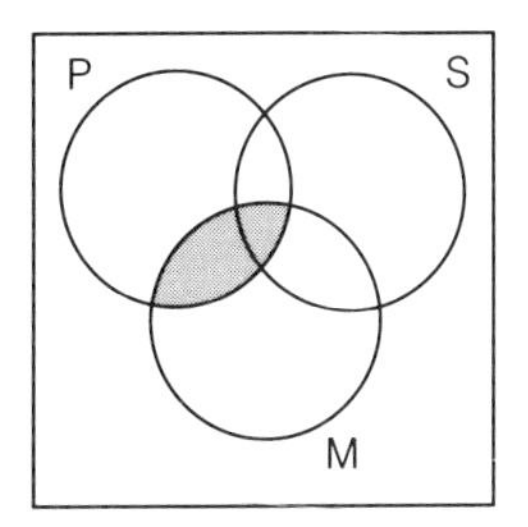

"No P are M."

4.

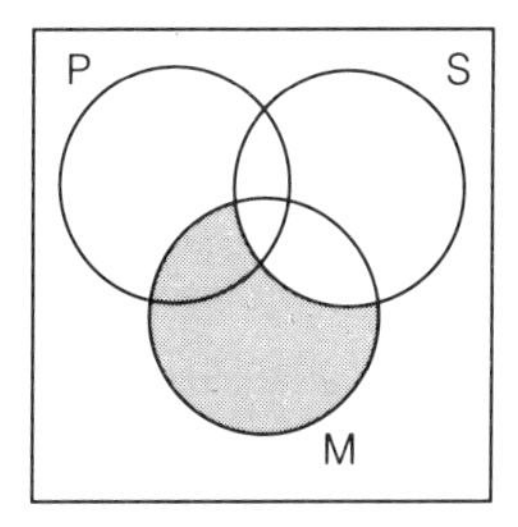

"All M are S."

MODULE 21, ***pages 220–223***

2. **Possible Conclusions**

All S are P	Invalid
No S are P	Invalid
Some S are P	Invalid
Not all S are P	Invalid

4. **Possible Conclusions**

All S are P	Invalid
No S are P	Invalid
Some S are P	Valid
Not all S are P	Invalid

UNIT THREE

MODULE 23.3, ***page 264***

Let 'S' = "The sun is at the center of the solar system."

Let 'E' = "Earth is a planet that has life."

Let 'M' = "The planet Mars orbits Jupiter."

Let 'R' = "Mars is the Red Planet."

Let 'O' = "The moon has rings."

Let 'C' = "Mercury is the closest planet to the sun."

Let 'P' = "Pluto is the closest planet to the sun."

2. $(S \vee E)$. 4. $(S \vee M)$. 6. $(R \supset O)$. 8. $(C \equiv R)$. 10. $(O \equiv P)$.

MODULE 23.4, ***page 266***

2. WFF (See footnote).
4. WFF.
6. No (missing right parenthesis).
8. WFF.
10. No (missing outside parentheses).
12. No ('&' is a connective that must be flanked on both sides by WFFs and enclosed in parentheses).

MODULE 23

Exercise 1, pages 276–277

2. K: I know what street Canada is on.
   $\sim K$
4. C: This will be a civilized country.
   S: We spend more money for books than we do for chewing gum.
   $(\sim C \vee S)$
6. D: The dead could return. W: War would end.
   $(D \supset W)$
8. C: You can guess. D: You dare.
   G: You ought to guess. H: You ought to choose.
   $((C \supset G)\ \&\ (D \supset H))$

10. T: We become two people—the suburban affluent and the urban poor, each filled with mistrust and fear of the other.
C: We shall effectively cripple each generation to come.
$(T \supset C)$

12. M: Heaven goes by merit. D: Your dog would go in.
Y: You would stay out. F: Heaven goes by favor.
$(M \supset (Y \,\&\, D))$
$\therefore F$

14. T: There was truth.
C: You cling to the truth even against the whole world.
M: You were mad.
$((T \,\&\, {\sim}T) \,\&\, (C \supset {\sim}M))$

Exercise 2, page 277

2. F: Free enterprise ended in the United States a good many years ago.
O: Big oil, big steel, big agriculture avoid the open marketplace.
P: Big corporations fix prices among themselves.
D: Big corporations drive out the small entrepreneur.
L: In their conglomerate forms, the huge corporations have begun to challenge the legitimacy of the state.
F
O
$(P \,\&\, D)$
$\therefore L$

4. D: We are drowning our children in violence, cynicism, and sadism piped into the living room and even the nursery.
L: The grandchildren of the kids who used to weep because the Little Match Girl froze to death now feel cheated if she isn't slugged, raped, and thrown into a Bessemer converter.
L
$\therefore D$

MODULE 24.3, *pages 283–284*

2. True.

4. True.

6. False (the antecedent is true, the consequent is false).

8. True.

10. True (both sides of the biconditional are false, so the whole biconditional is true).

12. True (the only time a conditional is false is when the antecedent is true and the consequent is false; here, the antecedent is false).

14. False (the antecedent is true, the consequent is false).

MODULE 24

Exercise 1, page 295

P	Q	~P	(P ⊃ Q)	(~P ∨ Q)
T	T	F	T	T
T	F	F	F	F
F	T	T	T	T
F	F	T	T	T

Exercise 2, pages 295–296

2.

D	∴(D & D)
T	T
F	F

VALID

4.

P	Q	(P ∨ Q)	∴(Q ∨ P)
T	T	T	T
T	F	T	T
F	T	T	T
F	F	F	F

VALID

6.

R	S	~R	~S	(R ⊃ S)	~S	∴~R
T	T	F	F	T	F	F
T	F	F	T	F	T	F
F	T	T	F	T	F	T
F	F	T	T	T	T	T

VALID

8.

A	B	(A ≡ B)	A	∴B
T	T	T	T	T
T	F	F	T	F
F	T	F	F	T
F	F	T	F	F

VALID

10.

S	K	B	(S ⊃ K)	(K ⊃ B)	∴(S ⊃ B)
T	T	T	T	T	T
T	T	F	T	F	F
T	F	T	F	T	T
T	F	F	F	T	F
F	T	T	T	T	T
F	T	F	T	F	T
F	F	T	T	T	T
F	F	F	T	T	T

VALID

12.

P	Q	~P	~Q	(P & Q)	~(P & Q)	∴(~P ∨ ~Q)
T	T	F	F	T	F	F
T	F	F	T	F	T	T
F	T	T	F	F	T	T
F	F	T	T	F	T	T

VALID

14.

P	Q	~P	~Q	(P & Q)	~(P & Q)	∴(~P & ~Q)	
T	T	F	F	T	F	F	
T	F	F	T	F	T	F	
F	T	T	F	F	T	F	✓
F	F	T	T	F	T	T	✓

INVALID, Rows 3 and 4

MODULE 25

Exercise 1, page 301

2.

A	B	(B ∨ A)	(A & (B ∨ A))	∴(B ∨ A)
T	T	T	T	T
T	F	T	T	T
F	T	T	F	T
F	F	F	F	F

VALID

4.

A	B	C	~A	~B	(B ∨ C)	(A ∨ (B ∨ C))	(~A & ~B)	∴C
T	T	T	F	F	T	T	F	T
T	T	F	F	F	T	T	F	F
T	F	T	F	T	T	T	F	T
T	F	F	F	T	F	T	F	F
F	T	T	T	F	T	T	F	T
F	T	F	T	F	T	T	F	F
F	F	T	T	T	T	T	T	T
F	F	F	T	T	F	F	T	F

VALID

6.

P	Q	R	S	(Q ⊃ R)	(P ⊃ (Q ⊃ R))	((Q ⊃ R) ⊃ S)	∴(P ⊃ S)
T	T	T	T	T	T	T	T
T	T	T	F	T	T	F	F
T	T	F	T	F	F	T	T
T	T	F	F	F	F	T	F
T	F	T	T	T	T	T	T
T	F	T	F	T	T	F	F
T	F	F	T	T	T	T	T
T	F	F	F	T	T	F	F
F	T	T	T	T	T	T	T
F	T	T	F	T	T	F	T
F	T	F	T	F	T	T	T
F	T	F	F	F	T	T	T
F	F	T	T	T	T	T	T
F	F	T	F	T	T	F	T
F	F	F	T	T	T	T	T
F	F	F	F	T	T	F	T

VALID

8.

A	H	B	~A	(H≡B)	(H⊃B)	(B⊃H)	((H⊃B)& (B⊃H))	(A⊃(H≡B))	~(A⊃(H≡B))	[~A&((H⊃B) &(B⊃H))]	∴(B⊃H)
T	T	T	F	T	T	T	T	T	F	F	T
T	T	F	F	F	F	T	F	F	T	F	T
T	F	T	F	F	T	F	F	F	T	F	F
T	F	F	F	T	T	T	T	T	F	F	T
F	T	T	T	T	T	T	T	T	F	T	T
F	T	F	T	F	F	T	F	T	F	F	T
F	F	T	T	F	T	F	F	T	F	F	F
F	F	F	T	T	T	T	T	T	F	T	T

VALID

10.

P	Q	R	S	(P ⊃ Q)	(R ⊃ S)	(P ∨ R)	(Q ∨ S)	((P ⊃ Q) & (R ⊃ S))	(P ∨ R)	∴(Q ∨ S)
T	T	T	T	T	T	T	T	T	T	T
T	T	T	F	T	F	T	T	F	T	T
T	T	F	T	T	T	T	T	T	T	T
T	T	F	F	T	T	T	T	T	T	T
T	F	T	T	F	T	T	T	F	T	T
T	F	T	F	F	F	T	F	F	T	F
T	F	F	T	F	T	T	T	F	T	T
T	F	F	F	F	T	T	F	F	T	F
F	T	T	T	T	T	T	T	T	T	T
F	T	T	F	T	F	T	T	F	T	T
F	T	F	T	T	T	F	T	T	F	T
F	T	F	F	T	T	F	T	T	F	T
F	F	T	T	T	T	T	T	T	T	T
F	F	T	F	T	F	T	F	F	T	F
F	F	F	T	T	T	F	T	T	F	T
F	F	F	F	T	T	F	F	T	F	F

VALID

12.

P	Q	(P ⊃ Q)	(P ⊃ P)	(P ≡ Q)	((P ⊃ Q) & (P ⊃ P))	∴(P ≡ Q)
T	T	T	T	T	T	T
T	F	F	T	F	F	F
F	T	T	T	F	T	F
F	F	T	T	T	T	T

INVALID, Row 3

14.

W	G	∴(W ∨ G)
T	T	T
T	F	T
F	T	T
F	F	F

VALID

Exercise 2, page 301

2.

P	~P	(P ∨ ~P)
T	F	T
F	T	T

TAUTOLOGY

4.

P	Q	(P ⊃ Q)	((P ⊃ Q) ⊃ P)	(((P ⊃ Q) ⊃ P) ⊃ P)
T	T	T	T	T
T	F	F	T	T
F	T	T	F	T
F	F	T	F	T

TAUTOLOGY (This is called "Peirce's Law.")

6.

P	Q	(P ⊃ Q)	((P ⊃ Q) ⊃ P)	(((P ⊃ Q) ⊃ P) ⊃ P)	~(((P ⊃ Q) ⊃ P) ⊃ P)
T	T	T	T	T	F
T	F	F	T	T	F
F	T	T	F	T	F
F	F	T	F	T	F

CONTRADICTION

8.

P	Q	~P	(P ⊃ Q)	(~P ⊃ (P ⊃ Q))
T	T	F	T	T
T	F	F	F	T
F	T	T	T	T
F	F	T	T	T

TAUTOLOGY

10.

P	S	Q	R	(P & S)	(Q ⊃ P)	(P ∨ R)	((P & S) ∨ (Q ⊃ P))	((P & S) ∨ ((Q ⊃ P) ≡ (P ∨ R)))
T	T	T	T	T	T	T	T	T
T	T	T	F	T	T	T	T	T
T	T	F	T	T	T	T	T	T
T	T	F	F	T	T	T	T	T
T	F	T	T	F	T	T	T	T
T	F	T	F	F	T	T	T	T
T	F	F	T	F	T	T	T	T
T	F	F	F	F	T	T	T	T
F	T	T	T	F	F	T	F	F
F	T	T	F	F	F	F	F	T
F	T	F	T	F	T	T	T	T
F	T	F	F	F	T	F	T	F
F	F	T	T	F	F	T	F	F
F	F	T	F	F	F	F	F	T
F	F	F	T	F	T	T	T	T
F	F	F	F	F	T	F	T	F

CONTINGENT

MODULE 26.2, *page 307*

2. WFF.
4. Not a WFF (small case 'n' is a statement letter).
6. Not a WFF (capital 'P' does not belong).
8. WFF.
10. WFF.

MODULE 26

Exercise 1, page 312

2.

a	b	Aba	KaAba	∴ Aba
T	T	T	T	T
T	F	T	T	T
F	T	T	F	T
F	F	F	F	F

VALID

4.

a	b	c	Na	Nb	Abc	AaAbc	KNaNb	∴ C
T	T	T	F	F	T	T	F	T
T	T	F	F	F	T	T	F	F
T	F	T	F	T	T	T	F	T
T	F	F	F	T	F	T	F	F
F	T	T	T	F	T	T	F	T
F	T	F	T	F	T	T	F	F
F	F	T	T	T	T	T	T	T
F	F	F	T	T	F	F	T	F

VALID

6.

p	q	r	s	Cqr	CpCqr	CCqrs	∴Cps
T	T	T	T	T	T	T	T
T	T	T	F	T	T	F	F
T	T	F	T	F	F	T	T
T	T	F	F	F	F	T	F
T	F	T	T	T	T	T	T
T	F	T	F	T	T	F	F
T	F	F	T	T	T	T	T
T	F	F	F	T	T	F	F
F	T	T	T	T	T	T	T
F	T	T	F	T	T	F	T
F	T	F	T	F	T	T	T
F	T	F	F	F	T	T	T
F	F	T	T	T	T	T	T
F	F	T	F	T	T	F	T
F	F	F	T	T	T	T	T
F	F	F	F	T	T	F	T

VALID

8.

a	h	b	Na	Ehb	Chb	Cbh	KChbCbh	CaEhb	NCaEhb	KNaKChbCbh	∴ Cbh
T	T	T	F	T	T	T	T	T	F	F	T
T	T	F	F	F	F	T	F	F	T	F	T
T	F	T	F	F	T	F	F	F	T	F	F
T	F	F	F	T	T	T	T	T	F	F	T
F	T	T	T	T	T	T	T	T	F	T	T
F	T	F	T	F	F	T	F	T	F	F	T
F	F	T	T	F	T	F	F	T	F	F	F
F	F	F	T	T	T	T	T	T	F	T	T

VALID

10.

p	q	r	s	Cpq	Crs	Apr	Aqs	KCpqCrs	Apr	∴ Aqs
T	T	T	T	T	T	T	T	T	T	T
T	T	T	F	T	F	T	T	F	T	T
T	T	F	T	T	T	T	T	T	T	T
T	T	F	F	T	T	T	T	T	T	T
T	F	T	T	F	T	T	T	F	T	T
T	F	T	F	F	F	T	F	F	T	F
T	F	F	T	F	T	T	T	F	T	T
T	F	F	F	F	T	T	F	F	T	F
F	T	T	T	T	T	T	T	T	T	T
F	T	T	F	T	F	T	T	F	T	T
F	T	F	T	T	T	F	T	T	F	T
F	T	F	F	T	T	F	T	T	F	T
F	F	T	T	T	T	T	T	T	T	T
F	F	T	F	T	F	T	F	F	T	F
F	F	F	T	T	T	F	T	T	F	T
F	F	F	F	T	T	F	F	T	F	F

VALID

12.

p	q	Cpq	Cqp	Epq	KCpqCqp	∴ Epq
T	T	T	T	T	T	T
T	F	F	T	F	F	F
F	T	T	F	F	F	F
F	F	T	T	T	T	T

VALID

14.

w	g	∴Awg
T	T	T
T	F	T
F	T	T
F	F	F

VALID

Exercise 2, page 312

2.

p	Np	ApNp
T	F	T
F	T	T

TAUTOLOGY

4.

p	q	Cpq	CCpqp	CCCpqpp
T	T	T	T	T
T	F	F	T	T
F	T	T	F	T
F	F	T	F	T

TAUTOLOGY (This is called "Peirce's Law.")

6.

p	q	Cpq	CCpqp	CCCpqpp	NCCCpqpp
T	T	T	T	T	F
T	F	F	T	T	F
F	T	T	F	T	F
F	F	T	F	T	F

CONTRADICTION

8.

p	q	Np	Cpq	CNpCpq
T	T	F	T	T
T	F	F	F	T
F	T	T	T	T
F	F	T	T	T

TAUTOLOGY

10.

p	s	q	r	Kps	Cqp	Apr	ECqpApr	AKpsECqpApr
T	T	T	T	T	T	T	T	T
T	T	T	F	T	T	T	T	T
T	T	F	T	T	T	T	T	T
T	T	F	F	T	T	T	T	T
T	F	T	T	F	T	T	T	T
T	F	T	F	F	T	T	T	T
T	F	F	T	F	T	T	T	T
T	F	F	F	F	T	T	T	T
F	T	T	T	F	F	T	F	F
F	T	T	F	F	F	F	T	T
F	T	F	T	F	T	T	T	T
F	T	F	F	F	T	F	F	F
F	F	T	T	F	F	T	F	F
F	F	T	F	F	F	F	T	T
F	F	F	T	F	T	T	T	T
F	F	F	F	F	T	F	F	F

CONTINGENT

MODULE 27

Exercise 1, pages 327–328

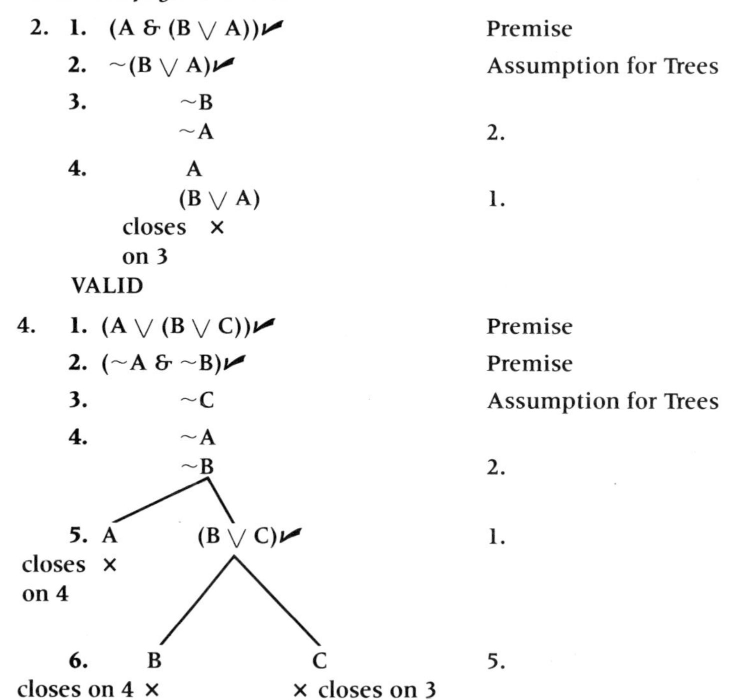

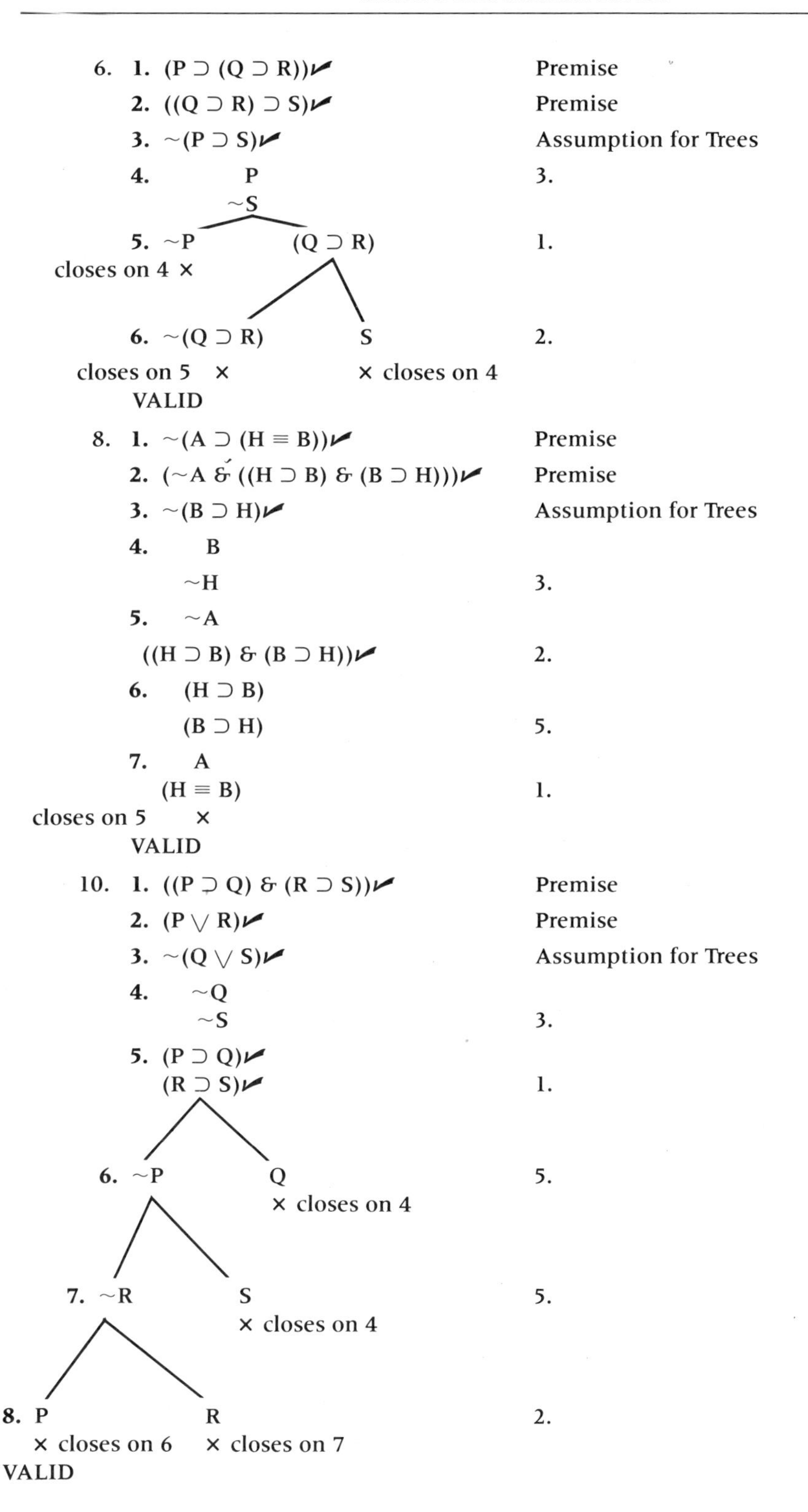
6. 1. (P ⊃ (Q ⊃ R))✓ Premise
2. ((Q ⊃ R) ⊃ S)✓ Premise
3. ~(P ⊃ S)✓ Assumption for Trees
4. P
~S 3.
5. ~P (Q ⊃ R) 1.
closes on 4 ×
6. ~(Q ⊃ R) S 2.
closes on 5 × × closes on 4
VALID
8. 1. ~(A ⊃ (H ≡ B))✓ Premise
2. (~A & ((H ⊃ B) & (B ⊃ H)))✓ Premise
3. ~(B ⊃ H)✓ Assumption for Trees
4. B
~H 3.
5. ~A
((H ⊃ B) & (B ⊃ H))✓ 2.
6. (H ⊃ B)
(B ⊃ H) 5.
7. A
(H ≡ B) 1.
closes on 5 ×
VALID
10. 1. ((P ⊃ Q) & (R ⊃ S))✓ Premise
2. (P ∨ R)✓ Premise
3. ~(Q ∨ S)✓ Assumption for Trees
4. ~Q
~S 3.
5. (P ⊃ Q)✓
(R ⊃ S)✓ 1.
6. ~P Q 5.
× closes on 4
7. ~R S 5.
× closes on 4
8. P R 2.
× closes on 6 × closes on 7
VALID

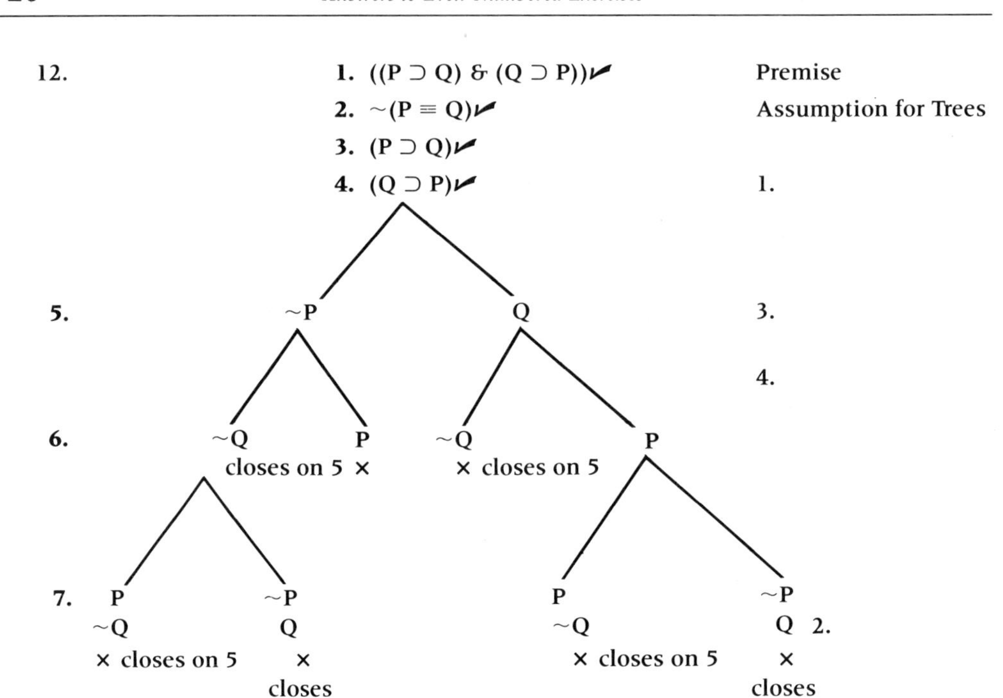

VALID

14. **1.** W Premise

2. ~(W ∨ G)✓ Assumption for Trees

3. ~W

~G 2.

closes on 1 ×

VALID

Exercise 2, page 328

2. **1.** ~(P ∨ ~P)✓ Assumption for Tautology

2. ~P

~~P 1.

closes on 1 ×

TAUTOLOGY

4. **1.** ~(((P ⊃ Q) ⊃ P) ⊃ P)✓ Assumption for Tautology

2. ((P ⊃ Q) ⊃ P)✓

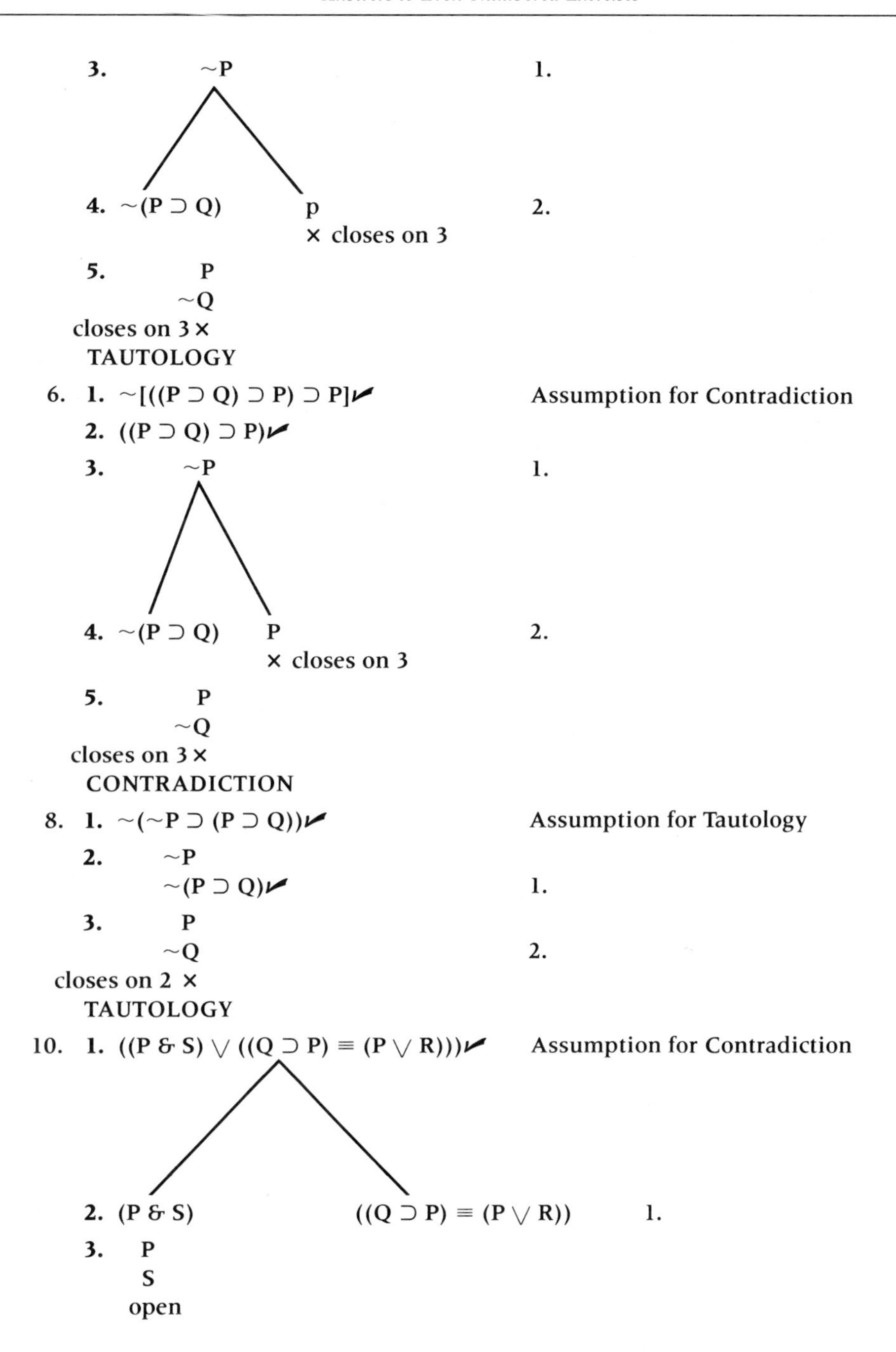

There is no need to work the right side of the branch since the left side is open—thus, this statement is *not* a contradiction. Now we need to work this statement on the "Assumption for Tautology," as below:

1. ~((P & S) ∨ ((Q ⊃ P) ≡ (P ∨ R)))✓ Assumption for Tautology

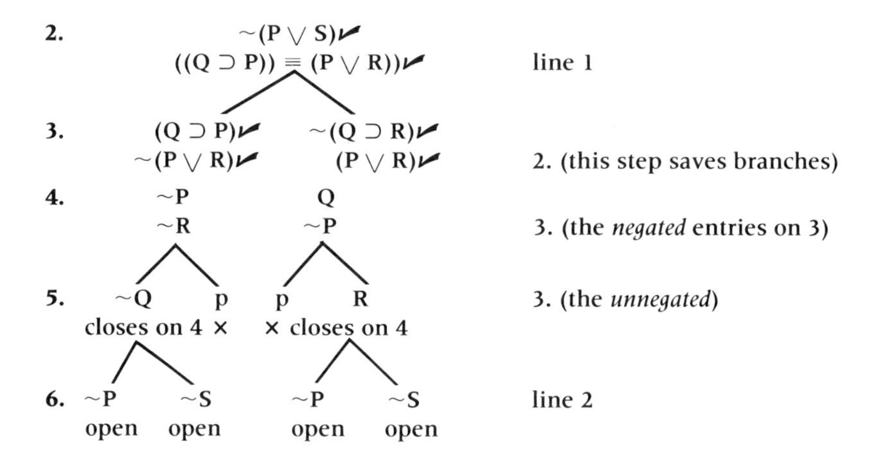

Since working a tree for each assumption leaves open branches, this statement is CONTINGENT.

Exercise 4, pages 329–330

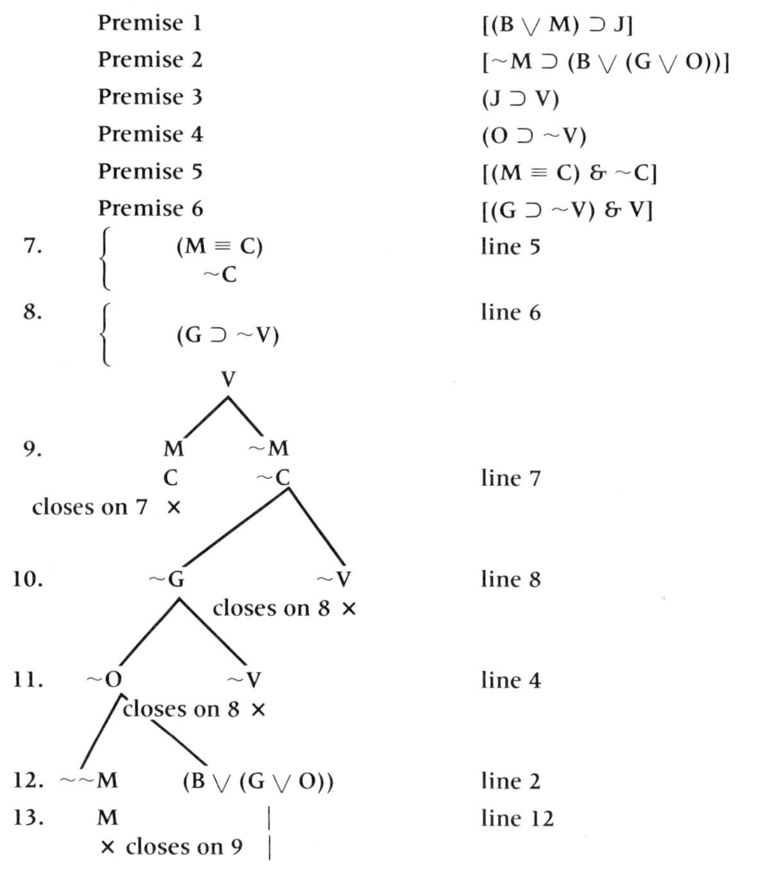

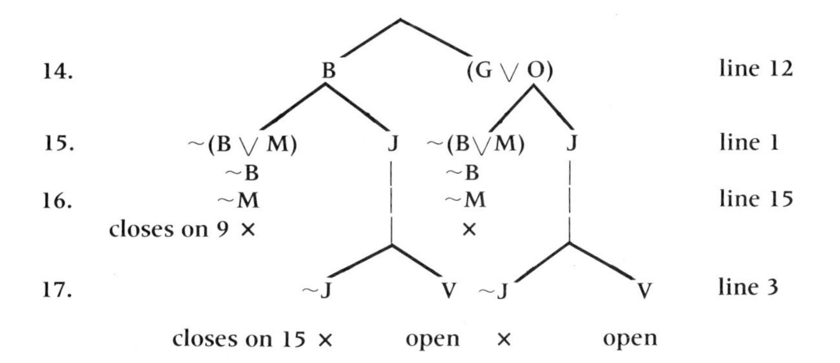

Conclusions:

'V' is true, 'J' is true, and 'B' is true.

Therefore, Bluebeard is in the vicinity, Jean Lafitt is drunk, and the treasure is buried on Barbados.

MODULE 28.2, *page 336*

2. $\sim\sim(\sim P \,\&\, \sim(Q \vee R))$
   $\therefore (\sim P \,\&\, \sim(Q \vee R))$

4. $(\sim P \,\&\, \sim(Q \vee R))$
   $(S \equiv T)$
   $\therefore ((\sim P \,\&\, \sim(Q \vee R)) \,\&\, (S \equiv T))$

6. $((\sim P \,\&\, \sim(Q \vee R)) \supset (S \equiv T))$
   $((S \equiv T) \supset \sim P)$
   $\therefore ((\sim P \,\&\, \sim(Q \vee R)) \supset \sim P)$

8. $\sim((\sim P \,\&\, \sim(Q \vee R)) \,\&\, (S \equiv T))$
   $\therefore (\sim(\sim P \,\&\, \sim(Q \vee R)) \vee (\sim(S \equiv T))$

10. $(\sim P \,\&\, \sim(Q \vee R))$
    $\therefore ((\sim P \,\&\, \sim(Q \vee R)) \vee (\sim P \,\&\, \sim(Q \vee R)))$

MODULE 29

Exercise 1, pages 349, 351

2. **DEDUCE:** $\sim R$

1.	$((Q \vee R) \supset \sim(A \,\&\, B))$	Premise
2.	$((\sim A \vee \sim B) \supset M)$	Premise
3.	$\sim M$	Premise
4.	$(\sim(A \,\&\, B) \supset M)$	2, DeM
5.	$((Q \vee R) \supset M)$	1, 4 HS
6.	$(\sim M \supset \sim(Q \vee R))$	5, Trans
7.	$\sim(Q \vee R)$	6, 3 MP
8.	$(\sim Q \,\&\, \sim R)$	7, DeM
9.	$(\sim R \,\&\, \sim Q)$	8, Com
10.	$\sim R$	9, Simp

Exercise 2, pages 352

2. DEDUCE: (A & (B & C))

1.	(A & B)	Premise
2.	(B ⊃ C)	Premise
3.	(B & C)	1, Com
4.	B	3, Simp
5.	C	2, 4 MP
6.	((A & B) & C)	1, 5 Conj
7.	(A & (B & C))	6, Assoc

4. DEDUCE: (T ∨ W)

1.	P	Premise
2.	(Q & S)	Premise
3.	((Q & P) ⊃ T)	Premise
4.	Q	2, Simp
5.	(Q & P)	1, 4 Conj
6.	T	3, 5 MP
7.	(T ∨ W)	6, Add

MODULE 30

Exercise 1, page 366

2. DEDUCE: ~B

1.	(~(M ⊃ S) & (B ⊃ L))	Premise
2.	(L ⊃ S)	Premise
3.	~~B	Assumption for IP
4.	~(M ⊃ S)	1, Simp
5.	((B ⊃ L) & ~(M ⊃ S))	1, Com
6.	(B ⊃ L)	5, Simp
7.	(~B ∨ L)	6, Imp
8.	L	3, 7 DS
9.	S	2, 8 MP
10.	~(~M ∨ S)	4, Imp
11.	(~~M & ~S)	10, DeM
12.	(~S & ~~M)	11, Com
13.	~S	12, Simp
14.	~B	3-13, IP

4. DEDUCE: R

1.	$(P \vee (P \& Q))$	Premise
2.	$(P \supset R)$	Premise
3.	$\sim R$	Assumption for IP
4.	$\sim P$	2, 3 MT
5.	$(P \& Q)$	1, 4 DS
6.	P	5, Simp
7.	R	3-6, IP

Exercise 2, page 366

2. DEDUCE: $(A \supset (B \supset C))$

1.	$((A \& B) \supset C)$	Premise
2.	A	Assumption for CP
3.	$(A \supset (B \supset C))$	1, Exp
4.	$(B \supset C)$	2, 3 MP
5.	$(A \supset (B \supset C))$	2-4, CP

4. DEDUCE: $(P \supset G)$

1.	$((\sim G \vee \sim B) \supset \sim P)$	Premise
2.	P	Assumption for CP
3.	$\sim\sim P$	2, DN
4.	$\sim(\sim G \vee \sim B)$	1, 3 MT
5.	$(\sim\sim G \& \sim\sim B)$	4, DeM
6.	$\sim\sim G$	5, Simp
7.	G	6, DN
8.	$(P \supset G)$	2-7, CP

Exercise 3, page 366

2. DEDUCE: $((\sim Q \supset \sim P) \supset (P \supset Q))$

1.	$(\sim Q \supset \sim P)$	Assumption for CP
2.	$(\sim\sim Q \vee \sim P)$	1, Imp
3.	$(Q \vee \sim P)$	2, DN
4.	$(\sim P \vee Q)$	3, Com
5.	$(P \supset Q)$	4, Imp
6.	$((\sim Q \supset \sim P) \supset (P \supset Q))$	1-5, CP

4. DEDUCE: $((P \supset (Q \supset R)) \supset ((P \supset Q) \supset (P \supset R)))$

1.	$(P \supset (Q \supset R))$	Assumption for CP
2.	$(P \supset Q)$	Assumption for CP
3.	P	Assumption for CP
4.	$(Q \supset R)$	1,3 MP
5.	Q	2,3 MP
6.	R	4,5 MP
7.	$(P \supset R)$	3-6, CP
8.	$((P \supset Q) \supset (P \supset R))$	2-7, CP
9.	$((P \supset (Q \supset R)) \supset ((P \supset Q) \supset (P \supset R)))$	1-8, CP

Exercise 4, pages 366–368

2. DEDUCE: C

1.	$(A \supset C)$	Premise
2.	$((C \supset B) \supset A)$	Premise
3.	$\sim C$	Assumption for IP
4.	$\sim A$	1,3 MT
5.	$\sim(C \supset B)$	2, 4 MT
6.	$\sim(\sim C \vee B)$	5, Imp
7.	$(\sim\sim C \,\&\, \sim B)$	6, DeM
8.	$\sim\sim C$	7, Simp
9.	C	3-8, IP

4. DEDUCE: $\sim I$

1.	$(I \supset \sim E)$	Premise
2.	$(E \vee C)$	Premise
3.	$\sim C$	Premise
4.	$\sim\sim I$	Assumption for IP
5.	I	4, DN
6.	E	2,3 DS
7.	$\sim\sim E$	6, DN
8.	$\sim I$	1, 7 MT
9.	$\sim I$	4-8, IP

6. DEDUCE: $(\sim M \supset T)$

1.	$(L \supset (T \vee M))$	Premise
2.	$(\sim M \supset L)$	Premise
3.	$\sim(\sim M \supset T)$	Assumption for IP
4.	$\sim(\sim\sim M \vee T)$	3, Imp
5.	$\sim(M \vee T)$	4, DN
6.	$(\sim M \& \sim T)$	5, DeM
7.	$\sim M$	6, Simp
8.	L	2, 7 MP
9.	$(T \vee M)$	1, 8 MP
10.	T	7, 9 DS
11.	$(\sim T \& \sim M)$	6, Com
12.	$\sim T$	11, Simp
13.	$(\sim M \supset T)$	3-12, IP

8. DEDUCE: T

1.	$(M \supset (S \supset T))$	Premise
2.	$(M \vee (S \supset T))$	Premise
3.	S	Premise
4.	$\sim T$	Assumption for IP
5.	$(\sim T \& S)$	3, 4 Conj
6.	$(\sim T \& \sim\sim S)$	5, DN
7.	$\sim(T \vee \sim S)$	6, DeM
8.	$\sim(\sim S \vee T)$	7, Com
9.	$\sim(S \supset T)$	8, Imp
10.	$\sim M$	1, 9 MT
11.	$(S \supset T)$	2, 10 DS
12.	T	4-11, IP

10. DEDUCE: W

1.	((~B & W) ∨ R)	Premise
2.	(~W ⊃ (B & R))	Premise
3.	~W	Assumption for IP
4.	(B & R)	2, 3 MP
5.	B	4, Simp
6.	~~B	5, DN
7.	(~~B ∨ ~W)	6, Add
8.	~(~B & W)	7, DeM
9.	~R	1, 8 DS
10.	(R & B)	4, Com
11.	R	10, Simp
12.	W	3-11, IP

12. DEDUCE: ((M & ~K) ⊃ (B ⊃ S))

1.	(M ⊃ B)	Premise
2.	(~S ⊃ K)	Premise
3.	~((M & ~K) ⊃ (B ⊃ S))	Assumption for IP
4.	~(~(M & ~K) ⊃ (B ⊃ S))	3, Imp
5.	(~~(M & ~K) & ~(B ⊃ S))	4, DeM
6.	~~(M & ~K)	5, Simp
7.	(M & ~K)	6, DN
8.	(~K & M)	7, Com
9.	~K	8, Simp
10.	~~S	2, 9 MT
11.	M	7, Simp
12.	B	1, 12 MP
13.	(~(B ⊃ S) & ~~(M & ~K))	5, Com
14.	~(B ⊃ S)	13, Simp
15.	~(~B ∨ S)	14, Imp
16.	(~~B & ~S)	15, DeM
17.	(~S & ~~B)	16, Com
18.	~S	17, Simp
19.	((M & ~K) ⊃ (B ⊃ S))	3-18, IP

14. DEDUCE: (~M ∨ ~T)

1.	(P ⊃ C)	Premise
2.	(~P ⊃ G)	Premise
3.	(G ⊃ ~T)	Premise
4.	(M ⊃ ~C)	Premise
5.	~(~M ∨ ~T)	Assumption for IP
6.	(~~M & ~~T)	5, DeM
7.	~~M	6, Simp
8.	M	7, DN
9.	(~~T & ~~M)	6, Com
10.	~~T	9, Simp
11.	~G	3, 10 MT
12.	~~P	2, 11 MT
13.	P	12, DN
14.	C	1, 13 MP
15.	~C	4, 8 MP
17.	(~M ∨ ~T)	5-15, IP

16. DEDUCE: ~L

1.	(S ≡ R)	Premise
2.	(L ⊃ S)	Premise
3.	((~S ∨ C) & ~R)	Premise
4.	~~L	Assumption for IP
5.	L	4, DN
6.	S	2, 5 MP
7.	((S ⊃ R) & (R ⊃ S))	1, Equiv
8.	(S ⊃ R)	8, Simp
9.	R	6, 8 MP
10.	(~R & (~S ∨ C))	3, Com
11.	~R	10, Simp
12.	~L	4-11, IP

MODULE 31.3, *pages 376–377*

2. ~(∃x)(Mx & Ix)   4. ~(∀x)(Mx ⊃ Px)   6. (∀x)(Hx ⊃ Lx)

8. (∀x)(Mx ⊃ Ex)   10. (∀x)(Mx ⊃ Fx)

12. Grammatically ambiguous. Conventions tell us to treat it as: ~(∃x)(Gx & Hx); but what is *meant* is: ~(∀x)(Gx ⊃ Hx).

14. (∃x)(Fx & (Ex & Sx))   16. ~(∃x)(Ix ⊃ Ex)   18. ~(∀x)(Mx ⊃ Gx)

20. ~(∃x)(Nx & Sx)   22. ~(∃x)~Lx   24. ~(∃x)~((Px & Cx) ⊃ Fx))

MODULE 31, *pages 384–385*

2. Ax: X is a backpacker
   Ex: x is bearbait
   Rx: x is a berry

$(\exists x)(Ax \,\&\, Ex)$
$\sim(\forall x)(Ex \supset Rx)$
$\therefore \sim(\forall x)(Ax \supset Rx)$

1.	$(\exists x)(Ax \,\&\, Ex)$✓	Premise
2.	$\sim(\forall x)(Ex \supset Rx)$✓	Premise
3.	$\sim\sim(\forall x)(Ax \supset Rx)$✓	Assumption for Trees
4.	$(Ax)(Ax \supset Rx)$✓	3
5.	$(Aa \,\&\, Ea)$✓	1
6.	$\sim(Eb \supset Rb)$✓	2
7.	$(Aa \supset Ra)$✓	4
8.	Aa Ea	5
9.	Eb $\sim Rb$	6
10.	$\sim Aa$ (x, 8) / Rb (open)	7

INVALID

4. Mx: x is a meander
   Fx: x is a fret
   Ox: x is an ornamental border

$(\exists x)(Mx \,\&\, Fx)$
$(\forall x)(Fx \supset Ox)$
$\therefore (\forall x)(Mx \supset Ox)$

1.	$(\exists x)(Mx \,\&\, Fx)$✓	Premise
2.	$(\forall x)(Fx \supset Ox)$✓	Premise
3.	$\sim(\forall x)(Mx \supset Ox)$✓	Assumption for Trees
4.	$(Ma \,\&\, Fa)$✓	1
5.	$(Fa \supset Oa)$✓	2
6.	$\sim(Mb \supset Ob)$✓	3
7.	Ma	
8.	Fa	4
9.	Mb $\sim Ob$	6
10.	$\sim Fa$ (x, 8) / Oa (open)	5

INVALID

6. Dx: x is a drove
Hx: x is a herd
Px: x is a pack

$(\forall x)(Dx \supset Hx)$
$\sim(\forall x)(Hx \supset Px)$
$\therefore \sim(\forall x)(Dx \supset Px)$

1.	$(\forall x)(Dx \supset Hx)$ ✓	**Premise**
2.	$\sim(\forall x)(Hx \supset Px)$ ✓	**Premise**
3.	$\sim\sim(\forall x)(Dx \supset Px)$ ✓	**Assumption for Trees**
4.	$(\forall x)(Dx \supset Px)$	**3**
5.	$\sim(Ha \supset Pa)$ ✓	**2**
6.	$(Da \supset Ha)$ ✓	**1**
7.	$(Da \supset Pa)$ ✓	**4**
8.	Ha ~Pa	**5**
9.	~Da / Pa (Pa: x 8)	**7**
10.	(under ~Da) ~Da open / Ha open	**6**

INVALID

8. Sx: x is a scorpion
Ax: x is an arachnid
Vx: x is a thing with a venomous sting

$(\forall x)(Sx \supset Ax)$
$(\exists x)(Ax \,\&\, Vx)$
$\therefore \sim(\forall x)(Sx \supset Vx)$

1.	$(\forall x)(Sx \supset Ax)$ ✓	**Premise**
2.	$(\exists x)(Ax \,\&\, Vx)$ ✓	**Premise**
3.	$\sim\sim(\forall x)(Sx \supset Vx)$ ✓	**Assumption for Trees**
4.	$(\forall x)(Sx \supset Vx)$ ✓	**3**
5.	$(Aa \,\&\, Va)$ ✓	**2**
6.	$(Sa \supset Aa)$ ✓	**1**
7.	$(Sa \supset Va)$ ✓	**4**
8.	Aa Va	**5**
9.	~Sa / Va	**6**
10.	(under ~Sa) ~Sa open / Aa open; (under Va) ~Sa open / Aa open	**5**

INVALID

10. Wx: x is a wisp
Sx: x is a snipe
Bx: x is a bevy

$(\forall x)(Wx \supset Sx)$
$(\exists x)(Bx \,\&\, Sx)$
$\therefore (\exists x)(Wx \,\&\, Bx)$

1.	$(\forall x)(Wx \supset Sx)$✔	Premise
2.	$(\exists x)(Bx \,\&\, Sx)$✔	Premise
3.	$\sim(\exists x)(Wx \,\&\, Bx)$✔	Assumption for Trees
4.	$(Ba \,\&\, Sa)$✔	2
5.	$(Wa \supset Sa)$✔	1
6.	$\sim(Wa \,\&\, Ba)$✔	3
7.	Ba Sa	4
8.	~Wa / ~Ba (x 7)	6
9.	~Wa (open) / Sa (open) [under ~Wa]	5

INVALID

12. Gx: x is a gam
Sx: x is a school
Hx: x is a shoal

$(\forall x)(Gx \supset Sx)$
$(\exists x)(Gx \,\&\, Hx)$
$\therefore (\exists x)(Sx \,\&\, Hx)$

1.	$(\forall x)(Gx \supset Sx)$✔	Premise
2.	$(\exists x)(Gx \,\&\, Hx)$✔	Premise
3.	$\sim(\exists x)(Sx \,\&\, Hx)$✔	Assumption for Trees
4.	$(Ga \,\&\, Ha)$✔	2
5.	$\sim(Sa \,\&\, Ha)$✔	3
6.	$(Ga \supset Sa)$✔	1
7.	Ga Ha	4
8.	~Ga (x 7) / Sa	6
	~Sa (x 8) / ~Ha (x 7) [under Sa]	5

VALID

14. Dx: x is a drift — $(\forall x)(Dx \supset Rx)$
Rx: x is a drove — $(\exists x)(Dx \,\&\, Hx)$
Hx: x is a hog — $\therefore (\exists x)(Rx \,\&\, Hx)$

1.	**$(\forall x)(Dx \supset Rx)$✔**	Premise
2.	**$(\exists x)(Dx \,\&\, Hx)$✔**	Premise
3.	**$\sim(\exists x)(Rx \,\&\, Hx)$✔**	Assumption for Trees
4.	**$(Da \,\&\, Ha)$✔**	2
5.	**$\sim(Ra \,\&\, Ha)$✔**	3
6.	**$(Da \supset Ra)$✔**	1
7.	**Da**	
	Ha	4
8.	**~Da** (x, 7) / **Ra**	6
	under Ra: **~Ra** (x, 8) / **~Ha** (x, 7)	5

VALID

MODULE 32

Exercise 1, page 404

2. " . . . polyadic quantification has two main benefits: (1) deductions that are very complicated and confusing in natural language very often can be worked more easily in polyadic symbolic form; and (2) symbolization, now as before, tends to resolve ambiguous expressions of natural language and give them precise formulation."

Exercise 2, pages 404–405

2. **Dx: x is a drowning man**
   Sx: x is a straw
   Cxy: x will catch at y
   $(\forall x)[Dx \supset (\forall y)(Sy \supset Cxy)]$
4. **Ox: x is a person**
   Cx: x is company
   Kxy: x is known by y
   $(\forall x)(Ox \supset (\exists y)(Cy \,\&\, Kxy))$
6. **Ox: x is a person**
   Wx: x is what one wishes
   Bxy: x believes y
   $(\forall x)(Ox \supset (\forall y)(Wy \supset Bxy))$
8. **Wx: x is a wish**
   Hx: x is a horse
   Bx: x is a beggar
   Rx: x is a rider
   $(\forall x)(Wx \supset Hx) \supset (\forall y)(By \supset Ry)$

10. Tx: x is a thing
    Wx: x is willing
    Mx: x is a mind
    Dxy: x is difficult for y
    $\sim(\exists x)(Tx \,\&\, (\forall y)((Wy \,\&\, My) \supset (Dxy))$

12. Bx: x is a bow
    Lx: x is long bent
    Sx: x is a thing's spring
    Oxy: x loses y
    $(\forall x)((Bx \,\&\, Lx) \supset (\exists y)(Sy \,\&\, Oxy))$

14. Lx: x is little
    Sx: x is stroke
    Gx: x is great
    Ox: x is an oak
    Fxy: x fells y
    $(\forall x)((Lx \,\&\, Sx) \supset (\exists y)((Gy \,\&\, Oy) \,\&\, Fxy))$

16. Hx: x is hope
    Dx: x is deferred
    Ex: x is a heart
    Sxy: x makes y sick
    $(\forall x)[(Hx \,\&\, Dx) \supset (\forall y)(Ey \supset Sxy)]$

18. Cx: x is a cap
    Px: x is a person
    Fxy: x fits y
    Wxy: x wears y
    $(\forall x)[Cx \supset (\forall y)((Py \,\&\, Fxy) \supset Wyx)]$

20. Px: x is a person
    a: some particular "lucky" person
    n: the Nile
    Ix: x is a fish
    Fxyz: x flings y in z
    Cxy: x comes up with y
    $(\forall x)[((Px \,\&\, Pa) \,\&\, Fxan) \supset (\exists y)(Iy \,\&\, Cay)]$

22. Px: x is psychology
    Ix: x is idleness
    Bxy: x is the beginning of y
    $(\forall x)(Px \supset (\exists y)(Iy \,\&\, Byx))$

24. Mx: x is a person
    Ex: x is an Englishman
    Sxy: x strives after y
    h: happiness
    $\sim(\forall x)[(Mx \,\&\, Sxh) \,\&\, (\exists y)((My \,\&\, Syh) \supset Ey)]$

Exercise 3, pages 405–406

2. Sx: x is a swarm
   Cx: x is a covey
   Bx: x is a bevy

   $\sim(\exists x)(Sx \,\&\, Cx)$
   $(\exists x)(Cx \,\&\, Bx)$
   $\therefore \sim(\forall x)(Bx \supset Sx)$

DEDUCE: $\sim(\forall x)(Bx \supset Sx)$

1.	$\sim(\exists x)(Sx \,\&\, Cx)$	Premise
2.	$(\exists x)(Cx \,\&\, Bx)$	Premise
3.	$(Ca \,\&\, Ba)$	2, EI
4.	$(\forall x)\sim(Sx \,\&\, Cx)$	1, QN
5.	$\sim(Sa \,\&\, Ca)$	4, UI
6.	$(\sim Sa \vee \sim Ca)$	5, DeM
7.	Ca	3, Simp
8.	$\sim\sim Ca$	7, DN
9.	$\sim Sa$	6, 8 DS
10.	$(Ba \,\&\, Ca)$	3, Com
11.	Ba	10, Simp
12.	$(Ba \,\&\, Sa)$	8, 11 Conj
13.	$\sim(\sim Ba \vee \sim\sim Sa)$	12, DeM
14.	$\sim(\sim\sim Ba \supset \sim\sim Sa)$	13, Imp
15.	$\sim(Ba \supset Sa)$	14, DN
16.	$(\exists x)\sim(Bx \supset Sx)$	15, EG
17.	$\sim(\forall x)(Bx \supset Sx)$	16, QN

4. Fx: x is a fall
   Cx: x is a covey
   Nx: x is a nide

$(\forall x)(Fx \supset Cx)$
$\sim(\exists x)(Cx \,\&\, Nx)$
$\therefore \sim(\exists x)(Nx \,\&\, Fx)$

DEDUCE: $\sim(\exists x)(Nx \,\&\, Fx)$

1.	$(\forall x)(Fx \supset Cx)$	Premise
2.	$\sim(\exists x)(Cx \,\&\, Nx)$	Premise
3.	$(\forall x)\sim(Cx \,\&\, Nx)$	2, QN
4.	$(Fa^1 \supset Ca^1)$	1, UI
5.	$\sim(Ca^1 \,\&\, Na^1)$	3, UI
6.	$(\sim Fa^1 \vee Ca^1)$	4, Imp
7.	$(\sim Ca^1 \vee \sim Na^1$	5, DeM
8.	$(\sim\sim Ca^1 \supset \sim Na^1)$	7, DeM
9.	$(Ca^1 \supset \sim Na^1)$	8, DN
10.	$(Fa^1 \supset \sim Na^1)$	4, 9 HS
11.	$(\sim Fa^1 \vee \sim Na^1)$	10, Imp
12.	$\sim(Fa^1 \,\&\, Na^1)$	11, DeM
13.	$(\forall x)\sim(Fx \,\&\, Nx)$	12, UG
14.	$\sim(\exists x)(Fx \,\&\, Nx)$	13, QN
15.	$\sim(\exists x)(Nx \,\&\, Fx)$	14, Com

6. Ex: x is an earth bike
   Sx: x is a speedy bike
   Tx: x is a ten-speed bike

   $(\forall x)(Ex \supset Sx)$
   $\sim(\exists x)(Sx \,\&\, Tx)$
   $\therefore \sim(\exists x)(Tx \,\&\, Ex)$

DEDUCE: $\sim(\exists x)(Tx \,\&\, Ex)$

1.	$(\forall x)(Ex \supset Sx)$	Premise
2.	$\sim(\exists x)(Sx \,\&\, Tx)$	Premise
3.	$(\forall x)\sim(Sx \,\&\, Tx)$	2, QN
4.	$\sim(Sa^1 \,\&\, Ta^1)$	3, UI
5.	$(Ea^1 \supset Sa^1)$	1, UI
6.	$(\sim Sa^1 \vee \sim Ta^1)$	4, DeM
7.	$(\sim\sim Sa^1 \supset \sim Ta^1)$	6, Imp
8.	$(Sa^1 \supset \sim Ta^1)$	7, DN
9.	$(Ea^1 \supset \sim Ta^1)$	5, 8 HS
10.	$(\sim Ea^1 \vee \sim Ta^1)$	9, Imp
11.	$(\sim Ta^1 \vee \sim Ea^1)$	10, Com
12.	$\sim(Ta^1 \,\&\, Ea^1)$	11, DeM
13.	$(\forall x)\sim(Tx \,\&\, Ex)$	12, UG
14.	$\sim(\exists x)(Tx \,\&\, Ex)$	13, QN

8. Hx: x is a hurricane
   Wx: x wreaked havoc in North Carolina
   Dx: x is potentially dangerous

   $(\exists x)(Hx \,\&\, Wx)$
   $(\forall x)(Hx \supset Dx)$
   $\therefore (\exists x)(Dx \,\&\, Wx)$

DEDUCE: $(\exists x)(Dx \,\&\, Wx)$

1.	$(\exists x)(Hx \,\&\, Wx)$	Premise
2.	$(\forall x)(Hx \supset Dx)$	Premise
3.	$(Ha \,\&\, Wa)$	1, EI
4.	$(Ha \supset Da)$	2, UI
5.	Ha	3, Simp
6.	Da	4,5 MP
7.	$(Wa \,\&\, Ha)$	3, Com
8.	Wa	7, Simp
9.	$(Da \,\&\, Wa)$	6, 8 Conj
10.	$(\exists x)(Dx \,\&\, Wx)$	9, EG

EXERCISES FOR UNIT THREE

Exercise 1, pages 409–410

2. Let 'M' = "New Wave Manhattan model Margaret (played by Ann Carlisle) serves as the primary attraction for a U.F.O. which lands atop her penthouse."

   Let 'H' = "She has heroin in her apartment."

Let 'S' = "Her sexual encounters provide the chemical nourishment that the aliens need."

Let 'R' = "The film was made by ex-patriot Russian immigrants who were denied such pleasures in their arid native land."

(M ≡ (H & S))

((H & S) ⊃ R)

R

∴ M

M	H	S	R	(H & S)	(M ≡ (H & S))	((H & S) ⊃ R)	R	∴ M	
T	T	T	T	T	T	T	T	T	
T	T	T	F	T	T	F	F	T	
T	T	F	T	F	F	T	T	T	
T	T	F	F	F	F	T	F	T	
T	F	T	T	F	F	T	T	T	
T	F	T	F	F	F	T	F	T	
T	F	F	T	F	F	T	T	T	
T	F	F	F	F	F	T	F	T	
F	T	T	T	T	F	T	T	F	
F	T	T	F	T	F	F	F	F	
F	T	F	T	F	T	T	T	F	✓
F	T	F	F	F	T	T	F	F	
F	F	T	T	F	T	T	T	F	✓
F	F	T	F	F	T	T	F	F	
F	F	F	T	F	T	T	T	F	✓
F	F	F	F	F	T	T	F	F	
M	H	S	R	(H & S)	(M ≡ (H & S))	((H & S) ⊃ R)	R	∴ M	

INVALID, Rows 11, 13, and 15

Exercise 2, page 410

2. (M ≡ (H & S))
   ((H & S) ⊃ R)
   R
   ∴ M

1.	(M ≡ (H & S))✓	Premise
2.	((H & S) ⊃ R)✓	Premise
3.	R	Premise

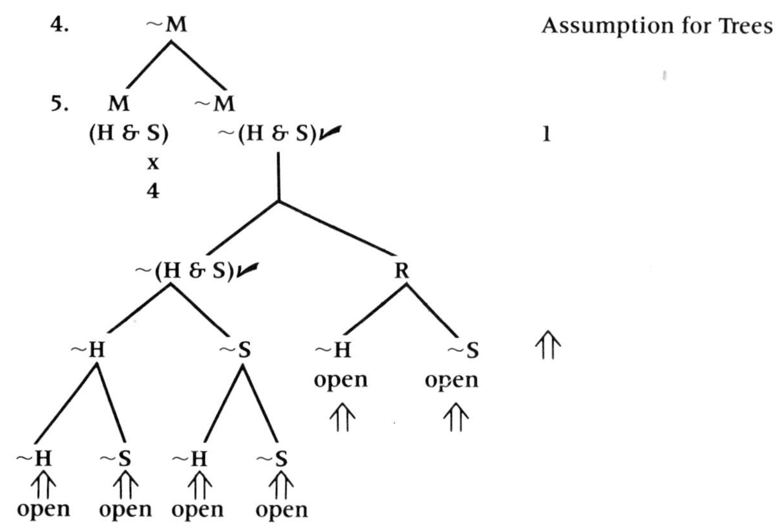

Reclaiming truth values:

The first open branch: ~H; ~M; R.
Thus, H = FALSE; M = FALSE; R = TRUE.

The second open branch: ~S, ~H, ~M, R.
Thus, S = FALSE; H = FALSE; M = FALSE; R = TRUE.

The third open branch is identical to the second.

The fourth open branch: ~S, ~M, R.
Thus, S = FALSE; M = FALSE; R = TRUE

The fifth open branch: ~H, ~S, R, ~M.

The fifth open branch is identical to the second.

The sixth open branch: ~S, R, ~M.
The sixth open branch is identical to the fourth.

Thus, the argument is invalid on rows 11, 13, and 15 of the truth table (where the rows have the shown values for the atomic statement letters traced from the open branches).

UNIT FOUR

MODULE 34

Exercise 2, pages 439–441

2. Appeal to Authority (the Bible).

4. Appeal to Ignorance (no complaints means no evidence one way or the other).

6. Appeal to Titles (Ph.D., D.D.S., M.L.S.).

8. *Ad Hominem.*

10. No Fallacy (good luck, Green Bay!).

12. Appeal to Popular Sentiment (the vote won't decide it).

14. Appeal to Fear.

16. *Tu Quoque* (you're one of them!).

18. Appeal to Popular People.

20. Poisoning the Well.

22. Appeal to Tradition.

24. Appeal to Force.

Exercise 3, pages 441–443

2. Appeal to Ignorance.

4. Appeal to Popular People.

6. *Ad Hominem.*

8. Appeal to Sympathy.

10. Appeal to Ignorance.

12. No Fallacy.

MODULE 35

Exercise 1, pages 458–459

2. No Fallacy.

4. *Post Hoc, Ergo Propter Hoc* (the timing was surely coincidence).

6. False Analogy (a government is only loosely like a ship).

8. Straw Man (this statement terribly mischaracterizes the Koran).

10. False Dilemma (there are other ways to make oneself heard).

12. Slippery Slope (it's a huge leap from watching television to such perdition!).

14. The Bureaucratic Fallacy ("if I make an exception for you . . . ").

16. Appeal to Superstition (a rabbit's foot has no effect on the turn of the cards).

Exercise 2, pages 459–461

2. *Post Hoc, Ergo Propter Hoc* (the dog's agitation doesn't establish the conclusion that it knew a disaster was about to happen).

4. False Analogy (compulsory seatbelts are a long way from Orwell's "Big Brother").

6. No Fallacy (this is a reasonable argument).

MODULE 36

Exercise 1, pages 475–476

2. Slanting (Mah gets Jong to believe him by emphasizing one aspect of an ambiguous photograph).

4. Ambiguity (the sentence is ambiguous; is the Dean making the announcement on a holiday, or is he preventing withdrawals on holidays?).

6. Composition (the bulldozer weighs tons when the parts are assembled).

Exercise 2, pages 476–477

2. Ambiguity (the prophecy was ambiguous: it seemed to Macbeth that the prophecy meant that it was impossible for him to lose).

4. Ambiguity (the instructor's comment was ambiguous; the fallacy is committed by the student for leaping to a particular interpretation when other, more plausible interpretations are possible).

MODULE 37, *pages 482–483*

2. Special Pleading (Groucho is violating his own generalization).

4. Denying the Antecedent. (If traffic lights were synchronized, then the automatic camera would be more useful. Traffic lights aren't synchronized. Therefore, the camera is more useful.)

6. Unequal Distribution.

8. Affirming the Consequent.

EXERCISES FOR UNIT FOUR, *pages 498–510*

2. (A) Attacking a Straw Man.

4. (C) *Ad Hominem.*

6. (D) Attacking a Straw Man.

8. (E) No fallacy at all.

10. (C) *Ad Hominem.*

12. (B) Irrelevant Appeal.

14. (B) Daniel Quayle.

16. (A) Lloyd Bentsen.

18. (C) The Questioner.

20. (C) Appeal to Sympathy.

22. (B) George Bush.

24. (C) The Questioner.

26. (C) The Questioner.

28. (B) False Dilemma.

30. (D) No fallacy.

32. (B) Ronald Reagan.

34. (A) Appeal to Ignorance.

36. (A) George Bush.

38. (C) Both (A) *Ad Hominem* and (B) Attacking a Straw Man.

40. (A) Leading Question.

42. (B) Slippery Slope.

44. (D) All of the above.

46. (D) All of the above.

UNIT FIVE

MODULE 40, *page 529*

He took his pulse, which is steady when a subject is at rest. Relying on one's pulse to tell time in this way was common practice before the invention of the clock.

MODULE 41, *pages 540–541*

2. "Tarzan's New York Adventure." Recall the occasional references to Tarzan in *Crocodile Dundee*.

4. Representative Felix Walker whose legendary "Speech to Buncombe County (NC)" in the early nineteenth century was reputed to be so bad that his colleagues in Congress came to refer to any self-serving bundle of nonsense as "buncombe" (later "bunkum"). The text of the speech has been lost. [If you find a copy, please notify me.]

6. Billy Sunday.

8. 'HAL' is a cryptogram for 'IBM'.

MODULE 42, *page 557*

Try feeding it a cookie. (Type in the word 'cookie' and hit "return.")

MODULE 43, *pages 570–571*

2. The solution is left to the reader.

4. The solution is left to the reader.

MODULE 44, *pages 590–591*

1. a woman $= 40/50 = 4/5$.
a man $= 10/50 = 1/5$

2. a. $(1/6) \times (1/6) \times (1/6) \times (1/6) \times (1/6) = 1/7776$
b. $(1/2) \times (1/2) = (1/4)$
c. $(1/6) + (1/6) + (1/6) = (3/6) = (1/2)$
d. $(13/52) \times (12/51) + (11/50) = (1716/132{,}600) = (11/850)$
e. $(1/2) \times (1/2) \times (1/2) = (1/8)$

4. $(13/52) \times (12/51) \times (11/50) = (1716/132{,}600) = (11/850)$

6. $(1/6) \times (1/6) = (1/36)$

8. $(1/2) \times (1/2) \times (1/2) \times (1/2) = (1/16)$

MODULE 45, *page 603*

Some suggestions: (If you haven't seen them, rent the videos.)

Michelangelo Antonioni's "Blow-Up" (1966)

Alfred Hitchcock's "Vertigo" (1958)

David Lynch's "Blue Velvet" (1986)

INDEX

Semantic Trees for the Quantificational Calculus

$(\forall X)\ \varphi$	$\sim(\exists X)\ \varphi$	$*(\exists X)\ \varphi$	$*\sim(\forall X)\ \varphi$
•	•	•	•
$\varphi\ (T/X)$	$\sim\varphi(T/X)$	$\varphi(T/X)$	$\sim\varphi(T/X)$

**Restrictions:* For '$(\exists X)\ \varphi$' and '$\sim(\forall X)\ \varphi$', 'T' must be a term new to everything on the pathway through the branches above it.

The Rules of Inference

1. *Modus Ponens* (MP)
   $(p \supset q)$
   p
   $\therefore q$
2. *Modus Tollens* (MT)
   $(p \supset q)$
   $\sim q$
   $\therefore \sim p$
3. Disjunctive Syllogism (DS)
   $(p \vee q)$
   $\sim p$
   $\therefore \sim q$
4. Simplification (Simp)
   $(p \& q)$
   $\therefore p$
5. Conjunction (Conj)
   p
   q
   $\therefore (p \& q)$
6. Addition (Add)
   p
   $\therefore (p \vee q)$
7. Hypothetical Syllogism (HS)
   $(p \supset q)$
   $(q \supset r)$
   $\therefore (p \supset r)$
8. Dilemma (Dil)
   $(p \supset q)$
   $(r \supset s)$
   $(p \vee r)$
   $\therefore (q \vee s)$
9. Absorption (Abs)
   $(p \supset q)$
   $\therefore (p \supset (p \& q))$
10. Repetition (Rep)
   p
   $\therefore p$

The Rules of Replacement

11. Double Negation (DN)	$p :: \sim\sim p$
12. Commutation (Com)	$(p \vee q) :: (q \vee p)$ $(p \& q) :: (q \& p)$
13. Association (Assoc)	$(p \vee (q \vee r)) :: ((p \vee q) \vee r)$ $(p \& (q \& r)) :: ((p \& q) \& r)$
14. De Morgan's Laws (DeM)	$\sim(p \& q) :: (\sim p \vee \sim q)$ $\sim(p \vee q) :: (\sim p \& \sim q)$
15. Distribution (Dist)	$(p \& (q \vee r)) :: ((p \& q) \vee (p \& r))$ $(p \vee (q \& r)) :: ((p \vee q) \& (p \vee r))$
16. Equivalence (Equiv)	$(p \equiv q) :: ((p \supset q) \& (q \supset p))$
17. Transposition (Trans)	$(p \supset q) :: (\sim q \supset \sim p)$